DICTIONARY OF PROCESS TECHNOLOGY

*English
German
French
Russian*

DICTIONARY OF PROCESS TECHNOLOGY

in four languages *English*
German
French
Russian

compiled by

Prof. Dr. sc. techn. Klaus Hartmann
Berlin, German Democratic Republic

Prof. Dr. sc. techn. Ernst-Otto Reher
Merseburg, German Democratic Republic

Dr.-Ing. Rolf Walde
Köthen, German Democratic Republic

ELSEVIER
Amsterdam – Oxford – New York – Tokyo
1989

Published in coedition with VEB Verlag Technik, Berlin 1989

This book is exclusively distributed in all non-socialist
countries with the exception of the Federal Republic of Germany,
West-Berlin, Austria and Switzerland by
Elsevier Science Publishers B. V.
Sara Burgerhartstraat 25
P.O. Box 211, 1000 AE Amsterdam, The Netherlands

Distributors for the United States and Canada
Elsevier Science Publishing Company, Inc.
655 Avenue of the Americas
New York, NY 10010

Library of Congress Cataloging in Publication Data

Dictionary of process technology.

  Bibliography: p.
  Includes index.
  1. Manufacturing processes—Dictionaries—Polyglot.
2. Chemical engineering—Dictionaries—Polyglot.
3. Dictionaries, Polyglot.   I. Hartmann, Klaus, Dr.
II. Reher, Ernst-Otto.   III. Walde, Rolf.
TS9.D54   1989   670'.3   88-30989
ISBN 0-444-98888-2

ISBN 0-444-98888-2

Printed in the German Democratic Republic by Grafischer Großbetrieb Völkerfreundschaft Dresden.

PREFACE

Among the modern engineering sciences, process technology is of a special importance. This is because industrial processes in which materials undergo changes in their physical or chemical state are an important factor in determining the effectiveness and level of industrial production as a whole.

Processes in which materials undergo changes are mainly found in the chemical, power, fuel and building material industries; however, they are also found in the glass and ceramics, metallurgical, food, pulp and paper, and wood and fibre industries, as well as in water management and many other areas. The principles of process technology are also commonly used in the manufacturing processes of the electrical and electronics, machine-building and automobile industries.

Process technology, as the branch of engineering that deals with the development of processes and the design and operation of plants in which materials undergo physical or chemical changes on an industrial scale, is a relatively young science. The development and recognition of process technology as an independent discipline, however, has varied greatly from country to country and the fields covered respectively by "Verfahrenstechnik", "process technology", "génie chimique" and "processy i apparaty" differ in structure. The German term "Verfahrenstechnik" was coined in 1935, and yet in the GDR, "Verfahrensingenieurwesen", a basic training course for chemical engineers, was only established as an independent discipline in 1968, while "process technology" in the United States and "processy i apparaty" in the USSR have long traditions.

During the past few years, a vast amount of specialized literature has been published, in textbooks, monographs and journals which have become indispensable aids and valuable sources of new knowledge for chemical engineers in research, industry, education and design. However, the dictionaries available to the student of foreign-language literature have been limited to the chemical-technological vocabulary (more than half of this vocabulary being chemical compounds and products) and a general polytechnical vocabulary. Thus, there is an urgent need for a specialized process technology dictionary, covering the vocabulary of modern process technology as it has evolved during the past 20 years.

This dictionary has been compiled by chemical engineers and translators from universities and academies who are actively involved in chemical engineering research and education as well as in foreign-language training.

The successful completion of the dictionary owes much to the support of the staff of the Dictionaries Department of the Verlag Technik publishing house. The authors would like to thank them for their friendly cooperation and critical review of the manuscript.

The authors and contributors realize it would be impossible not to have some shortcomings in the first edition of such a dictionary and, therefore, would like to ask all users to address any comments and suggestions aimed at improving the dictionary to: VEB Verlag Technik, Oranienburger Str. 13/14, Berlin, DDR-1020.

The Authors

DIRECTIONS FOR USE

1. Examples of alphabetization

addition
additional treatment
addition of energy
addition product
addition reaction

process group
processibility
process industry
processing direction
process intensification

Schmelze
Schmelzebruch
schmelzen
Schmelzen
Schmelzenthalpie

Temperaturhaltepunkt
Temperatur im Innern
Temperaturintervall
Temperaturmeßbereich
Temperaturmessung

gaz d'huile
gaz du four
gazéifiable
gaz étranger
gazoduc

limite de triage
limites/sans
limites de bilan
limites d'intégration
limite séparatrice

транспорт газа
транспортер
транспортирование
транспортируемый материал
транспорт масла

фильтрат
фильтрационная вода
фильтрация газа
фильтр грубой очистки
фильтр для масла

2. Signs and abbreviations

( ) absorbing (absorption) apparatus = absorbing apparatus *or* absorption apparatus
[ ] matrix [form] = matrix *or* matrix form
/   limites/sans = sans limites
( ) these brackets contain explanations
s. = see
s. a. = see also

English – German – French – Russian

A

	English	German	French	Russian
	ability to flow	s. F 282		
A 1	above-ground storage tank	Hochbehälter m, Über-Tage-Speicher m	réservoir m surélevé	надземный бак-хранилище
A 2	abrasion	Abriebwirkung f	effet m abrasif	абразивное действие
	abrasion	s. a. A 5		
A 3	abrasion resistance	Abriebfestigkeit f	résistance f à l'abrasion	прочность на истирание
A 4	abrasive test	Abriebversuch m	test m abrasif, essai m de l'abrasion	испытание на истирание
A 5	abrasive wear, abrasion	Abnutzung f, Abrieb m, Abreibung f	abrasion f, usure f [par abrasion]	истирание, износ, изнашивание
A 6	absolute delay	absolute Verzögerung f	retard m absolu	абсолютная задержка
A 7	absolute deviation	absolute Abweichung f	écart m absolu	абсолютное отклонение
A 8	absolute filter	absolutes Filter n (Stabilisation der Luft)	filtre m absolu	абсолютный фильтр
A 9	absolute gas constant	allgemeine Gaskonstante f	constante f des gaz générale	универсальная газовая постоянная
A 10	absolute heating effect	absolute Wärmetönung f, absoluter Wärmeeffekt m	effet m calorifique absolu, chaleur f de réaction absolue	абсолютная теплотворная способность, абсолютный тепловой эффект
A 11	absolute humidity	absolute Feuchtigkeit f	humidité f absolue	абсолютная влажность
A 12	absolute moisture content	absoluter Feuchtegehalt m	teneur m en humidité absolu, taux m d'humidité absolu	абсолютное влагосодержание
A 13	absolute pressure	absoluter Druck m, Absolutdruck m	pression f absolue	абсолютное давление
A 14	absolute sensitivity	absolute Empfindlichkeit f	sensibilité f absolue	абсолютная чувствительность
A 15	absolute system	absolutes System n	système m absolu	абсолютная система
A 16	absolute temperature	absolute Temperatur f	température f absolue	абсолютная температура
A 17	absolute viscosity	absolute Viskosität f	viscosité f absolue	абсолютная вязкость
A 18	absorb/to	absorbieren, aufsaugen, aufnehmen, anziehen	absorber	поглощать, абсорбировать, всасывать
	absorbability	s. A 57		
A 19	absorbable	absorbierbar, aufnahmefähig	absorbable	поглощаемый
A 20	absorbed horsepower	aufgenommene Leistung f	puissance f absorbée	потребляемая мощность
	absorbency	s. A 57		
A 21	absorbent, absorbing medium	Absorptionsmittel n, Absorbens n	absorbant m, agent m absorbant	абсорбент, поглотитель, поглощающая (абсорбирующая) среда
A 22	absorbent charging	Einfüllen n von Absorbens	chargement m de l'absorbant	зарядка системы абсорбентом, наполнение абсорбента
A 23	absorbent filter, absorption filter	Absorptionsfilter n	filtre m d'absorption	поглощающий (абсорбционный) фильтр
A 24	absorbent oil, absorption oil	Waschöl n	huile m de lavage	поглотительное (абсорбционное, промывное, скрубберное) масло
A 25	absorber, absorbing (absorption) apparatus	Absorber m, Absorptionsapparat m	absorbeur m, installation f d'absorption	абсорбер, поглотитель
A 26	absorber bottom	Absorbersumpf m	partie f inférieure de la colonne d'absorption	низ абсорбера
A 27	absorber cooler	Absorptionskühlaggregat n, Absorptionskühlmaschine f	machine f à froid à absorption	холодильник абсорбера (поглотителя)
A 28	absorber head	Absorberkopf m	tête f de colonne d'absorption	верх абсорбционной колонны
A 29	absorber off-gas	Absorberabgas n	gaz m de dégagement de l'absorbeur	отходящий газ из абсорбционной колонны
A 30	absorber washer	Absorber m, Absorptionswäscher m	laveur m, absorbeur m	абсорбер, скруббер
A 31	absorbing	Absorbieren n, Einsaugen n	absorption f, aspiration f	поглощение, абсорбирование, всасывание, подсасывание
	absorbing apparatus	s. A 25		
	absorbing capacity	s. A 57		
A 32	absorbing column, absorption column, absorbing (absorption) tower	Absorptionskolonne f, Absorptionsturm m, Waschturm m, Absorptionssäule f	colonne f d'absorption, tour f de lavage, absorbeur m, tour d'absorption, colonne liquide	поглотительная (абсорбционная) колонна, абсорбер, поглотительная (абсорбционная) башня
	absorbing medium	s. A 21		
A 33	absorbing temperature	Absorptionstemperatur f	température f d'absorption	температура абсорбции
	absorbing tower	s. A 32		
A 34	absorption	Absorption f, Aufsaugung f, Aufnahme f, Einsaugung f	absorption f	поглощение, абсорбция
	absorption	s. a. A 46		
	absorption apparatus	s. A 25		
A 35	absorption capacity	Absorptionskapazität f	pouvoir m absorbant, puissance f d'absorption	абсорбционная способность
A 36	absorption cell	Absorptionszelle f	cellule f d'absorption	абсорбционная кювета
A 37	absorption chiller	Absorptionskühler m	refroidisseur m à absorption	абсорбционный охладитель
	absorption coefficient	s. A 44		

	absorption column	s. A 32		сечение поглощения
A 38	absorption cross section	Absorptionsquerschnitt m	section f absorbante	сечение поглощения
A 39	absorption curve	Absorptionskurve f	courbe f d'absorption	кривая абсорбции
A 40	absorption dehumidifier	Absorptionstrockner m	sécheur m à absorption	абсорбционный влагопоглотитель
A 41	absorption equilibrium	Absorptionsgleichgewicht n	équilibre m d'absorption	равновесие абсорбции, абсорбционное равновесие
A 42	absorption experiment	Absorptionsversuch m	essai m d'absorption	испытание на поглощение
A 43	absorption extraction	Absorptionsextraktion f	extraction f absorptive	извлечение поглощением
A 44	absorption factor, absorption coefficient	Absorptionsfaktor m, Absorptionskoeffizient m, Absorptionsindex m	coefficient m d'absorption	коэффициент (показатель) поглощения
	absorption filter	s. A 23		
A 45	absorption line	Absorptionslinie f	raie f d'absorption	линия поглощения
A 46	absorption line, absorption [process]	Absorptionsverlauf m	processus m d'absorption	режим (протекание) абсорбции
A 47	absorption liquid	Absorptionsflüssigkeit f	liquide m absorbant	поглотительная (абсорбирующая) жидкость
A 48	absorption loss	Aufsaugverlust m	perte f par imbibage initial	абсорбционные потери, потеря абсорбции
	absorption oil	s. A 24		
A 49	absorption plant, absorption system	Absorptionsanlage f	installation f d'absorption	абсорбционная установка
A 50	absorption process	Absorptionsprozeß m, Absorptionsverfahren n	procédé m d'absorption	абсорбционный процесс
	absorption process	s. a. A 46		
A 51	absorption refrigerating machine	Absorptionskältemaschine f	machine f à froid à absorption	абсорбционная холодильная машина
A 52	absorption refrigeration	Absorptionskühlung f	refroidissement m par absorption	абсорбционное охлаждение
	absorption system	s. A 49		
	absorption tower	s. A 32		
A 53	absorption tube	Absorptionsrohr n	tuyau m d'absorption	абсорбционная (поглотительная) трубка
A 54	absorption unit	Absorptionseinheit f	unité f d'absorption	абсорбционная установка
A 55	absorption velocity	Absorptionsgeschwindigkeit f	vitesse f d'absorption	скорость абсорбции
A 56	absorptive	absorptiv, saugfähig	absorptif	поглощающий, абсорбирующий
A 57	absorptive power, absorptivity, absorbability, absorbing capacity, absorbency	Absorbierbarkeit f, Absorptionsvermögen n, Aufnahmefähigkeit f, Saugfähigkeit f, Absorptionskraft f, Aufsaugvermögen n	pouvoir m absorbant, absorptivité f, puissance (faculté, capacité) f d'absorption	абсорбционная (поглотительная, абсорбирующая, впитывающая) способность, всасываемость
A 58	abundance	Häufigkeit f	fréquence f	повторяемость, частота повторения, частота
A 59	abundance curve	Häufigkeitskurve f	courbe f des densités, courbe de fréquence	кривая повторяемости (частоты)
A 60	acceleration	Beschleunigung f	accélération f	ускорение
A 61	acceleration cost	Beschleunigungskosten pl	frais mpl d'accélération	расходы на ускорение
A 62	acceleration of settling rate	Sedimentationsbeschleunigung f	accélération f de la sédimentation	ускорение седиментации (оседания, осаждения)
A 63	acceleration time	Anlaufzeit f, Beschleunigungszeit f	durée f de démarrage	время разгона
A 64	accelerator	Beschleuniger m	accélérateur m	ускоритель
A 65	acceptance control	Übernahmekontrolle f	contrôle m à la réception, contrôle d'acceptation	приемный контроль, приемное испытание, приемка
A 66	acceptance of equipment	Ausrüstungsabnahme f	réception f du matériel	приемка оборудования
A 67	acceptance of erected plant, quality inspection	Bauabnahme f	réception f, acceptation f	приемка постройки
	acceptance power	s. P 305		
A 68	acceptance test	Abnahmeprüfung f, Abnahmeversuch m	essai m de réception	приемное (приемо-сдаточное) испытание
A 69	acceptor	Akzeptor m	accepteur m	акцептор
A 70	access duct	Eintrittskanal m	conduit m d'accès, canal m d'entrée, canal d'admission	входной канал
A 71	accessibility	Zugänglichkeit f, Erreichbarkeit f	accessibilité f	доступность
A 72	access of air	Luftzutritt m	entrée f d'air	доступ воздуха
	accessory gas	s. F 321		
A 73	access road, driveway	Zufahrt f, Zufahrtsweg m	chemin m d'accès, abord m, accès m	подъезд, подъездная дорога
A 74	accident prevention	Unfallschutz m	sécurité f industrielle, protection f contre les accidents	аварийная защита, техника безопасности
A 75	accumulated cold	akkumulierte Kälte f	froid m accumulé	аккумулированный холод
A 76	accumulation, heaping, aggregation, agglomeration	Häufung f, Anhäufung f, Aggregation f	accumulation f, agrégation f	аккумуляция, накопление, скопление, скучивание, собирание
	accumulation	s. a. 1. I 175, 2. S 697		
A 77	accumulation of mud	Verschlammung f	accumulation f de boue, colmatage m, envasement m	заиление, заиливание

A 78	**accumulative crystallization**	Anreicherungskristallisation f, Sammelkristallisation f, Kornvergrößerung f	cristallisation f d'enrichisse- ment	собирательная кристаллиза- ция
A 79	**accumulator,** run-down tank	Destillatsammelgefäß n	réservoir m de recette	приемный бак
A 80	**accumulator tank**	Sammelbehälter m, Akkumu- latorkasten m	récipient m collecteur, caisse f d'accumulateur	сборник, аккумуляторный бак
A 81	**accuracy of measurement**	Meßgenauigkeit f	exactitude f de mesurage, précision f des mesures	точность измерения
A 82	**accuracy of reading**	Ablesegenauigkeit f	précision f des lectures	точность отсчета
A 83	**acetylene converter**	Acetylenkonverter m	convertisseur m d'acétylène	ацетиленовый конвертер, конвертер для ацетилена
A 84	**acid bath,** pickle	Säurebad n	bain m acide	кислотная ванна
A 85	**acid content**	Säuregehalt m	teneur m en acide	кислотность
A 86	**acid elevator (feeder)**	Säureaufgabevorrichtung f, Säureheber m	alimentateur m d'acide, si- phon (feeder) m d'acide	подъемник для кислот, кис- лотный монжус (мон- тежю)
A 87	**acid-free**	säurefrei	exempt d'acides, neutre	не содержащий кислоты, бескислотный
A 88	**acidification**	Säuerung f, Ansäuern n	acidification f	подкисление
A 89	**acidimetry**	Säuremessung f	acidimétrie f, acétométrie f	ацидиметрия
A 90	**acidity,** acid value	Säuregrad m	degré m d'acidité, acidité f	кислотность
A 91	**acid plant**	Säureanlage f	installation f d'acide	кислотная установка, кис- лотный завод
A 92	**acid-proof,** acid-resistant	säurefest, säurebeständig	antiacide, résistant aux acides, à l'épreuve des acides	кислотоупорный, кислото- устойчивый, кислотостой- кий
A 93	**acid pump**	Säurepumpe f	pompe f à l'acide	насос для кислот, кис- лотный насос
A 94	**acid reaction**	Säurereaktion f	réaction f acide	кислая реакция
A 95	**acid receiver**	Säurebehälter m	réservoir m d'acide	кислотосборник, сборник для кислот
	acid recovery plant	s. A 97		
A 96	**acid resistance**	Säurefestigkeit f	résistance f à l'acide	кислотоупорность, кислото- стойкость
	acid-resistant	s. A 92		
A 97	**acid restoring plant,** acid recovery plant	Säurewiederaufbereitungsan- lage f, Säurerückgewin- nungsanlage f	installation f de régénération d'acide, installation de ré- cupération de l'acide	установка для регенерации кислоты
A 98	**acid seal**	Säureverschluß m, Säure- dichtung f	fermeture f résistante aux acides	кислотный затвор
A 99	**acid separator**	Säureabscheider m	séparateur m d'acide	отделитель кислот
A 100	**acid stability**	Säurebeständigkeit f	acidorésistance f, résistance f aux acides	кислотостойкость
A 101	**acid tower**	Säureturm m	tour f pour acides	поглотительная башня, оро- шаемая кислотой; кис- лотная башня
	acid value	s. A 90		
A 102	**action; effect**	Wirkung f, Einwirkung f, Ef- fekt m	effet m, action f	действие, воздействие, эф- фект, влияние, реакция
A 103	**action principle**	Wirkprinzip n	principe m d'action	принцип действия
A 104	**action turbine**	Aktionsturbine f	turbine f d'action	активная турбина
A 105	**activate/to**	aktivieren, anregen, erregen	activer	активировать, возбуждать
A 106	**activated carbon,** active car- bon	Aktivkohle f	charbon m activé	активный (активированный) уголь
A 107	**activated carbon filter**	Aktivkohlefilter n	filtre m à charbon activé	активированный угольный фильтр
A 108	**activated sludge**	Belebtschlamm m	boue f activée	активный ил
A 109	**activated sludge chamber**	Belebtschlammbecken n	bassin m à boue activée	аэротэнк, аэротенк, отстой- ник, аэротенк с активным илом
A 110	**activated sludge installation (plant)**	Belebtschlammanlage f	installation f de boue activée	установка для очистки сточных вод активным илом
A 111	**activated sludge process**	Belebtschlammverfahren n	procédé m de boues activées	аэротенк-метод
A 112	**activated sludge water**	Belebtschlammwasser n	eau f de boue activée	вода для активирования ила
A 113	**activation**	Aktivierung f, Anregung f, Aktivität f	activation f, activité f	активация, активность
	activation energy	s. E 155		
A 114	**activation heat**	Aktivierungswärme f	chaleur f d'activation	теплота активирования
A 115	**activation mechanism**	Aktivierungsmechanismus m	mécanisme m d'activation	механизм активации
A 116	**activation speed**	Aktivierungsgeschwindigkeit f	vitesse f d'activation	скорость активации
A 117	**activator**	Aktivator m	activeur m, activant m	активатор
	active carbon	s. A 106		
A 118	**active mass**	wirksame Masse f	masse f active	действующая масса
A 119	**active surface**	wirksame (aktive) Oberfläche f	surface f active	активная поверхность
	activity	s. R 101		
A 120	**activity coefficient**	Aktivitätskoeffizient m	coefficient m d'activité	коэффициент активности
A 121	**actual capacity**	tatsächliche (effektive) Lei- stung f	puissance f effective	действительная производи- тельность

A 122	**actual coefficient of performance**	tatsächlicher Wärmekoeffizient *m*	coefficient *m* effectif de performance	действительный холодильный коэффициент
A 123	**actual cooling surface**	tatsächliche Kühlfläche *f*	surface *f* effective de refroidissement	действительная поверхность охлаждения
A 124	**actual cycle**	realer Prozeß *m*	cycle *m* actuel, processus *m* réel	реальный цикл
A 125	**actual displacement**	tatsächlicher Förderstrom *m*	débit *m* effectif, volume *m* de transport réel	действительный объемный поток
A 126	**actual efficiency**	tatsächlicher Wirkungsgrad *m*	rendement *m* effectif	действительный коэффициент полезного действия
A 127	**actual gas,** real gas	reales Gas *n*	gaz *m* réel	реальный газ
A 128	**actual output**	Ist-Leistung *f*	puissance *f* réelle, production *f* effective	фактическая мощность, эффективная производительность
A 129	**actual output,** useful performance, effective power (output)	Nutzleistung *f*	effet *m* utile, travail *m* utile, puissance *f* utile	полезная мощность, полезная отдача
A 130	**actual plate number**	wirkliche Bodenzahl *f*	nombre *m* réel des plateaux	действительное число тарелок
A 131	**actual process temperature**	tatsächliche Prozeßtemperatur *f*	température *f* réelle de procédé	температура действительного процесса
A 132	**actual reflux**	tatsächlicher Rücklauf *m*	reflux *m* effectif	фактическое орошение, фактическая флегма
A 133	**actual refrigerating capacity**	tatsächliche Kälteleistung *f*	puissance *f* effective de réfrigération	действительная холодопроизводительность
A 134	**actual refrigeration process**	tatsächlicher Kühlprozeß *m*	processus *m* frigorifique réel	действительный процесс охлаждения
A 135	**actual state registration**	Ist-Zustandserfassung *f*	détection *f* de l'état actuel, enregistrement *m* de l'état donné	сбор актуальных переменных состояния
A 136	**actuating pressure**	Stelldruck *m*	pression *f* de réglage	давление сервопривода
	adaptability	s. U 63		
A 137	**adaptable**	verwendbar	utilisable	годный, применимый
	additament	s. A 140		
A 138	**addition**	Zuschlag *m*, Flußmittel *n*	fondant *m*, poudre *f* fondante	добавка, флюс
A 139	**addition**	Anlagerung *f*	addition *f*, réticulation *f*	присоединение
A 140	**addition,** additive, admixture, additament	Zusatz *m*, Zusatzmittel *n*, Zusatzstoff *m*, Additiv *n*	addition *f*, ajoutage *m*, additif *m*, produit *m* d'addition, agent *m*	добавка, присадка, вспомогательное вещество, примесь, аддитив
	additional plant	s. S 772		
	additional pressure	s. O 165		
A 141	**additional treatment,** subsequent treatment	Nachbehandlung *f*, Nachbearbeitung *f*	retraitement *m*, traitement *m* ultérieur (additionnel)	окончательная обработка, отделка, выдержка
	addition compound	s. A 143		
A 142	**addition of energy**	Energiezufuhr *f*, Energieeinspeisung *f*	alimentation *f* d'énergie, alimentation en énergie	подвод (питание, прирост) энергии
A 143	**addition product,** addition compound	Anlagerungsprodukt *n*, Anlagerungsverbindung *f*	produit *m* par addition	продукт присоединения
A 144	**addition reaction**	Anlagerungsreaktion *f*	réaction *f* par addition	реакция присоединения
	additive	s. A 140		
A 145	**additive property**	Additivitätseigenschaft *f*	propriété *f* d'additivité	аддитивное свойство
A 146	**additivity**	Additivität *f*, additive Eigenschaft *f*	additivité *f*	аддитивность
A 147	**adherence, adhesion**	Adhäsion *f*, Anhaftung *f*, Anziehung *f*, Haftfähigkeit *f*, Haftvermögen *n*	adhérence *f*, adhésion *f*	прилипание, сцепление, адгезия
A 148	**adhesion coefficient**	Adhäsionsbeiwert *m*	coefficient *m* d'adhésion	коэффициент сцепления (адгезии)
A 149	**adhesion heat**	Adhäsionswärme *f*, Adhäsionsenthalpie *f*	chaleur *f* d'adhésion	теплота прилипания (сцепления, адгезии)
A 150	**adhesion resistant,** resistant to adhesion	adhäsionsresistent	antiadhérent	антиадгезионный
A 151	**adhesive**	Kleber *m*, Haftmittel *n*, Klebstoff *m*	colle *f*, adhésif *m*, agglutinant *m*, matière *f* adhésive	клей, адгезив
	adhesive	s. a. F 191		
A 152	**adhesiveness**	Adhäsionsfähigkeit *f*, Adhäsionsvermögen *n*, Haftvermögen *n*	pouvoir *m* adhérent	сцепляемость, адгезионная способность, приставание к поверхности, способность к склеиванию
A 153	**adhesiveness**	Adhäsionsgewicht *n*	poids *m* d'adhésion	сцепной вес
A 154	**adhesive power**	Adhäsionskraft *f*, Haftvermögen *n*, Haftfestigkeit *f*	force *f* d'adhérence	сила сцепления, адгезионная способность
A 155	**adhesive separator**	Adhäsionsabscheider *m*	séparateur *m* par adhésion	отделитель, работающий по принципу адгезии
A 156	**adiabatic**	adiabat, adiabatisch	adiabatique	адиабатный, адиабатический
A 157	**adiabatic absorption**	adiabatische Absorption *f*	absorption *f* adiabatique	адиабатическая абсорбция
A 158	**adiabatic change of condition**	adiabatische Zustandsänderung *f*	changement *m* d'état adiabatique	адиабатическое изменение состояния
A 159	**adiabatic column**	adiabatische Kolonne *f*	colonne *f* adiabatique	адиабатическая колонна
A 160	**adiabatic compression,** isentropic compression	adiabate Verdichtung *f*	compression *f* adiabatique (isentropique)	адиабатическое сжатие

A 161	adiabatic compression temperature	adiabatische Verdichtungstemperatur f	température f de compression adiabatique	температура адиабатического сжатия
A 162	adiabatic compression work	adiabatische Verdichtungsarbeit f	travail m de compression adiabatique	адиабатическая работа сжатия
A 163	adiabatic cooling	adiabatische Kühlung f	refroidissement m adiabatique	адиабатическое охлаждение
A 164	adiabatic curve, adiabatic line	Adiabate f	adiabate f, courbe f adiabatique	адиабата, адиабатная кривая
A 165	adiabatic curve of expansion	Expansionsadiabate f	courbe f adiabatique d'expansion, adiabate f d'expansion	адиабата расширения
A 166	adiabatic desorption	adiabatische Desorption f	désorption f adiabatique	адиабатическая десорбция
A 167	adiabatic diagram	Adiabatendiagramm n	diagramme m adiabatique	адиабатная диаграмма
A 168	adiabatic dryer	adiabatischer Trockner m	sécheur (séchoir) m adiabatique	адиабатная сушилка
A 169	adiabatic efficiency	adiabatischer Wirkungsgrad m	rendement m adiabatique	адиабатический коэффициент полезного действия
A 170	adiabatic expansion	adiabatische Entspannung (Expansion) f	détente f adiabatique	адиабатическое расширение
A 171	adiabatic exponent, isentropic exponent	Adiabatenexponent m, Isentropenexponent m	exposant m adiabatique (isentropique)	адиабатный показатель
A 172	adiabatic heat drop	adiabatische Wärmeabgabe f	émission f de chaleur adiabatique	адиабатическая отдача тепла
A 173	adiabatic high temperature	adiabate Höchsttemperatur f	température f maximum adiabatique	максимальная адиабатическая температура
A 174	adiabatic horsepower	adiabatische Leistung f	puissance f adiabatique	адиабатическая мощность
	adiabatic line	s. A 164		
A 175	adiabatic membrane	adiabatische Membran f	membrane f adiabatique	адиабатная оболочка
A 176	adiabatic process	adiabatisches Verfahren n, adiabatischer Prozeß m	procédé m adiabatique	адиабатный процесс
A 177	adiabatic reactor	adiabatischer Reaktor m	réacteur m adiabatique	адиабатический реактор
A 178	adiabatic saturation	adiabatische Sättigung f	saturation f adiabatique	адиабатическое насыщение
A 179	adiabatic saturation process	adiabatischer Sättigungsprozeß m	processus m de saturation adiabatique	процесс адиабатического насыщения
A 180	adiabatic temperature curve	adiabatische Temperaturkurve f	courbe f de température adiabatique, adiabate f de température	адиабатическая кривая температуры
A 181	adiabatic temperature rise	adiabatische Temperaturerhöhung f	augmentation f de température adiabatique	адиабатическое повышение температуры
A 182	adiabatic throttling	adiabatische Drosselung f	étranglement m adiabatique	адиабатическое дросселирование
A 183	adiabatic wall	adiabatische Wand f	paroi f adiabatique, mur m adiabatique	адиабатная оболочка (стена)
A 184	adjacency	Adjazenz f	adjacence f	смежность, соседство
A 185	adjacency list	Adjazenzliste f	liste f d'adjacence	список смежностей
A 186	adjacency matrix	Adjazenzmatrix f	matrice f associée	смежная матрица
A 187	adjoining position	Nachbarposition f, benachbarte Lage f	position f voisine	смежное положение
A 188	adjoint computing technique	adjungierte Rechentechnik f	technique f de calcul adjointe	метод параллельных вычислений
A 189	adjoint matrix	adjungierte Matrix f	matrice f adjointe	сопряженная матрица
A 190	adjoint system	adjungiertes System n	système m adjoint	сопряженная система
A 191	adjust/to	justieren, einstellen	ajuster	юстировать
A 192	adjuster, final control element	Stellglied n	organe m de réglage	установочное звено, исполнительный элемент, регулирующий орган
A 193	adjusting device	Justiervorrichtung f, Nachstelleinrichtung f	dispositif m d'ajustage	приспособление для юстировки, регулировочное устройство
A 194	adjusting drive	Stellantrieb m	servomoteur m, moteur m de commande	сервопривод
A 195	adjusting screw	Justierschraube f	vis f d'ajustage	винт для юстировки
A 196	adjustment	Angleichung f, Regelung f, Anpassung f, Justierung f, Korrektur f	égalisation f, ajustage m, régulation f, réglage m	юстирование, регулирование, установка
A 197	adjustment control	Einstellelement n	élément m d'ajustement, élément de mise au point, élément de réglage	орган управления, установочный орган
A 198	adjustment time	Einstellzeit f	temps m d'ajustement	установочное время
	admission	s. I 188		
A 199	admission space	Füllungsraum m	volume m d'admission	объем наполнения
A 200	admission velocity	Eintrittsgeschwindigkeit f	vitesse f d'entrée	скорость на входе
	admixture	s. A 140		
A 201	adsorbability	Adsorbierbarkeit f	adsorbabilité f	адсорбирующая способность
A 202	adsorbable	adsorbierbar	adsorbable	адсорбируемый
A 203	adsorbate, sorbate	Adsorbat n, Adsorptiv n	adsorbat m, substance f adsorbée	[ад]сорбированное вещество, адсорбат, продукт адсорбции
A 204	adsorbed film	Adsorptionshaut f, Adsorptionsfilm m, Adsorptionsschicht f	couche f d'adsorption, film m d'adsorption	адсорбционная пленка

A 205	adsorbed layer	Adsorptionsschicht f	couche f adsorbée	адсорбционный слой
	adsorbent	s. A 209		
A 206	adsorbent bed	Adsorptionsbett n	lit m adsorbant	слой адсорбента
A 207	adsorbent layer	Adsorptionsschicht f	couche f d'adsorbant, couche adsorbante	слой адсорбента
A 208	adsorber	Adsorber m	adsorbeur m, adsorbant m	адсорбер
A 209	adsorbing agent (material); adsorbent, adsorbing substance	Adsorptionsmittel n, Adsorbens n	substance f adsorbante, adsorbant m	адсорбент, адсорбирующий агент, адсорбирующее вещество
A 210	adsorbing power, adsorptive power, adsorption capacity	Adsorptionsvermögen n, Adsorptionsfähigkeit f	pouvoir m d'adsorption	адсорбционная способность
	adsorbing substance	s. A 209		
A 211	adsorption	Adsorbieren n, Adsorption f	adsorption f	адсорбция
A 212	adsorption, adsorption phenomenon	Adsorptionserscheinung f, Adsorptionsvorgang m	phénomène m d'adsorption	адсорбционное явление, адсорбционный процесс
	adsorption capacity	s. A 210		
A 213	adsorption catalysis	Adsorptionskatalyse f	catalyse f d'adsorption	адсорбционный катализ
A 214	adsorption curve	Adsorptionskurve f	courbe f d'adsorption	кривая адсорбции
A 215	adsorption dehumidification plant (system)	Adsorptionstrocknungsanlage f	installation f de séchage par adsorption	адсорбционная осушительная установка
A 216	adsorption dehumidifier	Adsorptionstrockner m, Adsorptionsentfeuchter m	déshumidificateur m à adsorption, séchoir m à adsorption	адсорбционный осушитель
A 217	adsorption effect	Adsorptionseffekt m	effet m d'adsorption	адсорбционный эффект
A 218	adsorption efficiency	Adsorptionswirkungsgrad m	rendement m d'adsorption	коэффициент полезного действия адсорбции
A 219	adsorption equilibrium	Adsorptionsgleichgewicht n	équilibre m d'adsorption	адсорбционное равновесие
A 220	adsorption exponent	Adsorptionsexponent m	exposant m d'adsorption	показатель адсорбции
A 221	adsorption film	Adsorptionsfilm m	film m d'adsorption	адсорбционная пленка
A 222	adsorption filtration	Adsorptionsfiltration f	filtration-adsorption f	фильтрация через адсорбирующий слой
A 223	adsorption heat, heat of adsorption	Adsorptionswärme f, Adsorptionsenthalpie f	chaleur f d'adsorption	теплота адсорбции (адсорбирования)
A 224	adsorption indicator	Adsorptionsindikator m	indicateur m d'adsorption	адсорбционный индикатор
A 225	adsorption isotherm	Adsorptionsisotherme f	isotherme f d'adsorption	изотерма адсорбции
	adsorption phenomenon	s. A 212		
A 226	adsorption plant	Adsorptionsanlage f	installation f d'adsorption	установка адсорбции
A 227	adsorption potential	Adsorptionspotential n	potentiel m d'adsorption	потенциал адсорбции
A 228	adsorption process	Adsorptionsprozeß m, Adsorptionsverfahren n	procédé m d'adsorption	адсорбционный способ (процесс)
A 229	adsorption refrigerating system	Adsorptionskälteanlage f	installation f frigorifique à adsorption	адсорбционная холодильная установка
A 230	adsorption stoichiometry	Adsorptionsstöchiometrie f	stœchiométrie f d'adsorption	стехиометрия адсорбции
A 231	adsorption unit	Adsorptionseinheit f	unité f d'adsorption	адсорбционная установка
A 232	adsorption vessel	Adsorptionsgefäß n	adsorbeur m, récipient m d'adsorption	адсорбер
A 233	adsorption water	Adsorptionswasser n	eau f d'adsorption, eau adsorbée	адсорбционная вода, адсорбированная вода
A 234	adsorptive capacity	Beladefähigkeit f	pouvoir m d'adsorption	адсорбционная емкость (способность)
A 235	adsorptive flocculation	adsorptive Flockung f	floculation f adsorptive	адсорбтивная коагуляция
	adsorptive power	s. A 210		
A 236	aeration, airing	Belüftung f, Lüftung f, Durchlüftung f	aération f, aérage m	аэрация, продувание воздухом, насыщение углекислым газом
A 237	aeration tank	Belüftungsbecken n	bassin m d'aérage	аэротэнк, аэротенк
A 238	aerator	Belüftungsanlage f, Belüftungsapparat m, Belüfter m	aérateur m, ventilateur m	насытитель, сатуратор, аэратор
A 239	aerial contamination	Luftverseuchung f	pollution f de l'air, contamination f de l'air	воздушная инфекция, загрязнение воздуха
A 240	aeriferous	lufthaltig, luftdurchsetzt	aéré, contenant de l'air	воздухосодержащий, газированный
A 241	aerify/to	begasen (mit Luft)	aérer, ventiler	смешивать с воздухом, нагнетать воздух, газировать
A 242	aerobic	aerob	aérobie	аэробный
A 243	aerodynamic, aerodynamical	strömungsgünstig, aerodynamisch, strömungstechnisch	aérodynamique	удобообтекаемый, аэродинамический
A 244	aerodynamic properties	aerodynamische Eigenschaften fpl	propriétés fpl aérodynamiques	аэродинамические характеристики (свойства)
A 245	aerodynamics	Aerodynamik f	aérodynamique f	аэродинамика
A 246	affination bottom product, outlet of affinage, gutter of affination	Affinationsablauf m	égout m d'affination	аффинационный оттек
A 247	affination centrifuge	Affinationszentrifuge f	centrifuge f d'affination	аффинационная центрифуга
A 248	affination stirring machine	Affinationsrührmaschine f	mélangeur-brasseur m d'affination	аффинационная мешалка
A 249	aftercharging, subsequent charging	Nachbeschickung f	charge f postérieure	дополнительная (добавочная) загрузка
A 250	aftercondenser	Nachkondensator m	condenseur m secondaire	вторичный конденсатор

A 251	aftercooled	nachgekühlt	refroidi ultérieurement	вторично охлажденный, переохлажденный
A 252	aftercooler	Nachkühler m	condenseur m complémentaire, réfrigérant m postérieur	вторичный холодильник
A 253	aftercooling	Nachkühlung f	refroidissement m ultérieur	вторичное охлаждение, переохлаждение
A 254	aftereffect	Nachwirkung f	effet m résiduel, effet ultérieur	последствие
A 255	afterexpansion	Nachentspannung f	expansion f postérieure	дополнительное расширение
A 256	afterfermentation	Nachgärung f	fermentation f secondaire, maturation f	дображивание
A 257	afterfiltration	Nachfiltration f	filtration f secondaire	последующая (контрольная) фильтрация
A 258	afterheat	Nachwärme f	chaleur f ultérieure	остаточное тепло
A 259	afterproduct; low product	Nachprodukt n, Nachlauf m	bas-produit m, dernier jet m	побочный (низший) продукт
A 260	afterpurification	Nachreinigung f	épuration f ultérieure	доочистка, последующая очистка
	afterrun	s. T 7		
A 261	agglomerate/to	agglomerieren, zusammenballen	agglomérer	агломерировать, спекаться, слипаться
A 262	agglomerate	Agglomerat n	agglomérat m, aggloméré m	агломерат
A 263	agglomerating agent	Agglomerierungsmittel n	agent m d'agglomération	фактор агломерации, агент желатинизации
A 264	agglomeration	Agglomeration f, Agglomeratbildung f, Zusammenballung f, Schwarmbildung f	agglomération f	агломерация, агломерирование, спекание, скопление
	agglomeration	s. a. A 76		
A 265	agglomeration apparatus	Agglomerierofen m	four m d'agglomération	агломерирующая печь, печь для агломерации
A 266	agglomerative	agglomerativ	aggloméré	агломератный
	agglutinant	s. F 191		
A 267	agglutinating, agglutination	Zusammenklebung f, Agglutination f	agglutination f	спекание, склеивание, слипание, агглютинация
A 268	agglutinative	klebrig, zusammenklebend	gluant, glutineux, collant	спекающий, склеивающий, слипающий, агглютинирующий
A 269	aggregate	Aggregat n, Anhäufung f	agrégat m, corps mpl par agrégation	агрегат
	aggregation	s. A 76		
A 270	aging	Altern n, Alterung f, Reifung f	maturité f, vieillissement m, altération f	вызревание, старение, созревание
A 271	aging coefficient	Alterungskoeffizient m	coefficient m de vieillissement	коэффициент старения
A 272	aging period	Alterungsperiode f, Alterungsphase f	période f de vieillissement, période d'atonie	фаза старения
A 273	aging process	Alterungsprozeß m, Alterungsverfahren n, Alterungsvorgang m	processus m de vieillissement, vieillissement m	процесс старения (созревания)
A 274	aging resistance	Alterungsbeständigkeit f	résistance f au vieillissement	устойчивость к старению
A 275	aging resistant	alterungsbeständig	inaltérable, non-vieillissant	неподверженный старению
A 276	aging test	Alterungsprüfung f, Alterungsversuch m	essai m de vieillissement (artificiel à l'étuve)	испытание на [ускоренное] старение
A 277	aging velocity	Alterungsgeschwindigkeit f	vitesse f d'altération	скорость старения
A 278	agitate/to	rühren, schütteln	mouvoir, remuer, agiter	мешать, перемешивать
A 279	agitated crystallizer	Rührkristallisator m	cristallisoir m agité	кристаллизатор с мешалкой
A 280/1	agitated tank	Rührtank m	cuve f agitée	смеситель, чан с мешалкой, бак для смешения
	agitating	s. A 282		
	agitating pump	s. C 263		
A 282	agitation, agitating	Rühren n, Umrühren n, Schütteln n; Bewegung f	agitation f, mouvement m de secousses	размешивание, перемешивание
	agitation	s. a. S 256		
A 283	agitation vat	Bottich m mit Rührer	cuve f à rincer, cuvier m mélangeur	чан с мешалкой, агитатор
A 284	agitator	Rührer m, Rührwerk n, Rührvorrichtung f, Rührapparat m	remueur m, mélangeur m, agitateur m	мешалка, аппарат с мешалкой
A 285	agitator arm, agitator van	Rührflügel m	ailette (aile) f agitatrice, agitateur m	лопасть мешалки
A 286	agitator autoclave	Rührwerksautoklav m	autoclave m agité	автоклав с мешалкой
A 287	agitator dryer	Rührtrockner m	sécheur m agité	сушилка с мешалкой
	agitator van	s. A 285		
A 288	agitator vessel, mixing kettle	Rührwerksbehälter m, Rührwerkskessel m	réservoir m agité, agitateur m	чан (котел, емкость) с мешалкой
A 289	air and steam blast	Luftdampfgebläse n	soufflerie f à air et à vapeur	паровоздушное дутье
A 290	air bath	Luftbad n	bain m d'air	воздушная баня
A 291	air blast	Wind m, Gebläseluft f, Heißwind m	vent m soufflé, vent, air m forcé	[воздушное] дутье

	air blast mill	s. A 382		
A 292	air bleed vent	Entlüftungsventil n	robinet m de purge, soupape f de dégagement d'air, ventouse f	воздушный кран
A 293	air blow	Luftblasen n, Durchblasen n mit Luft	soufflage m d'air	дутье, воздушное дутье, продувание воздухом
A 294/5	air blower	Luftgebläse n	soufflerie f d'air	воздуходувка
A 296	air-blown	mit Luft durchgeblasen	soufflé [d'air]	продутый воздухом
A 297	air bottle [of pump]	Windkessel m	boîte f à vent, réservoir m à air	воздушный колпак [насоса]
A 298	air bypass	Luftbypass m	by-pass m d'air	обводная линия для воздуха
A 299	air casing	Luftkühlmantel m	chemise f réfrigérante à air	воздушная охладительная рубашка
A 300	air chamber	Luftbehälter m, Windkessel m, Luftkammer f	chambre f d'air, réservoir m à air, cloche f à vent	воздушная камера, воздушный колпак
A 301	air channel, air vent	Luftkanal m	canal m d'air, évent m	воздуховод
A 302	air circulation, air cycle	Luftkreislauf m, Luftzirkulation f	circulation f d'air	воздушная циркуляция
	air classification	s. A 413		
A 303	air classifier	Windsichter m	cribleur m à air, séparateur m à vent	воздушный классификатор (сепаратор)
A 304	air cleaner, air filter	Luftreiniger m, Luftfilter n	filtre m d'air, filtre à air	воздухоочиститель, воздушный фильтр
A 305	air cleaning	Luftreinigung f	épuration f (dépoussiérage m) de l'air	очистка воздуха
A 306	air cleaning plant	Luftreinigungsanlage f	installation f d'épuration de l'air	воздухоочистительное устройство
A 307	air cleanliness	Luftreinheit f	pureté f d'air	чистота воздуха
A 308	air–coal dust mixture	Luft-Kohlenstaub-Gemisch n	mélange m de l'air et du charbon pulvérisé	пылевоздушная смесь, смесь угольного порошка с воздухом
A 309	air cock	Entlüftungshahn m	robinet m décompresseur, clapet m de ventilation	воздушный кран
A 310	air collector	Luftsammler m, Luftsammelkanal m, Lufthaube f	conduite f collectrice d'air	воздухосборник
A 311	air compressor	Luftkompressor m	compresseur m d'air	воздушный компрессор
A 312	air condensation	Luftkondensation f	condensation f d'air	воздушная конденсация
A 313	air conditioning	Klimatisierung f	climatisation f, conditionnement m d'air	кондиционирование воздуха
A 314	air-conditioning circuit	Klimatisierungskreislauf m	circuit m de conditionnement d'air	цикл кондиционирования воздуха
A 315	air-conditioning plant	Klimaanlage f	conditionneur m d'air, climatiseur m	установка для кондиционирования воздуха
A 316	air-conditioning process	Klimatisierungsprozeß m	processus m de conditionnement d'air	процесс кондиционирования воздуха
A 317	air conditions	Luftparameter mpl	paramètres mpl d'air, valeurs fpl caractéristiques de l'air	параметры воздуха
A 318	air conduit, air duct	Luftkanal m	conduite (tuyauterie) f d'air	воздухопровод, провод дутья
A 319	air consumption	Luftverbrauch m	consommation f d'air	потребление воздуха
A 320	air conveyor	pneumatischer Förderer m	installation f de transport pneumatique	пневматический транспортер
A 321	air-cooled	luftgekühlt	refroidi par air, à refroidissement de l'air	охлажденный воздухом, с воздушным охлаждением
A 322	air-cooled condensation	luftgekühlte Kondensation f	condensation f à air	конденсация с воздушным охлаждением
A 323	air-cooled heat exchanger	luftgekühlter Wärmeaustauscher m	échangeur m thermique refroidi par air	теплообменник с воздушным охлаждением
A 324	air cooler	Luftkühler m	refroidisseur m à air, réfrigérant m à air	воздушный холодильник
A 325	air cooling	Luftkühlung f	refroidissement m à air	воздушное охлаждение
A 326	air cooling refrigerating system	Luftkühlkältesystem n	installation f frigorifique d'air, installation de refroidissement d'air	холодильная установка для охлаждения воздуха
A 327	air cooling system	Luftkühlsystem n	système m de refroidissement à air	воздушная система охлаждения
A 328	air cooling zone	Luftkühlzone f	zone f de refroidissement d'air	зона воздушного охлаждения
A 329	air current, air flow (input)	Luftfördermenge f, Luftausstoß m, Luftdurchsatz m, Luftstrom m	courant m d'air, débit m d'air	количество подаваемого воздуха
A 330	air cycle, cold-air process	Luftprozeß m, Kaltluftprozeß m	cycle m d'air froid	воздушный цикл
	air cycle	s. a. A 302		
A 331	air dehydration	Lufttrocknung f	déshydratation f (séchage m) de l'air	воздушная сушка
A 332	air delivery requirements	Förderluftbedarf m	demande f de l'air primaire	потребность в подаче воздуха
A 333	air delivery temperature	Förderlufttemperatur f	température f de l'air primaire	температура подаваемого воздуха

A 334	air density	Luftdichte f	densité f d'air, densité atmosphérique	плотность воздуха
A 335	air diffusion	Luftdiffusion f	diffusion f d'air	диффузия воздуха
A 336	air discharge	Luftentladung f	décharge f dans l'air	выпуск воздуха
A 337	air distributor	Windverteiler m	distributeur m d'air	воздухораспределитель
A 338	air-dried, air-dry	lufttrocken, luftgetrocknet	séché à l'air, sec à l'air	воздушно-сухой
A 339	air drum	Luftkessel m, Druckluftbehälter m	réservoir m à air comprimé	воздухосборник
	air-dry	s. A 338		
A 340	air dryer	Lufttrockner m, Luftentfeuchter m	déshumidificateur (déshydrateur) m d'air	воздушная сушилка
A 341	air drying	Lufttrocknung f	séchage m à l'air, déshumidification f de l'air	воздушная сушка
A 342	air drying plant	Lufttrocknungsanlage f	installation f pour le séchage de l'air	установка для сушки воздуха, воздушная сушилка
	air duct	s. A 318		
A 343	air duct circuit	Luftführungssystem n	circuit m de gaines d'air, système m de conduites aériennes	схема воздушных каналов
	air ejector	s. V 22		
A 344	air ejector condenser	Dampfstrahlkondensator m	éjecteur-condenseur m	параструйный конденсатор
A 345	air elutriation	Windsichtung f	élutriation f	воздушная классификация, воздушная сепарация
A 346	air engine	Kompressor m, Luftstrahlgetriebe n	machine f pneumatique (aérostatique)	компрессор
A 347	air enthalpy	Luftenthalpie f	enthalpie f de l'air	энтальпия воздуха
A 348	air entrainment	Luftporenbildung f, Luftmitführung f	formation f des porosités, entraînement m de l'air	образование воздушных пор
A 349	air entry	Lufteintritt m	admission f de l'air, entrée f d'air	поступление воздуха
A 350	air escape	Luftablaß m, Luftauslaß m	sortie f (échappement m) d'air	утечка воздуха
A 351	air escape cock	Luftablaßhahn m	soupape f de décompression, robinet m de décharge	воздушный кран
A 352	air exhaust	Entlüftung f, Luftabpumpen n	aspiration f d'air, élimination f de l'air	отсос (откачка) воздуха
A 353	air exhaust	Luftabzug m, Luftabzugsloch n	évent m, sortie f d'air	канал для выпуска воздуха
A 354	air exhauster	Entlüfter m, Exhaustor m	désaérateur m, exhausteur m, ventilateur m aspirant	эксгаустер, вытяжной воздушный вентилятор
A 355	air exhaust throat	Luftabzugskanal m	gaine f d'exhaustion d'air	вытяжной канал
A 356	air extractor	Entlüfter m	désaérateur m, appareil m de ventilation	вентилятор высокого давления
A 357	air feed	Lufteinspeisung f, Luftzuführung f	alimentation (adduction) f d'air	подача воздуха
A 358/9	air feeder	Luftzuführungsstutzen m	tubulure f d'admission d'air	патрубок, подающий воздух
	air filter	s. A 304		
A 360	air filter shell	Luftfiltergehäuse n	corps m du filtre à air	корпус воздушного фильтра
A 361	air filtration	Luftfilterung f	filtration f d'air	фильтрование воздуха
A 362	air-floated powder	windgesichtetes Pulver n	poudre f élutriée	тонкий порошок, выделенный воздушной сепарацией
A 363	air flow, air stream	Luftstrom m, Luftströmung f	courant m d'air, écoulement m d'air	ток воздуха, воздушное течение
	air flow	s. a. A 329		
A 364	air flowmeter	Luftmengenmesser m	débitmètre m d'air	воздухомер, воздушный счетчик
A 365	air flow velocity, current velocity	Strömungsgeschwindigkeit f	vitesse f d'écoulement	скорость потока (течения)
A 366	air flue	Luftkanal m	conduit m à air, évent m, canal m d'air	воздушный ход, канал для подачи воздуха
	air for combustion	s. C 427		
A 367	air-free	luftfrei	exempt d'air	безвоздушный
A 368	air handler	Luftaufbereiter m	épurateur m pneumatique	камера обработки воздуха
A 369	air handling	Luftaufbereitung f, Luftbehandlung f	conditionnement m d'air, épuration f pneumatique	пневматическое обогащение, подготовка (обработка) воздуха
A 370	air handling plant	Luftaufbereitungsanlage f	installation f de traitement de l'air, installation de conditionnement d'air	установка для подготовки (обработки) воздуха
A 371	air heater	Lufterhitzer m, Winderhitzer m	réchauffeur m d'air, aérochauffeur m	воздушный подогреватель, калорифер, рекуператор
A 372	air heat load	Luftwärmelast f	charge f thermique d'air	тепловая нагрузка воздуха
A 373/4	air hoist	Drucklufthebezeug n	appareils mpl de levage à air comprimé	воздушный (пневматический) подъемник
	air holder	s. A 426		
A 375/7	air humidificator	Luftbefeuchter m	humidificateur m d'air, saturateur m	увлажнитель воздуха
	air humidity	s. A 389		

A 378	air injection (inlet, input)	Lufteinlaß m, Luftzufuhr f, Luftzuführung f	entrée (amenée, alimentation, adduction) f d'air	впуск (подача, подвод, ток, инжекция) воздуха
	air input	s. a. A 329		
A 379	air intake cooler	Ansaugluftkühler m	refroidisseur m de l'air aspiré	охладитель всасываемого воздуха
A 380	air intermediate cooler	Luftzwischenkühler m	refroidisseur m d'air intermédiaire	промежуточный воздушный холодильник
A 381	air jet	Luftdüse f	tuyère (buse) f d'air	диффузор
A 382	air jet mill, air blast mill	Aeromühle f, Luftstrahlmühle f	broyeur m à courant d'air	пневматическая мельница
A 383	air leakage	Luftaustritt m; Luftaustrittsstelle f	échappement m (sortie f) d'air	утечка воздуха
A 384	air lift	pneumatische Förderung f	transport m pneumatique	пневматическая подача, подъем жидкости смешением с воздухом
A 385	air lift pump	Druck[luft]heber m	dispositif m de levage pneumatique	эрлифт
A 386	air load	Luftbelastung f	charge f de l'air	нагрузка воздуха (по воздуху)
A 387	air mixing unit	Luftmischer m	mélangeur m d'air	смеситель воздуха
A 388	air mixture	Luftgemisch n	mélange m d'air	воздушная смесь
A 389	air moisture, air humidity	Luftfeuchte, f, Luftfeuchtigkeit f	humidité f de l'air	влажность воздуха
A 390	air motion	Luftbewegung f	mouvement m de l'air	движение воздуха
A 391	air outlet conduit	Abluftrohr n	tube m d'évacuation d'air	вытяжная труба, труба для отработанного воздуха
A 392	air outlet conduit	Abluftkanal m	gaine f d'air sortant	вытяжной канал
A 393	air partial pressure	partieller Luftdruck m	pression f partielle d'air, pression atmosphérique partielle	парциальное давление воздуха
A 394	air permeability	Luftdurchlässigkeit f	perméabilité f à l'air	воздухопроницаемость
A 395	air-permeable	luftdurchlässig	perméable à l'air	воздухопроницаемый
A 396	air pilot valve	pneumatisches Steuerventil n	soupape f de manœuvre pneumatique	[автоматический регулирующий] клапан с пневматическим управлением
A 397	air pipe	Luftleitung f	conduite f d'air, tuyau m d'aérage, tuyauterie f d'air	воздухопровод
	air pipe	s. a. H 185		
A 398	air pocket	Luftblase f, Lufteinschluß m	bulle (inclusion, poche) f d'air	воздушный мешок (карман), газовое включение, воздушная раковина
A 399	air pollutant	Luftschadstoff m	élément m polluant de l'air, constituant m nuisible de l'air	загрязняющее воздух вещество
A 400	air pollution	Luftverschmutzung f	pollution f de l'air	загрязнение воздуха
	air preheater	s. E 11		
A 401	air preheating	Luftvorwärmung f	réchauffage m préliminaire d'air	подогрев воздуха, предварительный нагрев воздуха
A 402	air pressure, pneumatic pressure	Luftdruck m	pression f atmosphérique (d'air)	давление воздуха
A 403	air-pressure regulator	Luftdruckregler m	régulateur m de la pression d'air	регулятор давления воздуха
A 404	air-pressure test	Luftdruckprüfung f	épreuve f à air comprimé, essai m de pression à air	испытание сжатым воздухом
A 405	air-processing system	Luftaufbereitungssystem n	système m de traitement de l'air	система обработки воздуха
A 406	air pump	Luftpumpe f	pompe f à air	воздушный насос
A 407	air rate	Luftdurchsatz m, Luftrate f	débit m d'air	расход (кратность) воздуха
A 408	air receiver	Luftherd m, Luftaufnehmer m	réservoir m d'air	воздухосборник, ресивер
A 409	air regulator	Luftregler m	régulateur m d'air	воздушный регулятор, регулятор подачи воздуха
A 410	air requirements	Luftbedarf m	quantité f d'air nécessaire	потребность в воздухе
A 411	air saturator	Luftbefeuchter m	saturateur m	увлажнитель воздуха, воздухонасытитель
A 412	air separation	Luftzerlegung f	fractionnement m d'air	разложение (разделение) воздуха
A 413	air separation, air classification	Windsichtung f	élutriation f, séparation f à air	воздушная сепарация (классификация)
A 414	air separation method	Luftzerlegungsverfahren n	méthode f de fractionnement d'air	метод разделения воздуха
	air separation plant	s. A 416		
A 415	air separation process	Luftzerlegungsprozeß m	processus m de fractionnement d'air	процесс разделения воздуха
A 416	air separator, air separation plant	Luftzerlegungsanlage f, Lufttrennungsanlage f	installation f pour la séparation de l'air, unité f de fractionnement de l'air	установка для разделения воздуха, воздухоразделительная установка
A 417	air separator	Luftabscheider m, Entlüfter m	séparateur m d'air, désaérateur m	воздухоочиститель, воздушный сепаратор
A 418	air separator mill	Mühle m mit Windsichtung	broyeur m avec séparateur à air	мельница с воздушной сепарацией

A 419	air shrinkage	Luftschwindung f, Trocken-schwindung f	retrait m dû au séchage	воздушная усадка
A 420	air source	Luftquelle f	source f d'air	источник воздуха
A 421	air speed	Luftgeschwindigkeit f	vitesse f d'air	скорость движения воздуха
	air-steam mixture	s. A 443		
A 422/4	air strainer	Luftfilter n	filtre m à air	воздушный фильтр
	air stream	s. A 363		
A 425	air supply	Luftzuführung f, Luftzufuhr f	alimentation (adduction) f d'air	подача воздуха
A 426	air tank, air holder	Luftaufnehmer m, Druckluft-behälter m	récipient m d'air [comprimé], réservoir m de pression	воздухосборник
A 427	air temperature	Lufttemperatur f	température f d'air	температура воздуха
A 428	air throttling damper	Luftdrosselklappe f	clapet m d'étranglement d'air, papillon m d'air	воздушный дроссельный клапан
A 429	air-tight	luftdicht	étanche à l'air, hermétique	воздухонепроницаемый
A 430	air-tight joint	luftdichte Verbindung f, her-metische Dichtung f	garniture f hermétique, jonc-tion f imperméable à l'air	герметическое соединение
A 431	air-tight seal	luftdichter Abschluß m	fermeture f hermétique	воздухонепроницаемый за-твор
A 432	air-to-air heat exchanger	Luft-Luft-Wärmeaustauscher m	échangeur m thermique air-air, aéroréfrigérant m d'air	воздухо-воздушный тепло-обменник
A 433	air-to-fluid heat exchanger	Luft-Flüssigkeits-Wärmeüber-trager m	échangeur m de chaleur air-liquide	воздушно-жидкостный теп-лообменник
A 434	air-to-liquid heat transfer	Luft-Flüssigkeits-Wärmeüber-tragung f	transfert m de chaleur air-li-quide	теплопередача от воздуха к жидкости
A 435	air trap	Entlüftungsvorrichtung f, Luftabscheider m	désaérateur m, dispositif m de ventilation, dispositif décompresseur	отделитель воздуха
A 436	air treatment	Luftabsaugung f	aspiration f d'air, exhaustion f d'air	очистка (откачка) воздуха, вентиляция
A 437	air tube	Druckluftschlauch m	tuyau m à air comprimé	пневматический рукав
A 438	air-tube radiator	Luftröhrenkühler m	réfrigérant m à tuyaux d'air	трубчатый радиатор с цир-куляцией воздуха по трубкам
A 439	air turbulence	Luftturbulenz f	turbulence f de l'air	турбулентность потока воз-духа
A 440	air-vapour circulation	Dampf-Luft-Gemischzirkula-tion f	circulation f d'air-vapeur	циркуляция паровоздушной смеси
A 441	air-vapour enthalpy	Enthalpie f des Dampf-Luft-Gemisches	enthalpie f de l'air-vapeur	энтальпия паровоздушной смеси
A 442	air-vapour entropy	Dampf-Luft-Gemischentropie f	entropie f de l'air-vapeur	энтропия паровоздушной смеси
A 443	air-vapour mixture, air-steam mixture	Dampf-Luft-Gemisch n	mélange m d'air-vapeur	паровоздушная смесь
A 444	air vent	Luftabzug m, Luftablaßventil n	évent m, volet m de sortie d'air	воздушник, отводный канал
	air vent	s. a. A 301		
A 445	air ventilation	Lüftung f	ventilation f, aération f, aérage m	вентиляция, вентилирова-ние
A 446	air ventilation system	Lüftungsanlage f, Belüftungs-anlage f	installation f de ventilation, installation d'aération, dis-positif m d'aération	вентиляционная установка, вентиляционное устрой-ство
A 447	air vessel	Windkasten m, Luftaufneh-mer m	réservoir m à air, boîte f à vent	воздушный колпак, возду-хосборник
A 448	air volume	Luftvolumen n	volume m d'air	объем воздуха
A 449	air volume flow rate	Luftmengenstrom m	débit m d'air	объемный поток воздуха
A 450	alarm signal	Alarmsignal n	signal m d'alarme	сигнал тревоги, аварийный сигнал
A 451	alarm system	Signalanlage f; Alarmanlage f	système m d'alarme, installa-tion f de signalisation	система аварийной сигна-лизации, сигнальная си-стема
A 452	alignment	Einstellen n, Ausrichten n	réglage m, positionnement m, mise f au point	установка, регулировка, на-ладка
A 453	aliphatic	aliphatisch	aliphatique	алифатический
A 454	alkalinity	Alkalität f, Alkalinität f, alkali-sche Eigenschaft f	alcalinité f	щелочность
A 455	alkaliproof	laugenbeständig	résistant à la lessive	щелочестойкий
A 456	allocation	Zuteilung f, Zuweisung f, Aufstellung f, Anordnung f	allocation f, allotissement m	размещение, распределе-ние
A 457	allocation plan	Belegungsplan m	plan m d'attribution	план загрузки, план распре-деления
	allocation problem	s. A 585		
A 458	allowable limits	Toleranzbereich m	marge (zone) f de tolérance	поле допуска
	allowance for wear	s. W 136		
	alteration	s. C 158		
A 459	alteration of density	Dichteänderung f	variation f de la densité, changement m de densité	изменение плотности
A 460	ambient	umgebend, einschließend; Umgebungs..., Neben...	ambiant	окружающий
A 461	ambient condition, ambient medium	Umgebungsbedingung f, Um-gebungsmedium n	condition f ambiante, milieu m ambiant	условие окружающей среды, окружающая среда

A 462	ambient heat	Umgebungswärme f	chaleur f d'ambiance	теплота окружающей среды
	ambient medium	s. A 461		
A 463	ambient temperature	Umgebungstemperatur f	température f ambiante	температура окружающей среды
A 464	ammonia absorber	Ammoniakabsorber m	absorbeur m d'ammoniac	аммиачный абсорбер
A 465	ammonia absorption	Ammoniakabsorption f	absorption f d'ammoniac	абсорбция аммиака
A 466	ammonia circulation	Ammoniakumlauf m	recirculation f d'ammoniac	циркуляция аммиака
A 467	ammonia compressor	Ammoniakverdichter m	compresseur m d'ammoniac	аммиачный компрессор
A 468	ammonia evaporation	Ammoniakverdampfung f	vaporisation f d'ammoniac	испарение аммиака
A 469	ammonia evaporator	Ammoniakverdampfer m	vaporisateur m d'ammoniac	аммиачный испаритель
A 470	ammonia receiver	Ammoniakbehälter m	réservoir m à ammoniac	аммиачный ресивер
A 471	ammonia synthesis	Ammoniaksynthese f	synthèse f d'ammoniac	синтез аммиака
	ammonia water	s. G 101		
A 472	amortization	Amortisation f	amortissement m	погашение, амортизация
	amount of compression	s. D 85		
A 473	amount of energy	Energiebetrag m, Energie-menge f	quantité f (montant m) d'énergie	количество энергии
A 474/5	amount of liquid	Flüssigkeitsmenge f	quantité f de liquide	количество жидкости
	amount of moisture	s. M 303		
A 476	amount of steam	Dampfmenge f	quantité f (débit m) de va-peur	количество пара
A 477	amount of total air	Gesamtluftmenge f	quantité f totale d'air	общее количество воздуха
A 478	amount of water	Wassermenge f	quantité f (volume m) d'eau	количество воды
A 479	amplification, strengthening, reinforcement, concentra-tion, enrichment	Verstärkung f	renforcement m, intensifica-tion f, amplification f	усиление, укрепление
A 480	anaerobic fermentation	anaerobe Fermentation f	fermentation f anaérobie	анаэробная ферментация
A 481	analogue computer	Analogrechner m	calculateur m analogique	аналоговая вычислительная машина, аналоговый ком-пьютер
A 482	analogue computing tech-nique	Analogrechentechnik f	technique f de calcul analo-gique	аналоговая вычислительная техника
A 483	analogue-digital computer	Analog-Digital-Rechner m	calculatrice f analogique-di-gitale	аналого-цифровая вычисли-тельная машина
A 484	analogue-digital simulation	analog-digitale Simulation f	simulation f analogique-digi-tale	аналого-цифровое модели-рование
A 485	analogue-digital technique	Analog-Digital-Technik f	technique f analogique-nu-mérique	аналого-цифровая техника
A 486	analogue hybrid computer	Analoghybridrechner m	calculateur m analogique hy-bride	гибридная вычислительная машина
A 487	analysis by boiling (distilla-tion)	Siedeanalyse f, Destillations-probe f	analyse f fractionnée, analyse par distillation fractionnée, essai m de distillation	анализ разгонкой
A 488	analysis of variance	Varianzanalyse f	analyse f dispersionnelle (de variance)	вариационный анализ
A 489	analytical structure	analytische Struktur f	structure f analytique	аналитическая структура
A 490	analytic expression	analytischer Ausdruck m	expression f analytique	аналитическое выражение
A 491	analytic method	analytisches Verfahren n, analytische Methode f	méthode f analytique	аналитический метод
A 492	analytic relationship	analytische Beziehung f	relation f analytique	аналитическое соотноше-ние
A 493	analytic simulation	analytische Simulation f	simulation f analytique	аналитическое моделирова-ние
A 494	analyzer	Analysengerät n	appareil m d'analyse	анализатор
A 495	anchor agitator (mixer)	Ankerrührer m	agitateur m à ancres	якорная мешалка
A 496	angle-dependent	winkelabhängig	dépandant de l'angle, en fonction de l'angle	зависящий от угла
A 497	angle of incidence	Einfallswinkel m	angle m d'incidence	угол нападения
A 498	angle of reflection	Ausfallswinkel m	angle m de réflexion	угол отражения
A 499	angle of repose	Schüttwinkel m, Rutschwin-kel m	angle m naturel de repos, an-gle de déversement (glisse-ment)	угол естественного откоса
A 500	angle of rotation	Drehungswinkel m, Drehwin-kel m, Rotationswinkel m	angle m de rotation	угол вращения
A 501	angle valve	Eckventil n	robinet m (soupape f) d'équerre	угловой вентиль
A 502	angular momentum	Drehimpuls m, Impulsmo-ment n	quantité f de mouvement an-gulaire	импульс вращения, момент импульса
A 503	angular velocity	Winkelgeschwindigkeit f	vitesse f angulaire	угловая скорость (частота)
A 504	anhydric, anhydrous	wasserfrei	anhydre	безводный
A 505	annual balance	Jahresbilanz f, Jahresab-schluß m	bilan m annuel	годовой баланс
A 506	annul/to	vernichten	annuler, éliminer, annihiler	уничтожать, разрушать
A 507	annular flow	Ringströmung f, Ringraum-strömung f	courant m annulaire, écoule-ment m circulaire (annu-laire)	кольцевой [пространст-венный] поток
A 508	annulus	Ring m; Ringrohr n, Mantel-rohr n	tube m annulaire	кольцо, кольцевая трубка

A 509	anomalous viscosity	anomale Viskosität f	viscosité f anomale	аномальная вязкость
A 510	anticlockwise motion	Gegenlauf m	antirotation f, rotation f antagoniste, mouvement m à marche contraire	вращение в противоположном направлении
A 511	anticorrosive	Korrosionsschutzmittel n	anticorrosif m, agent m anticorrosif	антикоррозионное средство
A 512	anticorrosive coat, protection coat	Korrosionsschutzschicht f	couche f anticorrosive	антикоррозионный защитный слой
A 513	antifoam, antifoaming agent	Entschäumer m	agent m antimousse	антипенное средство, пеноудаляющий реагент
A 514	antifouling medium	Antifoulingmittel n	agent m antifouling	противообрастающее средство
A 515	antifreeze	Gefrierschutzmittel n, Frostschutzmittel n, Gefrierschutz m	agent m antigel, antigel m	антифриз; вещество, понижающее температуру застывания жидкости
A 516	antifreezing property	Kältebeständigkeit f	résistance f au froid	морозостойкость, холодостойкость
A 517	antifroth, defoaming agent	Demulgator m, Entschäumer m, Antischaummittel n	agent m antimousse	пеноустраняющий (пеноуничтожающий) агент, противопенное средство
A 518	aperiodic regime	aperiodischer Betriebszustand m	régime m apériodique, état m de fonctionnement apériodique	апериодический режим
A 519	aperture angle	Öffnungswinkel m	angle m d'ouverture, ouverture f angulaire	угол раскрытия
	apparatus	s. J 16		
A 520	apparatus construction	Apparatebau m	construction f des appareils	аппаратостроение
A 521	apparatus design	Apparateauslegung f	dimensionnement m d'appareil, conception f des appareils	расчет аппарата
A 522	apparatus determination	Apparatefestlegung f	détermination f (choix m) d'appareils	выбор аппаратов
A 523	apparatus platform	Apparatepodest n	plate-forme f de l'appareil	площадка для аппарата
A 524	apparent viscosity	Scheinviskosität f	viscosité f apparente	кажущаяся вязкость
	appear/to	s. O 8		
	appliance	s. J 16		
A 525	applicability	Anwendungsmöglichkeit f, Verwendungsmöglichkeit f, Einsatzmöglichkeit f	possibilité f d'application, possibilité d'utilisation	возможность применения
A 526	application, employment	Anwendung f, Einsatz m, Aufbringen n	application f, utilisation f	применение, приложение, употребление, внесение
A 527	apply/to	anwenden, verwenden; gelten	utiliser	применять, использовать
A 528	appreciation, evaluation	Einschätzung f, Abschätzung f, Bewertung f	évaluation f	оценка
A 529	approach	Herangehen n, Annäherung f, Betrachtungsweise f	approche f	подход
	approach	s. a. P 494		
A 530	approach velocity	Annäherungsgeschwindigkeit f	vitesse f d'approche	скорость сближения
A 531	approximate calculation	Näherungsrechnung f	calcul m approché	приближенный расчет
A 532	approximate equation	Näherungsgleichung f	équation f approximative	приближенное уравнение
A 533	approximately	näherungsweise	approximatif	приблизительно
	approximate method	s. A 539		
A 534	approximate quantity	Näherungsgröße f	grandeur f approchée	приближенная величина
A 535	approximate value	Näherungswert m, Annäherungswert m	valeur f approchée	приближенное значение, приближенная величина
A 536	approximation	Annäherung f, Näherungslösung f	approximation f, approche f	приближение, аппроксимация
A 537	approximation degree	Annäherungsgrad m	degré m d'approximation	степень приближения
A 538	approximation formula	Näherungsformel f	formule f d'approximation	приближенная формула
A 539	approximation method, approximate method	Näherungsmethode f, Näherungsverfahren n	méthode d'approximation, méthode approximative	метод аппроксимации (приближения), приближенный метод
A 540	approximation theory	Approximationstheorie f	théorie f d'approximation	теория аппроксимации
A 541	aqueous	wasserartig, wasserhaltig, wäßrig verdünnt	aqueux	водный
A 542	aqueous phase	wäßrige Phase f	phase f aqueuse	водная фаза
A 543	aqueous solution	wäßrige Lösung f	solution f aqueuse	водный раствор
A 544	arc furnace	Lichtbogenofen m	four m à arc	дуговая печь
A 545	Archimedean screw	Archimedes-Schnecke f, Archimedes-Spirale f	vis f d'Archimède	шнек (спираль) Архимеда
A 546	Archimedes' principle	Archimedessches Prinzip n	principe m d'Archimède	принцип Архимеда
A 547	arc of circle	Kreisbogen m	arc m [de cercle]	дуга круга
A 548	arithmetic constant	arithmetische Konstante f	constante f arithmétique	арифметическая константа
A 549	arithmetic data	arithmetische Daten pl	données fpl arithmétiques	арифметические данные
A 550	arithmetic mean	arithmetisches Mittel n	moyenne f arithmétique	арифметическое среднее
A 551	arithmetic mean temperature	arithmetisches Temperaturmittel n	température f moyenne arithmétique	среднеарифметическая температура

A 552	aromatics extraction	Aromatenextraktion f	extraction f des aromatiques, extraction des corps à noyau benzénique	экстрагирование ароматических углеводородов
A 553	arrange/to	ordnen, anordnen, einordnen, einreihen, einrichten	arranger	располагать, устраивать
A 554	arrangement	Einordnung f	classement m, classification f	размещение
A 555	arrangement, graph, layout, plan	Anordnung f, Aufstellung f, Anlage f	disposition f, arrangement m, plan m	расположение, установка, распределение, схема, расстановка, план, распорядок
A 556	arrangement planning	Anordnungsplanung f, Aufstellungsprojekt n	planification f de disposition, projet m d'arrangement	планирование компоновки (размещения), планировка расположения
A 557	arrangement principle	Anordnungsprinzip n	principe m d'arrangement, principe de disposition	принцип компоновки, принцип размещения
A 558	Arrhenius law	Arrhenius-Gesetz n, Arrhenius-Beziehung f	loi f d'Arrhenius	закон Аррениуса
A 559	asbestos insulation	Asbestisolierung f	isolation f à l'amiante	асбестовая изоляция
A 560	ascending conveyor	Steigförderer m	transporteur m ascendant (mécanique)	вертикальный конвейер, транспортер
A 561	ascending pipe	Steigleitung f	tuyau m ascendant, conduite f montante (de refoulement)	нагнетательный трубопровод, стояк
A 562	ascent	Anstieg m, Steigung f, Neigung f, Gefälle n	pente f	наклон, подъем
A 563	ascent velocity	Aufstiegsgeschwindigkeit f, Steiggeschwindigkeit f	vitesse f d'ascension	скорость подъема
A 564	ash removal bin	Aschenabzug m, Aschenbehälter m	cendrier m	зольник, золоудалитель
A 565	aspirator	Saugapparat m, Exhaustor m	aspirateur m, extracteur m	аспиратор
	aspirator	s. a. J 14		
A 566	assemblage, fitting, assembling, installation	Montierung f, Montage f	montage m, assemblage m	монтаж, сборка, установка
A 567/8	assemble/to	montieren	monter, assembler, mettre en place	монтировать, собирать
A 569	assembling condition	Montagebedingung f	condition f de montage	монтажное условие
A 570	assembling process	Montageprozeß m	procédé m de montage	процесс сборки (монтажа)
A 571	assembling technology	Montagetechnologie f	technologie f de montage	технология монтажа, технология сборки
A 572	assembling unit	Montageeinheit f	unité f d'assemblage, sous-groupe m de montage	монтажная группа (единица)
	assembly	s. U 17		
A 573	assembly beam	Montageträger m	poutre f de montage	монтажная балка
A 574	assembly hole, manhole	Montageöffnung f	orifice m de montage	монтажное отверстие, лаз, люк
A 575	assembly instruction	Montagehinweis m	instruction f d'assemblage, instruction de montage	монтажная инструкция
A 576	assembly instrument	Montagemittel n	outillage m de montage	монтажное приспособление, монтажный инструмент
A 577	assembly method	Montagemethode f	méthode f de montage	метод монтажа (сборки)
A 578	assembly model	Montagemodell n	modèle m de montage	монтажная модель
A 579	assembly operation	Montageoperation f	opération f de montage	операция монтажа (сборки)
A 580	assembly organization	Montageorganisation f	organisation f de montage	организация монтажа (сборки)
A 581	assembly seam	Montagenaht f	soudure f de montage	монтажный шов
A 582	assembly section	Montageabteilung f	atelier m de montage	сборочный цех
A 583	assembly technique	Montagetechnik f	technique f de montage	техника монтажа (сборки)
A 584	assignment list	Zuordnungsliste f	liste f d'attribution, liste d'affectation	список смежности
A 585	assignment problem, allocation problem	Zuordnungsproblem n, Zuteilungsproblem n	problème m d'attribution, problème d'affectation	задача о назначении
A 586	associate/to	assoziieren, vereinigen, verbinden, in Verbindung setzen	associer, lier, combiner	связывать, соединять, соединяться с ...
A 587	association, combination, union	Assoziation f, Vereinigung f, Verbindung f	association f, réunion f, jonction f, assemblage m, fusion f	ассоциация, объединение, соединение
	assorting	s. S 436		
	assortment	s. S 436		
A 588	assumption, supposition, condition	Annahme f, Voraussetzung f	supposition f, hypothèse f	допущение, предположение, предпосылка
	asynchronous alternator	s. I 95		
A 589	atmospheric condenser	Berieselungskondensator m, Rieselkondensator m	condenseur m atmosphérique, condenseur à ruissellement	оросительный конденсатор
A 590	atmospheric oxygen	Luftsauerstoff m	oxygène m atmosphérique	кислород воздуха
A 591	atmospheric pressure	Außendruck m	pression f extérieure	атмосферное (внешнее) давление
	atomic energy	s. N 88		

A 592	atomic fuel	s. F 176		
A 592	atomic heat	Atomwärme f	chaleur f atomique (molécu-laire)	атомная теплоемкость
A 593	atomic pile, nuclear reactor (pile)	Kernreaktor m	réacteur m (pile f) nucléaire	атомный реактор
A 594	atomic polarization	Atompolarisation f	polarisation f atomique	атомная поляризация
A 595	atomic refraction	Atomrefraktion f	réfraction f atomique	атомное преломление
A 596	atomic volume	Atomvolumen n	volume m atomique	атомный объем
A 597	atomic weight	Atommasse f	poids m atomique	атомный вес
A 598	atomize/to	atomisieren, fein zerstäuben	atomiser	распылять, тонко измель-чать
A 599	atomizer, sprayer	Zerstäuber m	atomiseur m, pulvérisateur m	распылитель
A 600	atomizing	Zerstäuben n, Zerstäubung f	atomisation f, pulvérisation f	распыление, тончайшее из-мельчение
A 601	atomizing scrubber	Zerstäubungswäscher m	laveur m (tour f de lavage) à pulvérisation, scrubber m	распылительный скруббер
A 602	attachment, fastening	Befestigung f	fixation f, attache f, fixage m	закрепление, укрепление, крепление, прикрепление
	attendance	s. 1. M 41; 2. O 77		
A 603	attenuation factor	Dämpfungsfaktor m	constante f d'amortissement, facteur m d'amortissement	коэффициент затухания
A 604	attenuation region	Dämpfungsbereich m	domaine m d'amortissement, zone f d'amortissement	область затухания (заглуше-ния)
A 605	attractive force	Anziehungskraft f	force f d'attraction	сила притяжения
A 606	attrition product	Ausmahlprodukt n	produit m de convertissage	продукт помола (вымола)
A 607	autocatalysis	Autokatalyse f	autocatalyse f	автокатализ
A 608	autocatalytic reaction	autokatalytische Reaktion f	réaction f autocatalytique	автокаталитическая реакция
A 609	autoclave	Autoklav m, Druckkocher m	autoclave m	автоклав
A 610	automatic control	[automatische] Regelung f	contrôle m automatique, ré-gulation f, réglage m, tech-nique f du contrôle auto-matique	автоматическое регулиро-вание
A 611	automatic receiving bunker	automatischer Aufnahmebun-ker m	trémie (soute) f de réception automatique	автоматический приемный бункер
	automation-friendly design	s. D 188		
A 612	autooxidation	Autooxydation f	autooxydation f	самоокисление, автоокис-ление
A 613	auxiliary agents, auxiliary material	Hilfsstoffe mpl, Hilfsmaterial n	auxiliaires mpl, matériau m auxiliaire	вспомогательные материа-лы
A 614	auxiliary agent supply	Hilfsstoffversorgung f	approvisionnement m en ma-tières auxiliaires	снабжение вспомога-тельным материалом
A 615	auxiliary air	Hilfsluft f	air m additionnel (auxiliaire)	добавочный воздух
A 616	auxiliary circuit	Hilfskreislauf m	circuit m auxiliaire	вспомогательная схема, вспомогательный цикл
A 617	auxiliary condenser	Hilfskondensator m, Reserve-kondensator m	condenseur m auxiliaire	вспомогательный (ре-зервный) конденсатор
A 618	auxiliary current	Hilfsstrom m	courant (flux) m auxiliaire	оперативный ток
A 619	auxiliary engine	Hilfsmotor m	moteur m auxiliaire	вспомогательный двигатель
A 620	auxiliary equipment	Hilfsanlage f, Nebenanlage f	installation f auxiliaire	вспомогательная установка
A 621	auxiliary equipment	Zusatzausrüstung f	appareillage m complémen-taire, équipement m addi-tionnel (auxiliaire)	вспомогательное оборудо-вание
A 622	auxiliary force	Hilfskraft f	force f auxiliaire	вспомогательная сила
	auxiliary material	s. A 613		
A 623	auxiliary material stream	Hilfsstoffstrom m	débit (flux) m de matières se-condaires	вспомогательный мате-риальный поток
A 624	auxiliary operation	Hilfsoperation f	opération f auxiliaire	вспомогательная операция
A 625	auxiliary power	Hilfsenergie f	énergie f auxiliaire	вспомогательная энергия
A 626	auxiliary process	Hilfsprozeß m	processus m auxiliaire	вспомогательный процесс
A 627	auxiliary product	Hilfsprodukt n	produit m auxiliaire	вспомогательный продукт
A 628	auxiliary product for filtration	Hilfsmittel n für Filtration	agent m auxiliaire de filtra-tion	фильтровальная присадка
A 629	auxiliary quantity	Hilfsgröße f	grandeur (quantité) f auxi-liaire	вспомогательная величина
A 630	auxiliary series	Nebenreihe f	série f secondaire	вспомогательный (по-бочный) ряд
A 631	availability	Verfügbarkeit f	disponibilité f	готовность, доступность, пригодность
	average	s. C 857		
A 632	average ambient tempera-ture	mittlere Umgebungstempera-tur f	température f moyenne am-biante	средняя температура окру-жающей среды
A 633	average evaporation temper-ature	mittlere Verdampfungstempe-ratur f	température f moyenne d'évaporation	средняя температура испа-рения
A 634	average life	mittlere Lebensdauer f	durée f d'existence moyenne, longévité f moyenne	средняя продолжитель-ность жизни
A 635	average molecular weight	mittlere [relative] Molekül-masse f	masse f moléculaire moyenne	средний молекулярный вес
A 636	average performance	Durchschnittsleistung f	puissance f moyenne, rende-ment m moyen	средняя производитель-ность
A 637	Avogadro's number	Avogadrosche Zahl f	nombre m d'Avogadro	число Авогадро
	avoid/to	s. P 449		

A 638	axial annular flow	rotationssymmetrische Ringströmung f	écoulement m annulaire symétrique à l'axe	ротационно-симметричное кольцевое течение
A 639	axial centrifugal pump	Axialkreiselpumpe f	pompe f centrifuge à écoulement axial	осевой центробежный насос
A 640	axial dispersion	Axialdispersion f, Längsvermischung f	dispersion f axiale	продольная дисперсия, осевое перемешивание
A 641	axial flow	axiale Strömung f	écoulement m axial	осевой поток
A 642	axial flow compressor	Axialverdichter m	compresseur m axial	осевой компрессор
A 643	axial flow pump	Axialpumpe f	pompe f axiale	осевой насос
A 644	axial flow turbine	Axialstrahlturbine f	turbine f axiale	реактивный двигатель с осевым компрессором
A 645	axial mixing	Axialvermischung f	mélange m axial	осевое перемешивание
A 646	axial strain	Axialdehnung f	allongement m axial	осевое растяжение
A 647	axial thrust	Axialschub m	poussée (force) f axiale	осевой сдвиг
A 648	axisymmetric	achsensymmetrisch	symétrique par rapport à l'axe	осесимметрично
A 649	azeotropic column	Azeotropenkolonne f	colonne f azéotropique	азеотропная колонна
A 650	azeotropic distillation	azeotrope Destillation f, Azeotropdestillation f	distillation f azéotropique	азеотропная дистилляция
A 651	azeotropic fluid	azeotrope Flüssigkeit f, Azeotropgemisch n	liquide (mélange) m azéotropique	азеотропная среда, азеотропная жидкость
A 652	azeotropic mixture	azeotrope Mischung f	mélange m azéotropique	азеотропная смесь
A 653	azeotropic point	Azeotroppunkt m	point m azéotropique	азеотропная точка

B

B 1	back end, last runnings	Nachlauf m (Destillation)	repasse f, après-coulant m, résidu m de redistillation, queue f	хвостовой погон
B 2	back flow	Rückstrom m	courant m de retour, retour m de courant	обратный поток
B 3	back mixing	Rückvermischung f	mélange m en retour	обратное смешение
B 4	backmix reactor	Rückvermischungsreaktor m	réacteur m à remélangeage	реактор с обратным перемешиванием
B 5	back-pressure turbine	Gegendruckturbine f	turbine f à contre-pression	турбина с противодавлением
B 6	back radiation (reflection)	Rückstrahlung f	réflexion f, rétroflexion f	лучеотражение, рефлексия
B 7	back stroke	Rückschlag m	choc m en arrière, contrecoup m	отскок, отдача, обратный удар
B 8	back valve	Rückschlagventil n, Rückschlagklappe f	soupape f de retenue	обратный клапан
B 9	backward flow	Rückwärtsströmung f	écoulement m en retour, flux m arrière	обратный поток
	back water	s. W 66		
B 10	baffle [plate]	Leitblech n, Prallblech n, Führungsblech n, Prellplatte f, Ablenkplatte f	chicane f, tôle-guide f, plaque f de déviation	отбойная (отражающая) перегородка
B 11	bag filter	Sackfilter n, Schlauchfilter n	filtre m à sac (manche)	мешочный (рукавный) фильтр
B 12	bagging machine	Absackvorrichtung f	dispositif m d'ensachage	автоматический мешконаполнитель
	bagging-off facility	s. S 2		
B 13	bagging scales	Absackwaage f	balance f ensacheuse, balance d'ensachage	весовыбойный аппарат
B 14	Bagley [end] correction factor	Bagley-Korrektur f	correction f de Bagley	корректура Бэглея
B 15	balance/to	abgleichen, ausgleichen	équilibrer, égaliser, niveler	выравнивать
B 16	balance, equilibrium	Gleichgewicht n	bilan m, balance f, équilibre m	баланс, равновесие
B 17	balance boundaries	Bilanzhülle f, Bilanzierungsgrenzen fpl	limites fpl de bilan	границы баланса, пределы баланса
B 18	balance conditions	Gleichgewichtsbedingungen fpl	conditions fpl d'équilibre	условия равновесия
B 19	balance equation	Bilanzgleichung f	équation f de bilan	уравнение баланса, балансовое уравнение
B 20	balance period	Bilanzzeitraum m	période f de bilan	период составления баланса
B 21	balance quantities	Bilanzgrößen fpl	grandeurs fpl de bilan	параметры (величины) баланса
B 22	balance range (region)	Bilanzbereich m, Bilanzgebiet n	zone f (rayon m, domaine m) de bilan	область баланса, балансовая область
B 23	balance smoothing	Bilanzausgleich m	égalisation f (équilibrage m) de bilan	выравнивание (уравновешивание) баланса, сбалансирование
B 24	balance space	Bilanzraum m	domaine (secteur) m de bilan	балансовое пространство, пространство балансирования, объем составления баланса
B 25	balancing	Bilanzierung f	mise f en équations de bilan	составление баланса
B 26	balancing	Abgleichen n, Ausgleichen n	équilibrage m, alignement m, réglage m	выравнивание, уравновешивание, компенсирование

B 27	balancing of masses	Massenausgleich *m*	équilibrage *m* de masses	корректировка масс при балансировании
B 28	balancing vapour tank, equalizing vapour tank, receiver	Ausgleichsdampfbehälter *m*	réservoir *m* compensateur à vapeur	уравнительный паровой резервуар, ресивер
B 29	ballast	Ballast *m*	lest *m*, surcharge *f*	балласт
B 30	ball mill, pebble (stirred-bead) mill	Kugelmühle *f*	moulin (broyeur) *m* à boules, moulin (broyeur) à boulets	шаровая мельница
B 31	ball valve	Kugelventil *n*	soupape *f* à bille	шаровой клапан
B 32	Banbury mixer	Banbury-Mischer *m*	mélangeur *m* Banbury	смеситель Банбури
B 33	band-pass filter	Bandfilter *n*	filtre *m* à courroie	ленточный (полосный) фильтр
B 34	band spectrum	Bandenspektrum *n*	spectre *m* de bandes	полосатый спектр
B 35	bar/to	absperren, schließen, unterbrechen	arrêter, stopper, fermer, couper	запирать, преграждать, закрывать
B 36	bare-pipe evaporator	Glattrohrverdampfer *m*	évaporateur *m* à tubes lisses	гладкотрубный испаритель
B 37	bare-tube water chiller	Glattrohrwasserkühler *m*	refroidisseur *m* de l'eau à tubes lisses	водоохладитель с гладкими трубами
B 38	bare vessel	nichtisolierter Kessel *m*	chaudière *f* sans isolement	неизолированный (голый) котел
B 39	barometric leg	barometrisches Fallrohr *n*, Rohr *n* eines barometrischen Kondensators	colonne *f* barométrique	барометрическая труба конденсатора
B 40	barrel shell, basket shell	Trommelmantel *m*	manteau *m* de tambour	кожух барабана, решето барабанного грохота
B 41	bar screen	Siebrost *m*, Stangenrost *m*	grille *f* à barreaux, grille de crible	колосниковый грохот
B 42	base, starting point	Ausgangspunkt *m*	point *m* de départ, base *f*	исходная точка, базис
	base	*s. a.* S 844		
B 43	base charge	Grundbelastung *f*	charge *f* de base	основная нагрузка
B 44	base circuit, basic scheme, fundamental circuit	Grundschaltung *f*, Prinzipschaltung *f*	schéma *m* de principe	основная (главная) схема
B 45	base exchange	Basenaustausch *m*	échange *m* de bases	обмен основаниями
B 46	base exchanger	Basenaustauscher *m*, Ionenaustauscher *m*	échangeur *m* d'ions	фильтр ионного обмена
B 47	base former	Basenbildner *m*	basificateur *m*, agent *m* de basification	агент, образующий основание
B 48	base load	Grundlast *f*	charge *f* normale	нормативная нагрузка
B 49	base mixing tank	Laugenmischer *m*	mélangeur *m* de lessive	щелочная мешалка
B 50	base plate, subbase	Grundplatte *f*	plaque *f* d'assise, plaque de fondation	фундаментная плита
B 51	basic assembling procedure	Montagegrundverfahren *n*	procédé *m* fondamental de montage	основной (главный) монтажный метод
	basic Bessemer process	*s.* T 179		
B 52	basic equipment	Grundausrüstung *f*	équipement *m* de base	основное оборудование
	basic flow	*s.* M 27		
B 53	basic load	Einheitsbelastung *f*	charge *f* unitaire	стандартная нагрузка
	basic material	*s.* R 63		
B 54	basic process	Elementarvorgang *m*	processus *m* élémentaire	элементарный процесс
	basic rate system	*s.* F 256		
B 55	basic repair	Grundinstandsetzung *f*	remise *f* à neuf, remise en état	капитальный ремонт
	basic scheme	*s.* B 44		
B 56	basin, tank, boiler	Becken *n*, Blase *f*	bassin *m*, cuvette *f*, chaudron *m*, bouilleur *m*	резервуар, бассейн
	basin	*s. a.* R 300		
	basket shell	*s.* B 40		
	batch	*s.* C 181		
B 57	batch crystallization	Chargenkristallisation *f*	cristallisation *f* discontinue, cristallisation par charges périodiques	периодическая кристаллизация
	batch distillation	*s.* B 69		
B 58	batches/in	stoßweise	intermittent, pulsatoire	рывками, толчками
B 59	batch extraction	diskontinuierliche Extraktion *f*	extraction *f* discontinue	однократная (периодическая) экстракция
B 60	batching	Chargierung *f*, Dosierung *f*, Zumessung *f*	dosage *m*, chargement *m*	дозирование
B 61	batch liquid	Badflüssigkeit *f*, Elektrolyt *m*	liquide *m* de bain, électrolyte *m*	жидкий электролит гальванической ванны
B 62	batch-loaded cooler	periodisch beschickte Kühlkammer *f*	chambre *f* froide à chargement périodique, refroidisseur *m* chargé périodiquement	камера охлаждения с периодической загрузкой
B 63	batch mixer	diskontinuierlicher Mischer *m*, Chargenmischer *m*, periodisch arbeitendes Rührwerk *n*	mélangeur *m* discontinu (périodique)	смеситель (мешалка) периодического действия, порционная мешалка
B 64	batch operation, batch process	diskontinuierlicher Prozeß *m*, Chargenprozeß *m*, periodische Betriebsweise *f*, diskontinuierliche Fahrweise *f*, Chargenbetrieb *m*	procédé *m* (opération *f*) périodique, régime (procédé) *m* discontinu	циклический (периодический) режим, процесс периодического действия, прерывный процесс

	batch pressure	s. D 587		
	batch process	s. B 64		
B 65	batch reactor	diskontinuierlich arbeitender Reaktor m	réacteur m discontinu	реактор периодического действия
B 66	batch retort	Chargenautoklav m	autoclave m discontinu	автоклав
B 67	batch system	diskontinuierliches System n	système m à fonctionnement discontinu	периодическая (периодически работающая) система
B 68	batchwise	chargenweise	par charges périodiques	периодический
B 69	batchwise distillation, batch distillation	diskontinuierliche Destillation f, Blasendestillation f	distillation f discontinue	периодическая дистилляция
B 70	beaker, bucket	Becher m	godet m, coupe f	ковш, черпак
B 71	beam double refraction	Strahldoppelbrechung f	réfraction f double, biréfringence f	двойное лучепреломление
B 72	beam pump	Balkenpumpe f	pompe f à poutre	балансирный насос
B 73	become poor/to, to exhaust	verarmen	appauvrir	обеднять
B 74	bed	Schüttung f, Bett n, Schüttschicht f, Grundplatte f	lit m	завалка, засыпка
	bed	s. a. L 46		
B 75	begin/to, to put in	einsetzen, einleiten	amorcer, provoquer	вводить в действие, пускать
B 76	behaviour, performance, response, reaction	Verhalten n, Einstellung f, Fahrweise f	comportement m, tenue f	поведение, реакция
	behaviour	s. a. P 580		
	behaviour during breakdown	s. F 2		
B 77	bell, receiver, glass air	Glocke f	cloche f	колокол, колпак
B 78	bell filter	Glockenfilter n	filtre-cloche m	колокольный фильтр
B 79	bell furnace	Glockenofen m, Haubenofen m	four m à cloche	колпачковая печь
	belt	s. F 24		
	belt conveyor	s. C 686		
B 80	belt dryer	Bandtrockner m	sécheur m à bande	ленточная сушилка
B 81	belt screen, travelling-belt screen	Bandsieb n	tamis m à bande	ленточное сито
B 82	belt separator	Bandscheider m (magnetisch), Sortierband n	trieur m à tamis roulant, séparateur m [magnétique] à ruban	ленточный [электромагнитный] сепаратор, конвейер для сортировки
B 83	bench-scale equipment	Labormaßstabsausrüstung f, kleintechnische Ausrüstung f	installation f pilote (de laboratoire)	оборудование лабораторного масштаба
B 84	bench sink	Abflußbecken n	bassin m d'évier	раковина для слива жидкостей
B 85	bending, deflecting	Abbiegen n, Biegung f, Durchbiegung f	courbement m	загибание, отгибание, отбортовка
B 86	bending pressure	Biegedruck m	pression f de pliage (flexion)	напряжение при изгибе
B 87	bending property	Biegefähigkeit f, Biegbarkeit f	flexibilité f, souplesse f	упругость при изгибе
B 88	bending stress	Biegespannung f, Biegebeanspruchung f	tension (contrainte) f de flexion	напряжение при изгибе
B 89	benefit-cost analysis	Kosten-Nutzen-Analyse f	étude f des profits et des dépenses	анализ затрат и прибыли
B 90	Berl saddle	Berl-Sattel[körper] m (Füllkörper)	selle f de Berl	седло Берля
B 91	Bernoulli equation	Bernoullische Gleichung f	équation f de Bernoulli	уравнение Бернулли
B 92	biaxial stress	ebener (zweiachsiger) Spannungszustand m	état m de tension plane, tension f plane	плоское напряженное состояние
B 93	bilge pump, drain pump, lenzine	Lenzpumpe f	pompe f de cale, pompe d'épuisement	насос Ленца
B 94	bin, silo, bunker	Behälter m, Silo n, Bunker m, Schachtspeicher m	réservoir m, silo m, trémie f, soute f	емкость, силос, бункер
B 95	binary feed	Zweistoffeinspritzung f, Zweistoffeinspeisung f	alimentation f binaire	двухкомпонентный впрыск, двухкомпонентное питание
B 96	binary matrix	binäre Matrix f	matrice f binaire	двойная матрица
B 97	binary mixture	Zweikomponentengemisch n	mélange m binaire	двойная (бинарная) смесь
B 98	binary search method	binäres Suchverfahren n	méthode f binaire de recherche	двойной метод поиска
B 99	binary structure	binäre Struktur f	structure f binaire	двоичная структура
B 100	binary system	Zweistoffsystem n, binäres Gemisch n, Zweikomponentensystem n, Zweistoffgemisch n	système m binaire	двухкомпонентная (бинарная) система, бинарная смесь
	binder	s. F 191		
	binding	s. 1. C 470; 2. L 127		
B 101	binding energy, linking energy	Bindungsenergie f	énergie f de liaison	энергия связи
B 102	binding property	Bindefähigkeit f	pouvoir m agglutinant	способность к сцеплению, связующая способность
B 103	Bingham fluid	Binghamsches fluides Medium n	fluide m de Bingham	жидкая среда Бингама, Бингамовская жидкая среда
B 104	Bingham material	Binghamscher Körper m, B-Körper m	corps m de Bingham	Бингамовское тело, тело Бингама

B 105	**Bingham model**	Binghamsches Modell n	modèle m de Bingham	Бингамовская модель, модель Бингама
B 106	**Bingham viscoplastic body**	Binghamscher viskoplastischer Körper m	corps m plastique de Bingham	Бингамовское вязкопластичное тело, вязкопластичное тело Бингама
B 107	**Bingham viscosity,** plastic viscosity	plastische Viskosität f, Plastizität f	viscosité f plastique	пластическая (Бингамовская) вязкость, пластичность
B 108	**bioaeration,** sludge activation	Schlammbelebung f	activation f de boue	оживление ила
B 109	**bioaeration plant,** sludge activation plant	Belebtschlammanlage f, Schlammbelebungsanlage f	installation f à boue activée	аэротэнк
B 110	**biofilm reactor**	Biofilmreaktor m	bioréacteur m à film	пленочный реактор для биотехнологических процессов, биофильмреактор
B 111	**biofilter**	Biofilter n, biologisches Filter n	filtre m biologique	биологический фильтр
B 112	**biogas generator**	Biogaserzeuger m	digesteur m pour le production de biogaz	генератор биогаза, биогазовая установка
B 113	**biological fluid**	biologisches fluides Medium n	fluide m biologique	биологическая жидкая среда
B 114	**biological treatment process**	Verfahren n zur biologischen Behandlung	procédé m de traitement biologique	процесс биологической обработки
B 115	**black body**	schwarzer Körper m	corps m noir	черное тело
B 116	**Blake-Kozeny equation**	Blake-Kozenysche Gleichung f	équation f de Blake-Kozeny	уравнение Блэка-Коцени
B 117	**Blasius resistance law**	Blasiussches Widerstandsgesetz n	loi f de résistance de Blasius	закон сопротивления Блазиуса
	blast	s. B 147		
B 118	**blast air,** forced air	Gebläseluft f	air m soufflé, vent m de soufflage	дутье
B 119	**blast air supply**	Windzuführung f	amenée f de vent, amenée d'air	подводка дутья
B 120	**blast furnace**	Hochofen m, Blasofen m, Gebläse[schacht]ofen m	haut-fourneau m	доменная печь
B 121	**blast-furnace gas,** waste (top) gas	Gichtgas n	gaz m du gueulard, gaz de hauts-fourneaux	колосниковый (доменный, отходящий) газ
B 122	**blast-furnace process**	Hochofenprozeß m	processus m de haut-fourneau	доменный процесс
B 123	**blast heating apparatus,** hot blast stove	Winderhitzer m	réchauffeur m d'air, régénérateur m	рекуператор, подогреватель дутья, воздухонагреватель
	blast pipe	s. J 9		
B 124	**blast process**	Windfrischverfahren n	affinage m au vent, affinage par soufflage	конвертерный процесс
B 125	**blaze off/to**	abbrennen	se consumer par le feu, brûler, incendier	сжигать, отжигать, прокаливать, сгорать
B 126	**blazing-off**	Abbrennen n, Verbrennen n	brûlement m	опалка, опаливание, обгорание; догорание
B 127	**bleaching**	Bleichen n, Bleichverfahren n	blanchiment m	беление, отбелка
B 128	**bleaching bath**	Bleichbad n	bain m de blanchiment	отбеливающая ванна
B 129	**bleeder**	Abzapfventil n	vanne f de soutirage	спускной вентиль
B 130	**bleeder,** exhaust valve	Abblaseklappe f, Abblaseventil n	clapet m d'évacuation	продувочный клапан
B 131	**bleed hole**	Entlüftungsbohrung f, Entlüftungsloch n	trou m d'évacuation de l'air	отверстие для газового деления
	bleed off/to	s. B 317		
B 132	**bleed steam**	Anzapfdampf m	vapeur f soutirée	отобранный пар
B 133	**bleed steam heater**	Anzapfdampfvorwärmer m	réchauffeur m à vapeur soutirée	подогреватель паром промежуточного отбора
B 134	**blend/to,** to mingle, to mix	mischen, [ver]mengen, versetzen, durchmischen	mélanger, mêler, melaxer	смешивать, составлять смесь
	blend	s. M 264		
B 135	**blender,** proportioning pump	Mischpumpe f	pompe f mélangeuse	смесительный насос
	blender	s. a. M 241		
B 136	**blind/to,** to plug (choke, stop) up, to tamp	verstopfen, zustopfen	colmater, boucher, obstruer	затыкать, закупоривать, засорять
B 137	**blisting**	Blasenbildung f	formation f de bulles	образование пузырей
	blockage	s. O 7		
B 138	**block diagram**	Blockschema n	schéma m fonctionnel (bloc)	блочная схема
B 139	**block flow diagram**	Blockfließdiagramm n	schéma m de fonctionnement par blocs	блок-схема работы
	block formation	s. R 154		
B 140	**block-oriented simulation**	blockorientierte Simulation f	simulation f orientée sur les blocs	блочное моделирование, последовательный (модульный) расчет
B 141	**block-oriented structure**	blockorientierte Struktur f	structure f orientée sur les blocs	блочная структура
B 142	**block valve**	Absperrventil n	soupape f d'arrêt	запорный вентиль (клапан)
B 143	**blow-down piping**	Abblaseleitung f	conduite f d'évacuation, conduite de purge	продувочный трубопровод, продувательная труба

B 144	blow-down system	Entspannungssystem n, Abblasesystem n	système m de détente (décompression), système d'évacuation	система сбрасывания давления, система отдувки
B 145	blow-down tank	Abblasetank m	réservoir m de purge (détente)	продувочный бак (куб)
B 146	blow-down tower	blow-down-Kolonne f, Abblasekolonne f	colonne f de purge	колонна для быстрой отдувки
B 147	blower, fan, ventilator, blast	Ventilator m, Lüfter m, Gebläse n, Kompressor m	soufflerie f, ventilateur m, compresseur m	воздуходувка, вентилятор, компрессор, дутье
B 148	blower chamber	Gebläsekammer f	chambre f de soufflante, chambre à jet	вентиляционная камера
B 149	blower efficiency	Gebläsewirkungsgrad m	rendement m de la soufflerie, rendement du ventilateur	коэффициент полезного действия вентилятора
B 150	blower throat	Gebläsekanal m	gaine f de soufflage	канал вентилятора
B 151	blowing	Blasen n, Leerblasen n	soufflage m, vidange f par soufflage	продувка, продувание, выдувание, дутье
B 152	blowing pressure	Blasdruck m	pression f de soufflage	давление дутья
B 153	blow off/to, to exhaust	ausblasen	souffler, purger	выдувать, продувать
B 154	blow-off cock	Ablaßhahn m, Abblasehahn m	robinet m purgeur, purgeur m, robinet de vidange, valve f de purge	спускной (продувочный) кран
B 155	blow-off pipe	Abblaserohr n	tuyau m de purge, tuyau d'évacuation	[паро]спускная (продувочная, выдувная) труба
B 156	blow-off pressure	Abblasedruck m	pression f de purge	продувочное (спускное) давление
B 157	blow-off valve	Abblaseventil n, Ausblaseschieber m	soupape f d'évacuation, soupape de purge	продувочный клапан, выдувной вентиль
B 158	blow-over cooling	Anblaskühlung f	refroidissement m par soufflage	охлаждение внешним обдувом
	blowpipe	s. G 12		
B 159	blow valve	Dampfablaßventil n	soupape f de décharge de la vapeur	вентиль для выпуска пара
B 160	Bodenstein principle	Bodenstein-Prinzip n	principe m de Bodenstein	принцип Боденштейна
B 161	body-centered	raumzentriert	centré dans l'espace	объемноцентрированный
B 162	body force	Volumenkraft f, Raumkraft f	force f volumétrique	объемная сила
B 163	boiler	Boiler m, Dampferzeuger m; Wasserkocher m; Heißwasserspeicher m	réservoir m d'eau chaude, bouilleur m, chauffe-eau m	кипятильник, бойлер, паровой котел
	boiler	s. a. B 56		
B 164	boiler availability	Kesselverfügbarkeit f	disponibilité f de la chaudière	коэффициент использования котла
B 165	boiler bearer	Kesselblock m, Kesseleinheit f	unité f de chaudière	котельная ячейка
B 166	boiler body	Kesselkörper m	corps m de chaudron	корпус котла
B 167	boiler brickwork setting	Kesselausmauerung f, Kesselmauerwerk n	maçonnage m de chaudières	обмуровка котла
B 168	boiler casing	Kesselabdeckung f	revêtement m de chaudière	облицовка котла
	boiler charging	s. B 173		
B 169	boiler construction, boiler manufacture	Kesselbau m	construction f de chaudières	конструкция котла
B 170	boiler draft	Kesselzug m	tirage m de chaudière	газоход котла
B 171	boiler efficiency	Kesselwirkungsgrad m	rendement m de chaudière	коэффициент полезного действия котлоагрегата
B 172	boiler end	Kesselboden m	fond m de chaudière	днище барабана котла
B 173	boiler feeding, boiler charging	Kesselbeschickung f	alimentation f de chaudière	подача топлива в котел
B 174	boiler feed piping	Kesselspeiserohr n	tube m d'alimentation de chaudière	питательная труба котла
B 175	boiler feed water	Kesselspeisewasser n	eau f d'alimentation de chaudière	вода для питания котла
B 176	boiler fittings	Kesselausrüstung f	armatures fpl de chaudière	арматура котла
B 177	boiler furnace	Kesselfeuerung f, Kesselfeuerraum m	foyer m d'une chaudière	топка котла
B 178	boiler grate	Kesselrost m	grille f de la chaudière	колосниковая решетка
B 179	boiler heating	Kesselbeheizung f	chauffage m d'une chaudière	обогрев (топка) котла
B 180	boiler heating surface	Kesselheizfläche f	surface f chauffante de chaudière	поверхность нагрева котла
B 181	boiler house	Kesselhaus n	halle f des chaudières, chaufferie f	котельная
B 182	boiler jacket	Kesselhülle f, Kesselmantel m	enveloppe f de chaudière, coque f de la chaudière	обшивка котла
B 183	boiler maintenance	Kesselwartung f, Kesselbedienung f	surveillance f de la chaudière	обслуживание котла
	boiler manufacture	s. B 169		
B 184	boiler output	Kesselleistung f	performance f de la chaudière, puissance f de chaudière	производительность котла
B 185	boiler pipe, heating tube	Kesselsiederohr n	tube m de chaudière, tube bouilleur	кипятильная труба котла
B 186	boiler pressure	Kesseldruck m	pression f dans la chaudière	давление пара в котле
B 187	boiler scale generation	Kesselsteinbildung f	génération f d'incrustations, formation f d'incrustations	образование накипи в котле

B 188	boiler size	Kesselgröße f	grandeur f (dimensions fpl) de la chaudière	размер котла
B 189	boiler supporting steelwork	Kesselgerüst n	charpente f de la chaudière	каркас котла
B 190	boiler test	Kesselprüfung f, Kesselrevision f, Kesselüberwachung f	surveillance f des chaudières, épreuve (révision) f de chaudière	испытание котла
B 191	boiler testing	Kesselkontrolle f, Kesseluntersuchung f	visite f de chaudière	котлонадзор
B 192	boiler tube	Kesselrohr n	tube m de chaudière	котельная труба
B 193	boiler unit	Kesseleinheit f	unité f de chaudière	котельная единица
B 194	boiler water	Kesselwasser n	eau f de chaudière	котловая вода
B 195	boiling	Sieden n	ébullition f	кипение
B 196	boiling heat	Siedehitze f	chaleur (température) f d'ébullition	тепло (температура) кипения
B 197	boiling number	Siedezahl f	nombre f d'ébullition	коэффициент кипения
B 198	boiling point	Siedepunkt m	point m d'ébullition	точка кипения
B 199	boiling point curve	Siedekurve f	courbe f des points d'ébullition	кривая разгонки (кипения)
B 200	boiling pressure	Siededruck m	pression f d'ébullition	давление кипения
	boiling range	s. D 425		
B 201	boiling reactor	Verdampferreaktor m, Siedereaktor m	réacteur m à eau bouillante	кипящий реактор, реактор с испаряющимся теплоносителем (растворителем)
B 202	boiling temperature	Siedetemperatur f	température f d'ébullition	температура кипения
B 203	boiling water reactor	Siedewasserreaktor m	réacteur m bouilleur	реактор с кипящей водой
B 204	boil-off gas	Siededampf m	vapeur f d'ébullition	пар, выделяющийся при кипении
B 205	boil-off losses	Siedeverluste mpl	pertes fpl d'ébullition	потери при кипении
B 206	Boltzmann constant	Boltzmann-Konstante f	constante f de Boltzmann	постоянная Больцмана
B 207	Boltzmann's superposition principle	Boltzmannsches Superpositionsprinzip n	principe m de superposition de Boltzmann	принцип суперпозиции Больцмана
B 208	Boltzmann-Volterra equation	Boltzmann-Volterra'sche Gleichung f	équation f de Boltzmann et Volterra	уравнение Больцмана-Вольтерра
	bond	s. L 127		
B 209	Boolean matrix	Boolesche Matrix f	matrice f booléenne	булева матрица
B 210	Boolean operator	Boolescher Operator m	opérateur m logique (de Boole)	логический оператор, оператор Буля
B 211	Boolean polynom	Boolesches Polynom n	polynôme m booléen	булев многочлен
B 212	Boolean product	Boolesches Produkt n	produit m booléen (de Boole)	логическое произведение
B 213	Boolean system function	Boolesche Systemfunktion f	fonction f booléenne de système	булева функция системы
B 214	Boolean value	Boolescher Wert m	valeur f booléenne	булева величина
B 215	booster pump	Druck[erhöhungs]pumpe f, Booster-Pumpe f, Hilfspumpe f, Zwischenpumpe f	pompe f foulante (de pression, auxiliaire, intermédiaire, de booster)	бустерный (вспомогательный) насос
B 216	bottling machine	Abfüllmaschine f, Flaschenfüllmaschine f	machine f à remplir les bouteilles	разлив[оч]ная машина
B 217	bottling of liquefied gases	Abfüllen f von Flüssiggasen	mise f en bouteille de gaz liquéfiés	разливка жидких газов
B 218	bottom	Boden m, unterster Teil m, Sohle f	bas m, sole f, base f, fond m	низ, дно, днище, осадок, отстой
B 219	bottom blow valve	Bodenventil n, Entleerungsventil n	soupape f de fond (vidange)	донный клапан
B 220	bottom discharge valve	Bodenablaßventil n	robinet m de fond de cuve	затвор донного спуска
B 221	bottom of column	Kolonnensumpf m	bas m de la colonne	куб колонны
B 222	bottoms	Vakuumrückstand m	résidu m de distillation poussée, résidu sous vide	остаток от перегонки в вакууме
B 223	bounce	Abprall m, Rückprall m	rebond m, rebondissement m	отскок, отскакивание
B 224	bound/to	aufprallen	rebondir, heurter	сталкиваться, наскакивать
B 225	boundary, limit	Begrenzung f, Grenze f, Rand m	limite f, barrière f, frontière f	ограничение, предел
B 226	boundary condition	Grenzbedingung f, Randbedingung f	condition f limite (marginale, aux limites)	граничное (краевое) условие
B 227	boundary element	Randelement n	élément m marginal	краевой элемент
B 228	boundary layer, interface	Grenzschicht f, Trennschicht f	couche f limite, interface f	поверхность раздела, [по]граничный слой
B 229	boundary-layer momentum transport	Grenzschichtimpulstransport m	transport m de la quantité de mouvement de couche limite, transfert m de l'impulsion de couche limite	перенос импульса в пограничном слое
B 230	boundary-layer theory	Grenzschichttheorie f	théorie f de la couche limite	теория пограничного слоя
B 231	boundary lubrication	Grenz[schicht]schmierung f, Teilschmierung f	lubrification f limite	смазка пограничного слоя, частичная смазка
B 232	boundary surface	Grenzfläche f, Begrenzungsfläche f	surface f limite (de séparation, de délimitation)	граничная поверхность, поверхность раздела
B 233	boundary value	Randwert m	valeur f marginale	краевое (граничное) значение
B 234	bounding	Aufprallen n	impact m, choc m, secousse f	столкновение, соударение

B 235	boundless	unbegrenzt	illimité, sans limites, infini	неограниченный, бесконечный
B 236	bound moisture	gebundene Feuchtigkeit f	humidité f liée	связанная влага
B 237	bound water	gebundenes Wasser n	eau f latente	связанная вода
B 238	Box's complex method	Komplexverfahren n nach Box	méthode f de Box	метод комплекса по Боксу, комплексный метод Бокса
B 239	**Box-Wilson method**	Box-Wilson-Methode f	méthode f Box-Wilson	метод Бокса-Вильсона
B 240	bracing	Versteifung f	étayage m, étançonnage m, étaiement m	жесткое крепление
B 241	**Bragg ['s] law**	Braggsches Reflexionsgesetz n	loi f de réflexion de Bragg	закон отражения Брагга
B 242	**branch,** discharge piping	Ableitung f	dérivation f	отвод, разветвление
B 243	branch current	Teilstrom m	courant m partiel	частичный ток
	branch current	s. a. P 41		
B 244	branched polymer	verzweigtes Polymer n	polymère m ramifié	разветвленный полимер
B 245	branching, bypass	Verzweigung f	bifurcation f, branchement m	разветвление, ответвление, отвод, байпас
B 246	branching index	Verzweigungsindex m	indice m de ramification	индекс разветвленности
B 247	branching method	Verzweigungsmethode f	méthode f de branchement	метод разветвления
B 248	branch of manufacture	Betriebszweig m	branche f de fabrication	отрасль производства
B 249	branch piece	Stutzen m, Rohrstutzen m	tubulure f [de branchement], tuyau m de rallonge	штуцер
B 250	branch pipe	Nebenrohr n	tuyau m secondaire, conduite f latérale	отводная труба
B 251	**break/to,** to switch off	abschalten	mettre hors circuit, déconnecter	отключать, выключать; прерывать
B 252	**break**	Bruch m	cassure f, rupture f, brisure f	излом, поломка; разрушение; трещина
B 253	breakdown, disturbance, failure	Ausfall m	défaillance f, arrêt m	отказ, выход из строя, развал, авария
B 254	breakdown, operating trouble	Betriebsstörung f	interruption f du travail, arrêt m de service	авария производства, неполадки производства
B 255	breakdown puncture, slugging	Durchschlag m	poinçon m, perçoir m, broche f	пробой
B 256	breakdown voltage	Durchschlagspannung f	tension f de claque (rupture)	пробивное напряжение
B 257	breaker, cracker, crusher	Mahlwerk n, Brecher m	broyeur m, concasseur m	дробилка, бурун
B 258	breaking, buckling	Knickung f	flambage m	продольный изгиб
B 259	breaking capacity	Abschaltleistung f	puissance f de rupture	разрывная мощность
B 260	breaking down temperature	Abbautemperatur f, Zersetzungstemperatur f	température f de décomposition	температура распада (разложения)
B 261	breaking plant, crushing plant	Brechanlage f	installation f de concassage (broyage)	дробилка, дробильная установка
B 262	breaking resistance	Knickfestigkeit f	résistance f au flambage	сопротивление продольному изгибу
B 263	breeder [reactor]	Brutreaktor m	réacteur m à surrégénération, breeder m	реактор, размножитель
	brewing	s. D 39		
B 264	bricking, brick lining	Ausmauerung f	maçonnerie f réfractaire	обмуровка, облицовка, футеровка
B 265	bridge circuit	Brückenschaltung f	méthode f du pont, montage m en pont	мостовая схема
B 266	brine	Sole f, Lauge f, Salzlauge f, Salzlösung f, Salzsole f, Salzwasser n, Kühlsole f	saumure f, eau f salée, eau saline	рассол, тузлук, соляной рассол, раствор соли
B 267	brine cooler	Solekühler m	refroidisseur m de saumure	холодильник на рассоле
B 268	brine cooling	Solekühlung f	refroidissement m par saumure	охлаждение рассолом
B 269	brine pump	Solepumpe f	pompe f de saumure, pompe d'eau salée	солесос
B 270	**briquetting**	Brikettieren n, Brikettierung f	briquetage m	брикетирование
B 271	brittle fracture	Sprödbruch m, Sprödigkeitsbruch m	rupture f fragile	хрупкое разрушение
B 272	brittleness	Sprödigkeit f, Zerbrechlichkeit f	fragilité f	хрупкость, ломкость
B 273	**Brownian motion (movement)**	Brownsche Bewegung f	mouvement m brownien	броуновское движение
	bubble boiling	s. N 94		
B 274	bubble cap plate; bubble tray	Glockenboden m	plateau m à cloche	колпачковая тарелка
B 275	bubble chamber	Blasenkammer f	chambre f à bulles	пузырьковая камера
B 276	bubble climbing speed	Blasenaufstiegsgeschwindigkeit f	vitesse f ascensionelle de bulles	подъемная скорость пузырей
B 277	bubble column	Blasensäule f, Blasenkolonne f, Sprudelkolonne f	colonne f de (à) barbotage	пузырьковая (барботажная) колонна
B 278	bubble flow	Blasenströmung f	écoulement m à bulles	барботирующий поток
B 279	bubble plate	Rektifizierboden m	plateau m échangeur	тарелка ректификационной колонны
B 280	bubble point	Blasensiedepunkt m, Blasenpunkt m	point m de bulles	температура начала кипения
B 281	bubble rising	Blasenaufstieg m	montée f de bulles	подъем пузырей

	English	German	French	Russian
	bubble tray	s. B 274		
B 282	bubble tray column	Glockenbodenkolonne f	colonne f à plateaux à cloche	колпачковая тарельчатая колонна
B 283	bubbling, cavitation	Blasenbildung f (Wirbelschicht)	bullage m, formation f de bulles	выделение (образование) пузырьков, кавитация
B 284	bubbling fluidized bed	Blasenwirbelschicht f	lit m fluidisé à bulles	пузырчатый кипящий слой
	bucket	s. B 70		
B 285	bucket-conveyor extractor	Becherwerkextraktor m	extracteur m à godets	корзиночный экстрактор
B 286	bucket elevator	Becherwerk n	transporteur m (chaîne f) à godets	ковшовый транспортер (элеватор)
B 287	bucket pump	Schaufelpumpe f, Flügelpumpe f	pompe f à palettes (ailettes)	крыльчатый насос
B 288	Buckingham-Reiner equation	Buckingham-Reinersche Gleichung f	équation f de Buckingham-Reiner	уравнение Бакингэма-Рейнера
	buckling	s. B 258		
B 289	Bueche model	Buechesches Modell n	modèle m de Bueche	модель Бики
B 290	buffer	Pufferspeicher m	réservoir m tampon (intermédiaire)	буферный резервуар
B 291	buffer solution, regulator mixture	Pufferlösung f	solution f tampon[née]	буферный раствор
B 292	buffer time	Pufferzeit f	temps m tampon	буферное время
B 293	buffing machine	Schwabbelmaschine f	machine f à polir à meule flexible	машина для буфирования, шлифовальная машина
	building	s. C 604		
B 294	building-block principle	Baukastenprinzip n	conception f par bloc-éléments, conception en éléments démontables	принцип блочного построения
B 295	building-block system	Baukastensystem n	système m de construction par blocs	система блочного построения
B 296	building code	Bauordnung f	règlement m sur les constructions	строительные правила
B 297	building construction, structure design	Baukonstruktion f	construction f de bâtiment	строительная конструкция
B 298	building material	Baustoff m, Baumaterialien npl	matériaux mpl de construction	строительный материал
B 299	building measure	Baumaßnahme f	mesure f de construction, projet m de génie civil	строительство, строительное мероприятие
	building site	s. C 607		
B 300	building trade	Baugewerk n	métier m du bâtiment	строительное дело
B 301	building trade	Ausrüstungsgewerk n	métier m de bâtisse intérieure	работа по оснащению, работа по монтажу оборудования
B 302	built-in stirrer	Einbaurührer m	agitateur m incorporé	встроенная мешалка
	bulb	s. F 208		
B 303	bulb condenser	Kugelkühler m, Kugelaufsatz m	réfrigérateur m à boule, colonne f à boules	шаровой холодильник
B 304	bulk	Volumen n, Umfang m, Masse f, Hauptmasse f, Großteil m	masse f, volume m	масса, основная часть, объем, вместимость
B 305	bulk density	Raummasse f, Schüttdichte f	poids m spécifique, densité f apparente (en vrac)	объемный вес, насыпной вес, плотность насыпки
B 306	bulk factor	Schüttfaktor m, Füllfaktor m	coefficient m de déversement, facteur m de remplissage (volume)	коэффициент уплотнения
B 307	bulk flow	Volumenstrom m, Volumendurchsatz m	débit m volumique	объемный расход, объемное течение
B 308	bulk modulus	Kompressionsmodul m, Raummodul m	module m d'élasticité volumique	модуль сжатия, объемный модуль
B 309	bulk temperature	Volumentemperatur f, Temperatur f im Innern, mittlere Temperatur	température f interne (moyenne)	объемная (внутренняя, средняя) температура
B 310	bulk temperature	Formmassetemperatur f	température f de la matière à mouler	температура формуемого материала
B 311	bulk viscosity	Volumenviskosität f	viscosité f volumique	объемная вязкость
B 312	bulk volume	Schüttvolumen n, scheinbares Volumen n; Rohvolumen n	volume m apparent	насыпной (удельный) объем
	bunker	s. B 94		
B 313	buoyant effect	Auftriebswirkung f	effet m de portance	эффект Архимеда
B 314	buoyant force, lift force	Auftriebskraft f	force f portante	подъемная сила, сила Архимеда
	burden	s. C 181		
B 315	Burgers-Frenkel model	Burgers-Frenkelsches Modell n	modèle m de Burgers-Frenkel	модель Бургерса-Френкеля
B 316	Burke-Plummer equation	Burke-Plummersche Gleichung f	équation f de Burke-Plummer	уравнение Бурке-Пламмера
B 317	burn/to, to bleed off, to feare	abfackeln	brûler à la torche	сжигать факелом
	burn	s. C 426		
	burner	s. C 435		

B 318	burner nozzle	Brennerdüse f	buse f (bec m) de brûleur	сопло горелки
	burning	s. C 426		
	burning gas	s. F 273		
B 319	burning quality	Brennqualität f, Brenneigenschaft f	qualité f de brûlage	качественная характеристика горения
	burning rate	s. R 47		
	burnt gas	s. F 273		
B 320	bus bar	Sammelschiene f	barre f omnibus, collectrice f	собирательная (сборная) шина
	butterfly valve	s. T 198		
B 321	bypass/to	umgehen, umleiten	dériver, contourner	обводить, байпасировать
B 322	bypass, by-pass, bypass line	Umführung f, Umführungsleitung f, Umgehungsleitung f, Bypass m	by-pass m, dérivation f, conduite f de dérivation	обход, обвод, байпас, обводный трубопровод
B 323	bypass, by-pass	Nebenauslaß m, Bypass m	sortie f latérale	вспомогательный спуск, байпас
	by-pass...	s. a. bypass...		
B 324	bypass air	Bypassluft f	air m de by-pass	байпасируемый воздух
B 325	bypass connection	Bypass-Schaltung f, Umgehungsschaltung f	montage m de déviation, connexion f de by-pass	байпасное включение, включение байпаса, обходная (байпасная) схема
B 326	bypass damper	Bypass-Schieber m, Bypassklappe f	clapet m de by-pass	клапан байпаса
B 327	bypass factor	Bypassfaktor m	facteur m de by-pass	коэффициент байпасирования
B 328	bypassing duct	Umgehungskanal m	conduite f de contournement, canal m de dérivation, by-pass m	байпасирующий канал
	bypass line	s. B 322		
B 329	bypass scheme	Bypass-Schema n	schéma m by-pass	схема байпасирования
B 330	bypass system	Umgehungssystem n	système m de dérivation, système by-pass	обходная система
B 331	bypass valve	Bypassventil n	soupape f à by-pass, vanne f de déviation	обводный вентиль, перепускной клапан
B 332	by-product, residual product	Nebenprodukt n, Abfallerzeugnis n, [verwertbares] Abfallprodukt n	sous-produit m, produit m secondaire, déchet m	побочный продукт, отходы, отбросы

C

	cable network	s. C 4		
C 1	cable run	Kabelführung f, Kabeldurchführung f	passe-câble m, traversée f de câble	кабелепровод
C 2	cable shaft	Kabelschacht m	puits m à câbles	кабельный колодец
C 3	cable sheating	Kabelummantelung f	gaine f (revêtement m) de câble	обкладка кабеля
C 4	cable system, cable network	Kabelnetz n	réseau m de câbles, canalisation f électrique souterraine	кабельная сеть
C 5	cable terminal box	Kabelschrank m, Kabelkasten m	boîte f de jonction, boîte à câbles	кабельный шкаф
C 6	cable trough	Kabelgraben m	tranchée f (caniveau m) à câbles	кабельная траншея
	CAD	s. C 507		
C 7	cake	Sinterkörper m, Preßling m	matériel (corps) m fritté	отжатый осадок, брикет, брусок
C 8	cake resistance	Kuchenwiderstand m	résistance f de tourteau	сопротивление отжатого осадка
C 9	calculated refrigerating capacity	berechnete Kälteleistung f	capacité f frigorifique calculée	расчетная холодопроизводительность
C 10	calculate in advance/to, to predetermine	vorausberechnen	calculer préalablement	предварительно оценивать, калькулировать
C 11	calculation, computation	Rechnung m, Berechnung m	calcul m, calculation f	вычисление, подсчет, расчет, исчисление
C 12	calculation algorithm	Berechnungsalgorithmus m	algorithme m de calcul	алгоритм вычисления
C 13	calculation method	Berechnungsmethode f	méthode f de calcul	метод вычисления
C 14	calculation principle, computing principle	Berechnungsprinzip n	principe m de calcul	принцип вычисления (расчета)
C 15	calender	Kalander m	calandre f, cylindre m	каландр
	calender bowl	s. C 18		
C 16	calender effect	Kalandereffekt m	effet (grain) m de calandrage (résine synthétique)	каландровый эффект
C 17	calendering	Kalandrieren n, Kalandern n	calandrage m	каландрование
C 18	calender roll, calender bowl	Kalanderrolle, f Kalanderwalze f	cylindre m calendreur	вал каландра
C 19	calender roll cambering, roll-crown	Kalanderwalzenbombierung f	bombé m de cylindre calendreur	бомбировка валков каландров
C 20	calibrated nozzle	Meßdüse f	tuyère (buse) f de mesure	измерительное сопло
C 21	calibration	Eichung f, Eichen n	étalonnage m, jaugeage m	калибровка, тарирование, юстировка, эталонирование

C 22	**calibration accuracy**	Eichgenauigkeit *f*	précision *f* d'étalonnage	точность калибровки
C 23	**calibration chart**	Eichtabelle *f*	table *f* d'étalonnage	таблица калибровки
C 24	**calibration current,** calibration flow	Eichstrom *m*	courant *m* étalon	эталонный ток, эталонный поток
C 25	**calibration curve**	Eichkurve *f*	courbe *f* d'étalonnage	градуировочная кривая
	calibration flow	*s.* C 24		
C 26	**calibration instrument**	Eichgerät *n*	appareil *m* d'étalonnage, instrument *m* d'étalonnage, étalon *m*	эталонный прибор
C 27	**calibration point**	Eichpunkt *m*	point *m* d'étalonnage	точка калибровки
C 28	**caloric capacity,** thermal capacity	Wärmeinhalt *m*	capacité *f* calorique	теплосодержание, содержание тепла
	caloric conductibility	*s.* H 26		
	calorie	*s.* T 136		
C 29	**calorimetry**	Wärmemessung *f*	calorimétrie *f*	калориметрия, пирометрия
C 30	**canonical distribution**	kanonische Verteilung *f*	distribution *f* canonique	каноническое распределение
	cap	*s.* C 808		
C 31	**capacity**	Kapazität *f*, Leistungsvermögen *n*, Leistungsfähigkeit *f*, Fassungsvermögen *n*, Tragfähigkeit *f*, Tragvermögen *n*	capacité *f*, productivité *f*	производительность, мощность, выработка, емкость, объем, вместимость, литраж, способность
C 32	**capacity,** delivery, output	Förderleistung *f*	capacité *f* de transport, rendement *m*, tonnage *m* transporté	грузоподъемность транспортера, производительность подачи
C 33	**capacity controller**	Leistungsregler *m*	régleur *m* de rendement	регулятор мощности
C 34	**capacity demand**	Leistungsbedarf *m*	puissance *f* nécessaire	потребление мощности
C 35	**capacity factor**	Kapazitätsfaktor *m*	facteur *m* de capacité	фактор производственной мощности
C 36	**capacity line**	Leistungskurve *f*	caractéristique *f*, courbe *f* de puissance	кривая производительности
C 37	**capacity maximum**	Leistungsspitze *f*	pointe *f* de puissance	импульсная мощность, максимум мощности
C 38	**capacity range**	Leistungsbereich *m*	rayon *m* d'action, champ *m* de travail	диапазон производительности
C 39	**capacity rating**	Leistungsberechnung *f*	calcul *m* de rendement (puissance)	подсчет (расчет) мощности
C 40	**capillar constant**	Kapillarkonstante *f*	constante *f* capillaire	капиллярная постоянная
C 41	**capillary cell**	Kapillarzelle *f*	cellule *f* capillaire	капиллярная ячейка
C 42	**capillary column**	Kapillarsäule *f*	colonne *f* capillaire	капиллярная колонка
C 43	**capillary condensation**	Kapillarkondensation *f*	condensation *f* capillaire	капиллярная конденсация
C 44	**capillary dryer**	Kapillartrockner *m*	sécheur *m* capillaire	осушитель капиллярной трубки
C 45	**capillary flow**	Kapillarströmung *f*	écoulement *m* capillaire	капиллярное течение
C 46	**capillary flowmeter**	Kapillardurchflußmesser *m*	débitmètre *m* capillaire	капиллярный расходомер
C 47	**capillary force**	Kapillarkraft *f*	force *f* capillaire	капиллярная сила
C 48	**capillary fringe**	Kapillarzone *f*	zone *f* capillaire	капиллярная зона
C 49	**capillary pressure**	Kapillardruck *m*	pression *f* capillaire	капиллярное давление
C 50	**capillary system**	Kapillarsystem *n*	système *m* capillaire	система с капиллярной трубкой, капиллярная система
C 51	**capillary tube**	Kapillarröhre *f*	tube *m* capillaire	капиллярная трубка, капилляр
C 52	**capillary viscometer**	Kapillarviskosimeter *n*	viscosimètre *m* capillaire	капиллярный вискозиметр
	capital expenditure	*s.* I 327		
	capstan	*s.* W 169		
C 53	**carbonaceous**	kohlenstoffhaltig	contenant du carbone	углеродистый
C 54	**carbon black**	Ruß *m*	noir *m* de fumée (carbone)	сажа
C 55	**carbon-chain polymer**	Kohlenstoffkettenpolymer *n*	polymère *m* à chaîne carbonée	углеродный цепный полимер
C 56	**carbon content**	Kohlenstoffgehalt *m*	teneur *m* en carbone	содержание углерода
	carbon dioxide snow	*s.* D 532		
C 57	**carbonization**	Karbonisierung *f*, Verkokung *f*, Verkohlung *f*	carbonisation *f*, cuisson *f*, cokéfaction *f*	обугливание, науглероживание, карбонизация, коксование
C 58	**carbonize/to**	verschwelen	carboniser à basse température, distiller lentement	обугливать, коксовать
C 59	**carbonize/to**	karbonisieren	carburer, saturer d'acide carbonique	карбонизировать
	carbonize/to	*s. a.* C 360		
C 60	**carbonizing apparatus**	Karburierungsapparat *m*	carbonisateur *m*, appareil *m* à carburer	карбонизатор, карбонизирующий аппарат
C 61	**carbonizing stove**	Karbonisationsofen *m*	four *m* à cémenter aux gaz carburants	карбонизационная печь
C 62	**carbon residue**	Verkokungsrückstand *m*	résidu *m* charbonneux, résidu de cokéfaction	углистый остаток
	carburet/to	*s.* G 54		
C 63	**carburetted fuel**	Frischgas *n*	fuel *m* carburé, gaz *m* d'affinage	свежая горючая смесь

C 64	cargo space	Laderaum *m*	espace *m* à marchandises	грузовое помещение
C 65	Carnot efficiency	Carnot-Wirkungsgrad *m*	rendement *m* de Carnot	коэффициент полезного действия цикла Карно
C 66	Carnot ideal coefficient of performance	Carnot-Kälteleistungsziffer *f*	coefficient *m* de performance idéale Carnot	холодильный коэффициент цикла Карно
	Carnot process	*s.* C 68		
C 67	Carnot refrigeration cycle	Carnot-Kälteprozeß *m*	cycle *m* frigorifique de Carnot	обратный цикл Карно
C 68	Carnot's cycle, Carnot process	Carnotscher Kreisprozeß *m*	cycle *m* de Carnot	цикл Карно, круговой процесс Карно
C 69	carrier of reaction	Reaktionsträger *m*	porteur *m* de réaction	носитель реакции
C 70	carry/to	tragen, befördern	porter, transporter	нести, везти, проводить
C 71	carrying capacity	Belastbarkeit *f*, Tragkraft *f*	capacité *f* de charge, portée *f*	нагружаемость
C 72	carry-over	Mitreißen *n*	entraînement *m*	унос
C 73	cascade condensation	Kaskadenkondensation *f*	condensation *f* en cascade	каскадная конденсация
C 74	cascade cooler	Kaskadenkühler *m*	refroidisseur *m* à cascade	каскадный холодильник
C 75	cascade crystallizer	Kaskadenkristallisator *m*	cristallisoir *m* en cascade	каскадный кристаллизатор
C 76	cascade distillation	Kaskadendestillation *f*	distillation *f* en cascade	каскадная дистилляция
C 77	cascade evaporator	Kaskadenverdampfer *m*	évaporateur *m* en cascade	каскадный испаритель
C 78	cascade heat exchanger	Kaskadenwärmeübertrager *m*	échangeur *m* thermique en cascade	теплообменник в каскадной холодильной машине
C 79	cascade operation	Kaskadenbetrieb *m*	opération *f* en cascade	каскадный процесс
C 80	cascade Peltier cooler	Peltier-Kaskadenkühler *m*	refroidisseur *m* à cascade Peltier	каскадный термоэлектрический охладитель
C 81	cascade process	Kaskadenprozeß *m*	processus *m* en cascade	каскадный процесс
C 82	cascade refrigeration system	Kältekaskade *f*	système *m* de refrigération en cascade	каскадная система охлаждения
C 83	cascade scrubber	Kaskadenwäscher *m*	scrubber *m* en cascade, laveur *m* à cascade	каскадный промыватель, каскадная промывная установка
C 84	cascade system	Kaskadensystem *n*, Kaskadenschaltung *f*	système *m* en cascade	каскадная система
C 85	cascade tray	Kaskadenboden *m*	plateau *m* à cascade	многоступенчатая тарелка
C 86	case/to	verrohren	tuber	ставить трубы
C 87	case of disturbance	Störungsfall *m*	cas *m* de dérangement	авария, отказ, возмущение
C 88	casing, tubing	Verrohrung *f*	tubage *m*, montage *m* des tuyaux	прокладка труб, тюбинг
	casing	*s. a.* H 202		
C 89	casting	Gießen *n*	coulage *m*	литье
C 90	catalyst, catalyzer; contact	Katalysator *m*, Kontakt *m*, Kontaktmasse *f*	contact *m*, catalyseur *m*, masse *f* de contact	катализатор, контакт, соприкосновение
C 91	catalyst activity	katalytische Wirkung (Wirksamkeit) *f*; Katalysatorwirkung *f*, Katalysatoraktivität *f*	activité *f* catalytique	активность катализатора, каталитическое действие
	catalyst area	*s.* C 102		
C 92	catalyst arrangement	Katalysatoranordnung *f*	disposition *f* (arrangement *m*) de catalyseur	распределение (размещение) катализатора
	catalyst bed	*s.* C 104		
C 93	catalyst bulk	Katalysatormasse *f*	masse *f* de catalyseur	масса катализатора
C 94	catalyst effectivness factor	Katalysatoreffektivitätskoeffizient *m*, effektiver Katalysatorfaktor *m*	coefficient *m* de l'efficacité du catalyseur	фактор эффективности катализатора
C 95	catalyst inhibitor, catalyst poison	Katalysatorgift *n*, Kontaktgift *n*	toxique *m* du catalyseur, poison *m* de catalyseur	каталитический яд
C 96	catalyst life	Katalysatorlebensdauer *f*	durée *f* de service de catalyseur	продолжительность срока службы катализатора
C 97	catalyst load	Katalysatorfüllung *f*	chargement *m* de catalyseur	загрузка катализатора
C 98	catalyst loading	Katalysatorbelastung *f*	charge *f* imposée au catalyseur	нагрузка на катализатор
C 99	catalyst particle (pellet)	Katalysatorkorn *n*	particule *f* de catalyseur	зерно катализатора
	catalyst poison	*s.* C 95		
C 100	catalyst property	Katalysatoreigenschaft *f*	propriété *f* de catalyseur	свойство катализатора
C 101	catalyst support	Katalysatorträger *m*	support *m* du catalyseur	носитель катализатора
C 102	catalyst surface, catalyst area	Katalysatoroberfläche *f*	superficie *f* de catalyseur	поверхность катализатора
C 103	catalytic activity	katalytische Aktivität *f*	activité *f* catalytique	каталитическая активность
C 104	catalytic bed, catalyst bed	Katalysatorbett *n*, Kontaktbett *n*, Katalysatorschicht *f*, Kontaktschicht *f*	lit *m* catalytique, couche *f* de catalyseur	слой катализатора, засыпка катализатора, неподвижный слой катализатора
C 105	catalytic mass	Kontaktmasse *f*, Kontaktsubstanz *f*	masse *f* de contact	контактная масса
	catalytic method	*s.* C 619		
C 106	catalytic process	katalytischer Prozeß *m*	procédé *m* catalytique	каталитический процесс
	catalytic reactor	*s.* R 103		
	catalyzer	*s.* C 90		
C 107	catch basin, collecting vat	Auffangwanne *f*	fosse *f* de noyage, bac *m* de chute	приемник, поддон для улавливания
C 108	Cauchy-Green strain tensor	Cauchy-Greenscher Dehnungstensor (Spannungstensor) *m*	tenseur *m* [de déformation] de Cauchy-Green	тензор напряжения Коши-Грина

C 109	Cauchy's linear equation	Cauchysche Lineargleichung f	équation f linéaire de Cauchy	линейное уравнение Коши
C 110	cause of disturbance	Störungsursache f	cause (source) f de dérangement	причина отказа (аварии, возмущения)
C 111	caustic, corrosive, mordant	Beize f, Beizmittel n	caustique m, corrosif m, mordant m	травитель, протрава
C 112	caustic scrubbing	Laugenwäsche f	lavage m à lessive	щелочная промывка, щелочная очистка
C 113	cavern	Hohlraum m, Höhle f, Kaverne f	espace m vide, cavité f	каверна
	cavitation	s. B 283		
C 114	cavity radiation	schwarze Strahlung f, Hohlraumstrahlung f	radiation f du corps noir	излучение (радиация) черного тела
C 115	ceiling coil, overhead coil	Deckenkühlschlange f	batterie f des tuyaux plafonnières, serpentin m plafonnier	потолочная батарея
C 116	ceiling-mounted air cooler	Deckenluftkühler m	refroidisseur m d'air monté au plafond	потолочный воздухоохладитель
C 117	cell model	Zellenmodell n	modèle m cellulaire	ячеечная модель
C 118	cell number	Zellenzahl f	nombre f de cellules	количество ячеек
C 119	cellular cooler	Zellenkühler m	radiateur m à cellules	ячейковый радиатор
	cellular-type radiator	s. L 14		
C 120	central control room	Zentralmeßwarte f	salle f de contrôle centrale	центральный контрольно-измерительный пункт
C 121	central temperature	Mittelpunktstemperatur f	température f centrale (de centre)	температура в центре
C 122	centre, node, junction, intersection	Knotenpunkt m	nœud m d'assemblage, point m de jonction	узел, узловая точка, транспортный узел
C 123	centre of gravity	Schwerpunkt m	centre m de gravité	центр тяжести
C 124	centrifugal acceleration	Zentrifugalbeschleunigung f	accélération f centrifuge	центробежное ускорение
C 125	centrifugal air separator	Schleudersichter m	séparateur m à vent centrifuge	центробежный сепаратор
C 126	centrifugal ball mill	Fliehkraftkugelmühle f, Kreiselkugelmischer m	moulin m centrifuge à boulets, mélangeur m à boules centrifuge	ротационная (центробежная) шаровая мельница
C 127	centrifugal blower, rotary blower	Kreiselgebläse n, Schleudergebläse n, Rotationsgebläse n	ventilateur m (soufflerie f) centrifuge, soufflante f rotative	центробежный вентилятор, центробежная (ротационная) воздуходувка
C 128	centrifugal compressor	Zentrifugalkompressor m	compresseur m centrifuge	центробежный компрессор
C 129	centrifugal decanter	Dekantierzentrifuge f	décanteur m centrifuge	центробежный декантер
C 130	centrifugal drum	Schleudertrommel f	tambour m centrifuge	барабан центрифуги
	centrifugal dust collector	s. V 205		
C 131	centrifugal efficiency, centrifugal force	Schleuderwirkung f, Zentrifugalkraft f	effet m centrifuge, force f centrifuge	центробежное действие, центробежная сила
C 132	centrifugal extractor	Zentrifugalextraktor m	ectracteur m centrifuge	центробежный экстрактор
C 133	centrifugal filter	Schleuderfilter n, Filterzentrifuge f	filtre m centrifuge	центробежный фильтр, фильтровальная центрифуга
C 134	centrifugal flow mill, desintegrator, centrifugal mill	Schleudermühle f, Fliehkraftmühle f	broyeur m centrifuge, moulin m centrifuge (désintégrateur)	центробежная мельница, дезинтегратор
	centrifugal force	s. C 131		
C 135	centrifugal mill, desintegrator	Schleuderprallmühle f	broyeur m à impact	ударноотражательная мельница
	centrifugal mill	s. a. C 134		
C 136	centrifugal mixer	Schleudermischer m	mélangeur m centrifuge	центробежный смеситель
C 137	centrifugal pump	Kreiselpumpe f, Zentrifugalpumpe f	pompe f centrifuge	центробежный насос
C 138	centrifugal separator	Zentrifugalabscheider m	séparateur m centrifuge	центробежный сепаратор
	centrifugal supercharger	s. C 140		
C 139	centrifugal turbine	Schleuderturbine f, Zentrifugalturbine f	turbine f centrifuge	центробежная турбина
C 140	centrifugal ventilator, turboblower, centrifugal supercharger	Radiallüfter m, Kreisellüfter m, Radialgebläse n	ventilateur m centrifuge, soufflerie f centrifuge	центробежный вентилятор
C 141	centrifugating	Abschleudern n (Kristallisation)	essorage m	центрифугирование
C 142	centrifugation, centrifuging	Schleudern n, Schleuderverfahren n	centrifugation f, essorage m centrifuge	способ центрифугирования, центрифугирование
C 143	centrifuge	Zentrifuge f, Schleuder f	machine f centrifuge, centrifugeuse f, essoreuse f	центрифуга
	centrifuging	s. C 142		
C 144	ceramic packing	Keramikfüllung f	remplissage m céramique	керамиковое наполнение
C 145	cetane number improver	Zündbeschleuniger m	accélérateur m d'inflammation	присадка для ускорения самовоспламенения, ускоритель самовоспламенения
C 146	chain conveyor	Kettenaufzug m	élévateur m à chaînes	цепной конвейер
C 147	chain drive	Kettenantrieb m	commande f par chaîne, entraînement m par pignons et chaînes	цепная передача

C 148	chain dryer	Kettenbandtrockner m	séchoir m à chaîne	цепная туннельная сушилка
C 149	chain elevator	Kettenbecherwerk n	élévateur m à chaînes, chaîne f à godets	цепной ковшовый элеватор
C 150	chain pump	Kettenpumpe f	pompe f à chaîne	цепной насос
C 151	chain reaction	Kettenreaktion f	réaction f en chaîne	цепная реакция
C 152	chain-type carrier, transportation line	Transportkette f	transporteur m à chaîne	цепной транспортер, транспортная линия
C 153	chamber dryer	Kammertrockner m	sécheur m à chambre	камерная сушилка
C 154	chamber filter press	Kammerfilterpresse f	filtre-presse m à chambres, filtre-presse à plateaux	камерный фильтр-пресс
C 155	chamber kiln, retort furnace	Kammerofen m	four m dormant, fourneau m à chambre	камерная печь
C 156	chamber reactor	Kammerreaktor m	réacteur m à chambres	камерный реактор
C 157	change/to, to convert, to transform	verwandeln, wechseln, ändern	changer, transformer	превращать, изменять, переходить
C 158	change, alteration, variation, transformation, conversion	Veränderung f, Verwandlung f, Umwandlung f, Umformung f, Transformation f	changement m, variation f, transformation f, formage m, conversion f	изменение, переработка, преобразование, трансформирование, превращение
C 159	change of state	Aggregatzustandsänderung f, Zustandsänderung f	changement m d'état	изменение агрегатного состояния вещества, изменение состояния
C 160	change of structure	Strukturveränderung f	changement m de structure, variation f structurale	структурное изменение
C 161	change of volume	Volumenänderung f	changement m du volume	изменение объема
C 162	channel, drain, drainage, drain pipe	Sickerrinne f, Rinne f, Kanal m	rigole f de drainage	дрена, сточный лоток
C 163	channelling	Kanalbildung f, Bachbildung f	formation f de canaux, formation de ruissellements	каналообразование
C 164	Chapman-Enskog theory	Chapman-Enskogsche Theorie f	théorie f de Chapman-Enskog	теория Чапмана-Энскога
C 165	char/to	verkohlen	carboniser	обугливать
C 166	characteristic curve	charakteristische Kurve f, Kennlinie f	courbe f caractéristique, ligne f de fonctionnement	характеристическая кривая, характеристика
C 167	characteristic energy	Eigenenergie f	énergie f propre	собственная энергия
C 168	characteristic function, eigenfunction	Eigenfunktion f	fonction f caractéristique (propre)	собственная (характеристическая) функция
C 169	characteristic gas equation	allgemeine Gasgleichung f, charakteristische Gaszustandsgleichung f	équation f caractéristique des gaz	характеристическое уравнение состояния газа
C 170	characteristic graph	charakteristischer Graph m	graphe m caractéristique	характеристический граф
C 171	characteristic matrix	charakteristische Matrix f	matrice f caractéristique	характеристическая матрица
C 172	characteristic point	charakteristischer Punkt m	point m caractéristique	характеристическая точка
C 173	characteristics	Kenndaten pl, Kennwerte mpl	caractéristiques fpl, données fpl caractéristiques	[характеристические] данные
C 174	characteristic state	Eigenzustand m	état m propre	характеристическое состояние
C 175	characteristic temperature	Eigentemperatur f	température f propre	характеристическая температура
C 176	characteristic time	Ausgleichszeit f	période f transitoire	время выравнивания (температуры)
C 177	characteristic value, parameter	Kennwert m, Parameter m, Kenngröße f	valeur f caractéristique, paramètre m	показатель, параметр, коэффициент
C 178	character of the surface	Oberflächenbeschaffenheit f	constitution f de la superficie, état m de surface	состояние [обработанной] поверхности
C 179	charcoal adsorber	Aktivkohleadsorber m	adsorbeur m à charbon activé	адсорбер с активированным углем
C 180	char dust bin	Kohlenstaubbunker m	soute (trémie) f de charbon pulvérisé	бункер угольной пыли
C 181	charge, batch, load, burden	Lademenge f, Füllgut n, Einsatz m, Füllung f, Charge f, Einsatzprodukt n, Menge f, Partie f, Beschickung f	quantité f de chargement, charge f, alimentation f	количество погружаемого материала, загрузка, зарядка, партия [на выработку], кампания, навеска, закладка
C 182	chargeable	belastbar	capable d'être chargé	допускающий нагрузку
C 183	charge coefficient	Beschickungskoeffizient m, Füllungsgrad m	coefficient m de charge	степень (коэффициент) заполнения
C 184	charge density	Ladedichte f	densité f de charge	плотность заряда
C 185	charge level	Füllstand m	niveau m de remplissage	уровень
C 186	charge material, primary material	Ausgangsmaterial n	matière f brute, produit m de départ	сырье, исходный материал
	charge pump	s. F 46		
C 187	charge quantity	Chargengröße f	quantité f de charge	количество загрузки
C 188	charge stock, feed stock	Einsatzprodukt n	produit m d'alimentation	сырье, исходный (загружаемый) материал
	charging	s. 1. I 131; 2. L 197		
C 189	charging belt, conveyor	Beschickungsband n	convoyeur m de chargement	ленточный питатель
C 190	charging device	Fülleinrichtung f	dispositif m de chargement (remplissage), chargeur m	устройство для загрузки (зарядки)

C 191	charging equipment	Beschickungsanlage f	équipement m de charge-ment, installation f d'ali-mentation	загрузочное устройство, за-грузочный механизм
C 192	charging funnel	Einfülltrichter m	entonnoir m de remplissage, trémie f de chargement	засыпная воронка
C 193	charging machine	Chargiermaschine f	machine f de chargement	загрузочная (завалочная) машина
C 194	charging of the surface	Oberflächenbeladung f	chargement m de la surface	наполнение поверхности, поверхностный заряд
C 195	charging opening	Beschickungsöffnung f, Ein-satzöffnung f	ouverture f de chargement	завалочное (загрузочное) окно
C 196	charging period	Chargenzeit f, Beschickungs-zeit f	période f de charge, temps m de chargement	время выдержки (загрузки)
C 197	charging point	Einfüllstelle f	point m de chargement	точка зарядки (загрузки)
C 198	charging rate	Beschickungsgeschwindigkeit f	vitesse f de chargement	скорость загрузки
C 199	charging station	Einfüllstation f	station f de chargement	зарядная (загрузочная) станция
	Charles' law	s. G 108		
C 200	chart, scheme	schematische Darstellung f, Schema n	représentation f schémati-que, schéma m	схема
C 201	check, revision	Überprüfung f	révision f, vérification f	проверка, контроль
C 202	check analysis	Kontrollanalyse f	analyse f de contrôle, ana-lyse témoin	контрольный анализ
C 203	checker work	Kammerfüllung f, Kammer-auskleidung f	remplissage m de la chambre	насадка регенератора, вну-тренняя облицовка
C 204	checking apparatus, control device, monitor	Kontrollapparat m	appareil m de contrôle	контрольный аппарат
C 205	check list	Checkliste f, Prüfliste f	check-liste f	контрольный (проверочный) список
C 206	check measurement, check test	Kontrollmessung f	mesure f de contrôle	контрольное измерение
C 207	checkout, inspection test	Kontrollprüfung f	contrôle m	контрольное испытание
	check test	s. C 206		
C 208	check valve	Rückschlagventil n	soupape f de retenue	обратный клапан
C 209	chemical adsorption	chemische Adsorption f	adsorption f chimique	химическая адсорбция
C 210	chemical bond	chemische Bindung f	liaison f chimique	химическая связь, химиче-ское сцепление
C 211	chemical cooling	chemische Kühlung f	refroidissement m chimique	химическое охлаждение
C 212	chemical decomposition	chemische Zersetzung f	décomposition f chimique	химическое разложение
	chemical drying	s. C 228		
C 213	chemical energy	chemische Energie	énergie f chimique	химическая энергия
C 214	chemical engineer, process engineer	Ingenieur m für chemische Verfahrenstechnik, Verfah-renstechniker m, Verfah-rensingenieur m	ingénieur m de génie chimi-que	инженер-химик-технолог
	chemical engineering	s. P 504		
C 215	chemical equilibrium	chemisches Gleichgewicht n	équilibre m chimique	химическое равновесие
C 216	chemical equipment building	chemischer Maschinenbau (Apparatebau) m	construction f d'appareils chimiques	химическое машиностро-ение
C 217	chemical industry	chemische Industrie f	industrie f chimique	химическая промышлен-ность
C 218	chemical kinetics	chemische Kinetik f	cinétique f chimique	химическая кинетика
C 219	chemical petroleum crude preparation	chemische Rohölaufberei-tung f	traitement m chimique du pétrole brut	химическая обработка сырой нефти
C 220	chemical plant	Chemieanlage f	unité f chimique, fabrique f de produits chimiques	химическая установка
C 221	chemical potential	chemisches Potential n	potentiel m chimique	химический потенциал
C 222	chemical process engineer-ing	chemische Verfahrenstechnik f	génie m chimique, technique f des procédés chimiques	химическая технология, процессы и аппараты хи-мической технологии
C 223	chemical process system	verfahrenstechnisches Sy-stem n	système m technologique (de génie chimique)	химико-технологическая си-стема
C 224	chemical process industries, chemical processing indus-tries	Stoffwirtschaft f, chemische und artverwandte Industrie f, chemische Industrie f, Chemieindustrie f	industrie f chimique et para-chimique	химическая и перерабаты-вающая промышленность
C 225	chemical processing	chemische Verarbeitung f, Stoffwandlung f	transformation f chimique	химическая переработка
	chemical processing indus-tries	s. C 224		
	chemical process system	s. C 231		
C 226/7	chemical raffination	chemische Raffination f	raffinage m chimique	химическая очистка
C 228	chemical seasoning, chemi-cal drying	chemische Trocknung f	desséchement m chimique, séchage m chimique	химическая сушка
C 229	chemical treatment	chemische Behandlung f	traitement m chimique	химическая обработка
C 230	chemical treatment process	Verfahren n zur chemischen Behandlung	procédé m de traitement chi-mique	процесс химической обра-ботки
C 231	chemico-technological sys-tem, chemical process sys-tem	chemisch-technologisches System n, verfahrenstech-nisches System n	système m chimico-technolo-gique	химико-технологическая си-стема

C 232	chemisorption	Chemisorption f	chimisorption f	хемисорбция
C 233	chemistry-based plant	stoffwirtschaftlicher Betrieb m	usine f de l'industrie chimique	предприятия (завод) химико-технологического профиля
C 234	chief reaction, main reaction	Hauptreaktion f	réaction f principale	главная (основная) реакция
C 235	chilled commodity	Kühlgut n	marchandise f de refroidissement	охлажденный товар
C 236	chilled-water circuit	Kühlwasserkreislauf m	circuit m d'eau réfrigérante	схема циркуляции охлажденной воды
C 237	chilled-water cooler	Kühlwasserkühler m	refroidisseur à eau réfrigérante, réfrigérant m à eau froide	охладитель охлажденной воды
C 238	chilled-water jacketed tank	Behälter m mit Kühlwassermantel	tank m à enveloppe d'eau réfrigérante	бак с рубашкой для охлаждающей воды
C 239	chilled-water output	Kühlwassermenge f	débit m d'eau réfrigérante	количество (расход) охлажденной воды
C 240	chilling bath	Kühlbad n	bain m de réfrigération	охлаждающая ванна
C 241	chilling rate	Abkühlgeschwindigkeit f	vitesse f de réfrigération (refroidissement)	скорость охлаждения
C 242	chilling water rate	Kühlwasserstrom m	courant m d'eau de refroidissement, débit m d'eau froide	поток охлаждающей воды
C 243	chimney cooler	Kaminkühler m	réfrigérant m à cheminée	башенная градирня, башенный охладитель
C 244	chimney draught	Kaminzug m	carneau m de cheminée	тяга дымовой трубы
	chimney gas	s. F 273		
C 245	chlorinated alkali electrolysis	Chloralkalielektrolyse f	électrolyse f du chlorure de sodium	электролиз хлористых щелочей
C 246	chlorination	Chlorierung f	chloration f	хлорирование
C 247	chlorine generator	Chlorentwickler m	générateur m de chlore	аппарат для получения газообразного хлора
C 248	chlorine-hydrogen reaction	Chlor-Knallgas-Reaktion f	réaction f d'un mélange de chlorure et d'hydrogène à volumes égaux	реакция хлористоводородного гремучего газа
C 249	choice, selection	Auswahl f	choix m, alternative f, sélection f	выбор, отбор, альтернатива, подбор, набор
C 250	choke, throttle	Drosselklappe f	clapet m d'étranglement	дроссель, дроссельный клапан
	choke up/to	s. B 136		
C 251	choking, throttling	Abdrosselung f, Drosselung f	étranglement m	дросселирование, прикрытие клапаном
C 252	choop stress component	Tangentialspannungskomponente f	composante f de tension tangentielle	тангенциальная составляющая напряжения
C 253	chromatographic adsorption [analysis]	Adsorptionschromatografie f, chromatografische Adsorptionsanalyse f	analyse f d'adsorption	адсорбционный анализ
	cinder	s. S 364		
C 254	circuit, loop, wiring system	Stromkreis m, Kreislauf m, Kreis m	circuit m, circulation f	цепь тока, электрическая цепь, круговод, цикл
C 255	circuit diagram (drawing), wiring diagram	Schaltbild n	schéma m de couplage, plan m de montage, diagramme m des connexions	схема, распределительная (монтажная) схема, схема токопрохождения
C 256	circuit graph	Schaltungsgraph m	graphe m structural	граф структуры, структура схемы
C 257	circular matrix	Kreismatrix f	matrice f circulaire	матрица циклов (контуров)
C 258	circular tube	Kreisrohr n, Rohr n mit kreisförmigem Qerschnitt	tube m circulaire	труба (канал) с круглым поперечным сечением
C 259	circulating dryer	Umlufttrockner m	sécheur m à air circulé	сушилка с рециркуляцией части воздуха
	circulating flow	s. S 301		
C 260	circulating fuel reactor	Kreislaufreaktor m	réacteur m à boucle	реактор с циркуляцией, циркуляционный реактор
	circulation	s. C 894		
C 261	circulation circuit	Umlaufschema n	schéma m de circulation	схема циркуляции
	circulation gas	s. S 68		
C 262	circulation loop	Umlaufleitung f	conduite f de circulation	обводный трубопровод
C 263	circulation pump, agitating pump	Umwälzpumpe f	pompe f de circulation	циркуляционный насос
C 264	circulation system	Kreislaufschaltung f, Umlaufsystem n, Zirkulationssystem n	système (montage) m de circulation, circuit m fermé	система циркуляции, циркуляционная схема
C 265	circulatory lubrication	Umlaufschmierung f	graissage m par circulation	циркуляционная смазка
C 266	circumstance	Zustand m	état m, situation f	состояние, режим
C 267	cire-perdue process	Wachsausschmelzverfahren n	procédé m à la cire perdue	прецизионный метод литья по восковыплавляемым моделям
C 268	clarification	Klären n, Klärung f	clarification f	осветление, очистка
C 269	clarification tank	Klärtank m	réservoir m de décantation, tank m de repos	отстойный резервуар
C 270	clarified water washing	Klarwasserspülung f	rinçage m à l'eau pure	промывка чистой водой
C 271	clarifier	Klärbecken n	bassin m de décantation, réservoir m de curement	отстойный (осадочный) бассейн

C 272	clarifier	Klärkessel *m*	réservoir *m* de décantation, chaudière *f* à défécation	осветлительный котел
C 273	clarifying apparatus	Klärapparat *m*	défécateur *m*	отстойник
C 274	clarifying tank	Klärbehälter *m*	vase *m* clarificatoire, réservoir *m* de curement, vase à clarifier	отстойный чан, отстойник
C 275	class frequency	Klassenhäufigkeit *f*	fréquence *f* de classe	частота классов
C 276	classification	Stromklassierung f, Sortierung f, Klassierung *f*	criblage *m* hydraulique, triage *m*, hydroclassification *f*, criblage pneumatique, classement *m*	классификация, сортировка
	classification	*s. a.* S 436		
C 277	classifier, ore sorter	Klassierer m, Klassierapparat *m*	classeur m, appareil *m* classeur	отборщик, разборщик, разделитель по фракциям, сортировка, классификатор
C 278	classify/to	klassifizieren, klassieren, sortieren	classer, cribler, trier	классифицировать
	classifying	*s.* S 204		
C 279	classifying screen	Sortiersieb n, Klassierrost m, Klassiersieb *n*	crible *m* de triage, crible – classeur m, tamis *m* de classement	сортировочный грохот, сортировочное сито, классификатор
C 280	class interval	Klassenbreite *f*	amplitude *f* de la classe	ширина класса
C 281	Clausius-Rankine process	Clausius-Rankine-Prozeß *m*	procédé *m* Clausius-Rankine	Клаузиус-Ранкин-процесс
	cleading	*s.* I 258		
C 282	clean air	reine Luft *f*	air *m* purifié	чистый воздух
C 283	cleaned air	gereinigte Luft *f*	air *m* épuré	очищенный воздух
C 284	cleaning	Reinigung *f*	nettoyage m, purification f, épuration *f*	очистка, очищение
	cleaning	*s. a.* W 16		
C 285	cleanliness, cleanness	Reinheit *f*	pureté *f*	чистота
C 286	cleansing, mordanting	Abbeizung f, Beizen *n*	décapage m, dérochage *m* chimique	травление, протравливание
C 287	clean-up reactor	Reinigungsreaktor *m*	réacteur *m* de purification	реактор-очиститель, реактор для очистки
C 288	clearance	Zwischenraum m, Spielraum m, Abstand *m*	intervalle m, espace m, interstice *m*	промежуточное пространство (расстояние)
C 289	clear from mud/to	entschlammen	débourber	удалить шлам
	clinkering temperature	*s.* S 342		
C 290	cloak room	Aufbewahrungsraum *m*	dépôt m, magasin *m*	хранилище, камера хранения
C 291	closed cooling circuit	geschlossener Kühlkreislauf *m*	circuit *m* de refroidissement fermé	закрытая система охлаждения
C 292	closed cycle	geschlossener Kreislauf m, Rückführung *f*	recyclage m, circuit *m* fermé, cycle *m* fermé	замкнутый кругооборот (цикл), закрытый цикл
C 293	closed shell-and-tube condenser	Rohrbündelkondensator m, Bündelrohrkondensator *m*	condenseur *m* multitubulaire, condenseur à faisceau de tubes	кожухотрубный конденсатор
C 294	closed system	abgeschlossenes System *n*	système *m* fermé	закрытая (замкнутая) система
C 295	closed technological cycle	geschlossener technologischer Kreislauf *m*	cycle *m* technologique fermé	замкнутый технологический цикл
C 296/7	closing delay	Einschaltverzögerung *f*	retard *m* à l'enclenchement	замедление (задержка, выдержка) включения
	closing signal	*s.* S 909		
	clot	*s.* C 304		
	cloud chamber	*s.* W 168		
	cloudiness	*s.* O 46		
C 298	cloud point	Trübungspunkt *m*	point *m* de trouble	точка помутнения
C 299	clustered errors	Fehlerhäufung *f*	accumulation *f* des erreurs	накопление ошибок
C 300	clutch	Kupplung *f*	accouplement m, embrayage *m* (tuyaux)	соединение
C 301	coagulability, coagulating property	Koagulierbarkeit f, Gerinnbarkeit *f*	aptitude *f* à la coagulation, coagulabilité *f*	способность к коагуляции, свертываемость, коагулируемость
C 302	coagulable	koagulierbar	coagulable	свертываемый, способный к коагуляции
C 303	coagulate/to	koagulieren, gerinnen, verdicken	coaguler	коагулировать
C 304	coagulate, clot	Koagulat *n*	coagulat m, matière *f* coagulée	коагулят
	coagulating property	*s.* C 301		
	coagulation	*s.* F 218		
C 305	coagulation stability	Koagulationsstabilität *f*	stabilité *f* de coagulation	устойчивость к коагуляции
C 306	coal consumption	Kohlebedarf *n*	besoins *mpl* en charbon	потребность в угле
C 307	coal conversion	Kohleveredlung *f*	ennoblissement *m* du charbon	переработка угля (в более ценные продукты), повышение качества угля
C 308	coal dressing	Kohleaufbereitung *f*	préparation *f* (traitement *m*) du charbon	обогащение угля

C 309	coal dust	Kohlenstaub m	poussière f (poussier m) de charbon, charbon m pulvérisé	угольная пыль
C 310	coal dust burner, pulverized coal burner	Kohlenstaubbrenner m	brûleur m à charbon pulvérisé	пылеугольная горелка
C 311	coal dust explosion	Kohlenstaubexplosion f	coup m de poussières de charbon	взрыв угольной пыли
C 312	coal feed	Kohlebeschickung f	chargement m de charbon	подача угля
C 313	coaling, supplying with coal	Bekohlung f	chargement m de charbon, alimentation f en charbon	загрузка угля, снабжение углем
C 314	coal liquefaction	Kohleverflüssigung f	liquéfaction f du charbon	ожижение угля
C 315	coal mill (pulverizer)	Kohlenstaubmühle f	moulin m à pulvériser le charbon, pulvérisateur m de charbon	угольная мельница
C 316	coal screening and grading plant	Kohleaufbereitungsanlage f	atelier m de lavage et criblage des charbons	углеобогатительная установка
C 317	coal stock	Kohlevorräte mpl	provision f (stock m) de charbon	запасы угля
C 318	coal supply	Kohleversorgung f	approvisionnement m charbonnier (en charbon)	снабжение углем
C 319	coarse crusher	Grobbrecher m, Vorbrecher m	concasseur m, premier concasseur, préconcasseur m	дробилка грубого дробления
C 320	coarse filter	Grobfilter n	filtre m ordinaire	фильтр грубой очистки
C 321	coarse grain	Grobkorn n	gros grain m	крупное зерно
C 322	coarse grinding	Grobzerkleinerung f	broyage (concassage) m grossier	крупное дробление
C 323	coarseness	Grobkörnigkeit f	grossièreté f	крупнозернистость
C 324	coarse screen, coarse sieve	Gräpelsieb n, Gröpelsieb n, Grobsieb n	gros sas m, tamis m gros	грубое сито
C 325/6	coarse screening	Grobsieben n	criblage m en gros	грубый рассев, крупное грохочение
	coarse sieve	s. C 324		
C 327	coat, layer, cover, coating	Überzug m, Auftrag m, Auflage f, Belag m	revêtement m, couverture f, enduit m, chemise f, enveloppe f	покрытие, отложение, осадок, обкладка, пленка, покровный слой
	coat	s. a. L 46		
C 328	coating; support	Auflage f	placage m, revêtement m	настил, покрытие
C 329	coating, covering	Belegung f	couvrement m, métallisage m	покрытие
C 330	coating	Beschichtung f, Beschichten n	revêtement m, enduction f	покрытие, покрывание слоем
	coating	s. a. 1. C 327; 2. V 86		
C 331	coating compound	Beschichtungsmaterial n, Streichmasse f	mélange m d'enduction, matière f de revêtement	материал для нанесения на поверхность
C 332	co-condensation, mixed condensation	Mischkondensation f	condensation f à mélange	смешанная конденсация
C 333	coefficient of cubical thermal expansion	kubischer Wärmeausdehnungskoeffizient m	coefficient m volumique de dilatation thermique	коэффициент кубического теплового расширения
C 334	coefficient of elasticity	Elastizitätskoeffizient m	coefficient m d'élasticité, constante f élastique	коэффициент упругости
C 335/6	coefficient of elongation	Dehnungskoeffizient m	coefficient (module) m d'allongement	коэффициент растяжения
	coefficient of evaporation	s. E 280		
C 337	coefficient of expansion	Ausdehnungszahl f, Expansionsgrad m	coefficient m d'expansion, degré m de détente	коэффициент расширения
C 338	coefficient of friction	Reibungszahl f	coefficient m de frottement	коэффициент трения
C 339	coefficient of heat radiation	Wärmestrahlungskoeffizient m	coefficient m de radiation thermique	коэффициент теплоизлучения
C 340	coefficient of linear thermal expansion	linearer Wärmeausdehnungskoeffizient m	coefficient m linéaire de dilatation thermique	коэффициент линейного теплового расширения
C 341	coefficient of performance	Kältefaktor m	coefficient m de froid	холодильный коэффициент
C 342	coefficient of quality	Qualitätskoeffizient m	coefficient m de qualité	коэффициент (параметр) качества
C 343	coefficient of solubility	Löslichkeitskoeffizient m	coefficient m de solubilité	коэффициент растворимости
C 344	coefficient of tension	Spannungskoeffizient m	coefficient m de tension	коэффициент упругости (напряжения)
C 345	coefficient of thermal conductivity	Wärmeleitzahl f	coefficient m de conductivité thermique	коэффициент теплопроводности
	coefficient of thermometric conductivity	s. T 119		
C 346	coefficient of variation	Streuungskoeffizient m	coefficient m de variation	коэффициент дисперсии
C 347	coefficient of volumetric expansion	räumlicher Ausdehnungskoeffizient m	coefficient m d'expansion volumétrique	коэффициент объемного расширения
C 348	cohesion	Kohäsion f, Zusammenhalt m	cohésion f, cohérence f	когезия, сцепление
C 349	cohesion force	Kohäsionskraft f	force f de cohésion	сила когезии
C 350	cohesion pressure	Kohäsionsdruck m	pression f de cohésion	когезионное давление
C 351	cohesion strength	Kohäsionsfestigkeit f	résistance f de cohésion	прочность на отрыв (сцепление)
C 352	cohesive energy density	Bindungsenergiedichte f	densité f de l'énergie de liaison	энергетическая плотность связи

C 353	cohobation, redistillation	wiederholte Destillation f	distillation f nouvelle	многократная перегонка, когобация
C 354	coil	Rohrschlange f	serpentin m	змеевик
C 355	coil condenser, spiral condenser	Schlangenkondensator m, Schlangenkühler m	condenseur (refroidisseur) m à serpentins	змеевиковый конденсатор (холодильник)
C 356	coil condenser, tubular cooler	Röhrenkühler m	radiateur (réfrigérant) m tubulaire	трубчатый холодильник
C 357	coil cooler	Schlangenrohrkühler m	refroidisseur m à serpentins	змеевиковый охладитель
C 358	coiled heat exchanger	Schlangenrohrwärmeübertrager m	échangeur m thermique à serpentins	змеевиковый теплообменник
C 359	coil-type absorber	Schlangenabsorber m	absorbeur m à serpentins	змеевиковый абсорбер
C 360	coke/to, to carbonize	verkoken	cokéfier	коксовать
C 361	coke chamber	Kokskammer f	chambre f de carbonisation	коксовая камера
C 362	coke oven, retort	Koksofen m	four m à coke	коксовальная печь
C 363	coke oven gas	Kokereigas n	gaz m de cokerie (fours à coke)	коксовый газ
C 364	coking, high-temperature carbonization	Verkokung f, Verkoken n	cokéfaction f	коксование
	coking	s. a. G 48		
C 365	coking index	Verkokungsindex m	indice m de cokéfaction	коксовое число
C 366	coking plant	Verkokungsanlage f, Kokerei f	installation f de cokéfaction, cokerie f	коксовый завод
C 367	coking process	Verkokungsprozeß m	procédé m de cokéfaction	процесс коксования
	cold accumulator	s. C 381		
C 368	cold air	Kaltluft f	air m froid	холодный воздух
	cold-air process	s. A 330		
C 369	cold-air stream	Kaltluftstrom m	courant m d'air froid	поток холодного воздуха
C 370	cold application	Kälteanwendung f	application f du froid	применение холода
C 371	cold boiler	Vakuumkocher m	appareil de cuisson à vide	вакуум-кипятильник
C 372	cold chamber	Kältekammer f, Kälteraum m	chambre f du froid, frigorifère m	холодильная камера
C 373	cold conditions	Kältebedingungen fpl	conditions fpl à basse température	условия пониженной температуры
C 374	cold consumption	Kälteverbrauch m	consommation f de froid	расход холода
C 375	cold deformation	Kaltverformung f	écrouissage m, déformation f à froid	холодная обработка
C 376	cold extractor	Kaltextraktor m	extracteur m travaillant à basse température	холодный экстрактор
C 377	cold forming process	Kaltverformungsverfahren n	écrouissage m, écrouissement m, façonnage m à froid	холодный способ формирования, холодное формование
C 378	cold gas	Kaltgas n	gaz m froid	холодный газ
C 379	cold generation rate	Kälteerzeugungsintensität f	intensité f (taux m) de génération du froid	интенсивность производства холода
C 380	cold granulation	Kaltgranulierung f	granulation f à froid	холодное гранулирование
C 381	cold hold-over, cold accumulator	Kältespeicher m	accumulateur m de froid	аккумулятор холода
C 382	cold hydrogenation	Kalthydrierung f	hydrogénation f à froid	гидрирование при низких температурах
C 383	cold hydrogenation process	Kalthydrierungsprozeß m	procédé m d'hydrogénation à froid	процесс низкотемпературного гидрирования
C 384	cold hydrogenation unit	Kalthydrierungsanlage f	unité f d'hydrogénation à basses températures	установка низкотемпературного гидрирования
C 385	cold insulation	Kälteisolation f, Kälteschutzisolierung f	isolation f contre le froid, isolement m à froid, calorifuge m frigorifique	изоляция для низких температур
C 386	cold loop	Kältekreislauf m	circulation f de froid	холодильный контур, холодильная циркуляция
C 387	cold polymer	Kaltpolymerisat n	produit m de polymérisation froide	холодный полимер
C 388	cold polymerization	Kaltpolymerisation f	polymérisation f à froid	полимеризация на холоду, холодная полимеризация
C 389	cold preservation, cold storage	Kaltlagerung f	stockage m frigorifique	холодное хранение
	cold production	s. R 226		
C 390	cold reserve	kalte Reserve f	réserve f froide	холодный резерв
C 391	cold rolling	Kaltwalzen n	laminage m à froid	холодное вальцевание
	cold storage	s. C 389		
C 392	cold treatment	Kältebehandlung f	traitement m à froid	холод[иль]ная обработка
C 393	cold vapour	Kaltdampf m	vapeur f froide	холодный пар
C 394	cold working	Kaltverarbeitung f	façonnage (travail) m à froid	холодная переработка
C 395	collect/to	einfangen, [sich] sammeln, sich vereinigen	collecter	собирать[ся]
C 396	collecting chamber	Auffangkammer f	chambre f collectrice	верхняя камера уравнительного резервуара, приемная камера
C 397	collecting funnel (hopper)	Auffangtrichter m	entonnoir m récepteur	приемная (сборная) воронка
C 398	collecting pipe	Sammelrohr n, Sammelrohrleitung f	tuyau m collecteur, collecteur m	смесительный резервуар, сборный (магистральный) трубопровод

C 399	collecting screw conveyor	Sammelschnecke f	vis f sans fin collectrice	сборочный шнек
C 400	collecting tank	Sammelbehälter m	récipient (réservoir) m collecteur	сборник, коллектор
C 401	collecting vat, reservoir	Sammelbecken n, Behälter m	récipient m collecteur, réservoir m	сборный бассейн, резервуар
	collecting vat	s. a. C 107		
C 402	collector	Sammler m, Kollektor m	collecteur m, égout m collecteur	собиратель, коллектор, сборник
C 403	collision, impact	Prall m	choc m, bond m, contrecoup m	удар, коллизия
C 404	collision factor	Stoßfaktor m	coefficient m de chocs	фактор частоты
C 405	colloid	Kolloid n	colloïde m	коллоид
C 406	colloidal fuel	kolloidaler Brennstoff m	combustible m colloïdal	коллоидальное топливо
C 407	colloidal state	kolloidaler Zustand m	état m colloïdal	коллоидальное состояние
	column	s. S 527		
C 408	column distillation	Kolonnendestillation f, Säulendestillation f	distillation f dans une colonne	перегонка на колонне
C 409	column internal	Kolonneneinbauten pl	pièces fpl incorporées d'une colonne	встроенные в колонну устройства
C 410	column plate, column tray	Kolonnenboden m	plateau m [de colonne]	тарелка колонны
C 411	column pressure	Säulendruck m, Kolonnendruck m	pression f dans la colonne	давление столба (в колонне)
C 412	column section	Kolonnenschuß m	tronçon m de colonne	секция колонны
C 413	column shell	Kolonnenmantel m	enveloppe f de colonne	оболочка (кожух) колонны
C 414	column still	Destillierkolonne f	colonne f de distillation	дистилляционная колонна
C 415	column temperature	Kolonnentemperatur f	température f de colonne	температура колонны
	column tray	s. C 410		
C 416	column volume	Kolonnenvolumen n	volume m de colonne	объём колонны
	combination	s. A 587		
	combination column	s. C 420		
C 417	combination dryer	Verbundtrockner m	sécheur m combiné	составная сушилка
C 418	combination matrix	Verknüpfungsmatrix f	matrice f de composition (combinaison)	матрица связей химико-технологической системы
C 419	combination termination	Kombinationsabbruch m	terminaison f par combinaison	комбинационный отрыв
C 420	combination tower, combination column	Kombinationskolonne f	tour f combinée	комбинированная колонна
C 421	combined lifting gears	kombiniertes Hebezeug n	dispositifs mpl de levage combinés	комбинированный подъемный механизм
C 422	combined process	Mischverfahren n	processus m mixte, procédé m combiné	комбинированный метод
C 423	combustibility	Brennbarkeit f	combustibilité f, inflammabilité f	горючесть
C 424	combustible	[ab]brennbar	combustible, inflammable	горючий, топливный
C 425	combustible gas	Verbrennungsgas n	gaz m de la combustion	печной газ, газ сгорания
C 426	combustion, burning, burn	Verbrennung f	combustion f	горение, сжигание, сгорание
C 427	combustion air, air for combustion	Verbrennungsluft f	air m de combustion	воздух, необходимый для горения; воздух сгорания
C 428	combustion chamber	Verbrennungskammer f	étuve f à combustion	камера горения (сгорания)
C 429	combustion deposit	Verbrennungsrückstand m	résidu m de combustion	остаток сгорания
C 430	combustion efficiency	Verbrennungsgütegrad m	rendement m de combustion	степень полноты сгорания, коэффициент полезного действия процесса сжигания
C 431	combustion energy	Verbrennungsenergie f	énergie f de combustion	энергия сгорания
C 432	combustion enthalpy	Verbrennungsenthalpie f	enthalpie f de combustion	теплота (энтальпия) сгорания
C 433	combustion equation	Verbrennungsgleichung f	équation f de combustion	уравнение химической реакции при горении
C 434	combustion front	Verbrennungsfront f	front m de combustion	фронт горения
C 435	combustion furnace, burner	Verbrennungsofen m	incinérateur m, grille f à analyse	печь для сжигания (элементарного анализа)
C 436	combustion nozzle	Verbrennungsdüse f	buse f de combustion	сопло горелки (камеры сгорания)
C 437	combustion plant	Verbrennungsanlage f	unité f de combustion, installation f d'incinération	топочная установка
C 438	combustion pressure	Verbrennungsdruck m	pression f de combustion	давление сгорания
C 439	combustion process	Verbrennungsprozeß m	processus m de combustion	процесс сгорания
	combustion rate	s. 1. C 443; 2. R 47		
C 440	combustion space	Verbrennungsraum m	chambre f de combustion	объём топочной камеры
C 441	combustion system	Verbrennungssystem n	système m de combustion	топливная система
C 442/3	combustion temperature, temperature of combustion	Verbrennungstemperatur f, Brenntemperatur f	température f de combustion	температура горения (сжигания)
	combustion velocity	s. R 47		
C 444	commensurability, comparability	Vergleichbarkeit f	commensurabilité f, comparabilité f	сравнимость, сопоставимость
	commensurable	s. C 449		
	comminutor	s. D 366		
C 445	commitment	Einsatz m	emploi m	применение, эксплуатация
	communication	s. T 262		

	compact/to	s. C 545		
C 446	compact construction (design)	Kompaktbauweise f, raumsparende Ausführung f	construction f compacte	компактная конструкция, компактное исполнение
C 447	compacting	Kompaktieren n, Kompaktierung f, Preßverdichten n, Verdichten n	compression f, compactage m	уплотнение
	compaction	s. C 538		
C 448	compactness	Kompaktheit f	compacité f	компактность
	comparability	s. C 444		
C 449	comparable, commensurable	vergleichbar	commensurable, comparable	сравнимый, сопоставимый
C 450	comparison process	Vergleichsprozeß m	procédé m de comparaison	идеальный термодинамический цикл, эталонный процесс
C 451	compatibility	Verträglichkeit f, Kompatibilität f	compatibilité f	совместимость
C 452	compensating curve (line)	Ausgleichslinie f	courbe f égalisatrice	линия регулирования, кривая выравнивания
C 453	compensating process, balancing	Ausgleichsvorgang m, Ausgleichserscheinung f	phénomène m d'évolution	процесс выравнивания
C 454	compensating tank	Ausgleichsbehälter m	réservoir m de compensation, récipient m de détente	регулирующий (выравнивающий) резервуар
C 455	compensation method	Kompensationsmethode f	méthode f de compensation	компенсационный метод
C 456	compensation of errors	Fehlerausgleich m	compensation f des erreurs	компенсация (выравнивание) ошибок
C 457	complete/to, to finish	vervollkommnen, vervollständigen, vollenden	perfectionner, achever, compléter	усовершенствовать, дополнять, завершать, заканчать
C 458	complete evaporation	vollständige Verdampfung f	évaporation f complète	полное испарение
C 459	complete mechanization, full mechanization	Vollmechanisierung f	mécanisation f complète	полная (комплексная) механизация
C 460	completeness	Vollständigkeit f	complet m, intégralité f	полнота, комплектность
	complete plant	s. O 143		
C 461	complete reversibility	vollständige Umkehrbarkeit f	réversibilité f complète	полная обратимость
C 462	completion	Vollendung f, Vervollständigung f	complètement m, achèvement m	завершение, окончание
	completion degree	s. S 523		
C 463	complex	mehrteilig, komplex	multiple, à plusieurs parties, complexe	составной, сложной
C 464	complex chemical reaction	Komplexreaktion f	réaction f chimique complexe	сложная химическая реакция
C 465	complex formation	Komplexbildung f	complexation f	образование комплекса
C 466	complex production plant	komplexe Produktionsanlage f	installation f de production complexe	сложная производственная установка
C 467	component, constituent	Bestandteil m, Komponente f	constituant m, composant m, partie f intégrante, élément m	компонент, составная часть, составляющая
C 468	component of assembly	Montagebauteil n	élément m de montage	монтажный элемент
C 469	compound	[chemische] Verbindung f, Masse f, Vergußmasse f, Kabelmasse f, Verbund m, Zusammensetzung f, Gemisch n, Mischung f	composé m, combinaison f	смесь, состав, соединение
C 470	compound, binding, joining, connection	Verbindung f	jonction f, assemblage m, connexion f, composé m	соединение, сочетание, сочленение, связь, сопряжение
	compound	s. a. M 264		
C 471	compound connection	Verbundschaltung f	montage m d'interconnexion	смешанная схема
C 472	compound cycle	Verbundprozeß m	procédé m combiné	связанный процесс, двухступенчатый цикл, параллельная работа
C 473	compounding	Mischen n, Mischungsherstellung f, Vermengung f	mélange m	смешение, приготовление смесей
C 474	compound material	Verbundwerkstoff m	matériaux mpl composites	композиция, композиционный (смешанный, многослойный) материал
C 475	compound operation	Verbundbetrieb m	compoundage m, marche f en interconnexion	совместная эксплуатация, параллельная работа
C 476	compound system, interconnected (overall) system	Verbundsystem n, Gesamtsystem n	système m de liaison, système compound (total, tout entier)	объединенная (общая) система
C 477	comprehensible, intelligible, logical	verständlich	intelligible	понятый, логичный
	compress/to	s. C 545		
C 478	compressed air	Preßluft f, Druckluft f	air m comprimé	сжатый воздух
C 479	compressed-air bottle	Druckluftflasche f	bouteille f à air comprimé	баллон со сжатым воздухом
C 480	compressed-air respirator	Druckluftatemgerät n	appareil m de respiration à air comprimé	кислородный аппарат со сжатым воздухом
	compressed gas	s. H 156		

C 481	compressed-gas cooler	Druckgaskühler *m*	réfrigérant *m* pour gaz comprimé	холодильник для сжатого газа
C 482	compressed-gas vessel	Druckgasbehälter *m*	cylindre *m* à gaz comprimé, récipient *m* sous pression pour gaz comprimé	резервуар для сжатого газа
C 483	compressibility	Kompressibilität *f*	compressibilité *f*	сжимаемость
C 484	compressibility factor	Kompressionsfaktor *m*, Verdichtungsfaktor *m*	facteur *m* de compression	коэффициент (степень, фактор) сжатия
	compressible	s. C 526		
C 485	compressing	Verdichten *n*	compression *f*, condensation *f*	конденсирование, сжижение, сжатие
	compression	s. C 538		
C 486	compression chamber, compression space	Verdichtungsraum *m*, Druckkammer *f*, Verdichtungskammer *f*, Kompressionsraum *m*	chambre *f* de compression	пространство (камера) сжатия
C 487	compression cycle	Verdichtungsprozeß *m*, Verdichtungsperiode *f*, Verdichtungstakt *m*	cycle *m* de compression	цикл (период) сжатия
C 488	compression line	Verdichtungskurve *f*	ligne *f* de compression	кривая сжатия
C 489	compression load	Druckbelastung *f*, Druckbeanspruchung *f*	effort *m* (contrainte *f*) de compression	нагрузка при сжатии, сжимающая нагрузка
C 490	compression loss	Verdichtungsverlust *m*, Kompressionsverlust *m*	perte *f* de compression	потеря компрессии (при сжатии)
C 491	compression pressure	Verdichtungsdruck *m*	taux *m* de compression, compression *f*	давление сжатия
C 492	compression ratio, ratio of compression	Kompressionsverhältnis *n*, Verdichtungsverhältnis *n*, Druckverhältnis *n*, Verdichtungsgrad *m*	taux (degré) *m* de compression	коэффициент (степень) сжатия
	compression space	s. C 486		
C 493/4	compression stage	Verdichtungsstufe *f*	étage *m* de compression	ступень сжатия
C 495	compression strength	Druckfestigkeit *f*	résistance *f* à la compression	прочность на сжатие
C 496	compression stroke	Kompressionshub *m*	course *f* de compression	ход сжатия
C 497	compression work	Verdichtungsarbeit *f*	travail *m* de compression	работа сжатия
C 498	compressive force	Druckkraft *f*	force *f* de pression, force foulante	сила давления
C 499	compressive stress	Druckspannung *f*	effort *m* (tension *f*) de compression	напряжение сжатия
	compressor	s. C 504		
C 500	compressor casing	Verdichtergehäuse *n*	carter (boîtier) *m* de compresseur	кожух компрессора
C 501	compressor drive	Verdichterantrieb *m*	commande *f* de compresseur	привод компрессора
C 502	compressor house	Verdichterhalle *f*	salle *f* de compresseurs	компрессорный цех
C 503	compressor stage	Verdichterstufe *f*	étage *m* de compresseur	ступень нагнетателя (компрессора)
C 504	compressor station, condenser, compressor	Kompressoranlage *f*, Kompressorsatz *m*, Verdichter *m*	installation *f* de compression, moto-compresseur *m*, compresseur *m*	компрессорная установка, компрессор, нагнетатель
C 505	computability	Berechenbarkeit *f*	calculabilité *f*	возможность расчета, рассчитываемость, возможность вычисления
	computation	s. C 11		
C 506	computer	Rechenanlage *f*, Rechenmaschine *f*, Rechenautomat *m*, Computer *m*	calculateur *m*, computer *m*, calculatrice *f* électronique	вычислительная машина, компьютер
C 507	computer-aided design, CAD	automatisierte (rechnergestützte) Projektierung *f*	projection *f* assistée par ordinateur, projection automatique	автоматизированное проектирование, САПР
C 508	computer simulation	Rechnersimulierung *f*, Simulation *f* (auf EDV)	simulation *f* par calculateur	имитационное моделирование, расчет
	computing principle	s. C 14		
C 509	computing system	Berechnungssystem *n*	système *m* de calcul	система расчета
	concentrate/to	s. 1. R 259; 2. T 169		
C 510	concentrate	Aufbereitungsprodukt *n*	produit *m* de préparation, produit obtenu par traitement des minérais	продукт обогащения, обогащенный продукт
C 511	concentrated, reinforced, strengthened	verstärkt	renforcé, armé, fortifié	концентрированный, усиленный, армированный
C 512	concentrated	eingedickt, konzentriert	concentré	концентрированный, сгущенный
C 513	concentrated solution	konzentrierte Lösung *f*	solution *f* concentrée	концентрированный раствор
C 514	concentrating	Eindicken *n*	épaississement *m*, concentration *f*	сгущение, концентрирование, уваривание
C 515	concentration, thickening	Verdickung *f*, Konzentration *f*, Eindickung *f*	concentration *f*, épaississage *m*, épaississement *m*	утолщение, концентрирование, сгущение, выпаривание
	concentration	s. a. 1. A 479; 2. C 538		
C 516	concentration cell	Konzentrationselement *n*	pile *f* de concentration	концентрационный элемент
C 517	concentration change	Konzentrationsänderung *f*	changement *m* (variation *f*) de concentration	изменение концентрации

C 518	concentration course	Konzentrationsverlauf *m*	allure *f* de concentration	ход (течение, протекание) концентрации
C 519	concentration of a saturated solution	Sättigungskonzentration *f*	concentration *f* de saturation	концентрация насыщения (насыщенного раствора)
C 520	concentration of mass, mass concentration	Massenkonzentration *f*	concentration *f* de masse	массовая концентрация
C 521	concentration plant	Eindampfanlage *f*	épaississeur *m*, conditionneur *m*, évaporateur *m*	выпарная установка
C 522	concentration ratio	Konzentrationsverhältnis *n*	proportion *f* de la concentration, rapport *m* de concentration	соотношение концентрации
C 523	concentrator, thickener, evaporator	Eindicker *m*, Eindickungsapparat *m*	épaississeur *m*, conditionneur *m*	сгуститель, концентратор, установка для концентрирования
C 524	conclusion	Folgerung *f*	déduction *f*, conclusion *f*	следствие, заключение, вывод
C 525	concurrent flow, parallel flow	Parallellauf *m*, Parallelstrom *m*, Parallelströmung *f*	courant (écoulement) *m* parallèle	параллельная работа, параллельное течение, параллельный поток
C 526	condensable, compressible	verdichtbar, kondensierbar	condensable, compressible	конденсируемый, сжимаемый
	condensate/to	s. 1. C 545; 2. L 143		
C 527	condensate	Kondensat *n*, Kondensationsprodukt *n*	condensé *m*, produit *m* de condensation, condensat *m*	конденсат, флегма
C 528	condensate collector	Kondensatsammelbehälter *m*	collecteur *m* de condensé	конденсатоотводчик, коллектор конденсата
C 529	condensate cooler	Kondensatkühler *m*	refroidisseur *m* de l'eau condensée	охладитель конденсата
C 530	condensate drum	Kondensatbehälter *m*	réservoir *m* à condensat	бак для конденсата
C 531	condensate network	Kondensatnetz *n*	canalisation *f* au condensé	система трубопроводов для конденсата
C 532	condensate pipe	Kondensatleitung *f*	tube *m* de l'eau condensée	конденсатопровод, трубопровод конденсата
C 533	condensate pump	Kondensatpumpe *f*, Extraktionspumpe *f*	pompe *f* à vapeur condensé, pompe au condensé	конденсатный насос
C 534	condensate recycling	Kondensatrückführung *f*	retour *m* de condensé, recirculation *f* de l'eau condensée	возврат конденсата
C 535	condensate removal	Kondensatabfluß *m*	retour (déversoir) *m* de condensé	устройство для спуска конденсата
C 536	condensate separator	Kondensatabscheider *m*, Kondenswasserabscheider *m*	séparateur *m* d'eau de condensation	горшок конденсата, конденсационный горшок
C 537	condensate stripper	Kondensatstripper *m*	stripper *m* du condensé	отделитель конденсата
C 538	condensation, compaction, densification, compression, concentration	Verdichtung *f*, Kondensierung *f*	compression *f*, condensation *f*, concentration *f*	уплотнение, конденсация, концентрация, сжатие, сжижение
C 539	condensation chamber	Kondensationskammer *f*	chambre *f* de condensation	камера конденсатора
C 540	condensation loss	Kondensationsverlust *m*	perte *f* par condensation	потери при конденсации
C 541	condensation point	Kondensationspunkt *m*	point *m* de condensation	точка конденсации
C 542	condensation polymer	Kondensationspolymer *m*	polymère *m* de condensation	конденсационный полимер
C 543	condensation process	Kondensationsprozeß *m*	processus *m* de condensation	процесс конденсации, конденсация
C 544	condensation product	Kondensationsprodukt *n*	produit *m* (résine *f*) de condensation, polycondensat *m*	продукт конденсации
C 545	condense/to, to condensate, to compact, to compress	verdichten, kondensieren	condenser, comprimer	сжимать, уплотнять, конденсировать
C 546	condensed steam	kondensierter Dampf *m*	vapeur *f* condensée	сконденсированный пар
C 547	condensed water	Kondens[ations]wasser *n*	eau *f* condensée (de condensation)	конденсационная вода
C 548	condenser, liquefier	Kondensator *m*, Verflüssiger *m*, Kühler *m*	condenseur *m*, liquéfacteur *m*, réfrigérant *m*	конденсатор, дефлегматор, конденсор, холодильник
	condenser	s. a. C 504		
C 549	condenser arrangement	Anordnung *f* der Kondensatoren	arrangement *m* (disposition *f*) des condenseurs	расположение конденсаторов
C 550	condenser jacket	Kühlermantel *m*	enveloppe *f* du refroidisseur, corps *m* de réfrigérant	кожух холодильника
C 551	condenser plant	Kondensationsanlage *f*	installation *f* de condensation	конденсационная установка
C 552	condenser unit	Kondensatorelement *n*	élément *m* de condenseur	элемент конденсатора
C 553	condenser water	Brüdenwasser *n*	eau *f* condensée	конденсат сокового пара
	condensing coil	s. C 707		
C 554	condensing pressure	Kondensationsdruck *m*	pression *f* de condensation	давление конденсации
C 555	condensing temperature	Kondensationstemperatur *f*	température *f* de condensation	температура конденсации
C 556	condensing turbine	Kondensationsturbine *f*	turbine *f* à condensation	конденсационная турбина
C 557	condition/to	bedingen, in Bewegung setzen	entraîner, conditionner	обусловливать
	condition	s. A 588		
C 558	conditional stress, engineering stress	bedingte Spannung *f*	tension *f* conditionelle	условное напряжение

C 559	conditioned vapour	aufbereiteter Dampf *m*	vapeur *f* conditionnée	кондиционированный пар
C 560	conditioning	Konditionierung *f*	conditionnement *m*	кондиционирование
C 561	conditioning process	konditionierender Prozeß *m*	procédé *m* de conditionnement	подготовительный процесс
C 562	conditions of supply	Lieferungsbedingungen *fpl*	conditions *fpl* de livraison	условия поставки
C 563	conduction cooling	Leitungskühlung *f*	refroidissement *m* par conduction	охлаждение за счет теплопроводности
	conduction of heat	s. H 118		
C 564	conductivity	Leitfähigkeit *f*	conductibilité *f*, conductivité *f*, pouvoir *m* conducteur	проводимость
C 565	conduit	Abzug[skanal] *m*, Leitung *f*, Leitungsrohr *n*, Röhre *f*, Rohrleitung *f*	tuyau *m* de décharge, conduit *m*, canalisation *f*, conduite *f*	трубопровод, труба, канал, ход, провод
	conduit	s. a. O 112		
C 566	cone-and-plate viscometer, rheogoniometer	Rheogoniometer *n*	rhéogoniomètre *m*	реогониометр
C 567	cone crusher	Rundbrecher *m*, Kegelbrecher *m*	broyeur *m* à cônes	конусная дробилка
C 568	cone mill	Glockenmühle *f*	moulin *m* conique	коническая мельница (дробилка)
C 569	cone shaped	kegelförmig	cône, conique, conoide, en forme de cône	конический, конусный
C 570	cone turbine	Kegelturbine *f*	turbine *f* conique	коническая турбина
C 571	cone valve	Kegelventil *n*	soupape *f* (obturateur *m*) conique	клиновый вентиль, конический клапан
C 572	configuration	Anordnung *f*, Konfiguration *f*, Konstitution *f*, Gestaltung *f*	configuration *f*	расположение, конфигурация
	confining liquid	s. P 11		
C 573	confluence	Zusammenfluß *m*	confluent *m*	слияние потоков
C 574	conformal mapping (representation)	konforme (winkeltreue) Abbildung *f*	transformation (application) *f* conforme	конформное отображение
	congealing point	s. S 403		
C 575	conical breaker (crusher)	Kegelbrecher *m*	concasseur (broyeur) *m* à cônes	конусная дробилка
	conical mill	s. J 23		
C 576	conical plug valve	Kegeldrehschieber *m*	tiroir *m* rotatif conique	конический золотник
C 577	connecting link (member)	Bindeglied *n*, Verbindung *f*	lien *m*	связ\усщее звено
C 578	connecting pipe	Anschlußrohr *n*	tuyau *m* de raccordement	насадок, соединительная труба
C 579	connecting tube	Anschlußschlauch *m*	flexible *m* de raccordement	соединительный рукав
C 580/1	connection, coupling, link	Verbindungsstück *n*, Zwischenstück *n*	pièce *f* de jonction (raccord), raccord *m*	соединительная часть, фитинг, соединение, сцепление
	connection	s. a. 1. C 470; 2. S 908		
	connection in parallel	s. P 20		
C 582	connection list	Verknüpfungsliste *f*	liste *f* de connexion	список связей химико-технологической системы
C 583	connection piece	Anschlußstutzen *m*	ajutage *m*, ajutoir *m*	соединительный штуцер (патрубок)
C 584	connection point	Anschlußpunkt *m*	point *m* de connexion (jonction, raccord)	опорная точка, пункт примыкания
C 585	conservation, preservation	Erhaltung *f*, Konservierung *f*	conservation *f*, traitement *m* préservatif	сохранение, консервирование
C 586	conservation and transport equation	Erhaltungs- und Transportgleichung *f*	équation *f* de conservation et de transport	уравнение сохранения и передачи (массы, энергии, импульса)
C 587	conservation laws	Erhaltungssätze *mpl*	lois *fpl* de conservation	законы сохранения
C 588	conservation of energy	Erhaltung *f* der Energie	conservation *f* de l'énergie	сохранение энергии
C 589	conservation of mass (matter)	Massenerhaltung *f*, Erhaltung *f* der Masse	conservation *f* de la masse	сохранение массы
C 590	conservation of momentum	Impulserhaltung *f*	conservation *f* de la quantité de mouvement	сохранение импульса
C 591	consistency	Konsistenz *f*	consistence *f*	консистенция, густота
C 592	constancy	Konstanz *f*, Aufrechterhaltung *f*, Unveränderlichkeit *f*	constance *f*, maintien *m* à une valeur	постоянство
C 593	constancy of volume	Volumenkonstanz *f*	constance (conservation) *f* de volume	постоянство объема
	constant	s. 1. L 28; 2. S 553		
C 594	constant-pressure change	Volumenarbeit *f*	changement *m* d'état à pression constante	полная работа изменения объема
C 595	constant-pressure cycle	Gleichdruckprozeß *m*	processus *m* à pression constante	процесс при постоянном давлении
C 596	constant-volume process	isochorer Prozeß *m*, isochore Zustandsänderung *f*	processus *m* (changement *m* d'état) isochore	процесс при постоянном объеме
C 597	constant weight	Gewichtskonstanz *f*	poids *m* constant, constance *f* de poids	постоянство веса, постоянный вес
	constituent	s. C 467		
C 598	constitute/to	ausmachen, bilden, darstellen	constituer	составлять, представлять
C 599	constitutive equation	Konstitutivgleichung *f*; Materialgleichung *f*	équation *f* de consistance	уравнение состояния

C 600	constitutive equation of a body	konstitutive Körpergleichung f	équation f de consistance d'un corps	уравнение состояния среды
C 601	constraints	Nebenbedingungen fpl	conditions fpl secondaires (imposées)	дополнительные (краевые, граничные, вспомогательные) условия, ограничения
C 602	construct/to, to establish	herstellen	produire, fabriquer, préparer	производить, изготовлять
C 603	construction, realization, execution	Ausführung f, Konstruktion f, Realisierung f	construction f, exécution f, réalisation f	реализация, выполнение
C 604	construction, building	Bauausführung f	exécution f constructive	конструкция, выполнение строительных работ
	construction	s. a. 1. S 752; 2. T 364		
C 605	construction design, construction project	Bauentwurf m	projet m de construction	строительный проект
C 606	construction engineering design	bautechnische Gestaltung f	projet m de construction	строительное оформление, проект строительства
C 607	construction field, construction site, [building] site	Bauplatz m, Baugelände n, Baustelle f	terrain m à bâtir, emplacement m du bâtiment, site m, chantier m	строительная площадка, строительный участок
	construction project	s. C 605		
	construction site	s. C 607		
C 608	consumer, user	Verbraucher m, Anwender m	consommateur m, utilisateur m	потребитель
C 609	consumer goods industry, light industry	Leichtindustrie f	industrie f légère (de biens de consommation)	легкая промышленность
C 610	consumption, expenditure, expense	Verbrauch m	consommation f	потребление, расход
C 611	contact	Berührung f, Kontakt m	contact m	контакт, прикосновение
	contact	s. a. C 90		
C 612	contact cooler	Kontaktkühler m	refroidisseur m de contact	контактный охладитель
C 613	contact distillation	Kontaktdestillation f	distillation f par contact	контактная перегонка
C 614	contact drying	Kontakttrocknung f	séchage m par contact	контактная сушка
C 615	contact filter, packing material	Füllkörper m, Kontaktfilter n	lit (filtre) m de contact, garnissage m	насадка, контактный фильтр
C 616	contact filtration	Kontaktfiltration f	filtration f (traitement m) par contact	контактная очистка (фильтрация)
C 617	contactor	Kontaktgeber m	contacteur m	контактный датчик
C 618	contact pressure	Anpreßdruck m	pression f d'assise, pression de serrage au montage	прижимное усилие
C 619	contact process, catalytic method	Kontaktverfahren n, katalytisches Verfahren n	procédé m de contact	контактный метод
C 620	contact reactor	Kontaktofen m	four m de contact	контактная печь
	contact-series	s. E 54		
C 621	contact surface	Kontaktfläche f, Benetzungsfläche f, Berührungsfläche f	surface (superficie) f de contact	поверхность контакта (смачивания, соприкосновения)
C 622	contact time	Kontaktzeit f	durée f de contact	продолжительность (время) контакта
C 623	contain/to	enthalten, fassen, einschließen	contenir	содержать
C 624	contaminate/to, to pollute, to soil	verunreinigen	contaminer, salir, polluer	загрязнять
	contaminate/to	s. a. I 316		
C 625	contaminated	verschmutzt, verunreinigt, kontaminiert	encrassé, contaminé	загрязненный
C 626	contaminating gas	Fremdgas n	gaz m contaminant (étranger)	газообразная примесь
C 627	contamination, pollution, soil, fouling, impurity	Verschmutzung f, Verunreinigung f, Unreinheit f, Verseuchung f	encrassement m, pollution f, impureté f, contamination f	загрязнение
	contamination	s. a. P 240		
	continuity equation	s. E 216		
C 628	continuous blender	kontinuierlicher Mischer m	mélangeur m continu	непрерывный смеситель
C 629	continuous circulation	kontinuierlicher Umlauf m	circulation f continue	непрерывная циркуляция
C 630	continuous-coil evaporator	Einspritzverdampfer m	évaporateur m à injection	испаритель со впрыском
C 631	continuous cooling	kontinuierliche Kühlung f	refroidissement m continu	непрерывное охлаждение
C 632	continuous dryer	Durchlauftrockner m	sécheur m continu	непрерывная сушилка
C 633	continuous extraction	kontinuierliche Extraktion f	extraction f continue	непрерывное экстрагирование
C 634	continuous function	stetige Funktion f	fonction f continue	непрерывная функция
C 635	continuous load, continuous ratings	Dauerbelastung f	charge f permanente	длительная нагрузка
C 636	continuous medium	kontinuierliches Medium n	milieu m continu, continuum m	сплошная среда
C 637	continuous operation	Dauerbetrieb m	opération f continue (ininterrompue)	постоянный режим, длительная эксплуатация
C 638	continuous phase	kontinuierliche Phase f	phase f continue	непрерывная (сплошная, однородная) фаза
C 639	continuous pressure	Dauerdruck m	pression f permanente	постоянное давление (при продолжительной работе)
C 640	continuous process	kontinuierlicher Prozeß m	procédé m continu	непрерывный процесс

	continuous ratings	s. C 635		
C 641	continuous screening	kontinuierliche Siebung f	criblage m continu	непрерывное грохочение
	continuous stirred vessel	s. S 676		
C 642	continuum mechanics	Kontinuumsmechanik f	mécanique f de continu	механика сплошных сред
C 643	contraction	Kontraktion f, Schrumpfen n, Schwindung f	contraction f	сжатие, сужение
C 644	contraction coefficient	Kontraktionskoeffizient m	coefficient m de contraction	коэффициент сжатия
	contraction in volume	s. V 181		
C 645	contraction strain, shrinkage stress	Schrumpfspannung f	tension f de contraction, effort m de retrait	усадочное напряжение
C 646	contractor, supplying firm	Lieferbetrieb m, Auftragnehmer m	établissement m fournisseur	завод-поставщик, поставщик
	control	s. 1. O 77; 2. S 908		
C 647	control box, control device	Steuerblock m	unité f de commande	блок управления
C 648	control cabinet, switch cabinet	Schaltschrank m	armoire f de distribution, boîte f de commande	пульт управления, распределительный шкаф
C 649	control date	Kontrolltermin m	date m de contrôle	контрольный срок
	control desk	s. O 85		
	control device	s. 1. C 204; 2. C 647		
C 650	control element	Regelorgan n	élément m de réglage	регулирующий орган
C 651	control elements, controls	Bedienelemente npl	éléments mpl de commande	элементы управления (настройки, контроля)
C 652	control engineering	BMSR-Technik f, Regelungs- und Steuerungstechnik f	technique f de mesure et de régulation	контрольно-измерительная техника
C 653	control equipment	Regler m	mécanisme m régulateur, équipement m de réglage	регулятор, стабилизатор
C 654	controllability	Steuerbarkeit f	contrôlabilité f, aptitude f d'être contrôlé	управляемость
C 655	controlled condition	Regelgröße f	grandeur f réglée	регулируемая величина
C 656	controlled temperature	geregelte Temperatur f	température f contrôlée (réglée)	регулируемая температура
C 657	controlled waste dumping	geordnete Mülldeponie f	décharge f publique ordonnée	упорядоченная свалка отбросов
C 658	control medium, manipulated variable	Stellgröße f	grandeur f de réglage, grandeur réglante	регулирующая величина
C 659	control model	Steuerungsmodell m	modèle m de commande	модель управления
C 660	control optimization	Steuerungsoptimierung f	optimisation f de commande	оптимизация управления
C 661	control platform	Bedienungsbühne f	plate-forme f de service	служебная платформа, платформа управления
C 662	control quantity	Steuergröße f	grandeur f de commande	параметр управления
C 663	control range	Regelbereich m	étendue f (domaine m) de réglage	диапазон регулирования
	controls	s. C 651		
C 664	control valve	Regelventil n, Steuerventil n	vanne f de réglage	регулирующий клапан, регулировочный вентиль
C 665	convection	Konvektion f, Mitbewegung f	convection f	конвекция, перенос
C 666	convection dryer	Heißluftstromtrockner m	sécheur m à courant d'air chaud	сушилка с подачей горячего воздуха
C 667	convection drying	Konvektionstrocknung f	séchage m par convection	конвекционная сушка
C 668	convection heating	Konvektionsheizung f	chauffage m de convection	отопление конвекцией
C 669	convection heating area	Konvektionsheizfläche f	surface f chauffante convective	конвективная поверхность нагрева
C 670	convection zone	Konvektionszone f	zone f de convection	конвекционная зона
C 671	convective transport	Konvektionstransport m, konvektiver Transport m	transport m convectif	конвективный транспорт (перенос)
C 672	convergence acceleration	Konvergenzbeschleunigung f	accélération f de convergence	ускорение сходимости
C 673	convergence rate	Konvergenzgeschwindigkeit f	vitesse f de convergence	скорость сходимости
C 674	conversion, recalculation	Konversion f, Konvertierung f, Umrechnung f	conversion f	конверсия, пересчет
	conversion	s. a. 1. C 158; 2. T 349		
C 675	conversion factor	Umwandlungsfaktor m	facteur m de conversion, coefficient m de transformation	фактор превращения, коэффициент пересчета
C 676	conversion measurement	Umsatzmessung f	mesure f de conversion	измерение конверсии (степени превращения)
C 677	conversion rate	Stoffumsatzgeschwindigkeit f	taux m (vitesse f) de conversion	скорость превращения вещества
	convert/to	s. C 157		
C 678	converted gas	Konvertierungsgas n	gaz m de conversion	газ конвертирования
	converter	s. T 271		
C 679	converter process	Konverterverfahren n	procédé m Bessemer	конвертерный процесс
C 680	converter reactor	Konverterreaktor m	convertisseur-réacteur m	конвертерный реактор
	conveyance	s. T 264		
	conveying device	s. C 685		
C 681	conveying equipment, conveying plant	Förderausrüstung f	installation f de transport	конвейерное оборудование, транспортное оснащение
C 682	conveying mean	Fördermittel n	moyen m (installation f) de transport	транспортное средство
	conveying plant	s. C 681		

C 683	conveying weigher	Bandwaage f	bascule f intégratrice pour les bandes transporteuses, peseuse f sur courroie	конвейерные весы
C 684	conveying worm, spiral conveyor	Transportschnecke f, Förderschnecke f	vis f transporteuse, transporteur m à hélice	транспортный шнек, винтовой транспортер, шнековый конвейер
C 685	conveyor, feeder, conveying device	Fördergerät n, Förderanlage f, Fördereinrichtung f, Förderer m, Zubringer m	transporteur m, convoyeur m	конвейер, транспортер, питатель, транспортное устройство
	conveyor	s. a. C 189		
C 686	conveyor belt, belt conveyor	Transportband n, Fließband n	bande (courroie) f transporteuse, tapis m roulant	ленточный транспортер, конвейер[ная лента]
C 687	conveyor chain	Förderkette f	chaîne f transporteuse	передаточная (подъемная) цепь, цепной транспортер
C 688	conveyor chute	Förderrinne f	transporteur m à auge, déversoir m, rigole f de transport	транспортный (качающийся) желоб
C 689	conveyor system	Förderanlage f	installation f de transport	подъемное устройство, транспортное сооружение, транспортная установка
	coolant	s. C 701		
C 690	coolant jacket, cooling jacket	Kühlmantel m, gekühlter Wärmeschutzmantel m	chemise (enveloppe) f réfrigérante, enveloppe f de protection thermique	охлаждающая рубашка, кожух холодильника
C 691	coolant network (system)	Kältemittelnetz n	système m du liquide frigorifique	система проводов охладительных средств
C 692	cool-down-duration, cool-down-time	Abkühldauer f, Abkühlzeit f	durée f (temps m) de refroidissement	продолжительность (время) охлаждения
C 693	cooled air	gekühlte Luft f	air m réfrigéré	охлажденный воздух
C 694	cooled gas	gekühltes Gas n	gaz m refroidi	охлажденный газ
C 695	cooled liquid	gekühlte Flüssigkeit f	liquide m réfrigéré (refroidi)	охлажденная жидкость
C 696	cooled medium	gekühltes Medium n	milieu m refroidi	охлажденная среда
C 697	cooled solution	gekühlte Lösung f	solution f refroidie	охлажденный раствор
C 698	cooler, refrigerator	Kühler m, Kühlapparat m, Abkühler m	réfrigérant m, condenseur m, refroidisseur m	холодильник, конденсатор, градирня
C 699	cooler area	Kühlerfläche f	surface f du réfrigérant	площадь холодильника, поверхность охладителя
C 700	cooling, refrigeration	Kühlen n, Abkühlung f	refroidissement m, réfrigération f	охлаждение
	cooling	s. a. Q 22		
C 701	cooling agent, refrigerant, cooling medium, coolant	Kältemittel n, Kälteträger m, Kühlmittel n, Kühlmedium n	frigorigène m, agent m frigorifique, réfrigérant m, produit (fluide) m réfrigérant	охлаждающий (холодильный) агент, хлад[о]агент, охлаждающее средство, охлаждающая среда, охладитель
C 702	cooling agent cycle	Kühlmittelkreislauf m	circulation f de l'agent réfrigérant	циркуляция охлаждающей среды
C 703/4	cooling air	Kühlluft f	air m de réfrigération (refroidissement)	охлаждающий воздух
C 705	cooling by mixing	Mischkühlung f	refroidissement m à mélange	охлаждение смешением
C 706	cooling chamber	Kühlkammer f	chambre f frigorifique (de refroidissement)	холодильная камера
C 707	cooling coil, condensing coil, cooling spiral	Kühlschlange f	serpentin m refroidisseur	холодильный (охлаждающий) змеевик, змеевик холодильника
C 708	cooling curve	Abkühlkurve f	courbe f de refroidissement	кривая охлаждения
C 709	cooling degree	Kühlstufe f, Abkühlgrad m	degré m de refroidissement	степень охлаждения
	cooling-down cycle	s. C 717		
C 710	cooling drum	Kühltrommel f	tambour m de refroidissement	охлаждающий (холодильный) барабан
C 711	cooling element	Kühlelement n	élément m frigorifique	холодильный элемент
C 712	cooling fin	Kühlrippe f	ailette f de refroidissement, nervure f ventilée	охлаждающее ребро
C 713	cooling grid, cooling screen	Kühlrost m	grille f de refroidissement	стеллаж для охлаждения
	cooling jacket	s. C 690		
C 714	cooling limit	Kühlgrenze f	limite f de refroidissement	предел охлаждения
C 715	cooling losses	Kühlverluste mpl	pertes fpl de refroidissement	потери на охлаждение
	cooling medium	s. C 701		
C 716	cooling pipe	Kühlleitung f	tube m de refroidissement	трубопровод для подвода охлаждающей жидкости
C 717	cooling process, cooling-down cycle	Abkühlprozeß m, Kühlvorgang m, Kühlprozeß m	procédé (processus) m de refroidissement	цикл (процесс) охлаждения
C 718	cooling pump	Kühlpumpe f	pompe f de refroidissement	охлаждающий насос
	cooling screen	s. C 713		
	cooling spiral	s. C 707		
C 719	cooling surface	Kühlfläche f	surface f de refroidissement	поверхность охлаждения
C 720	cooling tank	Kühlbehälter m	compartiment (récipient) m frigorifique	охлаждающий сборник, изотермический контейнер
C 721	cooling test	Kühltest m, Abkühlversuch m	essai m de refroidissement	испытание на охлаждение

C 722	cooling tower	Kühlturm m	tour f de réfrigération	градирня, скруббер, башенный охладитель
C 723	cooling water	Kühlwasser n	eau f de refroidissement	охлаждающая вода
C 724	cooling water circuit	Kühlwasserkreislauf m, Kühlwasserumlauf m	circuit m d'eau réfrigérante	схема циркуляции охлаждающей воды
	cooling water need	s. C 728		
C 725	cooling water network	Kühlwassernetz n	réseau m de l'eau réfrigérante	система трубопроводов охлаждающей воды
C 726	cooling water pipe	Kühlwasserleitung f	tuyau m d'eau réfrigérante	трубопровод системы охлаждения
C 727	cooling water pump	Kühlwasserpumpe f	pompe f à eau froide (de refroidissement)	охлаждающий водяной насос
C 728	cooling water requirements, cooling water need	Kühlwasserbedarf m	besoins mpl en eau réfrigérante	потребность в охлаждающей воде
C 729	cooling water supply	Kühlwasserförderung f; Kühlwasserversorgung f	approvisionnement m d'eau réfrigérante	подача охлаждающей воды, снабжение охлаждающей водой
C 730	cooling water treatment	Kühlwasseraufbereitung f	traitement m d'eau de refroidissement	обработка охлаждающей воды
C 731	cooling water valve	Kühlwasserventil n	vanne f de l'eau réfrigérante	вентиль охлаждающей воды
C 732	coolness	Kältestufe f, Kühlgrad m	degré m de froideur (refroidissement), froid m, fraîcheur m	степень охлаждения
C 733	cool water	Kaltwasser n	eau f froide	холодная вода
C 734	copolymer	Mischpolymerisat n	copolymère m	сополимер
C 735	copolymerization	Kopolymerisation f, Mischpolymerisation f	copolymérisation f	сополимеризация
	copolymerization	s. a. H 138		
C 736	copying, imitation	Nachbildung f, Simulation f, Simulierung f	simulation f	копирование, имитация
C 737	core flow	Kernströmung f	écoulement m central	ядро течения
C 738	Coriolis force	Coriolis-Kraft f	force f de Coriolis	сила Кориолиса
C 739	cork, stopper	Pfropfen m	tampon m, bouchon m	пробка, втулка
C 740	correction	Berichtigung f	correction f	правка, исправление, корректура
C 741	corrective factor	Korrekturfaktor m	coefficient m de correction	поправочный коэффициент
C 742	corroding bath	Beizbad n	bain m de décapage	ванна для травления
C 743	corrosion; rusting	Verrostung f, Korrosion f, Verrosten n	rouillage m, enrouillement m	ржавление, коррозия
C 744	corrosion fatigue	Korrosionsermüdung f	fatigue f due à la corrosion	коррозионная усталость
C 745	corrosion inhibition	Korrosionsbekämpfung f	lutte f contre la corrosion	борьба с коррозией
	corrosion prevention	s. C 748		
C 746	corrosion products	Korrosionsprodukte npl	produits mpl de corrosion	продукты коррозии
C 747	corrosion-prone, susceptible to corrosion	korrosionsanfällig	apte à la corrosion	подверженный коррозии
	corrosion proof	s. C 750		
C 748	corrosion protection, corrosion prevention	Korrosionsschutz m	protection f de métaux, protection contre la corrosion	защита против коррозии
C 749	corrosion resistance	Korrosionsbeständigkeit f	résistance f à la corrosion	коррозионностойкость
C 750	corrosion-resistant, corrosion-proof	korrosionssicher, korrosionsbeständig	résistant à la corrosion, non corrosif	нержавеющий, коррозионностойкий
C 751	corrosion testing	Korrosionsprüfung f	essai m de corrosion	коррозионное испытание
C 752	corrosive, corrosive agent	Korrosionsbildner m, Korrosionsmittel n	agent m corrosif, corrodant m	корродирующий агент
	corrosive	s. a. C 111		
	corrosive agent	s. C 752		
C 753	corrosive wear	korrosiver Verschleiß m	usure f corrosive (par la corrosion)	износ под влиянием коррозии
C 754	corrugated tube	geriffeltes Rohr n	tuyau m cannelé	рифленная труба
C 755	cost balances	Kostenbilanzen fpl	bilans mpl de frais	расходные балансы
C 756	cost distribution factor	Kostenaufteilungsfaktor m	facteur m de distribution de frais	фактор распределения затрат
C 757	cost distribution method	Kostenaufteilungsmethode f	méthode f de la distribution de frais	метод распределения затрат
C 758	cost equation	Kostengleichung f	équation f des frais	уравнение для расчета стоимости (оборудование)
C 759	cost estimate	Kostenvoranschlag m	devis m estimatif	смета, оценка затрат
C 760	cost estimate of construction	Bauanschlag m	devis m de construction	смета постройки
C 761	cost estimation	Kostenabschätzung f, Kostenberechnung f	estimation f des frais	калькуляция стоимости, расчет затрат
C 762	cost factor	Kostenfaktor m	facteur m de frais	фактор затрат
C 763	cost of construction	Baukosten pl	frais mpl de construction	затраты на строительство
C 764	cost of damage	Schadenskosten pl	frais mpl de dommage	стоимость ущерба, убытки от аварии
C 765	cost of production	Erzeugungskosten pl, Herstellungskosten pl	frais mpl de fabrication	стоимость производства, производственные затраты
	couch	s. L 46		
C 766	Couette flow	Couette-Strömung f	écoulement m Couette	течение Куэтта
C 767	Couette-Hatschek viscometer	Couette-Hatscheksches Viskosimeter n	viscosimètre m de Couette-Hatschek	вискозиметр Куэтта-Гатшека

C 768	counterbalance/to	ausgleichen, kompensieren; ausbalancieren; auswuchten, austarieren	compenser, égaliser	компенсировать, выравнивать
C 769	counter connection	Gegenschaltung f	montage m en opposition	противовключение
C 770	countercurrent, counterflow	Gegenströmung f, Gegenstrom m	courant m opposé, contrecourant m	обратное течение, противоток
C 771	countercurrent condensation	Gegenstromkondensation f	condensation f à contre-courant	конденсация по схеме противотока
C 772	countercurrent condenser	Gegenstromkondensator m	condenseur m à contre-courant	противоточный конденсатор
	countercurrent condenser	s. a. C 775		
C 773	countercurrent construction	Gegenstrombauweise f	exécution (construction) f à contre-courant	противоточная конструкция
C 774	countercurrent construction, countercurrent type	Gegenstromausführung f	réalisation f à contre-courant	противоточное исполнение
C 775	countercurrent cooler, countercurrent condenser	Gegenstromkühler m	réfrigérant m contre-courant	противоточный холодильник
C 776	countercurrent cooling	Gegenstromkühlung f	refroidissement m à contre-courant	противоточное охлаждение
C 777	countercurrent distillation	Gegenstromdestillation f	distillation f à contre-courant	противоточная дистилляция
C 778	countercurrent distribution	Gegenstromverteilung f	distribution f contre-courant	распределение противотока
C 779	countercurrent drum dryer, countercurrent rotary dryer	Gegenstromtrommeltrockner m	sécheur m rotatif à contre-courant	противоточная сушилка барабанного типа
C 780	countercurrent extraction	Gegenstromextraktion f	extraction f à contre-courant	противоточное извлечение, противоточная экстракция
C 781	countercurrent flow	Gegenstromfluß m	écoulement m contre-courant	противоток
C 782/3	countercurrent flow dryer	Gegenstromtrockner m	sécheur m à contre-courant	противоточная сушилка
C 784	countercurrent furnace	Gegenstromofen m	four m à contre-courant	противоточная печь
C 785	countercurrent gas scrubbing process	Gegenstromgaswäsche f	procédé m d'épuration de gaz, lavage m de gaz à contre-courant	противоточная промывка газов
C 786	countercurrent heat exchange	Gegenstromwärmeaustausch m, Gegenstromwärmeübertragung f	transfert m de chaleur à contre-courant	противоточный теплообмен
C 787	countercurrent pipe heat exchanger	Gegenstromrohrwärmeaustauscher m, Gegenstromrohrwärmeübertrager m	échangeur m thermique à tuyaux à contre-courant	противоточный трубчатый теплообменник
C 788	countercurrent principle	Gegenstromprinzip n	principe m de contre-courant	принцип противотока, противоточный принцип
	countercurrent rotary dryer	s. C 779		
C 789	countercurrent system	Gegensinnschaltung f	montage m contre-courant	инверсная схема
	countercurrent type	s. C 774		
C 790	counterflange	Gegenflansch m	contre-bride f	контрфланец
	counterflow	s. C 770		
C 791	counterflow air refrigeration	Gegenstromluftkühlung f	réfrigération f à contre-courant d'air	охлаждение в противоточном потоке воздуха
C 792	counterflow apparatus	Gegenstromapparat m	appareil m contre-courant	противоточный аппарат
C 793	counterflow condenser	Gegenstromverflüssiger m	condenseur m à contre-courant	противоточный конденсатор
C 794	counterflow cooling tower	Gegenstromkühlturm m	tour f réfrigérante à contre-courant	противоточная градирня
C 795	counterflow heat exchanger	Gegenstromwärmeaustauscher m, Gegenstromwärmeübertrager m	échangeur m méthodique (de chaleur à contre-courant)	противоточный теплообменник
C 796	counterflow heat transfer	Gegenstromwärmeübertragung f	transfert de chaleur méthodique (à contre-courant)	теплопередача при противотоке
C 797	counterflow operation	Gegenstromprozeß m	opération f contre-courant	противоточный процесс
C 798	counterflow process	Gegenstromverfahren n	procédé m à contre-courant	противоточный способ
C 799	counterpressure, reaction pressure	Gegendruck m, Reaktion f	contre-pression f, réaction f	противодавление, реакция опоры, противодействие давлению
C 800	counterreaction	Gegenreaktion f	contre-réaction f	обратная реакция
C 801	countersteam	Gegendampf m	contre-vapeur f	контрпар
	counter-wall	s. F 409		
	coupling	s. C 580/1		
C 802	coupling ability	Kopplungsfähigkeit f	aptitude f à l'accouplement	способность к соединению (сопряжению)
C 803	coupling factor	Kopplungsfaktor m	facteur m de couplage, facteur d'accouplement	коэффициент связи
C 804	coupling matrix	Kopplungsmatrix f	matrice f de couplage	матрица связей (смежности)
	course	s. P 580		
C 805	course of process	Prozeßablauf m, Prozeßverlauf m	allur f de procédé, déroulement m du procédé	течение (протекание, ход) процесса
C 806	course of reaction	Reaktionsablauf m	cours m de la réaction, processus m réactionnel	течение (протекание, ход) реакции
C 807	covalent bond	kovalente Bindung f	liaison f atomique (covalente)	ковалентная связь
C 808	cover, cap, hood, lid, top	Deckel m, Abdeckung f, Schutzkappe f, Haube f	couvercle m, chapeau m, dôme m	крышка, колпак, декель

	cover	s. a. C 327		
	covering	s. C 329		
C 809	covering surface	Abdeckfläche f	surface f de recouvrement	покрытие
C 810	cover plate	Abdeckplatte f	plaque f de recouvrement	карнизная плита, плита для перекрытия
C 811	cracked, flawed, sprung	gespalten	crevassé, gercé, fendu	расщепленный
C 812	cracked fraction	Krackfraktion f	fraction f de cracking	крекинг-фракция
C 813	cracked gas	Krackgas n, Spaltgas n	gaz m de cracking (dissociation)	крекинг-газ, газ пиролиза
C 814	cracked gas cooler	Spaltgaskühler m	réfrigérant m du gaz de dissociation	холодильник пиролизного газа
C 815	cracked product	Krackprodukt n	produit m de craquage	крекинг-продукт
C 816	cracked residue	Krackrückstand m	résidu m de cracking	крекинг-остаток
C 817	cracker	Kracker m, Spalter m	unité f de cracking	крекинг-установка
	cracker	s. a. B 257		
C 818	cracking, fission, decomposition	Spaltung f	dissociation f, fission f, division f, scission f, clivage m	пиролиз, расщепление, крекинг
C 819	cracking	Spaltdestillation f	cracking m, craquage m	крекинг, перегонка с разложением
C 820	cracking capacity	Krackkapazität f	capacité f de cracking	пропускная способность установки крекинга
C 821	cracking column	Spaltkolonne f	colonne f de craquage	крекинг-колонна
C 822	cracking furnace	Krackofen m	réacteur m de cracking	крекинг-реактор
	cracking output	s. C 826		
C 823	cracking plant	Spaltanlage f	installation f de cracking	крекинг-установка
C 824	cracking reaction	Krackreaktion f	réaction f de craquage	реакция крекинга
C 825	cracking temperature	Kracktemperatur f, Abbautemperatur f, Zersetzungstemperatur f	température f de cracking (décomposition)	температура разложения (крекинга, распада)
C 826	cracking yield, cracking output	Krackausbeute f	rendement m de cracking	выход при крекинге
C 827	craze	Haarriß m	microtapure f, fissure f microscopique (capillaire)	микротрещина
C 828	crazing	Haarrißbildung f (Plast), Crazing-Effekt m	formation f de fissures capillaires	образование трещин
C 829	creep	Kriechen n, Fließen n	fluage m	течение, ползучесть
C 830	creep function	Kriechfunktion f	fonction f de fluage	функция ползучести
C 831	creeping flow	Kriechströmung f	mouvement m rampant (lent)	ползучее течение
C 832	creep strength, endurance strength	Dauerstandfestigkeit f	résistance f à la rupture sous charge permanente	предел ползучести (усталости при статической нагрузке), долговременное сопротивление
	criterion	s. O 1		
C 833	critical constant	kritische Konstante f	constante f critique	критическая константа (постоянная)
C 834	critical density	kritische Dichte f	densité f critique	критический удельный вес, критическая плотность
C 835	critical limit	Abfallgrenze f	limite f critique	критический предел, граница (предел) спада
C 836	critical mass	kritische Masse f	masse f critique	критическая масса
C 837	critical point	kritischer Punkt m	point m critique	критическая точка
C 838	ciritical pressure	kritischer Druck m	pression f critique	критическое давление
C 839	critical temperature	kritische Temperatur f	température f critique	критическая температура
	critical value	s. L 106		
C 840	cross air draft	Querluftstrom m	courant m d'air transversal	поперечный поток воздуха
C 841	cross beater mill	Schlagkreuzmühle f	broyeur m à battoirs (marteaux fixes en croix), moulin m à croisillons	ударная мельница с крестообразным билом
C 842	cross-counterflow heat exchanger	Quer-Gegenstrom-Wärmeübertrager m	échangeur m thermique à contre-courant croisé	поперечно-противоточный теплообменник
C 843	cross-current condenser	Kreuzstromkühler m	réfrigérant m à courants croisés	перекрестный холодильник
C 844	cross-current extraction	Kreuzstromextraktion f	extraction f à courants croisés	перекрестная экстракция
C 845	cross-current extractor	Kreuzstromextraktor m	extracteur m à courants croisés	перекрестный экстрактор
C 846	cross flow	Querströmung f, Querstrom m	courant m transversal (en travers)	поперечное течение, поперечный ток
	cross flow	s. a. C 862		
C 847	cross-flow blower	Querstromgebläse n	soufflante f de courant croisé	воздуходувка с поперечным потоком
C 848	cross-flow cooling tower	Querstromkühlturm m	tour f réfrigérante à courants croisés	поперечноточная градирня
C 849	cross-flow evaporator	Querstromverdampfer m	évaporateur m à courant croisé	поперечноточный испаритель
C 850	cross-flow heat exchanger	Querstromwärmeübertrager m	échangeur m thermique à courants croisés	теплообменник с поперечными потоками
C 851	cross-flow principle	Kreuzstromprinzip n, Querstromprinzip n	principe m des courants croisés	принцип перекрестного потока
C 852	crossing	Übergang m	changement m [d'état physique]	переход

C 853	cross-link	Vernetzungsstelle f, Brücke f, Querverbindung f	réticulation f, point m de réticulation	узел зацепления, поперечная связь
C 854	cross-link density	Vernetzungsdichte f	densité f de réticulation	плотность зацепления
C 855	cross-linking	Vernetzen n, Vernetzung f	réticulation f	образование сетчатых молекул
C 856	cross-point, intersection point	Schnittpunkt m	point m d'intersection	точка пересечения (разрыва)
C 857	cross section, average	Querschnitt m, Durchschnitt m	section f [transversale], moyenne f, coupe f	[поперечное] сечение, [поперечный] разрез, среднее значение, профиль
C 858	cross-sectional area	Querschnittsfläche f	superficie f, aire f	площадь поперечного сечения
C 859	cross-sectional load	Querschnittsbelastung f	charge f de la section transversale	поперечная нагрузка
C 860	cross-sectional variation	Querschnittsänderung f	changement m de section (la coupe transversale)	изменение поперечного сечения
C 861	cross section of exhausting	Abzugsquerschnitt m, Abführungssektion f	section f de dégagement	сечение вытяжного канала
C 862	cross stream, cross flow	Kreuzstrom m	courant (flux) m croisé	перекрестный (перекрещающийся) ток
C 863	cross viscosity	Querviskosität f	coefficient m de viscosité quadratique	поперечная вязкость
C 864	crucible melting process	Tiegelschmelzverfahren n	fondage m en creuset	процесс тигельной плавки, тигельный процесс
C 865	crude oil distillation	Erdöldestillation f	distillation f du pétrole brut	перегонка сырой нефти
C 866	crude oil preparation	Erdölaufbereitung f	traitement m préalable du pétrole brut	переработка сырой нефти
C 867	crude oil refinery	Mineralölraffinerie f, Erdölraffinerie f	raffinerie f [de pétrole]	нефтеперегонный завод
C 868	crude yield, yield of crude product	Rohausbeute f	rendement m brut	выход сырого продукта
C 869	crush/to, to grind, to mill, to pulverize	vermahlen	moudre, broyer	размалывать, дробить, раздавливать, измельчать
C 870	crusher	Schlagmühle f	broyeur m à barres, moulin m à battoirs	ударная мельница
	crusher	s. a. B 257		
C 871	crushing (operation)	Brechen n, Zerkleinern n, Quetschen n, Kollern n, Zerdrücken n	concassage m, broyage m	дробление, раздавливание
C 872	crushing cylinder, grinding drum	Mahltrommel f	tambour m de broyage	размалывающий барабан
	crushing plant	s. B 261		
C 873	crushing point	kritische Belastung f	charge f critique	критическая нагрузка
	crust	s. D 176		
C 874	cryodesiccation chamber	Sublimationskammer f, Gefriertrocknungskammer f	chambre f de sublimation (lyophilisation)	сублимационная камера
C 875	cryogenic extraction process	Kälteextraktionsverfahren n	procédé m d'extraction cryogénique	метод экстрагирования глубоким охлаждением
C 876	cryogenic fractionation process	Kältefraktionierungsverfahren n, Tieftemperaturfraktionierung f	fractionnement m cryogénique, procédé m de séparation par cryogénique	метод фракционирования глубоким охлаждением
C 877	crystallizable	kristallisationsfähig, kristallisierbar	cristallisable	кристаллизуемый
C 878	crystallization	Kristallisation f, Kristallbildung f	cristallisation f, grainage m [sucre]	кристаллизация
C 879	crystallization process	Kristallisationsprozeß m	procédé m de cristallisation	процесс кристаллизации
C 880	crystallisation sequence	Kristallisationsfolge f, Ausscheidungsfolge f	suite f de cristallisations, cristallisation fractionnée	последовательность кристаллизации
C 881	crystallizer	Kristallisator m	cristallisoir m	кристаллизатор
C 882	crystal structure	Kristallstruktur f	structure f du cristal	кристаллическая структура
C 883	cubical expansion	räumliche Ausdehnung f	expansion f volumétrique	объемное расширение
	cubicle	s. S 906		
C 884	cumulative down-time	kumulative Ausfallzeit f	temps m cumulatif hors de service	совокупное время отказа
C 885	cumulative operating time	kumulative Betriebszeit f	temps m de service cumulatif	совокупное производственное время, совокупное время производства
C 886	cumulative reserve time	kumulative Reservezeit f	temps m de réserve cumulé (cumulatif)	общее (совокупное) резервное время
C 887	current density	Stromdichte f	densité f de courant	плотность [по]тока
C 888	current impulse	Stromstoß m	coup m de courant	импульс (толчок) тока
C 889	current source	Stromquelle f	source f de courant	источник тока
	current velocity	s. A 365		
C 890	curve of adiabatic condensation	Kondensationsadiabate f	adiabate f de condensation	адиабата конденсации
C 891	curve of boiling, point diagram	Siedelinie f	courbe f des points d'ébullition	линия кипения
C 892	cut-off valve	Absperrventil n	soupape f d'arrêt	запорный клапан (вентиль)
C 893	cut size	Trenngrenze f	limite f séparatrice	граница разделения
	cut size	s. a. S 358		

C 894	cycle, cyclic process, circulation	Kreisprozeß m, Kreislauf m	cycle m, circulation f, circuit m fermé	циклический (круговой, замкнутый) процесс, циркуляция
C 895	cycle efficiency	Prozeßwirkungsgrad m	rendement m de cycle	коэффициент полезного действия цикла (процесса)
C 896	cycle principle	Kreislaufprinzip n	principe m de recirculation	принцип циркуляции (кругового процесса)
C 897	cycle ratio	Kreislaufverhältnis n	taux m de recirculation	циркуляционное соотношение, фактор рециркуляции
C 898	cycle temperature	Prozeßtemperatur f	température f de cycle	температура цикла
C 899	cycle time	Zykluszeit f	période f	время цикла
C 900	cyclic operation	zyklische Betriebsweise f	opération f cyclique	циклический способ производства
	cyclic process	s. C 894		
C 901	cyclization	Zyklisierung f	cyclisation f	циклизация
	cyclone	s. D 570		
C 902	cylinder dryer	Schachttrockner m	sécheur m à couloir	шахтная сушилка
C 903	cylindrical batch pasteurizer	Trommelerhitzer m	pasteuriseur m à tambour	барабанный пастеризатор
C 904	cylindrical dryer, rotary (drum) dryer	Trommeltrockner m	tambour m sécheur, sécheur m rotatif	барабанная сушилка, сушильный барабан
C 905	cylindrical sieve, revolving screen, trommel, drum sieve	Trommelsieb n	tamis m à tambour, tambour m cribleur	барабанный грохот, барабанное сито

D

D 1	Dalton's law	Daltons[ches] Gesetz n	loi f de Dalton	закон Дальтона
D 2	damage, injury	Schaden m, Schadbild n	dommage m, dégât m	ущерб, поражение
D 3	damaged place	Schad[ens]stelle f	endroit m de dommage	место повреждения
D 4	damage limit	Schädigungsgrenze f	limite f de dommage	граница (предел) повреждения
D 5	damage protection	Havarieschutz m	protection f contre des avaries	аварийная защита
D 6	dampening air	Befeuchtungsluft f	air m de l'humidification	увлажненный воздух
D 7	damper	Anfeuchter m	humidificateur m, humecteur m	увлажнитель
D 8	damping device (machine)	Anfeuchtapparat m	appareil m à humidifier, humecteur m, humidificateur m	увлажнитель
D 9	danger of pollution	Verschmutzungsgefährdung f, Verschmutzungsgefahr f	danger m de pollution	опасность загрязнения
D 10	Darcy's law	Darcysches Gesetz n	loi f de Darcy	закон Дарси
	data	s. 1. M 108/9; 2. T 106		
D 11	data bank (file), data store	Datenbank f, Datenspeicher m	banque f de données	магазин (набор, банк) данных
D 12	data gathering	Datenerfassung f, Datenaufzeichnung f	détection f de l'information, détection des données, enregistrement m d'information	сбор данных
D 13	data logger	Datenspeicher m	mémoire f, accumulateur m d'information	накопитель данных, запоминающее устройство
D 14	data network	Daten[übertragungs]netz n	réseau m de transmission de données	сеть передачи данных, информационная сеть
D 15	data of supply, time of delivery	Ablieferungstermin m, Liefertermin m	date f de livraison	срок сдачи
D 16	data processing	Datenverarbeitung f	traitement m des données	обработка данных
	data store	s. D 11		
D 17	daughter	Folgeprodukt n	produit m consécutif	дочернее вещество, дочерний продукт
D 18	deacidification	Entsäuerung f	désacidification f	нейтрализация, раскисление, удаление кислоты
D 19	deactivation	Desaktivierung f, Entaktivierung f	désactivation f	дезактивация
D 20	deactivation rate	Desaktivierungsgeschwindigkeit f	vitesse f de désactivation	скорость дезактивации
D 21	dead-burning	Totbrennen n	cuisson f à fond, calcination f à mort	обжигание до конца, обжигание намертво
D 22	dead zone	Totzone f	zone f morte	застойная зона
D 23	deaerator	Entlüfter m	désaérateur m, appareil m de ventilation	эксгаустер
D 24	Deborah number	Deborah-Zahl f	nombre m de Deborah	число Дебора
D 25	debris	Haufwerk n, Schüttgut n	matières fpl en vrac	насыпной (сыпучий) материал
D 26	debris cone, dejection cone	Schüttkegel m	cône m de déjection, talus m	конус выноса, насыпной конус
D 27	debutanizer	Entbutaner m	débutaniseur m	дебутанизатор
D 28	decalcifying	Entkalkung f	décalcification f	удаление извести

D 29	decantation	Dekantierung f, Dekantation f, Abgießen n, Absetzen n, Entschlammen n	décantation f, décantage m	декантация, переливание
	decantation	s. a. E 98		
D 30	decanter	Abklärgefäß n, Dekantiergefäß n, Abscheider m, Dekanter m	décanteur m	декантатор, приемник
D 31	decarbonization, decarburization	Entkohlung f, Kohlenstoffentzug m, Entkarbonisierung f	décarbonisation f, décarburation f	обезуглероживание, декарбонизация
	decay	s. D 43		
D 32	decay time	Abklingzeit f	temps m d'affaiblissement	время затухания
D 33	dechlorination	Entchlorung f	déchloration f, déchlorage m	дехлорирование
	decision	s. I 14		
D 34	decision model	Entscheidungsmodell n	modèle m de décision	модель принятия решений
D 35	decision rules	Entscheidungsregeln fpl	règles fpl de décision	правила принятия решений
D 36	decision space	Entscheidungsraum m	espace m des décisions	пространство решений
D 37	decision theory	Entscheidungstheorie f	théorie f de décision	теория принятия решений
D 38	decision variable	Entscheidungsvariable f	variable f de décision	переменная принятия решений
D 39	decoction, brewing, mordant	Sud m, Abkochung f	décoction f, soude m, brassin m	отвар
D 40	decolouration	Verfärbung f, Entfärben n, Ausbleichen n	décoloration f, altération f de la couleur	изменение цвета
D 41	decomposable	zerlegbar, zersetzlich, zersetzungsfähig	décomposable, séparable	разложимый
D 42	decomposer	Zersetzungsapparat m, Zersetzer m	appareil m de décomposition, décomposeur m	аппарат для разложения
D 43	decomposition, decay	Zerfall m	décomposition f, désagrégation f, désintégration f	распад
D 44	decomposition, reduction	Zersetzung f, Auflösung f	décomposition f, dissolution f	расщепление, растворение
D 45	decomposition	Dekomposition f, Zerlegung f	décomposition f, séparation f	декомпозиция
	decomposition	s. a. 1. C 818; 2. F 419		
D 46	decomposition process	Abbauprozeß, Zerfallsprozeß m, Dekompositionsverfahren n	procédé m de décomposition, processus m de désintégration	декомпозиционный процесс
D 47	decomposition product	Abbauprodukt n, Zerfallsprodukt n	produit m de décomposition (réduction, désintégration)	продукт разложения (распада, деструкции)
D 48	decomposition rate	Zerfallsgeschwindigkeit f	vitesse f de désintégration	скорость распада
D 49	decomposition reaction	Zerfallsreaktion f	réaction f de décomposition, décomposition f	реакция распада
D 50	decomposition voltage	Zersetzungspannung f	tension f de décomposition	напряжение разложения
D 51	decompression	Druckentlastung f, Entspannung f, Expansion f	détente f, relaxation f, décompression f	перепад (разгрузка) давления
D 52	decontamination, elimination of toxic constituents	Entgiftung f, Entseuchung f, Desinfektion f, Dekontamination f	elimination f des constituants toxiques, décontamination f, désinfection f	обезвреживание, дезактивация
D 53	decontamination plant	Entgiftungsanlage f, Dekontaminieranlage f	installation f de décontamination	очистительная установка, установка обезвреживания (дегазации)
D 54	deduster	Entstauber m	dépoussiéreur m	пылеуловитель
D 55	deep freezing	Tiefkühlung f, Einfrostung f	surgélation f	глубокое охлаждение (замораживание)
D 56	deep-freezing plant	Tiefkühlanlage f	installation f de surgélation	установка глубокого холода, установка из морозильных аппаратов для низкотемпературного замораживания
D 57	de-ethanizer	Entethaner m	dééthaniseur m	деэтанизатор
D 58	defect, failure	Defekt m, Panne f, Versagen n, Ausfallen n	détérioration f, défaut m, panne f	дефект, неисправность, отказ
	deficient	s. I 247		
D 59	defining equation	Bestimmungsgleichung f	équation f de définition, équation déterminative	определяющее уравнение
	deflagration	s. E 381		
D 60	deflectability	Ablenkbarkeit f	déviabilité f	отклоняемость
	deflecting	s. B 85		
D 61	deflecting chamber	Ablenkkammer f	chambre f de déviation	отклоняющая камера
D 62	deflection	Ablenkung f	déviation f, déflexion f	отклонение
D 63	deflector plate	Ablenkplatte f	plaque f de déviation, déflecteur m, chicane f	отклоняющая пластина
	defoaming agent	s. A 517		
D 64	deform/to, to shape	verformen, umformen, verzerren	déformer	деформировать
D 65	deformation	Formänderung f	déformation f	деформация, коробление, формоизменение
D 66	deformation condition	Deformationsbedingung f, Umformungsbedingung f	condition f de déformation	условие деформации
D 67	degasification	Entgasung f	dégazification f, dégazation f, dégazage m	дегазация, сухая перегонка
D 68	degasification process	Entgasungsprozeß m	procédé m de dégazage, processus m de l'extraction du gaz	процесс дегазации

D 69	degasification rate	Entgasungsgeschwindigkeit f	vitesse f de dégazéification (dégazage)	скорость дегазации
D 70	degasifier	Entgaser m	dégazeur m	деаэратор
D 71	degassing plant	Entgasungsanlage f	installation f de dégazage (dégazation)	деаэрационная установка
D 72	degeneration	Degeneration f, Entartung f	dégénération f	вырождение
D 73	degradation	Abbau m, Zerlegung f, Degradation f	dégradation f, décomposition f	разложение, деградация
D 74	degradation constant	Zerfallskonstante f	constante f de désintégration	постоянная превращения (распада)
D 75	degrease/to	abfetten, entfetten, von Fett reinigen	dégraisser	снимать поверхностный слой жира
D 76	degree	Stufe f, Sektion f	degré m, échelon m, étage m	ступень, степень, секция, каскад
D 77	degree	Grad m	degré m, taux m	градус, степень
D 78	degree of accuracy	Genauigkeitsgrad m	degré m de précision	степень точности
D 79	degree of adsorption	Adsorptionsgrad m	degré m d'adsorption	степень адсорбции
D 80	degree of aeration	Belüftungsgrad m, Belüftungsrate f	taux m d'aération	степень аэрации
D 81	degree of aging	Alterungsgrad m	degré m de vieillissement	степень старения (износа)
D 82	degree of air conditioning	Klimatisierungsgrad m	degré m du conditionnement d'air	степень кондиционирования воздуха
D 83	degree of association	Assoziationsgrad m	degré m de l'association	степень ассоциации
D 84	degree of automation	Automatisierungsgrad m	degré m de l'automation	степень автоматизации
D 85	degree of compression, amount of compression	Verdichtungsgrad m, Kompressionsgrad m, Kompressionsverhältnis n, Verdichtungsdruck m	degré (taux, rapport) m de compression	степень сжатия
D 86	degree of condensation	Kondensationsgrad m	degré m de condensation	степень конденсации
D 87	degree of consistency	Konsistenz f, Beschaffenheit f, Flüssigkeitsgrad m	consistance f	степень текучести
D 88	degree of contamination	Vergiftungsgrad m, Kontaminationsgrad m	taux m d'empoisonnement	степень загрязнения (заражения)
D 89	degree of cross linking	Vernetzungsgrad m	degré m de réticulation	степень сетчатости
D 90	degree of desiccation	Austrockungsgrad m	degré m de dessiccation	степень усушки
D 91	degree of dilution	Verdünnungsgrad m	degré m de dilution	степень разрежения
D 92	degree of dispersion	Dispersitätsgrad m, Dispersionsgrad m	degré m de dispersion	степень дисперсности
D 93	degree of dissociation	Dissoziationsgrad m	degré m de dissociation	степень диссоциации
D 94	degree of elasticity	Elastizitätsgrad m	degré m d'élasticité	степень эластичности
D 95	degree of evaporation	Verdampfungsgrad m	degré m de l'évaporation	степень испарения
D 96	degree of expansion	Expansionsstufe f	degré m d'expansion, degré de détente	ступень расширения
D 97	degree of extraction	Extraktionsgrad m	degré m d'extraction	степень экстрагирования
D 98	degree of freedom	Freiheitsgrad m	degré m de liberté	степень свободы
D 99	degree of grinding	Mahlgrad m, Mahlfeinheit f	degré m de mouture (broyage)	степень размола
D 100	degree of hardness, hardness	Härtegrad m, Härtestufe f, Härte f	dureté f, degré m de dureté	степень твердости
D 101	degree of heat	Hitzegrad m, Wärmegrad m	degré m de chaud (chaleur)	степень жара (нагрева)
D 102	degree of humidity	Feuchtigkeitsgrad m	degré m hydrométrique	степень влажности
D 103	degree of leakage	Durchlässigkeitsgrad m, Durchlässigkeitsbeiwert m	degré (coefficient) m de pénétrabilité	степень неплотности (проницаемости)
D 104	degree of loading, load coefficient	Belastungsgrad m, Ausnutzungsgrad m	degré (facteur) m d'utilisation	степень (коэффициент) нагрузки
D 105	degree of miscibility	Mischbarkeitsgrad m	degré m de miscibilité	степень смешиваемости
D 106	degree of pollution	Verschmutzungsgrad m	degré m de pollution	степень загрязнения
D 107	degree of polymerization	Polymerisationsgrad m	degré m de polymérisation	степень полимеризации
D 108	degree of purity, percentage purity	Reinheitsgrad m	degré m de pureté	степень чистоты
D 109	degree of reliability	Zuverlässigkeitsgrad m	degré m de fiabilité	степень надежности
D 110	degree of safety	Sicherheitsgrad m, Sicherheitsfaktor m	coefficient m de sécurité	запас прочности, степень запаса
D 111	degree of saturation	Sättigungsgrad m	degré m de saturation	степень насыщения
D 112	degree of sensitivity	Empfindlichkeitsgrad m	degré m de sensibilité	степень чувствительности
D 113	degree of stability	Stabilitätsgrad m	degré m de stabilité	степень устойчивости
D 114	degree of stretching	Verstreckungsgrad m	degré m d'étirage	степень вытяжки
D 115	degree of superheating	Überhitzungsgrad m	degré m de la surchauffage	степень перегрева
D 116	degree of turbidity	Trübungsgrad m	degré m néphélémétrique	степень мутности
D 117	degree of turbulence	Turbulenzgrad m	degré m de turbulence	степень турбулентности
D 118/9	degree of vacuum	Vakuumstufe f	degré m du vide	степень вакуума
	degree of wear	s. W 138		
D 120	degree of wetness	Befeuchtungsgrad m, Benetzungsgrad m	degré m de mouillage (l'humectation)	степень смачиваемости
D 121	degree Rankine	Rankine-Grad m	degré m Rankine	градус (степень) Ранкина
D 122	degression exponent	Degressionsexponent n	exposant m dégressif	экспонента уменьшения (дегрессии)
D 123	dehumidification	Entfeuchtung f, Trocknung f (von Gasen)	déshumidification f, déshydratation f, séchage m	сушка, осушение
D 124	dehumidification technique	Entfeuchtungstechnik f	technique f de déshumidification	технология осушки

D 125	dehumidificator	Entfeuchter *m*	dèshydrateur *m*	эксикатор, осушитель
D 126	dehumidifier	Lufttrockner *m*, Luftentfeuchter *m*	dessicateur *m* d'air, dèshumidificateur *m*	осушитель
D 127	dehumidifying air cooler	Entfeuchtungsluftkühler *m*	refroidisseur *m* d'air à dèshumidification	осушающий воздухоохладитель
D 128	dehumidifying capacity	Entfeuchtungsleistung *f*	capacitè *f* de dèshumidification	производительность по осушке
D 129	dehumidifying effect	Entfeuchtungseffekt *m*, Entfeuchtungswirkung *f*	effet *m* de dèshumidification	осушающее действие
D 130	dehumidifying plant	Entfeuchtungsanlage *f*	installation *f* de dèshydratation	установка для обезвоживания
D 131	dehydration	Wasserentzug *m*, Wasserabspaltung *f*, Entwässerung *f*, Wasserentziehung *f*, Dehydration *f*	dèshydratation *f*	дегидратация
	dehydration	*s. a.* D 489		
D 132	dehydration cycle	Entfeuchtungszyklus *m*	cycle *m* de dèshydratation	цикл обезвоживания
D 133	dehydration of gas	Gastrocknung *f*, Gasentfeuchtung *f*	sèchage *m* (dessication *f*) de gaz	обезвоживание (дегидратация) газа
D 134	dehydration period	Entfeuchtungsperiode *f*, Trocknungsperiode *f*, Trocknungszeit *f*	pèriode *f* de dèshydratation, temps *m* de sèchage	период осушения
	dehydration plant	*s.* D 490/1		
D 135	dehydration process	Trocknungsprozeß *m*, Entfeuchtungsprozeß *m*, Entfeuchtungsvorgang *m*	processus *m* de dèshydratation, procèdè *m* de sèchage	процесс обезвоживания
D 136	dehydrator	Wasserabspaltungsapparat *m*, Trockenapparat *m*, Entfeuchter *m*	dèshydrateur *m*, dessiccateur *m*	дегидратор
D 137	dehydrator	Trockenmedium *n*, Trockenmittel *n*	dessèchant *m*, siccatif *m*, dessiccateur *m*	осушающее вещество
D 138	dehydrogenation	Wasserstoffabspaltung *f*, Dehydrierung *f*, Dehydrieren *n*	dèshydrogènation *f*	дегидрогенизация
D 139	dehydrogenation conditions	Dehydrierungsbedingungen *fpl*	conditions *fpl* de dèshydrogènation	условия дегидрирования
D 140	dehydrogenation plant	Dehydrierungsanlage *f*	unitè (installation) *f* de dèshydrogènation	установка дегидрирования
D 141	dehydrogenation pressure	Dehydrierungsdruck *m*	pression *f* de dèshydrogènation	давление при дегидрировании
D 142	dehydrogenation process	Dehydrierungsprozeß *m*	procèdè *m* de dèshydrogènation	процесс дегидрирования
D 143	dehydrogenation reaction	Dehydrierungsreaktion *f*	rèaction *f* de dèshydrogènation	реакция дегидрирования
D 144	dehydrogenation reactor	Dehydrierungsreaktor *m*	rèacteur *m* de dèshydrogènation	реактор дегидрирования
D 145	dehydrogenation temperature	Dehydrierungstemperatur *f*	tempèrature *f* du dèshydrogènation	температура дегидрирования
	dejection cone	*s.* D 26		
D 146	delayed effect	Spätwirkung *f*	effet *m* retardè	последействие
D 147	delay in boiling	Siedeverzug *m*	retard *m* à èbullition, surchauffe *f*	депрессия температуры кипения
D 148	delivery	Abgabe *f*, Lieferung *f*	livraison *f*, dècharge *f*	отдача, передача
D 149	delivery, feeding, haulage, hauling	Fördern *n*, Förderung *f*	extraction *f*, transport *m*, manutention *f*	подача, транспортирование, подъем
	delivery	*s. a.* C 32		
D 150	delivery air	Förderluft *f*, geförderte Luft *m*	air *m* primaire (forcè, soufflè)	сжатый воздух
D 151	delivery capacity	Förderstrom *m*	dèbit *m*	производительность, подача
D 152	delivery head, lifting height, pumping head	Förderhöhe *f*, Pumphöhe *f*	hauteur *f* de levage (refoulement)	высота подъема (подачи, напора, подачи подъемника)
D 153	delivery head, discharge head	Druckhöhe *f*	hauteur *f* manomètrique, refoulement *m*, pression *f* de refoulement	гидравлический напор
D 154	delivery height	Fördercharakteristik *f*, Förderkennlinie *f* (einer Pumpe)	caractèristique *f* de refoulement	характеристика насоса
	delivery pipe	*s.* P 413		
D 155	delivery pressure	Abgabedruck *m*	pression *f* de dècharge (sortie)	давление на конечном участке
D 156	delivery speed	Austrittsgeschwindigkeit *f*	vitesse *f* d'ècoulement, vitesse de sortie	скорость на выходе
D 157	delivery system	Fördersystem *n*	système *m* d'alimentation, système de transport	система подачи
D 158	delivery temperature	Abgabetemperatur *f*	tempèrature *f* de dècharge (sortie), tempèrature à la sortie	выходная температура
	demand	*s.* R 290		
D 159	demethanizer	Entmethaner *m*	dèmèthaniseur *m*	деметанизатор

D 160	demulsibility	Entemulgierbarkeit f, Demulgierbarkeit f	désémulsibilité f	деэмульгируемость
D 161	demulsification	Demulgierung f, Entmulgierung f, Dismulgierung f	désémulsionnement m, désémulsification f	деэмульгирование
	densification	s. C 538		
D 162	density function	Dichtefunktion f	fonction f de densité	функция плотности, плотностная функция
D 163	density of charge	Ladungsdichte f	densité f de charge	ретикулярная плотность
	density of dust	s. D 574		
D 164	density of state	Zustandsdichte f	densité f de l'état	плотность состояний
D 165	density of volume	Raumdichte f	densité f en volume	объемная плотность
D 166	density of water	Wasserdichte f	densité f de l'eau	плотность воды
D 167	de-oiler	Fettabscheider m, Ölabscheider m	séparateur m de graisse, séparateur d'huile	обезжириватель, маслоотделитель
D 168	dependant on temperature	temperaturabhängig	dépendant (en fonction) de la température	зависимый от температуры
D 169	dephenolating, dephenolization, phenol removal	Entphenolung f	déphénolation f	обезфеноливание, удаление фенола
D 170	dephlegmation column	Dephlegmierkolonne f, Rücklaufkolonne f	colonne f à reflux	дефлегмационная колонна
D 171	dephlegmator, reflux condenser, separator	Rückflußkühler m, Rücklaufkondensator m, Dephlegmator m	déphlegmateur m, réfrigérant m à reflux	дефлегматор, обратный конденсатор
D 172	dephosphorization	Entphosphorung f	déphosphoration f	обесфосфоривание
D 173	depletion	Entleerung f	vidage m, vidange f, évacuation f	опорожнение, выпуск, спуск, разгрузка, истощение
D 174	depolymerization	Depolymerisation f, Entpolymerisation f	dépolymérisation f	деполимеризация
D 175	deposit, precipitate	Niederschlag m, Fällungsprodukt n, Abscheidungsprodukt n, Ausfällung f, Ausfall m	dépôt m [précipité], précipité m, sédiment m	осадок, отложение, осаждение, конденсат
D 176	deposit, crust	Ansatz m, Kruste f, Ablagerung f, Bodensatz m	dépôt m, déposition f	настыль, корка, накипь, нагар
	deposit	s. a. 1. P 387; 2. U 7		
D 177	deposition, separation, segregation, elimination	Abscheidung f, Ausscheidungsvorgang m, Trennung f	séparation f, ségrégation f, déposition f	выделение, отделение, осаждение
D 178	deposition potential	Abscheidungspotential n	potentiel m de déposition	потенциал осаждения
	depression	s. L 256		
D 179	depropanizer	Entpropaner m	dépropaniseur m	депропанизатор
D 180	desalination, desalting	Entsalzung f	déminéralisation f, dessalaison f, dessalement m, dessalage m	отдаление соли, обессоливание
D 181/2	desiccate/to, to dry	eintrocknen, austrocknen	sécher, dessécher	сушать, осушивать
D 183	desiccation ratio	Entfeuchtungsgrad m, Austrocknungsgrad m, Trockenheitsgrad m	taux m de dessiccation	степень обезвоживания
	desiccator	s. D 528		
D 184	design	Auslegung f, Konzipierung f, Entwurf m	conception f, dimensionnement m, projet m	чертеж, прокладка, проектирование
D 185	design	Bauart m, Bauweise f, Konstruktion f, Ausführung f, Aufbau m, Gestaltung f, Form f, Entwurf m, Bemessung f, Berechnung f, Muster n	projet m, projection f, construction f, dimensionnement m	план, проект, тип, конструкция, расчет, проектирование
D 186	design boundary conditions	Entwurfsrandbedingungen fpl	limites (normes) fpl de conception	краевые условия проектирования
D 187	design conditions, rating conditions	Auslegungsbedingungen fpl	conditions fpl de dessin (dimensionnement)	расчетные условия, расчетный режим
	designing	s. P 583		
D 188	design meeting the demands of automation, automation-friendly design	automatisierungsgerechte Gestaltung f, automatisierungsfreundlicher Entwurf m	projection f correspondant aux exigences de l'automatisation	оформление процесса, удобное для автоматизации
D 189	design method	Gestaltungsmethode f, Konstruktionsverfahren n, Entwurfsverfahren n	méthode f de projection, méthode f de construction	метод проектирования
D 190	design model	Entwurfsmodell n, Auslegungsmodell n	modèle m de conception	модель для расчета
D 191	design of chemical engineering systems	Entwurf m verfahrenstechnischer Systeme	conception f des systèmes du génie chimique	проектирование химико-технологических систем
D 192	design of experiments	Versuchsplanung f	planning m d'expériences	планирование экспериментов
D 193	design optimization	Auslegungsoptimierung f	optimisation f de conception	оптимизация основных параметров процесса или системы
D 194	design parameter	Entwurfsparameter n	paramètre m de projet	проектный параметр
D 195	design pressure	Auslegungsdruck m	pression f de conception	расчетное давление

D 196	design pressure, licence pressure	zulässiger Kesseldruck *m*, Genehmigungsdruck *m* (*Kessel*)	timbre *m* de la chaudière	разрешенное давление
D 197	design state	Auslegungszustand *m*	état *m* de conception	расчетное (проектное) состояние
D 198	design variable	Entwurfsvariable *f*, Berechnungsvariable *f*	variable *f* de conception, variable du projet	проектный параметр, проектная переменная, переменная при проектировании
	desintegrator	*s.* 1. C 134; 2. C 135		
D 199	desired pressure	erforderlicher Druck *m*	pression *f* prévue (nécessaire)	требуемое давление
	desired pressure	*s. a.* N 43		
	desired temperature	*s.* N 48		
D 200	desired value	Soll-Wert *m*	valeur *f* préscrite (de consigne)	заданное значение, заданная величина, заданный параметр
D 201	deslagging	Entschlackung *f*	séparation *f* de la crasse, évacuation *f* du laitier	удаление шлака
D 202	desorber	Desorber *m*	appareil *m* de désorption, désorbeur *m*	десорбер
D 203	desorption	Desorbieren *n*, Desorption *f*	désorption *f*	десорбирование, десорбция
D 204	desorption column	Abblasekolonne *f*, Desorptionskolonne *f*	colonne *f* de désorption	отпарная (десорбционная) колонна
D 205	desorption kinetics	Desorptionskinetik *f*	cinétique *f* de désorption	кинетика десорбции
D 206	destructive distillation	Zersetzungsdestillation *f*	distillation *f* destructive	деструктивная перегонка
D 207	desulfuration	Entschwefelung *f*	désulfuration *f*	обессеривание, десульфурация, удаление серы
D 208	desuperheater	Heißdampfkühler *m*, Dampfkühler *m*	désurchauffeur *m*	пароохладитель
	detailed drawings	*s.* W 193		
D 209	detailed model	ausführliches Modell *n*	modèle *m* détaillé	подробная (деталированная) модель
D 210	detecting element	Meßfühler *m*, Meßwertgeber *m*	capteur *m* de mesure	датчик измеряемой величины
D 211	detergent	waschaktiver Stoff *m*, Tensid *n*	détersif *m*, détergent *m*	детергент
D 212	deterioration	Abnutzung *f*	détérioration *f*	износ
D 213	determination method	Abgrenzungsmethode *f*	méthode *f* de délimitation	метод разграничения (определения границ)
D 214	determination of errors	Fehlerbestimmung *f*	détermination *f* des erreurs	определение ошибки (погрешности)
D 215	determining factor (variable)	Einflußgröße *f*	grandeur *f* d'influence	переменная
D 216	deterministic element	deterministisches Element *n*	élément *m* déterminé	детерминистический (определяющий) элемент
D 217	detonate/to, to explode	verpuffen	exploser, déflagrer	детонировать, взрывать[ся]
	detonation	*s.* E 381		
D 218	detritus pit	Vorklärbecken *n*	premier bassin *m* de décantation	первичный отстойник
D 219	devaporation	Entnebelung *f*	condensation *f*	рассеяние тумана, конденсация пара
D 220	development, evolution	Entwicklung *f*, Evolution *f*	développement *m*, mise *f* au point, évolution *f*	развитие, проявление, эволюция
D 221	development work	Entwicklungsarbeit *f*	mise *f* au point, travaux *mpl* de développement	экспериментальная работа
D 222	deviation	Ablenkung *f*, Abweichung *f*, Deviation *f*	déviation *f*, déflexion *f*	отвлечение, отклонение
	device	*s.* J 16		
D 223	devolatilization	Abnahme *f* (Entfernen *n*) der flüchtigen Bestandteile, Entlüftung *f*, Entgasung *f*	désaeration *f*, dégazage *m*	выделение летучих частей, дегазация
D 224	dewaxing	Entparaffinierung *f* (*Mineralöle*)	déparaffinage *m*	выделение парафина, депарафинирование
D 225	dew point	Taupunkt *m*	point *m* de rosée	точка росы, температура точки росы
D 226	dew-point method	Taupunktverfahren *n*	méthode *f* à point de rosée	метод определения количества влаги по точке росы
D 227	dialyzer	Dialysator *m*	dialyseur *m*	диализатор
D 228	diaphragm	Diaphragma *n*, Membran *f*, Trennwand *f*, Scheidewand *f*	diaphragme *m*, membrane *f*	диафрагма
D 229	diaphragm chamber	Membrankammer *f*	chambre *f* de diaphragme	мембранная камера
D 230	diaphragm governor, diaphragm regulator	Membranregler *m*	régulateur *m* à membrane	мембранный регулятор
D 231	diaphragm pump	Membranpumpe *f*, Diaphragmapumpe *f*	pompe *f* à diaphragme (membrane)	мембранный насос
	diaphragm regulator	*s.* D 230		
D 232	diaphragm-type compressor	Membrankompressor *m*	compresseur *m* à membrane (diaphragme)	мембранный компрессор

D 233	diathermancy	Wärmedurchlässigkeit f	diathermanéité f, diathermansie f	теплопроницаемость
	die	s. M 344		
D 234	die casting	Spritzguß m	moulage m sous pression	литье под давлением
D 235	dielectric heating	dielektrische Beheizung f, Dielektroheizung f, Hochfrequenzheizung f	chauffage m par pertes diélectriques	диэлектрический обогрев, нагрев током высокой частоты
D 236	die lip opening	Austrittsspalt m des Extruderkopfs (Plast)	lèvres fpl (orifice m) de filière	фильера головки экструдера
D 237	dies	Matrizen fpl, Preßformen fpl	matrices fpl	матрицы, прессформы
	difference in solubilities	s. S 413		
D 238	difference of temperature	Temperaturdifferenz f	différence f de température	разность температур
D 239	differential absorption method	Differentialabsorptionsmethode f	méthode f de l'absorption différentielle	метод дифференциального поглощения
D 240	differential circulating reactor	Differentialkreislaufreaktor m	réacteur m différentiel en circuit fermé	дифференциальный замкнутый реактор
D 241	differential flowmeter	Differentialströmungsmesser m	débitmètre m différentiel	дифференциальный измеритель потока
D 242	differential method	Differentialmethode f	méthode f différentielle	дифференциальный метод
D 243	differential pressure	Wirkdruck m	pression f effective (active)	перепад (разность) давления, эффективное давление
D 244	diffraction	Beugung f, Diffraktion f	diffraction f	преломление, дифракция
D 245	diffuse double layer	diffuse Doppelschicht f	couche f double diffuse	диффузный двойной слой
D 246	diffuser, nozzle	Zerstäuberdüse f	buse f, douille f	распылительная форсунка
D 247	diffuser	Diffusionsapparat m	appareil m de diffusion	экстрактор, диффузионный аппарат, диффузор
D 248	diffusibility	Diffusionsvermögen n	diffusibilité f	диффузионная способность
D 249	diffusion, flattening out	Ausbreitung f, Diffusion f	extension f, propagation f, diffusion f	диффузия, рассеяние, рассеивание
D 250	diffusion cloud chamber	Diffusionsnebelkammer f	chambre f de diffusion	диффузионная камера Вильсона
D 251	diffusion coefficient	Diffusionskoeffizient m	coefficient m de diffusion	коэффициент диффузии
D 252	diffusion column	Diffusionskolonne f	colonne f de diffusion	диффузионная колонна
D 253	diffusion constant	Diffusionskonstante f	constante f de diffusion	константа диффузии
D 254	diffusion-controlled	diffusionsgesteuert	contrôlé par la diffusion	контролируемый (определяемый) диффузией
D 255	diffusion-controlled chemical reaction	diffusionsbegrenzte chemische Reaktion f	réaction f chimique limitée par diffusion	химическая реакция, определяемая диффузией
D 256	diffusion cooling	Diffusionskühlung f	refroidissment m par diffusion	диффузионное охлаждение
D 257	diffusion current	Diffusionsstrom m	courant m de diffusion, flux m diffusionnel	диффузионный поток
D 258	diffusion method	Diffusionsmethode f	méthode f de diffusion	диффузионный метод
D 259	diffusion model	Diffusionsmodell n	modèle m de diffusion	диффузионная модель
D 260	diffusion potential	Diffusionspotential n	potentiel m de diffusion	диффузионный потенциал
D 261	diffusion process	Diffusionsprozeß m	processus m de diffusion	процесс диффузии
D 262	diffusion pump	Diffusionspumpe f, Diffusorpumpe f	pompe f à diffusion	диффузионный насос
D 263	diffusion rate	Diffusionsrate f, Diffusionsgeschwindigkeit f	vitesse f (taux m) de diffusion	скорость диффузии
D 264	diffusion resistance	Diffusionswiderstand m	résistance f à la diffusion	сопротивление диффузии
D 265	diffusion separating column	Diffusionstrennkolonne f	colonne f de séparation par diffusion	диффузионная разделительная колонна
D 266	diffusion separation method	Diffusionstrennverfahren n	méthode f de séparation par diffusion	метод разделения диффузией
D 267	diffusion water	Diffusionswasser n	eau f de diffusion	диффузионная вода
D 268	diffusivity	Diffusität f	diffusité f	диффузность, коэффициент диффузии
D 269	digestion	Digerieren n, Aufschluß m, Aufschließen n, Faulung f	attaque f, dissolution f, désagrégation f	дигерирование, вываривание, переваривание
D 270	digital-analogue converter	Digital-Analog-Umwandler m	convertisseur m numérique-analogique, décodeur m	преобразователь дискретных данных в непрерывные
D 271	digital computer	Digitalrechner m	calculatrice f digitale, calculateur m numérique, ordinateur m	цифровой компьютер, цифровая вычислительная машина
D 272	digital converter	Digitalumsetzer m, Digitalumwandler m	convertisseur m digital	цифровой преобразователь
D 273	digital simulator	Digitalsimulator m	simulateur m digital	цифровое моделирующее устройство
D 274	digital technique	Digitaltechnik f	technique f digitale	цифровая техника
D 275	dilatability	Ausdehnungsfähigkeit f, Dehnbarkeit f	ductilité f, extensibilité f	растяжимость
D 276	dilatation, expansion	Expansion f, Ausdehnung f, Wärmedehnung f	expansion f, détente f, dilatation f [thermique]	расширение, растяжение
D 277	diluent	Streckmittel n, Verdünnungsmittel n, Verdünner m	charge f, diluant m, extendeur m	разбавитель, средство растяжения, разжижитель
D 278	dilute solution	verdünnte Lösung f	solution f diluée	разбавленный раствор
D 279	dilution	Verdünnen n, Verdünnung f	dilution f	разжижение, разбавление, разведение, разрежение

D 280	dilution coefficient	Verdünnungskoeffizient m	coefficient m de dilution	коэффициент разбавления
D 281	dilution factor	Verdünnungsfaktor m	facteur m de dilution	степень разбавления (раз- режения)
D 282	dilution law	Verdünnungsgesetz n	loi f de dilution	закон разбавления
D 283	dimensional stability	Dimensionsstabilität f, Form- beständigkeit f	résistance f à la déformation, stabilité f de forme, stabi- lité de dimension	постоянство (устойчивость) формы
D 284	dimensional tolerance	Maßtoleranz f	tolérance f dimensionnelle (de mesure)	допуск *(размера)*
D 285	dimension analysis	Dimensionsanalyse f	analyse f dimensionnelle	теория (анализ) размерно- стей
D 286	dimensioning programme	Dimensionierungsprogramm n	programme m de dimension- nement	программа определения размерностей
D 287	dimensionless	dimensionslos	sans dimension, non dé- nommé	безразмерный
D 288	dimensionless characteristic	dimensionslose Kennzahl f	nombre m sans dimension	безразмерный показатель
D 289	dimensionless coefficient	dimensionsloser Koeffizient m	coefficient m sans dimension	безразмерный коэффици- ент
D 290	dimensionless number	dimensionslose Zahl (Größe) f	chiffre m adimensionnel, nombre m sans dimension	безразмерная величина, безразмерное число
D 291	diminished capacity, reduced capacity	Minderleistung f, reduzierte Leistung f	rendement (débit) m infé- rieur, performance f ré- duite	недостаточная мощность
D 292	diminution	Abnahme f, Verminderung f, Abfall m	diminution f	убывание
D 293	dipper, plunger, sinker	Eintaucher m, Tauchkörper m	plongeur m	черпак
	dipping	s. S 768		
D 294	dipping method, dipping pro- cess	Tauchprozeß m, Tauchver- fahren n	méthode f d'immersion, pro- cédé m par trempage, pro- cédé d'immersion	формование маканием, способ формования по- гружением
D 295	dipping plant	Tauchanlage f	appareil m de trempage, in- stallation f d'immersion	устройство для замочки, промывная установка
	dipping process	s. D 294		
D 296	Dirac delta function	Diracsche Deltafunktion f	fonction f [impulsion unité] de Dirac	дельта-функция Дирака
D 297	direct-contact cooling	Direktkontaktkühlung f	refroidissement m par contact direct	непосредственное контакт- ное охлаждение
D 298	direct-contact heat ex- changer	Direktkontaktwärmeübertra- ger m	échangeur m thermique à contact direct	контактный теплообменник
D 299	direct cooling	Direktkühlung f	refroidissement m direct	непосредственное охла- ждение
D 300	direct-cooling system	Direktkühlsystem n	système m de refroidisse- ment direct	система непосредствен- ного охлаждения
D 301	direct digital control	direkte digitale Regelung f	contrôle m digital direct	прямое цифровое управле- ние
D 302	directed graph	gerichteter Graph m	graphe m dirigé	ориентированный граф
D 303	direct evaporation	direkte Verdampfung f	évaporation f directe	непосредственное испаре- ние
D 304	direct evaporation system	Direktverdampfungssystem n	système m d'évaporation di- recte	система прямого выпарива- ния
D 305	direct-expansion coil	Direktkühlschlange f	serpentin m refroidisseur di- rect	змеевик испарителя
D 306	direct-expansion cycle	Direktkühlprozeß m, Direkt- verdampfungsprozeß m	processus m de refroidisse- ment direct, processus de l'évaporation directe	цикл непосредственного охлаждения (испарения)
D 307	direct-expansion evaporator, direct-feed evaporator	Direktverdampfer m	évaporateur m à détente di- recte	испаритель непосредствен- ного охлаждения
D 308	direct-expansion refrigera- tion	Direktverdampfungskühlung f	refroidissement m par évapo- ration directe, réfrigération f à expansion directe	охлаждение непосред- венным испарением хо- лодильного агента
D 309	direct feed	direkte Zuführung f	alimentation f directe	непосредственная подача
	direct-feed evaporator	s. D 307		
D 310	direct flow-extractor	Durchflußextraktor m	extracteur m à passage di- rect	непрерывный экстрактор
D 311	direct injection	direkte Einspeisung f, Direkt- einspritzung f	injection f directe	непосредственный впрыск
	direction	s. I 252		
D 312	direction for use	Bedienungsvorschrift f, Ge- brauchsanweisung f	instruction f de service, mode m d'emploi	инструкция эксплуатации
D 313	direction of flow	Strömungsrichtung f	direction f de l'écoulement	направление течения (по- тока)
D 314	direct liquid refrigeration	direkte Flüssigkeitskühlung f	réfrigération f directe à li- quide	прямое охлаждение жидко- сти
D 315	directly operated valve	direkt betätigtes Ventil n, Di- rektregler m	soupape f à action directe	регулятор прямого дейст- вия
D 316	direct reaction	Hinreaktion f	réaction f aller	прямая реакция
D 317	direct-reading instrument	Ablesegerät n	instrument m de lecture	отсчетное устройство, ин- дикатор
D 318	direct simulation	direkte Simulation f	simulation f directe	непосредственное модели- рование

D 319	dirt trap	Schmutzfilter n, Schmutzfänger m, Reiniger m	pare-boue m, purgeur m, purificateur m	грязеуловитель
	disaggregation	s. F 419		
D 320	disaster propagation	Störungsausbreitung f, Störungsfortpflanzung f	propagation f de perturbations	распространение аварии (повреждения)
D 321	disaster susceptability	Störanfälligkeit f	sensibilité f aux perturbations	чувствительность к помехам
D 322	discharge, discharge funnel	Auslauftrichter m	entonnoir m d'écoulement, trémie f de sortie	разгрузочная воронка
D 323	discharge	Entladung f	déchargement m, décharge f	разряд
D 324	discharge	Austrag m, Ausschüttung f	déversement m, décharge f, sortie f	разгрузка
D 325	discharge	Ausströmen n, Auswurf m	émission f, évacuation f, échappement m	вытекание, истекание, выброс
	discharge	s. a. 1. D 488; 2. L 58; 3. O 111; 4. U 40		
D 326	discharge capacity	druckseitiger Förderstrom m, druckseitige Förderleistung f	débit m de refoulement	расход на стороне нагнетания, объем нагнетаемых паров
D 327	discharge chamber	Druckkammer f, Kompressionsraum m	chambre f de compression (refoulement)	нагнетательная камера
D 328	discharge cock	Abflußhahn m, Ablaßhahn m, Ablaßventil n	robinet m de décharge, purgeur m de vidange	расходный (спускной) кран
D 329	discharge coefficient	Durchflußzahl f	coefficient m de débit	коэффициент сопротивления течению
D 330	discharge cycle	Entleerungszyklus m, Entleerungsperiode f	période f de vidange, cycle m de déchargement	цикл нагнетания
D 331	discharge device, emptying device	Entleerungsvorrichtung f, Entladungsapparat m	dispositif m de vidange, dispositif d'évacuation, déchargeur m	приспособление для опоржнения (выгрузки, разгрузки), разгрузочный аппарат
D 332	discharge drum	Endabscheider m	réservoir m de décharge, séparateur m final	сепаратор
D 333	discharged water	Ablaufwasser n	eau f d'écoulement	сливная (сточная) вода
D 334/5	discharge flue	Entlüftungsrohr n, Abtußrohr n, Ausgangsrohr n, Auslaßrohr n	tube m d'évacuation, tuyau m de décharge	вытяжная труба
	discharge funnel	s. D 322		
D 336	discharge head	Förderhöhe f (Pumpe)	hauteur f de refoulement	напор, высота подъема
	discharge head	s. a. D 153		
D 337	discharge inclination	Ablaufschräge f, Abflußneigung f	pente f d'écoulement	наклон стока (сливного устройства), угол скоса
D 338	discharge knee	Abflußkrümmer m	coude m de décharge	обводное колено
D 339	discharge line	Abtußleitung f	décharge f, conduit m de décharge, tuyau m de descente	отводной трубопровод
	discharge line	s. a. P 413		
D 340	discharge liquor, sewage	Abwasser n	eaux fpl usées (d'égout)	сточная вода
D 341	discharge manifold	Hauptförderleitung f	conduite f principale	главный нагнетательный трубопровод
D 342	discharge pipe	Ausgangsrohr n, Abfallrohr n, Abflußrohr n, Ablaßstutzen m, Abfluß m	tube m d'écoulement, tuyau m de décharge (descente), déchargeoir m	отводная (канализационная, сточная) труба
	discharge pipe	s. a. P 413		
	discharge piping	s. B 242		
D 343	discharge pressure	Enddruck m, Förderdruck m, Ausgangsdruck m	pression f finale (de refoulement), pression à la sortie	конечное давление
D 344	discharge pump	Ablaufpumpe f	pompe f d'évacuation	насос для перекачивания оттеков
D 345	discharge quantity	Ausflußmenge f, Abflußmenge f	quantité f sortante, débit m de décharge (sortie)	количество истечения
D 346	discharge side	Druckseite f	côté f de refoulement	сторона нагнетания (высокого давления)
D 347	discharge temperature	Austrittstemperatur f, Ausgangstemperatur f, Ausflußtemperatur f	température f à la sortie, température de décharge	температура нагнетания, выходная температура
D 348	discharge valve	Abflußventil n	soupape f de décharge (tête, trop plein), soupape d'émission	выпускной вентиль, выпускной клапан
D 349	discharge velocity	Ausflußgeschwindigkeit f	vitesse f de sortie (décharge)	скорость истечения
D 350	discharging, drawing off	Ablassen n, Ausstoßen n	décharge f, évacuation f, émission f, vidange f	продувка, спуск, слив
D 351	discharging device	Entleerungseinrichtung f	dispositif m d'évacuation, dispositif de vidange	приспособление для опоржнения
D 352	discharging screw	Austragsschnecke f	vis m d'extraction	разгрузочный шнек
D 353	discharging stream	Austrittsstrom m	courant (débit) m de sortie, écoulement m sortant	выходящий поток
D 354	discharging time	Entladungsdauer f	durée f de décharge	продолжительность разгрузки
D 355	disconnection	Abschaltung f	déconnexion f	выключение, отключение
D 356	disconnection element	Abschaltelement n	élément m de coupure	отключающее устройство

D 357	discontinuity	Diskontinuität f, Ungleichmäßigkeit f	discontinuité f	дискретность, дисконтинуитет, разрыв
D 358	discontinuous control	diskontinuierliche Regelung f	contrôle m discontinu	прерывистое регулирование
D 359	discontinuous operation	diskontinuierliche Betriebsweise (Arbeitsweise) f	mode m opératoire discontinu	периодический способ производства
D 360	discrete maximum principle	diskretes Maximumprinzip n	principe m de maximum discret	дискретный принцип максимума
D 361	discrete quantity	diskrete (unstetige) Größe f	grandeur f discrète	дискретная величина
D 362	discrete representation	diskrete Darstellung f	représentation f discrète	дискретное представление
D 363	disintegration	Trennung f, Zerlegung f	séparation f, désintégration f, triage m	отделение, разделение
D 364	disintegration energy	Zersetzungsenergie f, Zerfallsenergie f	énergie f de décomposition (désintégration)	энергия распада
D 365	disintegration rate	Zerfallsrate f, Zerfallsgeschwindigkeit f	taux m (vitesse f) de désintégration	скорость распада
D 366	disintegrator, comminutor	Zerkleinerer m, Brechmaschine f	désintégrateur m	дробилка, дробильная машина
D 367	disk	Schaltpult n	pupitre m de commande	пульт управления, коммутационный пульт
D 368	disk attrition mill	Scheibenmühle f	broyeur m à disques	дисковая мельница
D 369	disk crusher	Scheibenbrecher m, Tellerbrecher m	concasseur m à disques	дисковая дробилка
D 370	disk filter, rotary disk filter	Scheibenfilter n	filtre m rotatif à disques	дисковый фильтр
D 371	disk mill	Tellermühle f, Scheibenmühle f	broyeur m à disques	дисковая дробилка
D 372	dislocation	Versetzung	engorgement m, accrochage m	перемещение
D 373	dismantling, dismounting	Demontage f, Zerlegen n, Abbau m	démontage m	демонтаж, разборка
D 374/5	disperse	dispers, fein verteilt (zerteilt), dispergiert	dispersé	диспергированный, рассеянный
D 376	dispersed water	dispergiertes Wasser n	eau f dispersée	дисперсная вода
D 377	disperse phase	disperse Phase f	phase f dispersée	дисперсная фаза
D 378	dispersing agent	Dispergator m, Dispersionsmittel n	agent m de dispersion, dispersant m	диспергатор
D 379	dispersion	Dispergieren n, Dispersion f, Streuung f	dispersion f	диспергирование, дисперсия
	dispersion	s. a. L 59		
D 380	dispersion index	Dispersionsgrad m, Dispersitätsgrad m	degré m de dispersion	показатель рассеяния (дисперсии)
D 381	displace/to, to eject, to push out	verdrängen, versetzen, verschieben	déplacer, supprimer	замещать, вытеснять
D 382	displacement	Verlagerung f, Verschiebung f, Verdrängung f, Ersetzung f	déplacement m, transposement m	перемещение, вытеснение, замена
D 383	displacement compressor	Verdrängungsverdichter m	compresseur m à déplacement	объемный компрессор, компрессор статического сжатия
D 384	displacement desorption	Verdrängungsdesorption f	désorption f par déplacement	вытеснительная десорбция
	displacement flow	s. P 171		
D 385	displacement reaction	Verdrängungsreaktion f	réaction f de déplacement	реакция замещения
D 386	displacer	Verdränger m	dispositif m déplaceur	вытеснитель, плунжер
D 387	disposal	Entsorgung f, Abfallbehandlung f	traitement m des déchets	удаление, устранение (отходов)
D 388	disposal plant	Abfallbeseitigungsanlage f	installation f de l'évacuation de déchets	установка для удаления отходов
	disregard/to	s. N 9		
D 389	dissipation	Dissipation f, Zerstreuung f	dissipation f	диссипация
	dissipation	s. a. 1. D 435; 2. L 235; 3. P 301		
D 390	dissipative mix-melting	dissipatives Mischschmelzen n	mélange-fusion f dissipative	диссипативное расплавление смеси
D 391	dissociation	Dissoziation f, Zerfall m, Aufspaltung f	dissociation f	диссоциация
D 392	dissociation constant	Dissoziationskonstante f	constante f de dissociation	постоянная диссоциации
D 393	dissociation energy	Dissoziationsenergie f	énergie f de dissociation	энергия диссоциации
D 394	dissociation equilibrium	Dissoziationsgleichgewicht n	équilibre m de dissociation	диссоциационное равновесие
D 395	dissociation pressure	Dissoziationsdruck m	pression f de dissociation	давление диссоциации
D 396	dissociation product	Dissoziationsprodukt n	produit m de dissociation	продукт диссоциации
D 397	dissociation temperature	Dissoziationstemperatur f	température f de dissociation	температура диссоциации
D 398	dissolution	Auflösung f, Auflösen n, Lösung f, Lösungsvorgang m	dissolution f, décomposition f, désagrégation f	растворение
	dissolution kinetics	s. D 402		
D 399	dissolved refrigerant	gelöstes Kältemittel n	réfrigérant m dissous	растворенный холодильный агент
D 400	dissolved substance	gelöster Stoff m	substance f dissoute	растворенное вещество
D 401	dissolving apparatus	Löseapparat m	appareil m de dissolution	аппаратура для растворения
D 402	dissolving kinetics, dissolution kinetics	Auflösungskinetik f	cinétique f de dissolution	кинетика растворения

D 403	**dissolving power,** solubilizing power	Lösungsvermögen n, Löslichkeit f, Auflösungsvermögen n	pouvoir m dissolvant, solubilité f	растворяющая способность
D 404	**distance control**	Fernbedienung f	télécommande f, commande f à distance	дистанционное обслуживание
D 405	**distil/to,** to distil off	abdestillieren, destillieren	éliminer par distillation, distiller	отгонять
D 406	**distillable**	destillierbar	distillable	перегоняемый
D 407	**distillate**	Destillat n	distillat m	дистиллят, погон
D 408	**distillate fuel oil**	Destillatheizöl n	fuel-oil m distillé	дистиллятное топливо
D 409	**distillate hydrogenation**	Destillathydrierung f	hydrogénation f du distillat	гидрирование дистиллята
D 410	**distillate of mineral oil, distillate of petrol**	Erdöldestillat n	distillat m du pétrole	нефтяной дистиллят
D 411	**distillate processing**	Destillataufbereitung f	traitement m du distillat	переработка дистиллята
D 412	**distillation**	Destillieren n, Destillation f	distillation f	перегонка, дистиллирование
D 413	**distillation apparatus,** distilling apparatus, still	Destillierapparat m	appareil m distillatoire (de distillation)	дистилляционный аппарат, перегонный аппарат (куб)
D 414	**distillation capacity**	Destillationskapazität f	capacité f de distillation	производственная мощность дистилляции
D 415	**distillation characteristic**	Destillationsverhalten n	caractéristique f de distillation	дистилляционная характеристика
D 416	**distillation column**	Destillationskolonne f, Trennungskolonne f, Trennsäule f, Destillationssäule f	colonne f de distillation	перегонная (дистилляционная) колонна
D 417	**distillation cooler**	Destillierkühler m, Destillatkühler m	refroidisseur m à distillat, réfrigérant m distillatoire	холодильник дистиллята (перегонных кубов)
D 418	**distillation curve**	Destillationskurve f, Destillationsablauf m	courbe f de distillation	кривая разгонки
D 419	**distillation in tray column**	Bodendestillation f	distillation f avec plateau	дистилляция в тарельчатой колонне
D 420	**distillation loss**	Destillationsverlust m	perte f par distillation	потери при разгонке
D 421	**distillation period**	Destillationsperiode f	période (durée) f de distillation	продолжительность дистилляции
D 422	**distillation plant**	Destillieranlage f, Destillationsanlage f	unité f de distillation	дистилляционная установка
D 423	**distillation product**	Destillationsprodukt n	produit m de distillation	продукт дистилляции (перегонки)
D 424	**distillation product**	Schwelprodukt n	produit m de la distillation à basse température	продукт швелевания (полукоксования)
D 425	**distillation range,** boiling range	Destillationsbereich m	zone f de distillation, intervalle m d'ébullition	границы кипения, пределы кипения фракции
D 426	**distillation tray**	Destillationsboden m, Austauschboden m, Rektifizierboden m	plateau m échangeur (de colonne de distillation)	тарелка [ректификационной колонны]
D 427	**distillation tube,** distilling pipe	Destillationsrohr n, Destillierrohr n	ampoule f à distiller, tuyau m de distillation	пароотводная труба
D 428	**distillation under pressure,** pressure distillation	Überdruckdestillation f	distillation f sous pression	перегонка при высоком давлении
D 429	**distilled water**	destilliertes Wasser n	eau f distillée	дистиллированная вода
	distilling apparatus	s. D 413		
D 430/1	**distilling chamber**	Destillationskammer f, Schwelkammer f	four m de distillation sèche, chambre f à lente distillation	камера деструктивной дистилляции
	distilling flosk	s. D 432		
	distilling furnace	s. D 433		
D 432	**distilling head,** distilling flosk, distilling vessel	Destillationskolben m, Destillationsgefäß n, Destillierkolben m, Destillierblase f	alambic m, vase m distillatoire, matras m à distillation	перегонная колба
	distilling pipe	s. D 427		
D 433	**distilling stove,** distilling furnace	Destillierofen m, Destillationsofen m	four m de distillation	дистилляционная печь
	distilling vessel	s. D 432		
	distil off/to	s. D 405		
D 434	**distil over/to**	[über]destillieren	distiller	отгонять
D 435	**distribution,** dissipation	Verteilung f	répartition f, distribution f	распределение
	distribution coefficient	s. P 65		
D 436	**distribution curve**	Verteilungskurve f	caractéristique (courbe) f de distribution	кривая распределения
D 437	**distribution function**	Verteilungsfunktion f	fonction f de distribution	функция распределения
D 438	**distribution law,** partition law	Verteilungsgesetz n	loi f de distribution	закон распределения
D 439	**distribution model**	Verteilungsmodell n	modèle m de distribution	модель распределения
D 440	**distribution of time**	Zeitverteilung f, Zeitaufteilung f	distribution f de temps	распределение времени
D 441	**distribution ratio**	Verteilungsverhältnis n, Aufteilungsverhältnis n	coefficient m de partage, rapport m de distribution	отношение распределения
D 422	**distribution time**	Verteilzeit f	temps m de distribution	время распределения
D 443	**distributor,** spreader	Verteiler m, Aufgabevorrichtung f, Aufgeber m, Verteilanlage f	distributeur m, installation f de distribution	распределитель[ное устройство]

D 444	disturbance	Störung f, Zwischenfall m, Panne f	trouble m, dérangement m, panne f, perturbation f, incident m	возмущение, авария, отказ, помеха
	disturbance	s. a. 1. B 253; 2. D 449		
D 445	disturbance analysis, trouble shooting	Störungsauswertung f, Störungsanalyse f, Störungsaufklärung f	analyse f des pannes (dérangements, perturbations), exploitation f des perturbations	подведение результатов причин аварии (отказа, возмущения), анализ причин аварии, отыскание причин аварии (отказа)
D 446	disturbance registration, trouble recording	Störungserfassung f	enregistrement m de troubles (pannes, perturbations)	регистрация аварии (возмущения, отказа)
D 447	disturbance response	Störverhalten n	comportement m au dérangement, comportement en cas de trouble	реакция на возмущение (нарушение режима)
D 448	disturbance time	Störungsdauer f	durée f de dérangement, période f de perturbation	продолжительность отказа (аварии, возмущения)
D 449	disturbance variable, disturbing quantity, disturbance	Störgröße f	grandeur f perturbatrice	возмущение
D 450	disturbing quantity analysis	Störgrößenanalyse f	analyse f des grandeurs perturbatrices	анализ возмущений
D 451	divider head	Verteilerkopf m, Teilerkopf m	tête f de distributeur	головка распределителя
D 452	dividing wall	Trennwand f, Zwischenwand f	paroi f de séparation, cloison f	перегородка
	diving	s. S 768		
D 453	dosage, dosing	Dosierung f, Zuteilung f, Zumessung f	dosage m, dose f	дозирование, дозировка
	dosage	s. a. I 131		
D 454	dosimeter	Dosimeter n, Dosismesser m, Geiger-Zähler m	dosimètre m, quantimètre m	дозиметр
	dosing	s. D 453		
D 455	dosing plant	Dosieranlage f	installation f de dosage	дозировочная машина, дозатор
D 456	dosing pump	Dosierpumpe f	pompe f doseuse	дозировочный насос
D 457	double cell furnace	Doppelkammerofen m	four m à deux chambres	двухкамерная печь
D 458	double-cone crusher	Doppelkegelbrecher m	concasseur (broyeur) m à deux cônes	двухконусная дробилка
D 459	double-diaphragm pump	Doppelmembranpumpe f	pompe f à double membrane	двойной мембранный насос, диафрагменный насос двойного действия
D 460	double-extraction process	Doppelextraktionsprozeß m	procédé m de double extraction	двухступенчатое экстрагирование
D 461	double jet, double nozzle	Doppeldüse f	buse f double	двойное сопло
D 462	double layer	Doppelschicht f	couche f double	двойной слой
	double nozzle	s. D 461		
D 463	double pipe	Doppelrohr n	tuyau m double, tube f concentrique	двойная концентрическая труба типа «труба в трубе»
D 464	double-pipe condenser	Doppelrohrkondensator m	condenseur m à tubes concentriques	двухтрубный конденсатор
D 465	double-pipe cooler	Doppelrohrkühler m	refroidisseur m à tubes concentriques	двухтрубный охладитель
D 466	double-pipe evaporator	Doppelrohrverdampfer m	évaporateur m à tubes concentriques	двухтрубный испаритель
D 467	double-pipe gas cooler	Doppelrohrgaskühler m	refroidisseur m de gaz à tubes concentriques	двухтрубный охладитель газа
D 468	double-pipe heat exchanger	Doppelrohrwärmeaustauscher m, Doppelrohrwärmeübertrager m	échangeur m de chaleur à tubes concentriques	теплообменник типа «труба в трубе»
D 469	double rectification column	Kolonne f für zweifache Rektifikation	colonne f pour double rectification	колонна двукратной ректификации
D 470	double refraction	Doppelbrechung f	réfraction f double, biréfringence f	двойное преломление
D 471	double stage desalting	zweistufige Entsalzung f	déminéralisation f à deux étages	двухступенчатое обессоливание
D 472	double stage process	Zweistufenprozeß m	procédé m à deux étages	двухступенчатый процесс
D 473	double stage reaction	zweistufiges Reaktionssystem n, zweistufige Reaktion f	réaction f à deux étages	двухступенчатая реакция
D 474	double wall, jacketed wall	Doppelmantel m	double enveloppe f	двойная рубашка
D 475	downcomer	Ablaufstutzen m	tubulure f de décharge	спускной патрубок
D 476	downcomer	Ablaufrohr n (von Glockenböden), Abfallrohr n	tuyau m de décharge	сливная труба
D 477	downstream	Abwärtsströmung f	écoulement m descendant	нисходящее течение
D 478	downstream	stromab, in Richtung der Strömung	en aval	вдоль течения
D 479	downstream component	untere (hintere) Komponente f	composante f de dessous	нижний (отдаленный) компонент
D 480	downtime, down-time	Ausfallzeit f, Abschaltzeit f, Stillstandzeit f	temps m de mise hors de service, temps d'arrêt	время простоя
D 481	downtime cost	Ausfallkosten pl	frais mpl de mise hors de service	стоимость простоя

D 482	**downward pressure**	Abtrieb m, negativer Auftrieb m	dépression f, force f descensionelle, portance f négative	смыв, снос, слон, отрицательная плавучесть
D 483	**draft tube,** draught tube	Abzugsrohr n	tuyau m de soutirage	вытяжная труба
D 484	**drag**	Luftwiderstand m, Strömungswiderstand m	résistance f aérodynamique (hydrodynamique)	сопротивление воздуха (течения)
D 485	**drag flow**	Widerstandsströmung f, Schleppströmung f, Couette-Strömung f	écoulement m Couette (à résistance)	куэттовское течение
D 486	**drag force**	hydrodynamische Widerstandskraft f	force f de résistance hydrodynamique	гидродинамическая сила сопротивления
D 487	**drain,** waste channel	Abflußkanal m	rigole f (canal m) d'écoulement	сточный канал (водовод), сточная канава
	drain	s. a. 1. C 162; 2. L 58; 3. O 111		
D 488	**drainage,** discharge, flowing off, leakage	Abfließen n	écoulement m, décharge f	стекание, оттекание, обтекание
D 489	**drainage,** water removal, sewerage, dehydration	Entwässerung f	drainage m, assèchement m	обезвоживание
D 490/1	**drainage,** dehydration plant	Entwässerungsanlage f	installation f d'assèchement, drainage m	водоотливная установка
	drainage	s. a. C 162		
D 492	**draining ditch**	Abflußgraben m	rigole f	сточная канава
D 493	**draining funnel**	Ablauftrichter m	entonnoir m de sortie (décharge)	спускная воронка, разгрузочная воронка
D 494	**drain line**	Abflußleitung f, Ablaßleitung f, Abzugsleitung f, Abwasserleitung f, Dränageleitung f, Entwässerungsleitung f	décharge f, conduit m de décharge	выпускной трубопровод, сточный водовод, сливная (дренажная) труба
D 495	**drain off**	Abzapfen n, Abziehen n, Abfüllen n	soutirage m	ответвление
D 496	**drain pipe**	Ableitungsrinne f	fossé m de dérivation	отводный желоб
	drain pipe	s. a. C 162		
	drain pump	s. B 93		
D 497	**drain valve**	Abflußventil n, Abscheideventil n	soupape f de décharge	выпускной вентиль (клапан)
D 498	**draught**	Tauchung f, Tauchen n	plongée f	погружение, осадка
	draught tube	s. D 483		
	drawing off	s. D 350		
D 499	**draw-off rate**	Ablaßgeschwindigkeit f	vitesse f (débit m) de décharge	скорость выхода
D 500	**dressing,** finishing	Appretur f	apprêt m, finissage m	аппретура, аппретирование
D 501/2	**dressing,** recovery, processing	Aufbereitung f, Behandlung f	traitement m, préparation f	обогащение, подготовка
	dressing	s. a. P 377		
	dressing process	s. D 527		
D 503	**drift of water,** water drift	Wasserverlust m	perte f de l'eau	утечка воды
D 504	**drip nozzle**	Tropfdüse f	tube m d'égouttage	капельное сопло, капельница
	dripping water	s. D 506		
D 505	**drip screen**	Abtropfsieb n	tamis m égouttoir	обезвоживающий грохот
D 506	**drip water,** dripping (splash) water	Tropfwasser n	eau f d'égouttage	вода, стекающая каплями
	driveway	s. A 73		
D 507	**driving force**	Triebkraft f	force f motrice	движущая сила
	driving force	s. a. D 509		
D 508	**driving moment**	Antriebsmoment n	couple m moteur (d'entraînement)	движущий (приводящий) момент
D 509	**driving power,** driving force	Antriebsbelastung f, Antriebskraft f	force f motrice	нагрузка привода
D 510	**driving time**	Antriebszeit f	temps m d'entraînement	время работы двигателя
	drop analysis	s. S 494		
D 511	**drop-counter**	Tropfenzähler m	compte-gouttes m	счетчик капель, капельница
D 512	**drop evaporation**	Tropfenverdunstung f	évaporation f de gouttes	капельное испарение
D 513	**droplet formation**	Tröpfchenbildung f	formation f des gouttelettes	образование капель
D 514	**droplet size**	Tröpfchengröße f	dimension f des gouttelettes	размер капель
D 515	**droplet stretching**	Tropfenverformung f	déformation f de goutte	вытягивание (деформация) капли
D 516	**dropping funnel**	Tropftrichter m	entonnoir m séparateur	капельная воронка
D 517	**dropwise condensation**	Tropfenkondensation f	condensation f sous forme de gouttes	капельная конденсация
	dross	s. S 364		
D 518	**drum,** roller	Trommel f	tambour m, trommel m	барабан
D 519	**drum cooler**	Trommelkühler m	refroidisseur m à tambour	барабанный охладитель
	drum dryer	s. C 904		
D 520	**drum mixer**	Trommelmischer m	malaxeur m à tambour, mélangeur m à trommel	барабанный смеситель
	drum pump	s. R 388		
	drum sieve	s. C 905		
	dry/to	s. D 181/2		
D 521	**dry blend**	Trockenmischung f	mélangeage m à sec	сухая смесь

D 522	dry cell	Trockenelement n	pile f sèche, élément m sec	сухой элемент
D 523	dry cleaning	Trockenreinigung f	épuration f sèche, nettoyage m à sec	сухая очистка
D 524	dry content	Trockengehalt m, Trockensubstanzanteil m	teneur m en matière sèche	степень сухости
D 525	dry crushing	Trockenmahlung f	broyage m à sec	сухой размол
D 526	dry distillation	trockene Destillation f, Trockendestillation f, Trockenentgasung f	distillation f sèche, dégazage m sec	сухая перегонка
D 527	dry-dressing process, dressing process	Trockenaufbereitung f	préparation f (traitement m) à sec	сухое обогащение
D 528	dryer, drying apparatus, desiccator	Trockner m, Trockeneinrichtung f, Trockenapparat m, Trockenkammer f, Trocknungsanlage f	sécheur m, séchoir m, installation f de séchage, dessiccateur m	сушильный аппарат
D 529	dry extract	Trockenextrakt m, Trockenauszug m	extrait m sec	сухой экстракт
D 530	dry friction	Trockenreibung f	frottement m à sec	сухое трение
D 531	dry gas	Trockengas n, getrocknetes Gas n	gaz m sec	сухой газ
D 532	dry ice, carbon dioxide snow	Trockeneis n	carboglace f, glace f sèche, acide m carbonique solide	сухой лед
D 533	drying, desiccation	Trocknen n, Trocknung f, Austrocknen n, Entwässern n, Austrocknung f	séchage m, dessèchement m, dessiccation f	высушивание, высыхание, [y]сушка
	drying agent	s. S 299		
D 534	drying air	Trockenluft f	air m sec (de séchage)	осушающий воздух
	drying apparatus	s. D 528		
D 535/6	drying bin, drying chamber	Trockenkammer f, Trocknungskammer f	chambre f de séchage, étuve f, séchoir m	сушильный бак
	drying by evaporation	s. E 284		
D 537	drying capacity	Trocknungsleistung f	capacité f de dessiccation, puissance f de séchage	производительность по осушке
	drying chamber	s. D 535/6		
D 538	drying cupbord, hot-air cabinet	Trockenschrank m	armoire f à sécher, séchoir m	сушильный шкаф
D 539	drying operation, drying process	Trocknungsverfahren n, Trockenprozeß m	opération f (procédé m, processus m) de séchage	способ сушки
D 540	drying oven	Trockenofen m	four m à sécher, sécheur m	сушильная печь
D 541	drying pipe	Trockenrohr n	tuyau m de séchage	сушильная труба
	drying process	s. D 539		
D 542	drying tower	Trockenturm m	tour f de séchage, colonne f sécheuse	сушильная колонна, угольная башня
D 543	drying tray	Trockenhorde f, Trockenboden m	plateau m de dessiccation (séchage)	сушильная кассета, сушильный лоток
D 544	drying zone	Trocknungsbereich m	zone f de séchage	зона сушки
D 545	dry residue	Trockenrückstand m	résidu m sec	сухой остаток
D 546	dry screening	Trockensiebung f, Trockensieben n	criblage m à sec	сухое просеивание (грохочение)
D 547	dry separator	Trocken[ab]scheider m	séparateur m à sec	сухой сепаратор
D 548	dry solid matter, dry substance	Trockengut n, Trockensubstanz f	matière f sèche	сухой материал
D 549	dry steam	trockener Dampf m	vapeur f sèche (surchauffée)	сухой пар
	dry substance	s. D 548		
D 550	dry system	Trockensystem n, Trocknungssystem n	système m de séchage	сухая система
D 551	dry weight, net weight	Trockengewicht n	poids m à l'état sec	вес сухого материала, сухой вес
D 552	dual-flow tray	zweiflutiger Boden m	plateau m à deux passes	сдвоенная тарелка
D 553	dual-screw extruder, twin-screw extruder	Doppelschneckenextruder m	extrudeuse f bi-vis	двухшнековый экструдер
	ductility	s. E 37		
D 554	duct wall	Rohrwand f	paroi f de tube, plaque f tubulaire	стена трубы
D 555	duct work	Leitungssystem n, Leitungsnetz n	réseau m de gaines (distribution)	система каналов
D 556	dulling agent	Trübungsmittel n	agent m troublant	глушитель
D 557	dump, dumping [ground], tip	Deponie f, Ablagerung f	décharge f, dépôt m, dépotoir m	захоронение отходов, свалка отходов, хранилище для отходов
D 558	dump drum	Entleerungstrommel f	tambour m de vidange	разгружающий барабан
D 559	dumping	Schütten n, Entladung f	déversement m, déchargement m	засыпка, загрузка
	dumping	s. a. D 557		
	dumping ground	s. D 557		
D 560	dumping height	Schütthöhe f	hauteur f de déchargement (déversement)	высота насыпаемого материала
D 561	durability	Haltbarkeit f, Dauerhaftigkeit f, Steifigkeit f, Beständigkeit f	durabilité f, stabilité f, inaltérabilité f	стойкость, прочность, крепкость

	English	German	French	Russian
D 562	**durability,** life, life time	Lebensdauer f, Nutzungszeitraum m, Einsatzdauer f	durée f utile (de service, de vie), vie f utile, durabilité f	срок службы, эксплуатационный срок, продолжительность жизни
D 563	**duration of action**	Einwirkungsdauer f, Wirkungsdauer f	durée f de l'attaque, durée de l'influence, durée d'action	продолжительность воздействия
D 564	**duration of cycle**	Zyklusdauer f, Zykluszeit f	durée f de cycle	продолжительность цикла
D 565/6	**duration of effect**	Wirkungsdauer f	durée f d'action	время воздействия
D 567	**duration of load [application]**	Belastungsdauer f	durée f de charge	продолжительность нагрузки
	duration of test	s. T 211		
	dust catcher	s. D 569		
D 568	**dust collection**	Staubabscheidung f, Entstaubung f	depoussiérage m, déposition f de poussière	пылеотделение, пылеулавливание
D 569	**dust collector,** dust catcher (precipitator)	Staubabscheider m, Staubsammler m, Staubsack m, Staubfänger m	séparateur m de poussière, dépoussiéreur m	пылеуловитель
D 570	**duster,** dust separator, cyclone	Staubabscheider m, Zyklon m	séparateur m de poussière, dépoussiéreur m [à cyclone]	пылеотделитель, пылесборник, пылеосадитель, циклон
D 571	**duster**	Staubbehälter m, Staubsammelbehälter m	collecteur m de poussière	пылесборник
D 572	**duster,** sprayer	Verstäuber m, Flüssigkeitssprüher m	pulvérisateur m de liquides	распылитель
D 573	**dust filter**	Staubfilter n, Entstaubungsfilter n	filtre m antipoussière (dépoussiéreur), dépoussiéreur m	фильтр для пыли
D 574	**dustiness,** density of dust	Staubdichtheit f	étanchéité f à la poussière	плотность пыли
D 575	**dust-laden gas**	staubhaltiges (staubbeladenes) Gas n, Staubgas n	gaz m poussiéreux (chargé de poussière)	газ, насыщенный пылью
D 576	**dust load**	Staubgehalt m, Staubbeladung f	charge f de poussière	содержание пыли
	dust precipitator	s. D 569		
D 577	**dust removal**	Gasentstaubung f, Gasreinigung f	épuration f (purification f, lavage m) du gaz	газоочистка
D 578	**dust removal plant**	Entstaubungsanlage f	installation f de dépoussierage	пылеуловительная установка
D 579	**dust screening**	Staubsiebung f	tamisage m de poussière	просеивание через частое сито
	dust separator	s. D 570		
D 580	**dust treating**	Staubaufbereitung f	traitement m de poussière	приготовление пыли
D 581	**dust treating plant**	Staubaufbereitungsanlage f	équipement m pour le traitement de poussière	установка для приготовления пыли
D 582	**dynamic behaviour**	dynamisches Verhalten n	comportement m dynamique	динамическое поведение, динамика
D 583	**dynamic characteristic**	dynamische Charakteristik f	caractéristique f dynamique	динамическая характеристика
D 584	**dynamic equilibrium**	dynamisches Gleichgewicht f	balance f (équilibre m) dynamique	динамическое равновесие
D 585	**dynamic flow**	dynamische Zähigkeit f	viscosité f dynamique	динамическая вязкость
D 586	**dynamic hold-up**	dynamisches Holdup n	rétention (retenue) f dynamique	динамическая задержка (в колоннах)
D 587	**dynamic pressure,** batch pressure	Staudruck m	pression f dynamique	динамическое давление, скоростной напор, повышенное давление воздуха
D 588	**dynamic programming**	dynamische Programmierung f	programmation f dynamique	динамическое программирование
D 589	**dynamic resistance**	dynamischer Widerstand m	résistance f dynamique	динамическое сопротивление
D 590	**dynamic sensitivity**	dynamische Empfindlichkeit f	sensibilité f dynamique	динамическая чувствительность
D 591	**dynamic state**	dynamischer Zustand m	état m dynamique	динамическое состояние
D 592	**dynamic system**	dynamisches System n	système m dynamique	динамическая система

E

	English	German	French	Russian
E 1	**early failure**	Frühausfall m	défaillance f prématurée	преждевременный отказ
E 2	**easily soluble**	leicht löslich	facilement soluble	легко растворимый
E 3	**eccentric**	Exzenter m, Exzenterscheibe f, Taumelscheibe f	roue f excentrique, excentri-que m	эксцентрик
E 4	**Eckert number**	Eckertzahl f, Eckertsche Zahl f	nombre m d'Eckert	число Эккерта
E 5	**economic coefficient**	ökonomische Kennzahl f	valeur f caractéristique économique	экономический показатель (критерий)
E 6	**economic design criteria**	ökonomische Entwurfskriterien npl	critères mpl de conception économiques	экономические критерии проектирования
E 7	**economic efficiency,** economy	Wirtschaftlichkeit f	économie f, efficience f, rentabilité f	экономичность, рентабельность
E 8	**economic evaluation**	ökonomische Bewertung f	évaluation f économique	экономическая оценка

E 9	economic evaluation quantity	ökonomische Bewertungs- größe f	grandeur f d'évaluation éco- nomique	величина (показатель) эко- номической оценки
E 10	economic optimization	ökonomische Optimierung f	optimisation f économique	экономическая оптимизация
E 11	economizer, air preheater, waste gas feeder	Abgasvorwärmer m, Rauch- gasvorwärmer m, Röhren- vorwärmer m mit Heizung durch Abgase, Ekonomiser m, Speisewasservorwärmer m, Luftvorwärmer m	économiseur m, réchauffeur m d'air	экономайзер, подогрева- тель воздуха, котел-ути- лизатор
	economizer	s. a. 1. F 58; 2. W 39		
	economy	s. E 7		
	eddy current	s. W 1		
E 12	eddy diffusivity	turbulente Diffusion f	diffusion f turbulente	турбулентная диффузия
E 13	eddy energy	Turbulenzenergie f	énergie f de turbulence	вихревая энергия
E 14	eddy flow, vortex (rotational) flow	Wirbel[quell]strömung f, ver- wirbelte Strömung f	courant m tourbillonnaire, écoulement m turbulent	вихреобразное (вихревое, турбулентное) течение
E 15	eddy formation	Wirbelbildung f	tourbillonnement m, forma- tion f de tournants	завихрение
E 16	eddy thermal conductivity	turbulente Wärmeleitfähig- keit f	conductibilité f thermique turbulente	турбулентная теплопровод- ность
E 17	eddy viscosity	scheinbare (turbulente) Vis- kosität f	viscosité f apparente (turbu- lente)	кажущаяся (эффективная, турбулентная) вязкость
E 18	educt	Abbauprodukt n, Zerset- zungsprodukt n	produit m de décomposition (dégradation)	продукт разложения
E 19	eduction	Abführung f, Auslaß m (Dampf, Gase)	éduction f	отвод
	effect	s. A 102		
E 20	effective calculating	effektive Rechenzeit f	temps m de calcul effectif	эффективное машинное (расчетное) время
E 21	effective compression work	effektive Verdichtungsarbeit f	travail m effectif de compres- sion	эффективная работа сжа- тия
E 22	effective cross section	Wirkungsquerschnitt m	section f efficace	эффективное поперечное сечение
E 23	effective diffusivity	effektive Diffusion f, Koeffi- zient m der effektiven Dif- fusion	coefficient m de diffusion ef- fective	эффективный коэффици- ент диффузии
E 24	effective electromotive force	elektromotorische Kraft f	force f électromotrice	электродвижущая сила
	effective output	s. A 129		
	effective power	s. A 129		
E 25	effective resistance	Wirkwiderstand m	résistance f effective	эффективное активное со- противление
E 26	effective value	Effektivwert m	valeur f efficace (effective)	эффективное значение
E 27	effective water energy (load)	Wasserlast f, effektiver Was- serdruck m	charge f d'eau effective	полезная нагрузка от воды
E 28	efficiency of the combustion	Verbrennungswirkungsgrad m	rendement m de combustion	эффективность сгорания
E 29	efficiency per unit	Einzelwirkungsgrad m	rendement m individuel	частичный коэффициент полезного действия
E 30	effluent	abfließendes Produkt n, Aus- fluß m, Abgang m	effluent m	эффлюент
E 31	effluent disposal	Abwasserbeseitigung f	élimination f des eaux usées	очистка сточных вод
E 32	effluent stream	Abwasserstrom m, abgehen- der Strom m	effluent m, écoulement m de sortie	поток сточных вод
E 33	efflux	Ausströmung f, Ausfluß m	évacuation f, décharge f	вытекание, спуск, истека- ние, выпуск, утечка
	efflux	s. a. I 189		
	eigenfunction	s. C 168		
	eject/to	s. D 381		
E 34	ejector	Ejektor m, Saugstrahlpumpe f, Strahlpumpe f	pompe f à jet, injecteur m, éjecteur m	эжектор, струйный насос
	ejector	s. a. J 9		
E 35	ejector air-conditioning plant	Dampfstrahlklimaanlage f	installation f de conditionne- ment d'air type éjecteur	пароэжекторная установка кондиционирования воз- духа
E 36	ejector refrigerating machine	Dampfstrahlkältemaschine f	machine f frigorifique à éjec- teur	эжекторная холодильная машина
E 37	elasticity, ductility, malleabil- ity	Dehnbarkeit f, Elastizität f, Streckbarkeit f	ductilité f, extensibilité f, élasticité f	растяжимость, тягучесть, вязкость, ковкость, эла- стичность
E 38	elastic recovery	elastische Erholung (Rückver- formung) f, Rückfederung f	portée (reprise, restitution) f élastique	обратное восстановление
	elastic recovery ratio	s. S 902		
E 39	elastomer	Elastomer n, Elast m	élastomère m	эластомер
E 40	elastomer processing	Elastverarbeitung f	transformation f d'élasto- mères	переработка эластомеров
	elaterometer	s. S 641		
E 41	electrical conductance	elektrische Leitfähigkeit f	conductivité f électrique	электрическая проводи- мость
E 42	electrical gas cleaning equip- ment	elektrische Gasreinigung f	épuration f du gaz électrique	электрическая газоочистка (очистка газа)
E 43	electrical installation	elektrische Anlage f, Elektro- installation f	installation f électrique	электроустановка, электро- оборудование

E 44	electric charge	elektrische Ladung *f*	charge *f* électrique	электрический заряд
E 45	electric controller	elektrischer Regler *m*	régulateur *m* électrique	электрический регулятор
E 46	electric double layer	elektrische Doppelschicht *f*	couche *f* double électrique	электрический двойной слой
E 47	electric humidifier	elektrischer Befeuchter *m*	humidificateur *m* électrique	электроувлажнитель
E 48	electric power	elektrische Leistung *f*	puissance *f* électrique	электрическая мощность
E 49	electric separation	elektrische Aufbereitung *f*	séparation *f* électrique	электрическое обогащение
E 50	electroanalysis	Elektroanalyse *f*	électro-analyse *f*, analyse *f* électrique	электроанализ
E 51	electrochemical equivalent	elektrochemisches Äquivalent *n*	équivalent *m* électrochimique	электрохимический эквивалент
E 52	electrochemical process	elektrochemisches Verfahren *f*	processus *m* électrochimique	электрохимический процесс
E 53	electrochemical reaction	elektrochemische Reaktion *f*	réaction *f* électrochimique	электрохимическая реакция
E 54	electrochemical series, electromotive chain, contact-series	Spannungsreihe *f*	série *f* de contact (tension), chaîne *f* des forces électromotorices	ряд напряжений
E 55	electrochemistry	Elektrochemie *f*	électrochimi *f*	электрохимия
E 56	electrode characteristic	Elektrodenkennlinie *f*	caractéristique *f* d'électrode	характеристика электрода
E 57	electrode current	Elektrodenstrom *m*	courant *m* d'électrode	ток электрода
E 58	electrodeposition	elektrolytische Abscheidung *f*	déposition *f* électrolytique	электролитическое осаждение (отложение)
E 59	electrodialysis	Elektrodialyse *f*	électrodialyse *f*	электродиализ
E 60	electrodispersion	Elektrodispersion *f*	dispersion *f* électrique	электродисперсия
E 61	electrokinetic potential	elektrokinetisches Potential *n*	potentiel *m* électrocinétique	электрокинетический потенциал
E 62	electrolysis	Elektrolyse *f*	électrolyse *f*	электролиз
	electromotive chain	s. E 54		
E 63	electronic information processing system	elektronisches Informationsverarbeitungssystem *n*	système *m* de traitement électronique des informations	электронная система обработки информации
E 64	electroplate/to	galvanisieren	galvaniser	наносить слой металла, покрывать гальванически, гальванизировать
E 65	electroplating	Galvanisieren *n*	électrogalvanisage *m*, électroplacage *m*, galvanostégie *f*	гальванизация, гальванопокрытие
E 66	electrostatic coating	elektrostatisches Beschichten *n*	enduction *f* électrostatique	электростатическое нанесение
E 67	electrostatic filter (precipitator)	Elektrostatikfilter *n*, Elektrofilter *n*, Elektroabscheider *m*	filtre *m* électrostatique, séparateur *m* électrostatique, électrofiltre *m*	электрофильтр
E 68	electrothermal, electrothermic	elektrothermisch	électrothermique	электротермический
E 69	elementary analysis, ultimate analysis	Elementaranalyse *f*	analyse *f* élémentaire (ultime)	элементарный анализ
E 70	elementary building block	Elementarbaustein *m*	bloc *m* fonctionel élémentaire	элементарный блок
E 71	elementary effective quantum	elementares Wirkungsquantum *n*	quantum *m* d'action élémentaire	элементарный квант действия
E 72	elementary function	Grundfunktion *f*	fonction *f* élémentaire	элементарная функция
E 73	elementary iteration procedure	elementares Iterationsverfahren *n*	méthode *f* d'itération élémentaire	метод элементарных итераций
E 74	elementary lattice cell	Elementarzelle *f*	cellule *f* élémentaire	базисная клетка, элементарная ячейка
E 75	elementary mesh	Elementarmasche *f*	maille *f* élémentaire	элементарная ячейка
E 76	elementary particle	Elementarteilchen *n*	particule *f* élémentaire	элементарная частица
E 77	elementary process	Elementarprozeß *m*	procédé *m* élémentaire	элементарный (первичный) процесс
E 78	elementary quantum of electricity	elektrisches Elementarquantum *n*	quantum *m* élémentaire électrique	элементарное количество электричества
E 79	elementary reaction	Elementarreaktion *f*	réaction *f* élémentaire	элементарная реакция
E 80	elementary way	Elementarweg *m*	voie *f* élémentaire	элементарный путь (способ)
E 81	element efficiency	Elementwirkungsgrad *m*	rendement *m* élémentaire (de l'élément)	коэффициент полезного действия элемента, элементный коэффициент полезного действия
E 82	element redundancy	Elementredundanz *f*	redondance *f* élémentaire	элементный избыток
E 83	element with distributed parameters	Element *n* mit verteilten Parametern	élément *m* à constantes distribuées	элемент с распределенными параметрами
E 84	elevated tank	Hochbehälter *m*	réservoir *m* surélevé	надземный напорный резервуар, напорный бак
E 85	elimination, separation, precipitation	Ausscheidung *f*, Abscheidung *f*, Ablagerung *f*	séparation *f*, dégagement *m*, précipitation *f*	выделение, осаждение
E 86	elimination	Eliminierung *f*, Beseitigung *f*, Entfernung *f*	élimination *f*	исключение, удаление
	elimination	s. a. D 177		
	elimination of toxic constituents	s. D 52		

E 87	eliminator plate	Tropfenabscheider m	plaque f d'élimination, séparateur m de gouttes	каплеотделитель
E 88	**Ellis fluid**	Ellissches fluides Medium n	fluide m d'Ellis	жидкость Эллиса
E 89	**Ellis model**	Ellissches Modell n	modèle m d'Ellis	модель Эллиса
E 90	**elongation**	Verlängerung f, Elongation f, Längung f, Dehnung f	élongation f	расширение, удлинение
E 91	**elongational flow**	Dehnungsfluß m	flux m d'allongement, flux d'elongation	течение растяжения
E 92	**elongational viscosity**	Dehnungsviskosität f	viscosité f à l'élongation	вязкость растяжения
E 93	**elongation of rupture**	Bruchdehnung f	allongement m de rupture	удлинение при разрыве
E 94	**elongation resistance** elution	Streckfestigkeit f s. W 17	résistance f à l'allongement	прочность на растяжение
E 95	**elution process**	Waschprozeß m, Eluierung f	procédé m de lavage, élution f	процесс промывки
E 96	**elutriate/to**	abschlämmen, entschlammen	débourber, élutrier	отмучивать
E 97	**elutriating**	Abschlämmung f	élutriation f	удаление грязи, продувка
E 98	**elutriation,** washing, decantation	Schlämmen n, Schlämmung f	décantage m, lavage m, élutriation f	отмучивание, промывка, декантация, приготовление шлама
E 99	**elutriator**	Schlämmgerät n, Schlämmapparat m	installation f d'élutriation	аппарат для отмучивания
E 100	**embedding,** potting	Einbetten n, Einfügen n	insertion f, encastrement m	заделка, заливка, вставление, укладка, монтирование, погружение
E 101	**emergency blowdown tank**	Notentspannungstank m	réservoir m de détente de secours	танк для аварийного сброса
E 102	**emergency cutout**	Notausschalter m	disjoncteur m de secours	аварийный выключатель
E 103	**emergency measure**	Sofortmaßnahme f	mesure f d'urgence	неотложное мероприятие
E 104	**emergency shut-down**	Notabschaltung f	arrêt m d'urgence	аварийное отключение
E 105	**emission analysis**	Emissionsanalyse f	analyse f par émission	эмиссионный анализ
E 106	**emission band**	Emissionsbande f	bande f d'émission	полоса спектра пропускания
E 107	**emission capability**	Emissionsfähigkeit f	pouvoir m émissif	способность излучения
E 108	**emission coefficient**	Emissionskoeffizient m	coefficient m d'émission	коэффициент излучения
E 109	**emission law**	Emissionsgesetz n	loi f d'émission	закон электронной эмиссии
E 110	**emission measurement technology**	Emissionsmeßtechnik f	technique f de mesure d'émission	эмиссионная измерительная техника
E 111	**emission source**	Emissionsquelle f	source f d'émission	источник эмиссии
E 112	**emission spectrum**	Emissionsspektrum n	spectre m d'émission	эмиссионный спектр
E 113	**emissive power,** emissivity emollient	Emissionsvermögen n s. P 207	pouvoir m émissif	эмиссионная способность
E 114	**empirical model,** statistical model	empirisches (statistisches) Modell n	modèle m empirique (statistique)	эмпирическая (статистическая) модель
	employment	s. A 526		
	emptying device	s. D 331		
E 115	**emulsifiability**	Emulgierfähigkeit f, Emulgiervermögen n	aptitude f à donner des émulsions, pouvoir m émulsionnant	эмульгирующая способность, способность к эмульгированию
E 116	**emulsification**	Emulsionsbildung f	formation f d'émulsion	образование эмульсии
E 117	**emulsification**	Emulgieren n	émulsionnement m	эмульгирование
E 118	**emulsifier**	Emulgiermaschine f	émulsionneuse f, émulseur m	аппарат для эмульгирования
E 119	**emulsifier,** emulsifying agent	Emulgator m, Emulsionsbildner m, Emulgierzusatz m	émulsionnant m, émulsifiant m, émulsifieur m	эмульгатор, эмульсор
E 120	**emulsion layer**	Emulsionsschicht f	couche f d'émulsion	эмульсионный слой
E 121	**emulsion polymerization**	Emulsionspolymerisation f	polymérisation f d'émulsion	эмульсионная полимеризация
E 122	**end group,** terminal group	Endgruppe f	groupe m terminal	конечная (концевая) группа
E 123	**endothermic process**	endothermer Prozeß m	processus m endothermique	эндотермический процесс
E 124	**end point**	Endpunkt m, Umschlagpunkt m	point m d'extrémité, point final de la réaction	конечная точка, конец процесса, точка перехода
	endurance strength	s. C 832		
E 125	**energetic efficiency**	energetischer Wirkungsgrad m	rendement m énergétique	энергетический коэффициент полезного действия, энергетическая отдача
E 126	**energetic regeneration**	energetische Regeneration (Regenerierung) f	régénération f énergétique	энергетическая регенерация
E 127	**energy absorption**	Energieaufnahme f, Energieabsorption f	absorption f d'énergie	поглощение энергии
E 128	**energy accumulation**	Energiestau m	accumulation f d'énergie	накопление энергии
E 129	**energy analyzer**	Energieanalysator m	analyseur m d'énergie	энергетический анализатор
	energy application	s. E 173		
E 130	**energy balance**	Energiegleichgewicht n, Energiebilanz f	balance f d'énergie, équilibre m (bilan m, balance f) énergétique	энергетический баланс
E 131	**energy barrier**	Energieschranke f, Energieschwelle f	seuil m énergétique (d'énergie)	энергетический барьер
E 132/3	**energy carrier**	Energieträger m	porteur m d'énergie	носитель энергии, теплоемник
E 134	**energy consumption**	Energieverbrauch m, Energieaufwand m	consommation (dépense) f d'énergie	потребление (расход) энергии, энергозатраты

E 135	energy-containing	energiehaltig	contenant de l'énergie	энергосодержательный, энергосодержащий
E 136	energy content	Energieinhalt m	contenu m énergétique	содержание энергии, энергосодержание
E 137	energy conversion, energy transformation	Energieumwandlung f, Energieumsetzung f	transformation f d'énergie	превращение (преобразование) энергии
E 138	energy converter	Energieumformer m	transformateur (convertisseur) m d'énergie	преобразователь энергии
E 139	energy cycle	Energiekreislauf m	cycle m d'énergie	энергетический цикл
E 140	energy decrease	Energieabnahme f, Energieverminderung f	décroissance f énergétique	уменьшение энергии
E 141	energy density	Energiedichte f	densité f énergétique	плотность энергии
E 142	energy distribution	Kraftverteilung f, Energieverteilung f	distribution f d'énergie, distribution de force motrice	распределение энергии
E 143	energy drop, energy loss	Energieverlust m	perte f d'énergie	уменьшение (потеря) энергии
E 144	energy equation	Energiegleichung f	équation f énergétique	уравнение энергии
E 145	energy equivalent	Energieäquivalent n	équivalent m d'énergie	энергетический эквивалент
E 146	energy exchange	Energieaustausch m	transfert (échange) m d'énergie	обмен энергии, энергообмен
E 147	energy flow	Energiefluß m, Energieströmung f, Energiestrom m	flux m d'énergie, flux énergétique	энергетический поток, поток энергии
E 148	energy flow sheet	Energieflußbild n	flow-sheet m énergétique (d'énergie)	схема энергетических потоков (процесса)
E 149	energy fluctuation	Energieschwankung f	variation f de l'énergie, fluctuation f énergétique	разброс по энергии
E 150	energy gain	Energiegewinn m	gain m d'énergie, rendement m énergétique	выигрыш энергии
E 151	energy input	zugeführte Energie f	énergie f introduite	входная энергия
E 152	energy jump	Energiesprung m	changement m énergétique brusque	скачок энергии
E 153	energy level, energy term	Energiestufe f, Energieniveau n	niveau m énergétique	уровень энергии, энергетический уровень
	energy loss	s. E 143		
E 154	energy model	Energiemodell n, energetisches Modell n	modèle m énergétique	энергетическая модель
E 155	energy of activation, activation energy	Aktivierungsenergie f	énergie f d'activation	энергия активации
E 156	energy of formation	Bildungsenergie f	énergie f de formation	энергия образования
E 157	energy of steam	Dampfenergie f	énergie f de vapeur	энергия пара
E 158	energy output	Energieabgabe f, Energielieferung f	fourniture (livraison) f d'énergie	выделение энергии, энергоотдача
E 159	energy-poor	energiearm	pauvre d'énergie	низкоэнергетический
E 160	energy prices	Energietarife mpl	tarifs mpl énergétiques	тарифы энергии, энергетические тарифы
E 161	energy release	Energiefreigabe f, Energiefreisetzung f	apport m d'énergie	выделение (отдача) энергии
E 162	energy reserve, energy storage	Energiereserve f, Energiespeicherung f	réserve f d'énergie, réserve énergétique, accumulation f d'énergie	запас энергии
E 163	energy rich, high-energy	energiereich	riche en énergie	высокоэнергичный, высококалорийный, с большой энергией
E 164	energy-saving	energiesparend, mit geringem Energieverbrauch	à faible consommation d'énergie	энергосберегающий
E 165	energy-saving technology	energiesparende Technologie f	technologie f économisatrice d'énergie	энергосберегающая технология
E 166	energy source	Energiequelle f	source f énergétique (d'énergie)	источник энергии
E 167	energy state	Energiezustand m, energetischer Zustand m	état m énergétique	энергетическое состояние
	energy storage	s. E 162		
E 168	energy store	Energiespeicher m	accumulateur m d'énergie	накопитель (аккумулятор) энергии
	energy term	s. E 153		
E 169	energy theorem	Energiesatz m, Energieerhaltungssatz m	loi f de la conservation d'énergie	закон сохранения энергии
E 170	energy transfer	Energieübertragung f	transmission f (transport m) de l'énergie	передача (перенос) энергии
	energy transformation	s. E 137		
E 171	energy transport	Energietransport m	transport m de l'énergie	передача энергии, энергопередача, энерготранспорт
E 172	energy unit	Energieeinheit f	unité f énergétique (d'énergie)	единица энергии
E 173	energy utilization, use of energy, energy application	Energieverwendung f, Energieanwendung f, Energieausnutzung f	utilisation f d'énergie, application f de l'énergie	утилизация (использование, применение) энергии
E 174	energy yield	Energieumsatz m, Energieausbeute f	transformation f d'énergie, gain m énergétique	оборот (выход по) энергии

E 175	engineering flow sheet	technologisches Fließbild n, Betriebsschema n	schéma m de procédé, flow-sheet m technologique	технологическая схема
	engineering fluid-mechanics	technische Strömungsmechanik f	mécanique (technique) f des fluides	техническая гидромеханика
E 176	engineering stress	s. C 558		
E 177	engine fuel	Brennstoff m, Kraftstoff m, Treibstoff m	carburant m	топливо, горючее
E 178	enquire drawing (specification)	Anfragezeichnung f	dessin m de projet, dessin de demande	чертежи по требованию, техническая документация по требованию
E 179	enriching with oxygen	Sauerstoffanreicherung f	concentration f (enrichissement m) d'oxygène	обогащение кислородом
E 180	enrichment (on the tray)	Bodenverstärkung f	renforcement m du plateau	обогащение тарелки
E 181	enrichment	Anreicherung f	enrichissement m, concentration f	обогащение, концентрирование
	enrichment	s. a. A 479		
E 182	enrichment factor	Anreicherungsfaktor m	facteur m d'enrichissement	коэффициент обогащения
E 183	enrichment ratio	Anreicherungsgrad m	degré (taux) m d'enrichissement	степень обогащения
E 184	entering section	Eingangsquerschnitt m	section f d'entrée	сечение на входе
E 185	entering temperature difference	Eintrittstemperaturdifferenz f	différence f de température d'entrée	разность температур на входе
E 186	enthalpy	Enthalpie f, Wärmeinhalt m	enthalpie f	энтальпия, теплосодержание
E 187	enthalpy balance equation	Wärmebilanzgleichung f	équation f de balance d'enthalpie, équation de bilan m thermique	уравнение теплового баланса
E 188	enthalpy change	Enthalpieänderung f	changement m (variation f) d'enthalpie	изменение энтальпии
E 189	enthalpy curve (diagram)	Enthalpiediagramm n, Enthalpiekurve f	diagramme m (courbe f) d'enthalpie	энтальпийная диаграмма, кривая энтальпии
E 190	enthalpy-entropy chart	Enthalpie-Entropie-Diagramm n	diagramme m d'enthalpie-entropie	энтальпийно-энтропийная диаграмма
E 191	enthalpy of activation	Aktivierungsenthalpie f	enthalpie f d'activation	энтальпия активации
E 192	enthalpy of bonding	Bindungsenthalpie f	enthalpie f de liaison	энтальпия связи
E 193	enthalpy of dilution	Verdünnungsenthalpie f	enthalpie f de dilution	энтальпия разбавления
E 194	enthalpy of mixing	Mischungsenthalpie f	enthalpie f de mélange	энтальпия (теплота) смешения
E 195	enthalpy-water content diagram	Enthalpie-Wassergehalt-Diagramm m	diagramme m d'enthalpie-teneur en eau	диаграмма энтальпия-влагосодержание
E 196	entrained air	mitgerissene Luft f	air m entraîné	унесенный воздух
E 197	entrained droplet	mitgerissener Tropfen m	gouttelette f entraînée	унесенная капля
E 198	entrained liquid	mitgerissene Flüssigkeit f	liquide m entraîné	унесенная жидкость
E 199	entrained vapour	mitgerissener Dampf m	vapeur f entraînée	унесенный пар
E 200	entrainment	Mitreißen n	entraînement m, primage m	увлечение, унос
E 201	entrance loss	Eintrittsverlust m	perte f d'entrée	потери на входе
E 202	entropy	Entropie f, Wärmegewicht n	entropie f	энтропия
E 203	entropy balances	Entropiebilanzen fpl	balances fpl entropiques, bilans mpl d'entropie	энтропийные балансы, балансы энтропии
E 204	entropy change	Entropieänderung f	changement m d'entropie	изменение энтропии
E 205	entropy curve, entropy diagram	Entropiediagramm n, Entropiekurve f, Entropieänderungskurve f	courbe f (diagramme m) d'entropie	кривая изменения энтропии, энтропийная диаграмма
E 206	entropy density	Entropiedichte f	densité f d'entropie	плотность энтропии
	entropy diagram	s. E 205		
E 207	entropy of activation	Aktivierungsentropie f	entropie f d'activation	энтропия активации
E 208	entropy of evaporation	Verdampfungsentropie f	entropie f d'évaporation	энтропия при парообразовании
E 209	entropy of mixing	Mischungsentropie f	entropie f de mélange	энтропия смешения
E 210	entropy principle	Entropiesatz m	principe m d'entropie	энтропийный принцип
E 211	entropy-temperature relationship	Entropie-Temperatur-Beziehung f	relation f entropie-température	соотношение энтропии и температуры
E 212	entry side, inlet	Eintrittsseite f	côté m d'entrée	сторона входа, входная сторона
E 213	environmental variable	Umweltvariable f, Umgebungsvariable f	variable f d'environnement	параметр (переменная) окружающей среды
E 214	equalized pressure	Ausgleichsdruck m	pression f égalisée	разность потенциалов
E 215	equalizing line	Ausgleichsleitung f	conduit m de compensation, conduite f de l'égalisation	уравнительная линия
	equalizing vapour tank	s. B 28		
E 216	equation of continuity, continuity equation	Kontinuitätsgleichung f	équation f de continuité	уравнение сплошности (непрерывности)
E 217	equation of decay	Zerfallsgleichung f	équation f de décomposition (désintégration)	уравнение распада
E 218	equation of energy	Energie[erhaltungs]satz m, Energiegleichung f	principe m de la conservation de l'énergie	закон сохранения энергии, уравнение энергии
E 219	equation of motion	Bewegungsgleichung f	équation f de mouvement	уравнение движения
E 220	equation of state, state equation	Zustandsgleichung f	équation f d'état	уравнение состояния
E 221	equation-oriented procedure	gleichungsorientierte Vorgehensweise f	méthode f basée sur des équations	расчет основанный на учете структуры уравнений

	equilibrium	*s.* B 16		
E 222	equilibrium chart	Gleichgewichtsdiagramm *n*	diagramm *m* d'équilibre	диаграмма равновесия
E 223	equilibrium concentration	Gleichgewichtskonzentration *f*	concentration *f* d'équilibre	равновесная концентрация
E 224	equilibrium conditions	Gleichgewichtszustand *m*	état *m* d'équilibre	равновесное состояние
E 225	equilibrium constant	Gleichgewichtskonstante *f*	constante *f* d'équilibre	константа (постоянная) равновесия
E 226	equilibrium curve	Gleichgewichtskurve *f*	courbe *f* d'équilibre	кривая равновесия
E 227	equilibrium data	Gleichgewichtsdaten *pl*	données *fpl* d'équilibre	равновесные данные
E 228	equilibrium distillation	Gleichgewichtsdestillation *f*	distillation *f* équilibrée	перегонка в равновесных условиях
E 229	equilibrium equation	Gleichgewichtsgleichung *f*	équation *f* d'équilibre	уравнение равновесия
E 230	equilibrium of evaporation	Verdampfungsgleichgewicht *n*	équilibre *m* d'évaporation	равновесие испарения
E 231	equilibrium pressure	Ausgleichsdruck *m*, Gleichgewichtsdruck *m*	pression *f* égalisée (d'équilibre)	равновесное давление, давление в равновесной системе
E 232	equilibrium stage extractor	Gleichgewichtsstufenextraktor *m*, Extraktor *m* mit idealen Böden	extracteur *m* aux étages idéals	экстрактор с идеальными (равновесными) тарелками
E 233	equilibrium state	Gleichgewichtszustand *m*	état *m* d'équilibre	состояние равновесия
E 234	equilibrium still	ideale Kolonne *f*, Gleichgewichtsdestillationskolonne *f*	colonne *f* de distillation équilibrée	равновесный перегонный аппарат, идеальная ректификационная колонна
E 235	equilibrium temperature	Gleichgewichtstemperatur *f*	température *f* d'équilibre	температура равновесия (выдержки)
E 236	equilibrium water content	Gleichgewichtsfeuchte *f*	humidité *f* d'équilibre	равновесная влажность
E 237	equipartition	Gleichverteilung *f*	équipartition *f*	равномерное распределение
E 238	equipment, fitting out	Ausrüstung *f*, Anlage *f*	équipement *m*, matériel *m*, installation *f*	оборудование, оснащение
E 239	equipment arrangement	Ausrüstungsanordnung *f*	disposition *f* d'équipement	компоновка (расположение, расстановка) оборудования
E 240	equipment block	Ausrüstungsblock *m*	bloc *m* d'équipement	блок (элементарный комплекс) оборудования
E 241	equipment group	Ausrüstungsgruppe *f*	groupe *m* d'équipement	узел, группа оборудования, [сложный] комплекс оборудования
E 242	equipotential line	Äquipotentiallinie *f*	ligne *f* équipotentielle	эквипотенциальная линия
E 243	equivalent conductivity	Äquivalentleitfähigkeit *f*	conductivité *f* équivalente	эквивалентная проводимость
E 244	equivalent diameter	äquivalenter Durchmesser *m*	diamètre *m* équivalent	эквивалентный диаметр
E 245	equivalent weight	Äquivalentmasse *f*	poids *m* équivalent	эквивалентный вес
	erecting crane	*s.* S 670		
E 246	erecting tool	Montageeinrichtung *f*	installation *f* (dispositif *m*) de montage	монтажное приспособление, устройство для монтажа
E 247	erection drawing	Montagedokumentation *f*, Montageunterlagen *fpl*	documentation *f* de montage	монтажная техническая документация
E 248	erection frame	Montagegerüst *n*	échafaudage *m* de montage	монтажные леса
E 249	erection mast	Montagemast *m*	poteau *m* de montage	монтажная мачта
E 250	erection project	Montageprojekt *n*	projet *m* de montage	монтажный проект
E 251	ercection site	Montageort *m*, Bauplatz *m*	chantier *m*, site *m*	монтажная площадка, место монтажа
E 252	erection time	Montageablaufzeit *f*, Montagedauer *f*	temps *m* (durée *f*) de montage	время реализации монтажа
E 253	Ergun equation	Ergunsche Gleichung *f*	équation *f* d'Ergun	уравнение Эргуна
E 254	error of construction	Konstruktionsfehler *m*	défaut (vice) *m* de construction	конструктивная ошибка, ошибка конструкции
	escape	*s.* W 51		
E 255	escape pipe	Ausblaseleitung *f*	tuyau *m* à purgeur	выдувная (выхлопная) труба
E 256	escaping, leakage	Entweichen *n*, Leckage *f*	fuite *f*, échappement *m*	утечка, улетучивание
	establish/to	*s.* C 602		
	establishment	*s.* P 188		
E 257	esterification	Esterbildung *f*	formation *f* d'ester	эстерификация, образование сложного эфира
	estimation	*s.* P 318/9		
E 258	etching	Abätzen *f*, Korrosion *f*, Ätzen *n*	corrosion *f*, mordançage *m*	выщелачивание, травление
E 259	ethylene	Ethen *n*, Ethylen *n*	éthylène *m*	этилен
E 260	ethylene manufacturing process	Ethylengewinnungsverfahren *n*	processus *m* de production de l'éthylène	способ получения этилена
E 261	ethylene production	Ethylengewinnung *f*	production *f* d'éthylène	получение этилена
E 262	Eucken formula	Euckensche Formel *f*	formule *f* d'Eucken	формула Ейкена
E 263	Euler equation	Eulersche Gleichung *f*	équation *f* d'Euler	уравнение Эйлера
	evacuate/to	*s.* P 608		
	evaluation	*s.* A 528		
E 264	evaluation factor	Bewertungsfaktor *m*	facteur *m* d'évaluation	оценочный фактор, фактор оценки
E 265	evaluation methods	Bewertungsmethoden *fpl*	méthodes *fpl* d'évaluation	методы оценки

E 266	**evaluation model**	Bewertungsmodell *n*	modèle *m* d'évaluation	оценочная модель, модель оценки
E 267	**evaluation quantity**	Bewertungsgröße *f*	grandeur *f* d'évaluation	величина (параметр) оценки
	evaporate/to	*s.* V 66		
	evaporating	*s.* V 68		
	evaporating apparatus	*s.* E 297		
E 268	**evaporating boiler**	Abdampfkessel *m*, Abhitzekessel *m*	chaudière *f* à vapeur d'échappement	выпарной (концентрационный) котел
	evaporating column	*s.* F 199		
E 269	**evaporating cycle**	Verdampfungszyklus *m*	cycle *m* d'évaporation	цикл испарения
E 270	**evaporating disk**	Abdampfschale *f*	écuelle *f* d'évaporation	выпарная чашка
E 271	**evaporating efficiency**	Verdampfungsleistung *f*	débit *m* évaporatif, puissance *f* d'évaporation	мощность выпаривания (испарения)
E 272	**evaporating liquid**	verdampfende Flüssigkeit *f*	liquide *m* s'évaporant	испаряющаяся жидкость
E 273	**evaporating rate**	Siedegeschwindigkeit *f*, Verdampfungsrate *f*	degré *m* d'évaporation, vitesse *f* d'ébullition	скорость испарения
E 274	**evaporating region**	Verdampfungsbereich *m*	région *f* d'évaporation	зона парообразования
E 275	**evaporating step**	Verdampfungsstufe *f*	stade *f* d'évaporation	ступень испарения
E 276	**evaporating surface**	Verdampfungsfläche *f*	surface *f* d'évaporation	поверхность испарения
E 277	**evaporating temperature**	Verdampfungstemperatur *f*	température *f* de vaporisation	температура испарения
E 278/9	**evaporation,** volatilization, vaporization	Verflüchtigung *f*, Verdunstung *f*, Verdampfung *f*	volatilisation *f*, évaporation *f* lente, vaporisation *f*	испарение, выпаривание, улетучивание
	evaporation apparatus	*s.* E 297		
E 280	**evaporation coefficient,** coefficient of evaporation	Verdampfungszahl *f*	coefficient *m* d'évaporation	коэффициент испарения
E 281	**evaporation condenser**	Verdunstungskondensator *m*	condenseur *m* à évaporation	испарительный конденсатор
E 282	**evaporation cooling**	Verdunstungskühlung *f*	refroidissement *m* par évaporation	охлаждение испарением
E 283	**evaporation curve**	Verdampfungskurve *f*	courbe *f* de vaporisation	кривая испарения
E 284	**evaporation drying,** drying by evaporation	Verdampfungstrocknung *f*, Verdunstungstrocknung *f*	séchage *m* par évaporation	сушка нагреванием до температуры испарения жидкости, сушка нагретым воздухом
E 285	**evaporation height**	Verdunstungshöhe *f*	hauteur *f* évaporatoire	высота слоя испарения
E 286	**evaporation loss,** loss by evaporation	Verdampfungsverlust *m*	perte *f* par évaporation	потеря от испарения
E 287	**evaporation nucleus**	Verdampfungsnukleon *n*	nucléon *m* d'évaporation	центр кипения
E 288	**evaporation number**	Verdunstungszahl *f*	facteur *m* d'évaporation	коэффициент испаряемости
E 289	**evaporation process**	Verdampfungsprozeß *m*	processus *m* d'évaporation	процесс испарения
E 290	**evaporation region**	Dampfbildungszone *f*, Verdampfungsbereich *m*	région *f* d'évaporation, zone *f* de formation de vapeur	зона парообразования
E 291	**evaporation residue**	Abdampfrückstand *m*	résidu *m* d'évaporation	остатки выпаривания
E 292	**evaporation surface**	Verdunstungsfläche *f*	surface *f* d'évaporation lente	поверхность испарения
E 293	**evaporation time,** time of evaporation	Verdunstungszeit *f*, Verdampfungszeit *f*	temps *m* d'évaporation, temps *m* de vaporisation	время испарения (выпаривания)
E 294	**evaporation unit**	Verdampfstation *f*, Eindampfeinheit *f*, Verdampferkörper *m*	station *f* d'évaporation, étage *m* évaporatoire	выпарная станция, выпарка
E 295	**evaporative capacity**	Verdampfungsfähigkeit *f*	aptitude *f* à l'évaporation	испаряемость
E 296	**evaporative crystallization**	Verdampfungskristallisation *f*	cristallisation *f* par évaporation	кристаллизация выпариванием
E 297	**evaporator,** vaporizer, evaporation (evaporating) apparatus	Verdampfer *m*, Eindampfer *m*, Verdampfapparat *m*, Eindampfapparat *m*, Verdampfanlage *f*	vaporisateur *m*, évaporateur *m*, appareil *m* évaporatoire (à évaporer)	выпарной аппарат, испаритель
	evaporator	*s. a.* 1. C 523; 2. R 133		
E 298	**evaporator body**	Verdampferkörper *m*	appareil *m* d'évaporation	корпус выпарной установки
E 299	**evaporator condenser**	Brüdenkondensator *m*	condenseur *m* à buées	конденсатор влажного пара
E 300	**evaporator crystallizer**	Verdampfungskristallisator *m*	cristallisoir *m* à évaporation	выпарной кристаллизатор
E 301	**evaporator furnace**	Verdampfungsofen *m*	four *m* vaporisateur	испарительная печь
E 302	**evaporator pressure**	Verdampferdruck *m*	pression *f* d'évaporateur	давление в испарителе
E 303	**evaporator station**	Verdampfanlage *f*	station *f* d'évaporation	выпарная установка, выпарка
E 304	**even flow**	gleichmäßige Strömung *f*	écoulement *m* uniforme	равномерный поток, равномерное течение
	evolution	*s.* D 220		
E 305	**excess air**	Luftüberschuß *m*	excès *m* d'air	избыток воздуха
E 306	**excess base**	Basenüberschuß *m*	excès *m* de bases	избыток оснований
E 307	**excess vapour**	Überschußdampf *m*	surplus *m* de vapeur	избыточный пар
E 308	**excess vapour pressure**	Dampfüberdruck *m*	surpression *f* de vapeur, pression *f* effective	избыточное давление
E 309	**exchange capacity**	Austauschleistung *f*, Übertragungsleistung *f*	puissance *f* d'échange	обменная способность (мощность), производительность обмена
E 310	**exchange coefficient**	Austauschkoeffizient *m*	coefficient *m* d'échange	коэффициент обмена (передачи, переноса)
E 311/2	**exchange column**	Austauschkolonne *f*	colonne *f* échangeuse (d'échange)	колонна для обменных процессов

E 313	exchange equipment, exchanger	Austauschapparat m, Übertragungseinrichtung f, Austauscher m, Übertrager m	appareil m d'échange, échangeur m	обменный аппарат, обменник
E 314	exchange zone	Austauschzone f	zone f d'échange	зона обмена
	execution	s. C 603		
E 315	exergetic efficiency, exergetic grade	exergetischer Wirkungsgrad m, Gütegrad m	rendement m exergétique	эксергетический коэффициент полезного действия
E 316	exergetic element efficiency	exergetischer Elementwirkungsgrad m	rendement m élémentaire exergétique	эксергетический коэффициент полезного действия элемента
	exergetic grade	s. E 315		
E 317	exergetic system efficiency	exergetischer Systemwirkungsgrad m	rendement m exergétique du système	эксергетический коэффициент полезного действия системы
E 318	exergy balance	Exergiebilanz f	balance f (bilan m) exergétique	баланс эксергии
E 319	exergy flow (flux)	Exergiefluß m	flux m d'exergie	поток эксергии
	exhaust/to	s. 1. B 73; 2. B 153		
E 320	exhaust air	Absaugluft f, Austrittsluft f	air m évacué (de sortie)	выпускаемый воздух
E 321	exhaust air	Abluft f	air m sortant (d'évacuation, d'échappement)	отходящий (отводимый, отработанный) воздух
	exhaust air blower	s. E 325/6		
E 322	exhaust air flow	Abluftstrom m	courant m d'air sortant	поток выпускаемого воздуха
E 323	exhaust air heat regeneration	Abluftwärmerückgewinnung f	récupération f de chaleur sur l'air extrait	утилизация тепла отходящего воздуха
E 324	exhaust duct, extract duct	Abzugsrohr n, Entlüftungsleitung f, Abgasleitung f	gaine f (tuyau m) d'échappement, tube m d'évacuation, tuyau de décharge	вытяжная (дымовая, отводная) труба
	exhaust duct	s. a. S 791		
E 325/6	exhaust fan, exhaust air blower	Sauglüfter m, Abluftgebläse n	ventilateur m d'évacuation d'air, exhausteur m	вытяжной вентилятор
	exhaust gas	s. W 33		
E 327	exhaust gas chimney, ventilation stock, waste gas chimney	Abgasschornstein m, Schornstein m	cheminée f [à gaz d'échappement]	дымовая труба для отходящих газов
E 328	exhaust gas collector, waste gas collector	Abgassammler m	canal m collecteur des gaz perdus	выхлопной коллектор, коллектор выхлопных газов
E 329	exhaust gas dryer	Abgastrockner m	sécheur m des gaz d'échappement	сушилка, работающая на отработавшем газе
E 330	exhaust gas plant	Abgasanlage f	installation f d'exhaustion	система выпуска газа, газовыпускная система
E 331	exhaust gas purification	Abgasreinigung f	épuration f des gaz d'échappement	очистка выхлопного газа
E 332	exhaust hood, exhaustion device, puller	Abzugseinrichtung f	installation f d'évacuation des gaz	съемный механизм, вытяжная установка
E 333	exhausting device	Absaugvorrichtung f	installation f d'aspiration, dispositif m d'exhaustion	отсасывающее устройство, отсасыватель
	exhaustion device	s. E 332		
E 334	exhaustion machine	Absaugmaschine f	machine f d'aspiration	машина с отсасыванием отработавших газов
E 335	exhaust method	Absaugeverfahren n	procédé m d'aspiration, processus m d'exhaustion	способ бурения с отсосом буровой пыли
E 336	exhaust passage	Dampfausgang m; Dampfaustritt m	sortie f de vapeur	выход (выпуск) пара
E 337	exhaust pump	Absaugpumpe f	pompe f d'exhaustion, pompe d'évacuation	откачивающий насос
E 338	exhaust steam, waste steam	Abgasbrüden m, Abdampf m, Ausblasedampf m	vapeur f chaude (épuisée)	сдувка, мятый (отработанный, отходящий) пар
E 339	exhaust steam	Rückdampf m	vapeur d'échappement	контрпар, мятый пар
E 340	exhaust steam absorption refrigerating machine	Abdampfabsorptionskältemaschine f	machine f frigorifique d'absorption à vapeur d'échappement	абсорбционная холодильная машина
E 341	exhaust steam inlet	Abdampfeintritt m	entrée f de la vapeur épuisée	вход отработанного пара
E 342	exhaust steam pipe, exhaust vent pipe	Abdampfrohr m	tube m de vapeur d'échappement	пароотводная труба, трубопровод отработавшего пара
E 343	exhaust steam piping	Abdampfrohrnetz n	conduite f de vapeur épuisée (d'échappement)	сеть паровыпускных труб
E 344	exhaust steam pressure	Abdampfspannung f, Abdampfdruck m	pression f de la vapeur épuisée	давление отработавшего пара
	exhaust valve	s. 1. B 130; 2. O 127		
E 345	exhaust vapour utilization	Brüdenverwertung f	utilisation f de vapeur chaude	использование сокового пара
E 346	exhaust ventilation	Saugentlüftung f	ventilation f à exhaustion	вытяжная вентиляция
	exhaust vent pipe	s. E 342		
	existence	s. P 387		
E 347	exit conditions	Austrittsparameter mpl, Austrittsbedingungen fpl	conditions fpl (paramètres mpl) de sortie	параметры на выходе
E 348	exit pipe, outlet pipe	Ausgangsleitung f	tube f de décharge, tuyau m de sortie	исходящая линия

E 349	exit section	Austrittsquerschnitt m	section f de sortie	сечение на выходе
E 350	exit temperature	Abzugstemperatur f, Austrittstemperatur f	température f de sortie, température d'évacuation	температура на выходе
E 351	exothermal	wärmeabgebend, exotherm	exothermique	экзотермический
E 352	exothermal reaction, exothermic reaction	exotherme Reaktion f	réaction f exothermique	экзотермическая реакция
E 353/4	exothermic process exothermic reaction	exothermer Prozeß m s. E 352	processus m exothermique	экзотермический процесс
E 355	expanded air	expandierte (entspannte) Luft f	air m détendu	расширившийся воздух
E 356	expanded gas	expandiertes (entspanntes) Gas n, Entspannungsgas n	gaz m détendu	расширившийся газ
E 357	expanded vapour	expandierter (entspannter) Dampf m, Entspannungsdampf m	vapeur f détendue	дросселированный пар
E 358	expanding heat exchanger	Entspannungswärmeaustauscher m, Entspannungswärmeübertrager m	échangeur m de chaleur de détente	детандерный теплообменник
E 359	expansibility	Expansionsfähigkeit f, Dehnbarkeit f, Ausdehnbarkeit f	expansibilité f, pouvoir m expansif	расширяемость
	expansion	s. D 276		
E 360	expansion cooler	Entspannungskühler m	refroidisseur m d'expansion	холодильник мгновенного испарения
E 361	expansion factor	Expansionskoeffizient m	facteur m de détente, coefficient m d'expansion	коэффициент расширения
E 362	expansion layer	Ausdehnungsschicht f	couche f d'expansion	расширительный слой
E 363	expansion process	Entspannung f, Expansion f, Entspannungsprozeß m	processus m de détente, détention f	процесс расширения
E 364	expansion ratio	Expansionsgrad m	degré m de détente, degré d'expansion	степень расширения
E 365	expansion refrigeration	Expansionskühlung f, Entspannungskühlung f	réfrigération f de détente, réfrigération par expansion	охлаждение расширением
E 366	expansion temperature	Expansionstemperatur f, Entspannungstemperatur f	température f de détente	температура расширения
E 367	expansion turbine	Entspannungsturbine f	turbine f de détente	турбодетандер
E 368	expansion valve	Entspannungsventil n, Drosselventil n	soupape f de détente	устройство для сброса давления
E 369	expansion vessel	Expansionsbehälter m, Ausdehnungsgefäß n	vase m de détente, vase d'expansion	расширительный бак
E 370	expansion volume	Expansionsvolumen n	volume m de détente	расширяющийся объём
E 371	expansion work	Expansionsarbeit f, Entspannungsarbeit f	travail m de détente	работа расширения
E 372	expectation value	Erwartungswert m	valeur f escomptée	ожидаемое значение
	expenditure	s. C 610		
E 373	expenditure of force	Kraftaufwand m	dépense f d'énergie, énergie f dépensée	расход силы
E 374	expenditure of work	Arbeitsaufwand m	travail m nécessaire	затрата на работу
	expense	s. C 610		
	experiment/to	s. T 319		
E 375	experimental conditions	experimentelle Bedingungen fpl	conditions fpl expérimentales	экспериментальные условия
E 376	experimental data	experimentelle Daten pl	données fpl expérimentales	экспериментальные данные
	experimental model	s. P 147		
	experimental unit	s. P 148		
E 377	experimental value	experimenteller Wert m	valeur f expérimentale	опытные данные, результат опыта
E 378	experimental work	experimentelle Arbeit f	travail m expérimental	экспериментальная работа
E 379	experimental work	experimentelle Untersuchung f	étude f expérimentale	экспериментальное исследование
	explode/to	s. D 217		
E 380	exploitation, mining loss	Abbauverluste mpl	pertes fpl d'exploitation	потери полезного ископаемого
E 381	explosion, detonation, deflagration	Explosion f, Detonation f, Verpuffung f	explosion f, détonation f, déflagration f	взрыв, детонирование, вспышка
E 382	explosion chain	Explosionskette f	chaîne f d'explosion	взрывная цепь
E 383	explosion limit	Explosionsgrenze f	limite f d'explosibilité	предел взрыва
E 384	explosion pressure	Explosionsdruck m	pression f de (à) l'explosion	взрывное давление
E 385	explosion-proof	explosionssicher, explosionsgeschützt	inexplosible, à l'épreuve de l'explosion	взрывобезопасный
E 386	explosion protection	Explosionsschutz m	protection f antidéflagrante (contre les explosions)	взрывозащита
E 387	explosion range	Explosionsbereich m	intervalle m d'explosibilité	область взрывоопасности
E 388	explosion reaction	Explosionsreaktion f	réaction f d'explosion	взрывная реакция
E 389	explosive	Explosivstoff m, Sprengstoff m, Sprengmittel n	explosif m, matière f explosive	взрывчатое вещество
E 390	explosive mixture	Explosionsgemisch n, zündfähiges Gemisch n	mélange m détonant (explosif, explosible)	взрывчатая (взрывоопасная) смесь
E 391	explosiveness	Explosionsgefahr f	danger m d'explosion	взрывоопасность
E 392	exponential distribution	Exponentialverteilung f	distribution f exponentielle	показательное распределение
E 393	exponential function	Exponentialfunktion f	exponentielle f	показательная функция

E 394	extended circular matrix	erweiterte Kreismatrix f	matrice f circulaire étendue	расширенная матрица циклов
E 395	extensional viscosity	Dehnungsviskosität f	viscosité f d'allongement	вязкость растяжения
E 396	external airflow	Oberflächenluftstrom m	courant m d'air extérieur	внешний поток воздуха
E 397	external cooling	äußere Kühlung f, Außenkühlung f	refroidissement m extérieur	наружное охлаждение
E 398	external corrosion	Außenkorrosion f	corrosion f extérieure	внешняя коррозия
E 399	external energy	äußere Energie f	énergie f extérieure	внешняя энергия
E 400	external equalizer line	äußere Druckausgleichsleitung f	conduit m d'égalisation de pression extérieur	наружная уравнительная линия
E 401	extinguisher	Löschgerät n, Feuerlöschgerät n	outillage m des pompiers, extincteur m, boîte f à extinction	огнетушитель
	extract duct	s. E 324		
E 402	extracting agent	Extraktionsmittel n	moyen m d'extraction, solvant m pour extraire	экстракционное средство
E 403	extracting plant	Extraktionsanlage f	installation f d'extraction	экстракционное устройство
E 404	extracting tower	Extraktionsturm m, Extraktionskolonne f	colonne (tour) f d'extraction	экстракционная колонна
E 405	extraction, stripping	Extrahieren n, Auslaugung f, Extraktion f	extraction f	экстрагирование, извлечение
E 406	extraction apparatus	Extraktionsapparat m	appareil m d'extraction	экстрактор, экстракционный аппарат
E 407	extraction coefficient	Extraktionskoeffizient m	coefficient m d'extraction	коэффициент извлечения
E 408	extraction constant	Extraktionskonstante f	constante f d'extraction	константа извлечения
E 409	extraction filter, filter for insertion	Einsatzfilter n	filtre m d'extraction	вставной фильтр
E 410	extraction of grease	Entfettung f	dégraissage m	обезжиривание
E 411	extraction process	Extraktionsprozeß m, Extraktionsverfahren n	procédé m d'extraction	процесс экстрагирования
E 412	extraction system	Extraktionssystem n	système m d'extraction	система вытягивания, экстракционная система
	extractive	s. E 414		
E 413	extractive distillation	Extraktivdestillation f	distillation f extractive	экстрагирующая перегонка, отгон адсорбата из жидкости
E 414	extractive matter, extractive	Extraktivstoff m, Extrakt m, Auszug m	matière f extractive, extrait m	экстракт, экстракт[ив]ное вещество
E 415	extractive stripper	Extraktivstripper m	stripper m pour désorption d'extrait	колонна для десорбирования экстракта
E 416	extractor	Extraktor m, Extrakteur m, Extraktionsapparat m	extracteur m	экстрактор, экстракционный аппарат
E 417	extrudate	Extrudat n, Strangpreßling m	produit m d'extrusion, extrudat m	экструдат
E 418	extruder, extruding press	Strangpresse f, Extruder m, Schneckenpresse f	presse f à filage (extrusion), extrudeuse f, boudineuse f	червячный (прутковый, шнековый) пресс, пресс, непрерывно выдавливающий изделие, экструдер
E 419	extruder head	Extruderkopf m, Strangpressenkopf m	tête f d'extrudeuse, tête de boudineuse	формующий инструмент экструдера
	extruding press	s. E 418		
E 420	extrusion	Strangpressen n, Extrusionsverfahren n	extrusion f, filage m	прессование на червячном прессе, прутковое прессование, выдавливание, экструзия
	extrusion	s. a. P 392		
E 421	Eyring kinetic theory	Eyringsche kinetische Theorie f	théorie f cinétique d'Eyring	кинетическая теория Эйринга
E 422	Eyring model	Eyringsches Modell n	modèle m d'Eyring	модель Эйринга

F

F 1	fabricating technique	Herstellungsverfahren n	procédé m de fabrication	метод производства, способ изготовления
	facility	s. J 16		
	factory	s. P 188		
	failure	s. 1. B 253; 2. D 58		
F 2	failure behaviour, behaviour during breakdown	Ausfallverhalten n	comportement m de défaillance	характеристики потери работоспособности
F 3	failure probability	Ausfallwahrscheinlichkeit f	probabilité f de défaillance	вероятность отказа
F 4	failure probability density distribution	Ausfallwahrscheinlichkeitsdichteverteilung f	distribution f probable des défaillances	распределение плотности вероятности отказа
F 5	failure rate	Ausfallrate f	taux m de panne (défaillance)	частота отказа
F 6	falling film	fallender Film n, Fallfilm m, Rieselfilm m	film m descendant	падающая пленка
F 7	falling film column, wetted-wall tower	Fallfilmkolonne f, Rieselfilmkolonne f	tour (colonne) f à parois humectées	пленочная колонна
F 8	falling film cooler	Rieselkühler m, Fallfilmkühler m	refroidisseur m à ruissellement (film descendant)	холодильник с падающей пленкой
F 9	falling film vaporizer	Fallfilmverdampfer m	évaporateur m à film descendant	испаритель с падающей пленкой

	English	German	French	Russian
	fan	s. B 147		
F 10	fan air cooler	Gebläseluftkühler m	refroidisseur m d'air à ventilateur	воздухоохладитель с вентилятором
F 11	fan-assisted fin coil	Gebläserippenkühler m	serpentin m à ailettes avec ventilateur	змеевик из оребренных труб с вентиляторным обдувом
F 12	fan cooler	Ventilatorkühler m	refroidisseur m à ventilateur	охладитель с вентилятором
F 13	fan cooling	Gebläsekühlung f	refroidissement m à ventilation	воздушное охлаждение с помощью вентилятора
F 14	fan efficiency	Lüfterwirkungsgrad m	rendement m de ventilateur	коэффициент полезного действия вентилятора
F 15	fan pressure	Lüfterdruck m, Gebläsedruck m	pression f de ventilateur (soufflage)	напор вентилятора
F 16	fan system	Lüfteranlage f	installation f de ventilation	вентиляторная установка
F 17	fast breeding reactor	schneller Brüter m, Schnellbrutreaktor m	pile f couveuse rapide	реактор с расширенным воспроизводством на быстрых нейтронах
	fastening	s. A 602		
F 18/20	fatigue of material	Materialermüdung f	fatigue f de matériel	усталость материала
	feare/to	s. B 317		
F 21	feed amount	Aufgabemenge f	quantité f (débit m) d'alimentation	производительность, дебит, расход, подача
F 22	feed apparatus	Aufgabeapparat m, Speiser m	appareil m alimentateur	механизм подачи, питатель
F 23	feedback	Rückkopplung f, Rückleitung f	réaction f, couplage m en arrière	обратная связь, обратное питание
F 24	feed belt, belt, feeder	Aufgabeband n	bande f de chargement	ленточный питатель, загрузочная лента
F 25	feed bin	Aufgabebunker m	trémie f de chargement, soute f d'alimentation	бункер-питатель, загрузочный (расходный) бункер
F 26	feed chamber, inlet chamber	Einlaufkammer f	chambre f d'entrée	входная камера, аванкамера
F 27	feed chute, feed spout	Speiserinne f	rigole f d'alimentation	питающий (питательный) желоб
F 28	feed composition	Zulaufzusammensetzung f	composition f d'alimentation	состав исходного сырья (продукта)
F 29	feed concentration	Zulaufkonzentration f	concentration f d'entrée, concentration d'alimentation	концентрация питания, состав питающего потока
F 30	feed controller	Aufgaberegler m, Zulaufregler m	système m régulateur de chargement	регулятор загрузки, автоматический питатель
F 31	feed device, feeder	Zuteileinrichtung f, Zuteiler m, Aufgabevorrichtung f, Aufgeber m, Eintragvorrichtung f, Beschickungsvorrichtung f, Einspeisevorrichtung f, Speiser m, Feeder m, Dosiervorrichtung f	dispositif m d'alimentation, appareil m d'alimentation, alimentateur m, feeder m, dispositif de dosage	питатель, загрузочное устройство, дозатор
	feeder	s. a. 1. C 685; 2. F 24; 3. L 51		
F 32	feedforward control	feedforward-Steuerung f	commande f à l'aide d'un modèle	управление с помощью модели
F 33	feedforward optimization (optimizing)	„Vorwärts"-Optimierung f	optimisation f feed-forward	оптимизация с помощью модели
F 34	feed gas	Ausgangsgas n	gaz m d'alimentation	исходный газ
F 35	feed gas dryer	Speisegastrockner m	sécheur m de gaz d'alimentation	сушилка для питательного газа
F 36	feed hopper, feeding hopper	Speisetrichter m, Einlauftrichter m, Einwurftrichter m, Aufgabetrichter m	entonnoir m d'alimentation, trémie f de chargement	питающая (загрузочная) воронка
F 37	feeding	Einspeisung f, Beschickung f, Materialzuführung f, Einspeisen n, Einfüllen n	charge f, chargement m, alimentation f, arrivée f	питание, подача [материала], загрузка, приток
	feeding	s. a. D 149		
F 38/9	feeding disk	Aufgabeteller m	table f de chargement	загрузочная тарелка
	feeding hopper	s. F 36		
F 40	feeding mixture	Beschickungsgemisch n, Einspeisemischung f	mélange m d'alimentation	питательная смесь
F 41	feeding point	Einspeisungsstelle f, Zulaufstelle f	point m d'alimentation	место ввода (питания, введения исходных веществ)
	feeding screw	s. F 50		
	feeding system	s. F 53		
F 42	feed injector	Kesselspeisepumpe f	pompe f d'alimentation de chaudière	питательный насос для котла
F 43	feed line	Zuleitung f, Speiseleitung f	conduite f d'alimentation, tube m d'arrivée, feeder m, alimentateur m	питающий трубопровод, питательная труба
F 44	feed plate	Zulaufboden m, Einlaufboden m	plateau m d'alimentation	тарелка для подачи, питающая тарелка колонны
F 45	feed pressure	Zulaufdruck m	pression f de l'alimentation, pression d'entrée	давление питания (исходного продукта)

F 46	**feed pump,** charge pump	Speise[wasser]pumpe f, Zulaufpumpe f	pompe f alimentaire (d'alimentation, d'injection)	питающий (нагнетательный, питательный) насос
F 47	**feed pump**	Förderpumpe f	pompe f de circulation	подъемный насос
F 48	**feed rate**	Durchsatz m, Einsatzmenge f, Einsatzrate f, Zulaufrate f	débit m, vitesse f (débit) d'alimentation	расход, скорость питания
F 49	**feed screen**	Aufgabesieb n	tamis m de chargement	грохот-питатель
F 50	**feed screw,** feeding screw	Speiseschnecke f, Zulaufschnecke f, Aufgabeschnecke f	vis f alimentaire (d'alimentation)	шнековый питатель
F 51	**feed solvent**	Speiselösungsmittel n	solvant m d'alimentation	питательный растворитель
	feed spout	s. F 27		
	feedstock	s. R 63		
F 52	**feedstock utilization,** material utilization (economy)	Stoff[aus]nutzung f	utilisation f de matière	использование вещества (материала), эффективность использования веществ
F 53	**feed system,** feeding system	Beschickungssystem n, Speisevorrichtung f	système m de charge, dispositif m d'alimentation	система подачи
F 54	**feed tank**	Speisebehälter m	tank m d'alimentation	питательный бак
F 55	**feed temperature**	Einsatzprodukttemperatur f, Zulauftemperatur f	température f de l'alimentation	температура исходного продукта
F 56	**feed tube**	Beschickungsrohr n	goulotte f d'alimentation	питательная труба
F 57	**feed water**	Speisewasser n	eau f d'alimentation	питательная вода
F 58	**feed-water heater,** economizer	Speisewasservorwärmer m	réchauffeur m d'eau d'alimentation	подогреватель питательной воды, экономайзер
F 59	**feed-water heating**	Speisewasservorwärmung f	réchauffement m de l'eau d'alimentation	предварительный подогрев воды для парогенератора
F 60	**feed zone**	Beschickungszone f	zone f d'alimentation	зона загрузки (питания)
F 61	**female mold method**	Negativverfahren n	procédé m à moule négatif	негативное формование
F 62	**ferment**	Ferment n	ferment m, enzyme m	фермент, энзим
F 63	**fermentation**	Fermentation f, Fermentierung f	fermentation f	ферментация
F 64	**fermentation method**	Fermentierung f, Fermentationsverfahren n, Faulverfahren n	procédé m de fermentation	метод ферментации (брожения)
F 65	**fertilizer**	Düngemittel n, Dünger m, Mineraldünger m	engrais m [minéral]	удобрение
F 66	**fibre pattern**	Faserstruktur f	structure f de fibre	волокнистая структура, структура волокна
F 67	**fibrous material**	Faserstoff m	matière f textile (fibreuse)	волокнистый материал
F 68	**Fick's diffusion law,** Fick's law of diffusion	Ficksches Gesetz (Diffusionsgesetz) n	loi f [de diffusion] de Fick	закон [диффузии] Фикка
F 69	**filled polymer,** loaded polymer	gefülltes Polymer n, gefüllter Plast m	polymère (plastique) m chargé	наполненный полимер
F 70	**filler, filling**	Füllmittel n, Füllstoff m	matériel m de remplissage, matière f inerte, charge f	наполнитель
F 71	**filling,** topping up	Auffüllung f, Füllen n	remplissage m	наполнение
F 72	**filling control**	Füllkontrolle f	contrôle m de remplissage	контроль наполнения
F 73	**filling pressure**	Fülldruck m	pression f de remplissage	давление наполнения
F 74	**filling station**	Abfüllanlage f	station (installation) f de remplissage	заправочный пункт
F 75	**film,** sheeting	Folie f	feuille f, lamelle f, film m	пленка
F 76	**film boiling**	Filmverdampfung f	ébullition f de film	пленочное кипение
F 77	**film casting**	Filmgießen n	coulage m de film	отливание пленки
F 78	**film column**	Filmkolonne f	colonne f à film	пленочная колонна
F 79	**film condensation,** filmwise condensation	Filmkondensation f	condensation f en film	пленочная конденсация
F 80	**film cooling**	Filmkühlung f	refroidissement m à film	пленочное охлаждение
F 81	**film cooling tower**	Filmkühlturm m	tour f réfrigérante à évaporation en film	пленочная градирня
F 82	**film drying**	Filmtrocknung f	séchage m de film	сушка поверхностных слоев
F 83	**film evaporator**	Filmverdampfer m	évaporateur m à film	пленочный испаритель
F 84	**film former**	Filmbildner m	matière f formant la pellicule	пленкообразователь
F 85	**film forming**	filmbildend	filmogène	пленкообразующий
	film reactor	s. W 158/9		
F 86	**film thickness**	Filmdicke f	épaisseur m de couche	толщина слоя (пленки)
F 87	**film-type flow**	Filmströmung f	écoulement m en forme de film	пленочный поток
	filmwise condensation	s. F 79		
F 88	**filter/to**	filtern, filtrieren, [durch]seihen	filtrer	фильтровать
F 89	**filter**	Filter n, Filtereinsatz m	filtre m	фильтр
F 90	**filter aid**	Filterhilfsmittel n, Filterzusatz m	adjuvant m de filtration	фильтровальное вспомогательное вещество, вспомогательный порошок для фильтрования
F 91	**filter bed**	Filterschicht f, Filterbett n	couche f filtrante	фильтрующий слой
F 92	**filter bottom**	Filterboden m	plateau m de filtre	дно фильтра
	filter box	s. F 96		
F 93	**filter cake,** press cake	Filterkuchen m	tourteau m de filtration	уплотненный осадок на фильтре, лепешка, кек, кухен, отжатый осадок

F 94	**filter candle**	Filterkerze *f*	chandelle *f* filtrante	фильтровальная свеча
F 95	**filter cell**	Filterelement *n*	élément *m* filtrant	ячейка фильтра
F 96	**filter chamber,** filter box	Filterzelle, *f*, Filterkammer *f*	cellule *f* filtrante, chambre *f* de filtre	ячейка фильтра
F 97	**filtered,** strained	abgefiltert, filtriert	filtré	фильтрованный
F 98	**filtered air**	gefilterte Luft *f*	air *m* filtré	фильтрованный воздух
F 99	**filter efficiency**	Filterwirkungsgrad *m*	rendement *m* de filtre	эффективность фильтра, коэффициент полезного действия фильтра
F 100	**filter element**	Filtereinsatz *m*	élément *m* filtrant	вставная часть фильтра
F 101	**filter equipment**	Filterausrüstung *f*	installation *f* (équipement *m*) de filtration	оборудование фильтра
	filter flask	*s.* S 793		
	filter for insertion	*s.* E 409		
F 102	**filter frame**	Filterrahmen *m*	cadre *m* de filtre	фильтровальная рама
F 103	**filter gravel**	Filterkies *m*	gravier *m* filtrant	фильтровальный гравий
F 104	**filtering,** filtration	Filtrieren *n*, Filtration *f*, Filterung *f*	filtration *f*, filtrage *m*	фильтрование
F 105	**filtering aids**	Filterhilfen *fpl*	moyens *mpl* de secours pour le filtrage, adjuvant *m* de filtration	вспомогательные средства фильтрации
F 106	**filtering area,** filtering surface	Filtrierfläche *f*, Filterfläche *f*	surface *f* filtrante	площадь фильтрации, фильтрующая площадь
F 107	**filtering cloth**	Filtertuch *n*	tissu *m* filtrant	салфетка фильтрпресса, фильтровальная ткань
F 108	**filtering material,** filter medium	Filtermaterial *n*, Filterstoff *m*	masse *f* filtrante	фильтрующий (фильтрационный) материал
F 109	**filtering off**	Abfiltrieren *n*	séparation *f* par filtration, filtration *f*	отфильтрование
F 110	**filtering sieve**	Filtersieb *n*	tamis *m* filtrant, crible *m* filtrant	сетка фильтра
	filtering surface	*s.* F 106		
F 111	**filtering time**	Filterzeit *f*, Filtrierzeit *f*	temps *m* de filtrage	время фильтрации
F 112	**filter mass**	Filtermasse *f*	masse *f* filtrante	фильтрующая масса
	filter medium	*s.* F 108		
F 113	**filter mud**	Filterschlamm *m*	boues *fpl* de filtration	шлам с фильтра
F 114	**filter paper**	Filtrierpapier *n*, Filterpapier *n*	papier *m* fluant (à filtre, emporéthique)	фильтровальная бумага
F 115	**filter plate**	Filterplatte *f*	plaque *f* filtrante (de filtre)	фильтрующая пластина, пластина фильтрпресса
F 116	**filter press,** plate filter	Filterpresse *f*, Plattenfilter *n*	filtre-presse *m*, presse *f* à filtres, filtre *m* à plaques	фильтрпресс
F 117	**filter pressure**	Filterdruck *m*, Filtrierdruck *m*	pression *f* au filtre, pression de filtration	напор на фильтр
F 118	**filter pressure drop**	Filterdruckverlust *m*	chute *f* de pression au filtre	потеря напора на фильтре
F 119	**filter pump**	Filterpumpe *f*	pompe-filtre *f*	фильтровальный (водоструйный) насос
F 120	**filter run**	Filterlaufzeit *f*, Filterzyklus *m*	période *f* entre deux lavages d'un filtre	время пробега фильтра, цикл работы фильтра
F 121	**filter screen**	Filtersieb *n*, Filtergewebe *n*, Filterstoff *m*	tissu *m* filtrant, étoffe *f* à filtrer	сетка фильтра
F 122	**filter tower**	Filterturm *m*	tour *f* à filtre	фильтровальная башня
F 123	**filter wash**	Filterwäsche *f*	lavage *m* du filtre	промывание на фильтре
F 124	**filtrate**	Filtrat *n*	filtrat *m*, produit *m* filtré	фильтрат
F 125	**filtrate liquid (water)**	Filterwasser *n*, Filtratwasser *n*	eau *f* de filtration	фильтрационная вода
	filtration	*s.* F 104		
F 126	**filtration plant**	Filteranlage *f*	installation *f* de filtrage	фильтровальная установка, установка для фильтрования
F 127	**filtration rate**	Filtriergeschwindigkeit *f*, Filtergeschwindigkeit *f*	vitesse *f* de filtration (passage)	скорость фильтрации
F 128	**filtration residue**	Filterrückstand *m*	résidu *m* de filtrage	остаток после фильтрации
F 129	**final boiling point**	Siedeende *n*, Siedeschluß *m*	fin *f* d'ébullition	конец (температура конца) кипения
F 130	**final concentration**	Endkonzentration *f*	concentration *f* finale	конечная концентрация
F 131	**final condenser**	Endkühler *m*	refroidisseur *m* final	конечный конденсатор, конденсатор, поставленный после дефлегматора
	final control element	*s.* A 192		
F 132	**final conversion**	Endumsatz *m*	transformation *f* finale, taux *m* de conversion final	конечная степень превращения
F 133	**final cooling**	Endkühlung *f*, Schlußkühlung *f*	refroidissement *m* final	конечное охлаждение
F 134	**final filtering**	Nachfiltration *f*	filtration *f* ultérieure, postfiltrage *m*	дополнительная фильтрация
F 135	**final humidity (moisture)**	Endfeuchtigkeit *f*, Endfeuchte *f*	humidité *f* finale	конечная влажность
F 136	**final pressure,** output pressure	Abnahmedruck *m*	pression *f* de réception	приемное (отдаваемое) давление

F 137	final temperature	Endtemperatur f	température f finale	конечная температура
F 138	final test	Endabnahme f	réception f finale	конечное приемное испытание, конечный контроль при приемке
F 139	fin chiller	berippter Kühler m, Rippenkühler m	refroidisseur m muni d'ailettes	охладитель с оребренными трубами
F 140	fin configuration	Rippenanordnung f	arrangement m d'ailettes	конфигурация ребер
F 141	fine crusher	Mühle f, Feinzerkleinerer m, Feinbrecher m	moulin m, broyeur m [fin], pulvérisateur m	дробилка тонкого (для мелкого) измельчения
F 142	fine crushing, fine grinding	Feinzerkleinerung f, Feinbrechen n, Feinmahlung f	concassage m, broyage m fin	тонкое измельчение
F 143	fin efficiency	Rippenwirkungsgrad m	efficacité f d'ailette	коэффициент эффективности ребра
F 144	fine filtration	Feinfiltrierung f	filtration f fine	тонкая фильтрация
F 145	fine fraction	Feinfraktion f, Feingut n	fraction f fine	тонкая фракция
F 146	fine grain	Feinkorn n, Feinzeug n	grain m fin	мелкое зерно
F 147	fine-grained	feinkörnig	à grain fin (serré)	мелкозернистый
	fine grinding	s. F 142		
F 148	fine jigging	Feinkornsetzarbeit f	sédimentation f des grains fins	отсадка мелкого материала
F 149	fine-meshed	feinmaschig, engmaschig	à mailles fpl serrées, à petites mailles	с мелкими ячейками
F 150	fineness of grain	Kornfeinheit f	finesse f de grain	мелкозернистость
F 151	fine ore bin	Feinerzbunker m	trémie f pour fines de minérai	бункер рудной мелочи
F 152	fine screening	Fein[ab]siebung f	tamisage (criblage) m fin	тонкий отсев (рассев)
F 153	Finger strain tensor	Fingerscher Spannungstensor m	tenseur m de déformation de Finger	тензор напряжения по Фингеру
F 154	fining agent	Klärmittel n	clarifiant m	осветлитель, осветляющее вещество
	finish/to	s. C 457		
F 155	finishing, subsequent treatment	Nachbereitungsoperation f	traitement m ultérieur	последующая (окончательная) обработка, отделка
	finishing	s. a. D 500		
F 156	finishing machine	Feinmühle f (Papier)	broyeur m, moulin m fin	рафинер
F 157	finned coil cooler	Rippenrohrkühler m	batterie f des tuyaux ailettés	змеевик из оребренных элементов
F 158	finned element	Rippenelement n	élément m ailetté	оребренный элемент
F 159	finned evaporator	Rippenrohrverdampfer m, berippter Verdampfer m	évaporateur m à tuyaux à ailettes, évaporateur muni d'ailettes	испаритель с оребренными трубами
F 160	finned heat exchanger	Rippenrohrwärmeübertrager m	échangeur m thermique à tuyaux à ailettes	теплообменник из оребренных труб
F 161	finned pipe	Rippenrohr n	tube (tuyau) m à ailettes	оребренная труба
F 162	finned surface	berippte Oberfläche f, Rippenfläche f	surface f ailettée	оребренная поверхность
F 163	fin spacing	Rippenabstand m	fente f des ailettes	зазор между ребрами
F 164	fin thickness	Rippenbreite f	épaisseur m d'ailette	толщина ребра
F 165	fire and explosion hazard	Feuer- und Explosionsgefahr f	danger m d'incendie et d'explosion	опасность пожара и взрыва, пожаро- и взрывоопасность
F 166	fire fighting installation	Feuerschutzanlage f	installation f antifeu (ignifuge)	противопожарное оборудование
F 167	fireproof, refractory	feuerfest, feuersicher	réfractaire, ignifuge, ininflammable, résistant au feu	огнеупорный, жаростойкий
F 168	fireproofness	Feuerbeständigkeit f	résistance f au feu, fixité f	огнестойкость
F 169	fire protection	Brandschutz m	protection f contre le feu, protection contre les incendies	пожарная охрана, противопожарные мероприятия
	fire surface	s. H 110		
F 170	fire tube, heater flue, hot tube	Heizrohr n	tube m de fumée, tuyau m de chauffage	дымогарная труба
F 171	fire-water distribution system	Löschwasserverteilungssystem n	système m de distribution pour l'eau à éteindre l'incendie	распределительная система пожарного водоснабжения
	first coat	s. P 482		
F 172	first evaporator	Erstverdampfer m	évaporateur m premier	первый корпус выпарки
F 173	first law of thermodynamics	erster Hauptsatz m der Thermodynamik	première loi f principale de la thermodynamique	первый закон термодинамики
F 174	first run[nings], fore-runnings	Vorlauf m (Alkoholdestillation)	avant-coulant m, eau-de-vie f première	головной погон
F 175	Fischer-Tropsch-synthesis	Fischer-Tropsch-Synthese f	synthèse f de Fischer-Tropsch	синтез Фишера-Тропша
F 176	fissile material, atomic fuel	Spaltmaterial n	matière f fissile	материал расщепления
F 177	fissility	Spaltbarkeit f	fissilité f, aptitude f au dédoublement, aptitude de clivage	расщепляемость, спайность
	fission	s. C 818		
F 178	fission energy	Spaltungsenergie f	énergie f de fission	энергия расщепления
F 179	fission process	Spaltprozeß m	procédé m de fission	процесс разделения (расщепления)

F 180	**fission product,** product of fission	Spaltprodukt *n*	produit *m* de fission	продукт расщепления (крекинга, деления)
F 181	**fission yield**	Spaltungsausbeute *f*	gain *m* de fission	выход продуктов деления
	fitting	*s.* A 566		
	fitting out	*s.* E 238		
F 182	**fittings**	Armaturen *fpl*	garnitures *fpl*, robinetterie *f*	арматуры
F 183	**fixed bed**	Festbett *n*	lit *m* fixe	неподвижный слой
F 184	**fixed-bed absorber**	Festbettabsorber *m*	absorbeur *m* à lit fixe	абсорбер с неподвижным слоем
F 185	**fixed-bed adsorption**	Festbettadsorption *f*	adsorption *f* à lit fixe	адсорбция на неподвижном слое
F 186	**fixed-bed catalyst**	ortsfester Katalysator *m*, Festbettkatalysator *m*	catalyseur *m* fixé	неподвижный (стационарный) катализатор
F 187	**fixed-bed process**	Festbettprozeß *m*	procédé *m* à lit fixe	метод с неподвижным слоем катализатора
F 188	**fixed-bed reactor,** packed-bed reactor	Festbettreaktor *m*	réacteur *m* à lit fixe	реактор с неподвижным слоем [катализатора]
F 189	**fixed catalyst**	Feststoffkatalysator *m*	catalyseur *m* solide	твердый катализатор
F 190	**fixed layer**	Festschicht *f*	couche *f* fixe	неподвижный слой
F 191	**fixing agent,** agglutinant, binder, adhesive	Bindemittel *n*, Andicker *m*	adhésif *m*, matière *f* agglutinante, liant *m*	связующее вещество
	fixture	*s.* J 16		
F 192	**flameproof**	unbrennbar, unentflammbar	ininflammable, non inflammable, non-combustible	огнестойкий
F 193/4	**flame temperature**	Flammentemperatur *f*	température *f* de flamme	температура пламени
	flammability	*s.* I 133		
F 195	**flammable**	[ent]flammbar	inflammable	воспламеняемый, воспламеняющийся
F 196	**flange**	Flansch *m*, Kragen *m*, Bund *m*	rebord *m*, collet *m*, bride *f*, collerette *f*	фланец
	flare	*s.* T 226		
F 197	**flare device (system)**	Fackelanlage *f*, Fackelsystem *n*	installation *f* (système *m*) de torche	факельная установка
	flash/to	*s.* V 66		
F 198	**flash boiler,** high-speed evaporator	Schnellverdampfer *m*	chaudière *f* à vaporisation rapide	выпарной аппарат с интенсивной циркуляцией
F 199	**flash column,** evaporating column, flash tower	Entspannungskolonne *f*, Verdampfungssäule *f*	colonne *f* flash (de distillation momentanée par détente)	испарительная колонна
F 200	**flash cooler,** jet cooler	Einspritzkühler *m*	refroidisseur *m* de flash, condenseur *m* à jet	холодильник со впрыскиванием (смешением)
F 201	**flash evaporation**	augenblickliche Verdampfung *f*	évaporation *f* instantanée	мгновенное испарение
F 202	**flash fire**	schnellaufflammendes Feuer *n*, Aufflammen *n*, Auflodern *n*, Entflammung *f*	inflammation *f* [rapide]	вспышка, быстрое (мгновенное) сгорание
F 203	**flashing vessel** (in a refrigeration loop)	Verdampfer *m*	vaporisateur *m* de détente	испаритель
F 204	**flash point**	Flammpunkt *m*, Entflammungstemperatur *f*	point *m* d'inflammation, point d'éclair	точка вспышки, температура вспышки
F 205	**flash tank**	Entspannungsbehälter *m*	réservoir *m* de détente	сосуд для мгновенного испарения
	flash tower	*s.* F 199		
F 206	**flash vaporization process**	Entspannungsverdampfung *f*	évaporation *f* par détente	процесс мгновенного испарения
F 207	**flash vapour**	Drosseldampf *m*	vapeur *f* d'étranglement	пар, образующийся при дросселировании жидкого холодильного агента
F 208	**flask,** bulb	Kolben *m*, Destilliergefäß *n*	alambic *m*, ballon *m*, flacon *m*	колба
F 209	**flat screen**	Plansieb *n*, Flachsieb *n*	tamis *m* plat	плоский грохот
	flattening out	*s.* D 249		
	flawed	*s.* C 811		
F 210	**Fletcher-Reeves method**	Fletcher-Reeves-Verfahren *n*	méthode *f* de Fletcher-Reeves	метод Флетчера-Ривса
F 211	**flexibility**	Flexibilität *f*, Biegbarkeit *f*	souplesse *f*, flexibilité *f*	подвижность, текучесть, гибкость
F 212	**flexible-chain polymer**	flexibel-kettiges Polymer *n*	polymère *m* à chaîne flexible	гибко-цепной полимер
F 213	**flight-conveying**	Flugförderung *f*, pneumatischer Transport *m*	transport *m* pneumatique	пневматический транспорт
F 214	**floatation**	Flotieren *n*, Flotation *f*, Schwimmaufbereitung *f*	flottation *f*	флотация
F 215	**floatation agent**	Flotationsmittel *n*	agent *m* de flottation	флотационный агент
F 216	**floatation plant**	Flotationsanlage *f*	installation *f* de flottation	флотационная фабрика
F 217	**flocculant**	Flockungsmittel *n*	flooulant *m*	коагулирующее средство, коагулянт
F 218	**flocculation,** coagulation	Flockung *f*, Ausflockung *f*, Gerinnen *n*, Koagulation *f*	floculation *f*, coagulation *f*	коагуляция, свертывание
F 219	**flocculation plant**	Flockungsanlage *f*	installation *f* de floculation	установка для коагулирования
F 220	**flocculent**	flockig, ausgeflockt	floculeux	хлопьевидный
F 221	**flock-coating**	Beflocken *n*, Beflockung *f*	flocage *m*	флокирование, покрывание пухом

F 222	flooded condenser	überfluteter Kondensator m	condenseur m noyé	затопленный конденсатор
F 223	flooded evaporator	überfluteter Verdampfer m	évaporateur m inondé	затопленный испаритель
F 224	flooded fluid cooler	überfluteter Flüssigkeitskühler m	refroidisseur m de liquide noyé	затопленный охладитель с непосредственным охлаждением жидкости
F 225	flooded shell-and-tube cooler	überfluteter Bündelrohrverdampfer m	refroidisseur m multitubulaire noyé	затопленный кожухотрубный охладитель
F 226	flooding, overflowing	Beflutung f, Überflutung f	inondation f	затопление, наводнение
F 227	flooding nozzle	Berieselungsdüse f	tuyère f à ruissellement, buse f de ruissellement	орошающее сопло, форсунка
F 228	flow behaviour, flow conditions	Strömungsverhältnisse fpl, Strömungsbedingungen fpl	conditions fpl d'écoulement	режим потока, гидродинамические условия
F 229	flow behaviour	Fließverhalten n	visco-élasticité f, écoulement m	текучесть
F 230	flow component	Strömungskomponente f	composante f d'écoulement	составляющая потока
	flow conditions	s. F 228		
F 231	flow cross section	Strömungsquerschnitt m, Durchlaßquerschnitt m	section f de passage	поперечное сечение потока, живое сечение
F 232	flow curve	Fließkurve f	courbe f d'écoulement	кривая течения
F 233	flow density	Strömungsdichte f	densité f de flux	плотность потока
	flow diagram	s. F 250		
	flow drag	s. F 262		
F 234	flow energy	Strömungskraft f	force f d'écoulement	гидроэнергия течения
F 235	flow field	Strömungsfeld n	champ m de courant	поле скоростей потока
F 236	flow heater	Durchflußerhitzer m	appareil m de chauffage à passage	проточный электронагреватель
F 237	flow indicator	Durchflußanzeiger m	indicateur m du débit	расходомер
F 238	flow-induced crystallization	Zwangskristallisation f, strömungsinduzierte Kristallisation f	cristallisation f causée par l'écoulement	принудительная кристаллизация, кристаллизация, наведенная потоком
F 239	flowing	Durchfluß m, Durchsickern n	circulation f, passage m, percolation f	протекание, поток, расход
	flowing off	s. D 488		
F 240	flow-in pressure	Einströmdruck m	pression f d'entrée, pression d'afflux	давление (напор потока) на входе
F 241	flow-in stream	Einströmung f	admission f, introduction f	втекание, приток
F 242	flow layer	Strömungsschicht f	couche f de courant, couche d'écoulement	слой потока
F 243	flow loss	Strömungsverlust m	perte f d'écoulement	гидравлическая потеря
F 244	flow measurement	Strömungsmessung f	mesure f du courant, mesure d'écoulement	измерение течения
F 245	flow meter	Strömungsmengenmesser m, Mengenmesser m, Strömungsmesser m, Durchflußmesser m, Mengenmeßgerät n, Durchflußmeßgerät n	rhéomètre m, ailette f hydrométrique, débitmètre m, indicateur m de débit, appareil m à mesurer des quantités	измеритель течения, расходомер, счетчик расхода, реометр
F 246	flow mixer	Fließmischer m, Flüssigkeitsmischer m	mélangeur m de liquides	непрерывно действующий смеситель
F 247	flow model	Strömungsmodell n	modèle m hydraulique (d'écoulement)	модель потока (течения)
F 248	flow off/to, to leak off	abfließen	se décharger	стекать
F 249	flow path	Strömungsverlauf m	voie (forme) f de l'écoulement	путь потока
F 250	flow pattern, flow diagram	Fließdiagramm n, Strömungsbild n, Fließkurve f	image f des filets fluides, forme f des filets d'air, lignes fpl de courant	кривая (картина) течения
F 251	flow point, pour point	Fließpunkt m	point m d'écoulement, point de fusion (ramollissement)	точка ожижения (перехода в жидкое состояние), точка начала течения
F 252	flow pressure	Fließdruck m, Strömungsdruck m	pression f de fluage, pression d'écoulement, pression du courant	давление истечения, гидродинамическое давление
F 253	flow profile	Strömungsprofil n	lignes fpl du courant, représentation f d'écoulement	профиль скорости потока
F 254	flow property	Fließfähigkeit f, Fließeigenschaft f	capacité f de fluage, fluidité f	текучесть, свойство текучести
F 255	flow rate, rate of flow	Durchsatz[strom] m, Durchfluß[strom] m, Förderstrom m, Strom m, Fließgeschwindigkeit f, Strömungsgeschwindigkeit f, Durchflußgeschwindigkeit f, Durchlaufgeschwindigkeit f, Strömungsmenge f, Durchflußrate f, Durchsatzmenge f	débit m, vitesse f (taux m) de passage, vitesse d'écoulement	скорость расхода (течения, протекания, потока), подача, расход [потока]
F 256	flow rate control, basic rate system	Mengenregelung f	réglage (contrôle) m du débit	регулирование расхода
F 257	flow rate loss	Durchflußverlust m	perte f à la circulation	потеря расхода
F 258	flow rate profile	Durchsatzprofil n	profil m du débit	профиль расхода (производительности)

F 259	flow reactor	Durchflußreaktor m	réacteur m à passage	проточный реактор, реактор для непрерывного процесса
F 260	flow regulator	Durchflußregler m	régulateur m du débit	регулятор потока
F 261	flow resistance, resistance to flow	Fließwiderstand m, Fließfestigkeit f	résistance f d'écoulement, résistance de fluage	сопротивление течения, гидравлическое сопротивление
F 262	flow resistance, hydraulic resistance, flow drag	Strömungswiderstand m, hydrodynamischer (hydraulischer) Widerstand m	résistance f hydraulique	сопротивление течения, гидродинамическое (гидравлическое) сопротивление
	flow sheet	s. 1. P 502; 2. T 25		
F 263	flow stability	Fließstabilität f	stabilité f d'écoulement, stabilité de fluage	устойчивость течения
F 264	flow stream	Strömung f, Flüssigkeitsstrom m	écoulement m, courant m liquide	течение, поток
F 265	flow temperature	Fließtemperatur f	température f de fluage	температура ожижения (перехода в жидкое состояние, текучести)
F 266	flow-through cooler	Durchlaufkühler m	refroidisseur m à passage	проточный охладитель
F 267	flow time, processing time	Durchlaufzeit f	temps m (durée f) de passage	время протекания, продолжительность процесса
F 268	flow tube	Strömungsrohr n	tuyau m d'écoulement	трубка потока
F 269	flow valve	Durchflußventil n	vanne f de passage	проходной вентиль
F 270	flow volume	Strömungsvolumen n	volume m d'écoulement	объем потока
F 271	flue ash	Flugasche f	cendre f folle (volante)	летучая зола
F 272	flue dust	Flugstaub m	cendres fpl volantes, cendre f fine (de fumée)	зола, унесенная дымовыми газами
F 273	flue gas, smoke (burning, burnt, chimney) gas	Abgas n (von Feuerungen), Rauchgas n, Verbrennungsabgas n, Abzugsgas n, Kamingas n	gaz m de fumée (cheminée, combustion), gaz brûlé (perdu, d'évacuation)	топочный (дымовой) газ
F 274	flue gas boiler	Abgaskessel m, Abhitzekessel m	chaudière f chauffée par la chaleur perdue	запечный котел, котел-утилизатор
F 275	flue outlet	Abgasstutzen m	tubulure f d'échappement	выпускной патрубок
F 276	fluid bed, fluidized bed	Fließbett n, Wirbelschicht f, Wirbelbett n	couche f fluidisée, lit m fluidisé	ожиженный (кипящий) слой
F 277	fluid content, moisture content	Flüssigkeitsgehalt m	teneur m de liquide	содержание жидкости
F 278	fluid coolant	Kühlflüssigkeit f	liquide m de refroidissement	жидкий холодоноситель, охлаждающая жидкость
F 279	fluid dynamics	Strömungslehre f, Flüssigkeitsdynamik f, Hydrodynamik f	dynamique f des fluides	гидродинамика, динамика жидкостей
F 280	fluid fuel reactor	Flüssigbrennstoffreaktor m	réacteur m à carburant liquide	реактор на жидком топливе
F 281	fluid hydroforming	Wirbelschichtreformierverfahren n	procédé m hydroforming à lit fluidisé	процесс гидроформинга в псевдоожиженном слое
	fluidifying	s. F 284		
F 282	fluidity, ability to flow	Fließfähigkeit f, Fluidität f, Fließvermögen n	fluidité f, capacité f de fluage	текучесть, текучеспособность
F 283	fluidization	Verflüssigung f, Fluidisierung f	fluidisation f	ожижение, флюидизация
F 284	fluidization, fluidifying	Aufwirbelung f, Wirbelschichtbildung f	fluidisation f	завихрение, образование кипящего слоя
F 285	fluidization velocity	Fluidisierungsgeschwindigkeit f, Wirbel[punkt]geschwindigkeit f	vitesse f de fluidisation	скорость ожижения (флюидизации)
	fluidize/to	s. L 143		
	fluidized bed	s. F 276		
F 286	fluidized bed adsorption	Wirbelbettadsorption f	adsorption f à lit fluidisé	адсорбция в псевдоожиженном слое
F 287	fluidized bed catalyst	Fließbettkatalysator m, Wirbelbettkatalysator m	catalyseur m de lit fluidisé	псевдоожиженный катализатор
F 288	fluidized bed coating	Fließbettbeschichtung f, Wirbelsintern n	enduction f en lit fluidisé, procédé m d'enduction par frittage	нанесение покрытия в кипящем слое
F 289	fluidized bed cracking	Wirbelbettkrackung f	cracking m à lit fluidisé	крекинг в флюидизированном слое
F 290	fluidized bed dryer	Fließbetttrockner m	sécheur m à lit fluidisé	сушилка с кипящим слоем
F 291	fluidized bed furnace	Wirbelschichtofen m, Wirbelschichtfeuerung f	foyer m à lit fluidisé, installation f de chauffe à couche fluidisée	печь (топка) с псевдоожиженным слоем
F 292	fluidized bed gasification	Wirbelbettvergasung f	gazéification f à lit fluidisé	газификация в псевдоожиженном слое
F 293	fluidized bed gasifier	Wirbelschichtvergaser m, Wirbelschichtgaserzeuger m	gazéificateur m à couche fluidisée	газогенератор с псевдоожиженным слоем
F 294	fluidized bed principle	Wirbelschichtprinzip n	principe m de couche fluidisée	принцип кипящего слоя

F 295	fluidized bed process	Wirbelschichtprozeß m, Fließbettverfahren n, Fließbettprozeß m	procédé (processus) m à lit fluidisé	процесс в кипящем слое
F 296	fluidized bed reactor	Wirbelbettreaktor m, Wirbelschichtreaktor m, Fließbettreaktor m	réacteur m à lit fluidisé, réacteur à couche fluidisée	реактор с кипящим (псевдоожиженным) слоем
	fluidizing point	s. V 203		
F 297	fluid mechanics	Strömungsmechanik f	mécanique f des fluides	гидромеханика
F 298	fluid process	Wirbelschichtverfahren n	procédé m par couche fluidisée, fluidisation f	метод вихревого слоя, процесс в кипящем слое
F 299	fluid-solid reaction	Fluid-Feststoff-Reaktion f	réaction f solide-fluide	реакция между флюидом и твердым телом
F 300	fluid system	Flüssigkeitssystem n	système m liquide	жидкостная система
F 301	fluid tank	Flüssigkeitsbehälter m	réservoir m à liquide	бак для жидкости
F 302	fluid temperature	Fluidtemperatur f	température f de fluide	температура среды
	flush gas	s. S 68		
	flushing	s. W 17		
F 303	flux, fluxing agent	Flußmittel n, Fließmittel n	fondant m	флюс, плавень
F 304	foam drying	Schaumtrocknung f	séchage m de mousse	сушка пены (в вспененном состоянии)
F 305	foamer, frothing agent	Schaumbildner m, Verschäumer m, Schaummittel n	agent m moussant	пенообразователь, пенообразующее средство
F 306	foam extinguisher	Schaumlöscher m	extincteur m à mousse	пенный огнетушитель
F 307	foam fire extinguisher	Schaumlöschmittel n	extincteur m mousse, agent m moussant d'extinction du feu	пенистый огнетушитель
F 308	foam generator	Schaumgenerator m	générateur m de mousse	пеногенератор
F 309	foaming	Verschäumung f, Verschäumen n	moussage m	вспенивание
F 310	foaming apparatus	Schaumapparat m	appareil m à mousse	пенообразующий аппарат
F 311	foaming property	Schaumfähigkeit f, Schaumeigenschaft f	propriété f de mousse	пенообразующая способность
F 312	food packaging	Lebensmittelverpackung f	emballage m des denrées alimentaires	упаковка пищевых продуктов
F 313	food processing	Lebensmittelverarbeitung f	traitement m des denrées alimentaires	переработка пищевых продуктов
	forced air	s. B 118		
F 314	forced air evaporator	Gebläseverdampfer m	évaporateur m à l'air forcé	испаритель с принудительной конвекцией воздуха
F 315	forced circulation	Zwangsumlauf m, erzwungener Umlauf m	circulation f forcée	принудительная циркуляция
F 316	forced-circulation cooling tower	Ventilatorkühlturm m	tour f de réfrigération à circulation forcée	градирня с принудительной вентиляцией
F 317	forced-circulation evaporator	Zwangsumlaufverdampfer m, Druckumlaufverdampfer m	évaporateur m à circulation forcée	выпарной аппарат с принудительной циркуляцией
	forced-circulation generator	s. I 75		
F 318	forced convection	Zwangskonvektion f, Zwangsströmung f	convection f forcée, écoulement m forcé	вынужденная (принудительная) конвекция
F 319	forced highly elastic deformation	erzwungene hochelastische Deformation f	déformation f élastique forcée	вынужденная высокоэластичная деформация
	forcing pump	s. P 420		
F 320	forecast demand	Voraussageanforderung f, Prognoseanforderung f	demande f prévisionnelle	прогностический спрос, прогностическая потребность
	forecooling	s. P 333		
F 321	foreign gas, accessory gas	Begleitgas n	gaz m associé	сопутствующий газ
F 322	foreign substance	Beimengung f, Beimischung f, Fremdsubstanz f, Verunreinigung f	addition f, impureté f, substance f étrangère	примесь, примешивание
	fore-runnings	s. F 174		
F 323	formation constant	Bildungskonstante f	constante f de formation	константа образования
F 324	formation enthalpy	Bildungsenthalpie f	enthalpie f de formation	энтальпия образования
F 325	formation of azeotrope	Azeotropbildung f	formation f d'azéotrope	образование азеотропа
F 326	formation of drops	Tropfenbildung f	formation f de gouttes	каплеобразование
F 327	formation of layers	Schichtenbildung f	formation f de couches	образование слоев, наслоение
F 328	formation of mist	Nebelbildung f	formation f de brouillard (brume)	образование тумана, туманообразование
F 329	formation of steam	Dampfbildung f	vaporisation f, production f de vapeur	парообразование
	fouling	s. C 627		
F 330	fouling factor	Verschmutzungsfaktor m, Verkrustungsfaktor m	coefficient m d'encrassement	фактор загрязнения
F 331	Fourier's law of heat conduction	Fouriersches Gesetz n der Wärmeleitung	loi f de Fourier	уравнение теплопроводности Фурье
F 332	fractional condensation	Fraktionierkondensation f	condensation f fractionnée	фракционная конденсация
F 333	fractional distillation	Fraktionierdestillation f, Rektifikation f	distillation f fractionnée, rectification f	ректификация, фракционирование
F 334	fractional precipitation	fraktionierte Fällung f	précipitation f fractionnée	фракционированное осаждение

F 335	fractionating column	Fraktionieraufsatz m, Fraktionierkolonne f, Fraktioniersäule f, Trennkolonne f, Fraktionator m	tour (colonne) f de fractionnement	фракционирующая колонна
F 336	fractionating tray	Fraktionierboden m	plateau m d'une colonne de fractionnement	тарелка ректификационной колонны
F 337	fractionation	Fraktionierung f	fractionnement m	фракционирование
	fractive force	s. T 87		
F 338	frame filter	Rahmenfilter n	filtre m à cadres et à plateaux	рамный фильтр
F 339	free circulation	Naturumlauf m, freier Umlauf m	circulation f naturelle	естественная циркуляция
F 340	free convection	freie (natürliche) Konvektion f	convection f naturelle	свободная (естественная) конвекция
F 341	free discharge	freier Ablauf m	décharge f libre	свободное истечение
F 342	freedom from stress, stressless	Spannungsfreiheit f	état m exempt de tension	отсутствие [внутренних] напряжений
F 343	freedom of movement	Bewegungsfreiheit f	liberté f de déplacement (mouvement)	свобода движения, нестесненное движение
F 344	free elastic recovery	freie elastische Rückfederung f	reprise (relaxation) f élastique libre, ressort m libre	свободное эластическое (упругое) обратное восстановление
F 345	free energy	freie Energie f	énergie f libre	свободная энергия
F 346	free flow	freie Strömung f	courant (écoulement) m libre	свободный поток, свободное течение
F 347	free liquid knockout	Flüssigkeitsabscheider m	séparateur m de liquide	сепаратор жидкости
F 348	free molecule diffusion	Molekulardiffusion f, Molekularströmung f	mouvement m (diffusion f) moléculaire	[молекулярная] диффузия
F 349	free path [lenght]	freie Weglänge f	libre parcours m	свободный путь пробега
F 350	free vortex	freier Wirbel m	tourbillon m libre	свободный вихрь
F 351	free water knock-out	Vorabscheider m	séparateur m préliminaire	предварительный отстойник
F 352	freeze drying	Gefriertrocknung f	séchage m par congélation, lyophilisation f	сублимационная сушка, сушка при температуре ниже 0°C
F 353	freezing liquid	Gefrierflüssigkeit f	liquide m frigorifique (de congélation)	замораживающая жидкость
F 354	freezing mixture	Kältemischung f	mélange m frigorifique	охлаждающая смесь
F 355	freezing point	Gefrierpunkt m	point m de congélation	точка замерзания
F 356	freezing-point curve	Erstarrungskurve f	courbe f de solidification	кривая затвердевания
F 357	freezing point depression	Gefrierpunkterniedrigung f	abaissement m du point de congélation	понижение точки замерзания
F 358	freezing preservation	Kältekonservierung f	conservation f par le froid	холодильное хранение
F 359	freezing pressure	Gefrierdruck m	pression f de congélation	давление между морозильными плитами
F 360	freezing temperature	Gefriertemperatur f	température f de congélation	температура замерзания (застывания)
F 361	frequency distribution curve	Häufigkeitsverteilungskurve f	courbe f de Gauss (distribution de fréquence)	кривая распределения (частоты)
	frequency factor	s. I 39		
F 362	frequency function	Häufigkeitsfunktion f	fonction f de fréquence	функция частоты
F 363	frequency value	Häufigkeitswert m	valeur f de fréquence	величина частоты
F 364	fresh air cooler	Frischluftkühler m	refroidisseur m d'air frais	охладитель свежего воздуха
F 365	fresh air cooling	Frischluftkühlung f	refroidissement m à (de l')air frais	охлаждение свежим воздухом
F 366	fresh air make-up	Frischluftzusatz m	appoint m d'air frais	добавка наружного воздуха
F 367	fresh air volume	Frischluftmenge f, Frischluftvolumen n	volume m d'air frais	количество (объем) свежего воздуха
F 368	fresh water cooler	Frischwasserkühler m	refroidisseur m d'eau fraîche	охладитель свежей водой
F 369	fresh water cooling	Frischwasserkühlung f	refroidissement m à eau fraîche	охлаждение свежей (пресной) водой
F 370	frictional resistance	Reibungswiderstand m	résistance f de (due au) frottement	сопротивление трения
F 371	friction drag	Reibungswiderstand m (gegenüber einem flüssigen oder gasförmigen Medium)	résistance f de frottement, résistance de friction	сопротивление трения
F 372	friction factor	Reibungskoeffizient m	coefficient m de frottement	коэффициент трения
F 373	friction force	Reibungskraft f	force f de frottement	сила трения
F 374	friction head	Reibungsdruckverlust m	perte f de pression due au frottement	потери давления от трения, напор на преодоление трения
F 375	friction heat	Reibungswärme f	chaleur f de friction	теплота трения
F 376	friction horse power	Reibungsleistung f	puissance f de frottement	мощность трения
F 377	friction loss	Reibungsverlust m	perte f par friction	потеря трения
F 378	friction slope	Reibungsgefälle n	perte f de charge par frottement par unité de longueur	потеря напора на трение
F 379	friction welding	Reib[ungs]schweißen n	soudage m à (par) friction	сварка трением
F 380	frost-hardy, frost-resistant	kältebeständig	résistant au froid	морозостойкий, холодостойкий
F 381	frost resistance	Frostbeständigkeit f	résistance f au froid	морозостойкость

	frost-resistant	s. F 380		
F 382	froth flow	turbulente Blasenströmung f	écoulement m de bulles turbulent	пенистое течение, пенистый поток, турбулентное пузырчатое течение
	frothing agent	s. F 305		
F 383	Froude number	Froudesche Zahl f	nombre m de Froude	число Фруда
F 384	fuel, heating material	Heizstoff m, Heizmittel n, Brennstoff m	combustible m	топливо, горючее
F 385	fuel, motor fuel	Treibstoff m	carburant m, combustible m	горючее, жидкое топливо, моторное масло
F 386	fuel consumption	Kraftstoffverbrauch m	consommation f de carburant, consommation d'essence	расход [моторного] топлива
F 387	fuel gas	Heizgas n	gaz m chaud (combustible)	топочный (отопительный, печной, горючий) газ
F 388	fuel oil	Heizöl n	huile f à brûler, fuel-oil m, mazout m	нефть, мазут, жидкое топливо
F 389	fugacity	Flüchtigkeit f	volatilité f	летучесть
F 390	full load	Vollast f, Vollbelastung f	pleine charge f	полная нагрузка
F 391	full-load hours	Vollaststunden fpl	heures fpl à pleine charge	часы работы при полной нагрузке
	full-load rate	s. F 393		
F 392	full-load regime	Vollastbetrieb m	régime m à pleine charge	эксплуатация при полной нагрузке
	full mechanization	s. C 459		
F 393	full power, full-load rate	Vollförderung f	transport m maximum	полная подача
F 394	fully developed turbulence	entwickelte (vollausgeprägte) Turbulenz f	turbulence f développée	развитая турбулентность
F 395	fumigate/to, to gas	begasen	faire passer du gaz	окуривать, выделять газ
F 396	fumigation, gas emission	Ausgasung f, Entgasung f	dégagement m de gaz	газовыделение, выгазование
	function	s. M 288		
F 397	functional scheme	Funktionsschema n	schéma m fonctionnel	функциональная схема
F 398	functional structure	Funktionsstruktur f	structure f fonctionelle	функциональная структура
F 399	function of temperature, temperature variation	Temperaturgang m	allure (variation) f de température	ход температуры
	fundamental circuit	s. B 44		
F 400	fundamental law	Grundgesetz n	loi f fondamentale	основной закон
F 401	funnel	Fülltrichter m, Trichter m	entonnoir m, bec m d'alimentation, bec de remplissage, trémie f, cône m	приемная (загрузочная) воронка
	furnace	s. O 139		
F 402	furnace chamber	Ofenraum m, Ofenkammer f	chambre f du four	помещение печи, печное помещение
F 403	furnace charge	Ofenbeschickung f	fournée f, charge f du four	загрузка печи, шихта
F 404	furnace efficiency	Ofenleistung f, Ofendurchsatz m	capacité f du four, fournée f	мощность печи
F 405	furnace exhaust gas	Ofenabgas n	gaz m du four sortant	печной отходящий газ
F 406	furnace gas	Ofengas n	gaz m du four	обжигательный газ
F 407	furnace grate	Ofenrost m	grille f du four	колосниковая решетка печи
F 408	furnace installation	Ofenanlage f	four[neau] m, chauffe f, installation f de fours	печная установка, печь
F 409	furnace lining, counter-wall	Ofenfutter n	garnissage (revêtement) m de four	футеровка печи
F 410	furnace operation	Ofenbetrieb m	marche f d'un four, opération f d'un fourneau	режим (ход, работа) печи
F 411	furnace pressure	Ofendruck m	pression f de four	давление в печи
F 412	furnace shell	Ofenmantel m	paroi f du four	кожух печи
F 413	furnace unit	Ofeneinheit f	four[neau] m	печной агрегат
F 414	furnace wall	Ofenwandung f	paroi f du four	стенка печи
	fuse/to	s. M 162		
F 415	fusibility	Schmelzbarkeit f	fusibilité f	плавкость
F 416	fusing-in, melting-down	Einschmelzen n	fusion f, fondage m	плавление, расплавление, плавка
F 417	fusing temperature	Erweichungstemperatur f	température f de ramollissement	температура размягчения
F 418	fusing temperature	Schmelztemperatur f	température f de fusion	температура плавления
F 419	fusion, decomposition, disaggregation	Aufschluß m, Zerlegung f	désagrégation f, dissolution f	вскрытие, переведение в удобное для переработки состояние, переведение в растворимое состояние
	fusion	s. a. M 164		
F 420	fusion curve	Schmelzkurve f	courbe f de fusion	кривая плавления
	fusion point	s. M 172		
F 421	fusion zone	Schmelzzone f	zone f de fusion	пояс (зона) плавления

G

G 1	gain, output, produce	Ausbeute f	rendement m, profit m, gain m	выход, выработка, добыча

G 2	gap-type filter, space filter	Spaltfilter n	filtre m à fentes (disques)	пластинчатый фильтр
G 3	gap volume coefficient gas/to	Lückenvolumenkoeffizient m s. F 395	coefficient m de porosité	коэффициент пористости
G 4	gas absorption	Gasabsorption f, Gasaufnahme f	absorption f de gaz	абсорбция газа
G 5	gas adsorbate	gasförmiges Adsorbat n	substance f adsorbée, gaz m adsorbé	газообразный адсорбат
G 6	gas analysis	Gasanalyse f	analyse m du gaz	газовый анализ
G 7	gas analyzer	Gasanalysator m	appareil m d'analyse de gaz, analys[at]eur m de gaz	газоанализатор
G 8	gas binding	Gasansammlung f	coussin m gazeux, enrichissement m en gaz	газовая «подушка»
G 9	gas blower	Gasgebläse n	soufflerie f à gaz	газодувка
G 10	gas bubble	Gasblase f	bulle f de gaz	газовый пузырь
G 11	gas bubbler	Gaszylinder m	tube m à barbotage	газовый цилиндр
G 12	gas burner, gas jet torch, blowpipe	Brenner m (Gas)	brûleur m, bec m	сварочная горелка
G 13	gas catalysis	Gaskatalyse f	catalyse f de gaz	газовый катализ
G 14	gas cleaning process	Gasreinigungsprozeß m	procédé m d'épuration de gaz	процесс газоочистки
G 15	gas cleaning unit	Gasreinigungsanlage f	unité f d'épuration de gaz	газоочистительная установка
G 16	gas compressor	Gasverdichter m	compresseur m à gaz	газовый компрессор
G 17	gas constant	Gaskonstante f	constante f des gaz	газовая постоянная
G 18	gas cooler	Gaskühler m	refroidisseur m du gaz, réfrigérant m à gaz	газовый холодильник
G 19	gas cooling gas counter	Gaskühlung f s. G 62	refroidissement m du gaz	охлаждение газа
G 20	gas cylinder	Druckgasflasche f	bouteille (bonbonne) f à gaz	баллон для компримированного газа
G 21	gas density	Gasdichte f	densité f de gaz	плотность газа
G 22	gas dynamics gas emission	Gasdynamik f s. F 396	dynamique f des gaz	газовая динамика
G 23	gaseous decomposition products	Zersetzungsgase npl	produits mpl de décomposition gazeux	газообразные продукты разложения
G 24	gaseous form	gasförmige Phase f, Gasphase f	phase f gazeuse	газовая фаза
G 25	gaseous fuel	gasförmiger Brennstoff m	combustible m gazeux	газообразное топливо
G 26	gaseous mixture	gasförmiges Gemisch n	mélange m gazeux	газовая смесь
G 27	gaseous refrigerant	gasförmiges Kältegemisch (Kältemittel) n	frigorigène m (agent m frigorifique) gazeux	газообразный холодильный агент
G 28	gaseous refrigerant stream	Kältemitteldampfstrom m, gasförmiger Kältemittelstrom m	écoulement m de fluide frigorifique gazeux	поток газообразного холодильного агента
G 29	gas equation	Gasgleichung f	équation f des gaz	газовое уравнение
G 30	gas expansion refrigerating system	Gasexpansionskühlsystem n	système m frigorifique à détente de gaz	холодильная система газового расширения, газовая холодильная машина с детандером
G 31	gas explosion	Gasexplosion f	explosion (détonation) f de gaz	взрыв газа
G 32	gas filter	Gasfilter n	filtre m à gaz	газоочиститель, газовый фильтр
G 33	gas-fired furnace	Gasfeuerung f (Ofen mit Gasbrenner)	chauffage m au gaz, foyer m à gaz	газовая печь
G 34	gas-fired heat exchanger	gasbeheizter Wärmeübertrager m	échangeur m thermique chauffé au gaz	теплообменник с газовым нагревом
G 35	gas flow	Gasstrom m	courant (flux) m de gaz, courant gazeux	поток газа
G 36	gas formation	Gasbildung f	formation f de gaz	газообразование
G 37	gas fractionation	Gasfraktionierung f	séparation f (fractionnement m) des gaz	разделение газа, газовое фракционирование
G 38	gas fractionation plant	Gastrennanlage f	unité f de fractionnement de gaz	установка фракционирования газа
G 39	gas generation, gas production	Gaserzeugung f	production (génération) f de gaz	производство газа, газообразование
G 40	gas generator	Gaserzeuger m	gazogène m, générateur m de gaz	газогенератор
G 41	gas heater	Gaserhitzer m	réchauffeur m de gaz	нагреватель газа
G 42	gas heating	Gaserhitzung f	échauffement m de gaz	нагрев газа
G 43	gas holder, gasometer, gas tank	Gasometer m, Gasbehälter m	gazomètre m, réservoir m à gaz	газометр, газгольдер
G 44	gas hold-up	Gas-hold-up n	hold-up m de gaz	газосодержание, содержание газа
G 45	gasifiable, vaporizable	vergasbar	gazéifiable	газифицирующийся
G 46	gasification, gasifying, vaporizing	Vergasung f, Entgasung f, Gaserzeugung f	gazéification f	газообразование, газификация
G 47	gasification factor	Vergasungsfaktor m	facteur m de gazéification	фактор газификации
G 48	gasification of coal, coking	Kohlevergasung f	gazéification f du charbon	газификация угля
G 49	gasification plant	Vergasungsanlage f	unité f de gazéification	установка газификации
G 50	gasification process	Vergasungsverfahren n, Vergasungsprozeß m	procédé m de gazéification	процесс газификации

G 51	gasification reactor	Vergasungsreaktor *m*	réacteur *m* de gazéification	реактор газификации
G 52	gasification under pressure	Vergasung *f* unter Druck, Druckvergasung *f*	gazéification *f* sous pression	газификация под давлением
G 53	gasifier	Vergasungsapparat *m*, Gasgenerator *m*, Gaserzeuger *m*	gazéificateur *m*	газогенератор
G 54	gasify/to, to vaporize, to carburet	vergasen, Gas erzeugen	gazéifier, carburer	газифицировать, карбюрировать
	gasifying	*s.* G 46		
G 55	gasket, packing ring	Dichtungsring *m*	rondelle *f* de joint, anneau *m* de garniture, anneau d'étanchéité	прокладка, прокладочное кольцо
	gas jet torch	*s.* G 12		
G 56	gas leakage	Gasaustritt *m*, Gasaustrittsstelle *f*	fuite *f* de gaz	утечка газа
G 57	gas liquefaction plant	Gasverflüssigungsanlage *f*	installation *f* de liquéfaction de gaz	установка для сжижения газа
G 58	gas liquefaction process	Gasverflüssigungsprozeß *m*	procédé *m* de liquéfaction de gaz	процесс ожижения газа
G 59	gas-liquid-extraction	Gas-Flüssigkeits-Extraktion *f*	extraction *f* gaz-liquide	экстракция в системе газ-жидкость
	gas-liquid mixing	*s.* M 255		
G 60	gas-liquid reaction	Gas-Flüssigkeits-Reaktion *f*	réaction *f* gaz-liquide	реакция в системе газ-жидкость
G 61	gas make	Gasausbeute *f*	rendement *m* en gaz	выход газа
G 62	gas meter, gas counter	Gasmesser *m*, Gasuhr *f*	compteur *m* à gaz	газомер, газовый счетчик
G 63	gas mixture	Gasgemisch *n*	mélange *m* gazeux (des gaz)	газовая (горючая, газообразная) смесь
G 64	gas oil	Gasöl *n*	gas-oil *m*	газойль
G 65	gas oil separator	Gasölabscheider *m*	séparateur *m* de gas-oil	сепаратор газойля
	gasometer	*s.* G 43		
G 66	gas operated	gasbetätigt	commandé par gaz	с газовым исполнительным механизмом
G 67	gas percolation	Gasfiltration *f*	filtration *f* de gaz	фильтрация газа
G 68	gas pipeline	Erdgasleitung *f*, Ferngasleitung *f*	gazoduc *m*	трубопровод для природного газа
G 69	gas pressure	Gasdruck *m*	pression *f* de gaz, tension *f* gazeuse	давление газа
G 70	gas pressure regulator	Gasdruckregler *m*	régulateur *m* de pression de gaz	регулятор давления газа
G 71	gas producer	Gasgenerator *m*, Gaserzeuger *m*	générateur *m* de gaz	генератор газа
G 72	gas producting plant	Gaserzeugungsanlage *f*	installation *f* de production de gaz	газогенераторная установка
	gas production	*s.* G 39		
G 73	gas-proof	gasdicht	étanche aux gaz	газонепроницаемый
G 74	gas purification (purifying)	Gasreinigung *f*	épuration *f* (purification *f*, lavage *m*) du gaz	газоочистка, очистка газа
G 75	gas purifying mass	Gasreinigungsmasse *f*	masse *f* d'épuration de gaz	газоочистительная масса
G 76	gas reactor	Gasreaktor *m*	réacteur *m* à gaz	газонаполненный реактор
G 77	gas recovery	Gasrückgewinnung *f*	récupération *f* de gaz	газоулавливание
G 78	gas recovery process	Gasrückgewinnungsverfahren *n*	procédé *m* de récupération de gaz	процесс газоулавливания
G 79	gas recycle rate	Gasumlaufrate *f*	vitesse *f* (taux *m*) de recyclage de gaz	скорость циркуляции газа
G 80	gas reforming plant	Gasspaltanlage *f*	unité *f* de réformation de gaz	установка риформинга газа
G 81	gas re-liquefying	Gasrückverflüssigung *f*	reliquéfaction *f* de gaz	повторное сжижение газа
G 82	gas remover	Gasabscheider *m*	séparateur *m* de gaz	газоуловитель
G 83	gas scrubbing	Gaswaschen *n*, Gaswäsche *f*	lavage *m* de gaz	промывка газа
G 84	gas scrubbing column	Gaswäscher *m*, Gaswaschkolonne *f*	laveur *m* de gaz, scrubber *m*	скруббер для промывки газа
G 85	gas separation	Gastrennung *f*, Gaszerlegung *f*	séparation *f* (fractionnement *m*) de gaz	газоотделение, газоразделение
G 86	gas separation plant	Gaszerlegungsanlage *f*	installation *f* de séparation de gaz	установка для разделения газов
G 87	gassing	Entgasen *n*, Entgasung *f*	dégazonnage *m*, dégazonnement *m*	дегазация, выделение газа
G 88	gas storage	Gasspeicherung *f*	stockage *m* des gaz	подземное хранение газа
G 89	gas storage tank	Gasspeicher *m*	réservoir *m* à gaz	хранилище для газа, газометр, газохранилище
G 90	gas supply	Gaszuführung *f*	admission (arrivée) *f* du gaz	подводка (подача) газа, питание газом
G 91	gas supply	Gasförderung *f*, Gasversorgung *f*	transport *m* du gaz	подача газа
	gas tank	*s.* G 43		
G 92	gas temperature	Gastemperatur *f*	température *f* de gaz	температура газа
G 93	gas thermometer	Gasthermometer *n*	thermomètre *m* à gaz	газовый термометр
G 94	gas throttling	Gasdrosselung *f*	étranglement *m* de gaz	дросселирование газа
G 95	gas-tight	gasundurchlässig	étanche (imperméable) aux gaz	газонепроницаемый
G 96	gas transport	Gastransport *m*	transport *m* de gaz	транспорт газа

G 97	gas treatment	Gasaufbereitung f	traitement m de gaz	подготовка газа
G 98	gas turbulence	Gasturbulenz f	turbulence f de gaz	турбулентность потока газа
G 99	gas velocity	Gasgeschwindigkeit f	vitesse f de gaz	скорость газа
G 100	gas washing bottle	Gaswaschflasche f	épurateur m à gaz	склянка для промывания газов
G 101	gas water, ammonia water	Gaswasser n	eau f ammoniacale (de gazomètre)	газовая вода, аммиачный раствор
G 102	gate valve, slide valve	Schieber m	coulisse f, curseur m, tiroir m, vanne f	задвижка, заслонка, золотник, шибер
G 103	gate valve	Absperrklappe f	clapet m d'arrêt	запорный клапан
	gate valve	s. a. S 297		
G 104	gauge	Meßgerät n, Meßeinrichtung f	appareil (instrument) m mesureur, dispositif m de mesure	измерительный прибор
G 105	gauge, standard measure	Eichmaß n	jauge f, échantillon m, étalon m	нормальная мера, эталон
G 106	gauge pressure	Manometerdruck m	pression f manométrique	манометрическое давление
	Gaussian distribution	s. N 79		
	gauze	s. N 19		
G 107	gauze filter	Siebfilter n	filtre-tamis m, filtre m à tamis	сетчатый фильтр
G 108	Gay-Lussac's law, Charles' law	Gay-Lussac'sches Gesetz n	loi f de Gay-Lussac	закон Гей-Люссака
G 109	Gay-Lussac-tower	Gay-Lussac-Turm m	tour f de Gay-Lussac	башня Гей-Люссака
G 110	gear box	Getriebe[gehäuse] n	engrenage m, carter m	редуктор, корпус редуктора
G 111	gear pump	Zahnradpumpe f	pompe f à engrenages	зубчатый насос
G 112	gel formation	Gelbildung f	formation f de gel, gélification f	гелеобразование
	gel point	s. S 403		
G 113	general gas law	allgemeine Zustandsgleichung f	équation f d'état générale	уравнение состояния газа
G 114	generalized Newtonian fluid	verallgemeinertes Newtonsches fluides Medium n	fluide m newtonien généralisé	обобщенная Ньютоновская жидкая среда
	general plan	s. L 49		
G 115	generation of gases	Gasentwicklung f, Gasfreisetzung f	dégagement m de gaz	выделение газа, газообразование
G 116	generator-absorber	Austreiber-Absorber m	générateur-absorbeur m	генератор-абсорбёр
G 117	generator gas, producer gas	Generatorgas n	gaz m de gazéificateur (gazogène)	генераторный газ
G 118/9	Gibb's phase rule	Gibbssche Phasenregel m	règle f de Gibbs	правило фаз Гиббса
	gilled cooler	s. R 346		
G 120	gland vapour	Sperrdampf m	vapeur m de barrage	пар лабиринтовых уплотнений
	glass air	s. B 77		
G 121	glass filter pump, water jet aspirator (injector)	Wasserstrahlpumpe f	pompe f de (par) jet d'eau, injecteur m hydraulique	водоструйный насос
G 122	glass-transition temperature	Glasübergangstemperatur f, Glastemperatur f	température f de vitrification	[переходная] температура стеклования
G 123	global iteration block	globaler Iterationsblock m	bloc m global d'itération	глобальный итерационный блок
G 124	Glover tower	Glover-Turm m	tour f de Glover, dénitrificateur m	башня Гловера
G 125	graded crushing	schrittweise (stufenweise) Zerkleinerung f	broyage m étagé (pas à pas)	стадийное измельчение
G 126	grade efficiency	Trenngrad m	taux m de séparation	фактор разделения
G 127	gradientless	gradientfrei	sans gradient	безградиентный
G 128	grading	Korngrößentrennung f	séparation f (classement m) granulométrique	сортировка, сортирование
G 129	gradual condensation	Stufenkondensation f	condensation f graduelle	постепенная конденсация
G 130	graduated cylinder, measuring glass	Meßzylinder m	éprouvette f graduée	измерительный цилиндр, мерник
G 131	graduated flask, measuring flask	Meßkolben m	ballon m jaugé	мерная колба
	graduated jar	s. M 152		
	graduation	s. 1. S 56; 2. S 664		
G 132	graft copolymer	Pfropfkopolymer n	copolymère m greffé	привитой полимер
	graft flow	s. P 171		
	grain	s. G 154		
G 133	grain density	Korndichte f	densité f de grain	плотность зерен
	graining	s. G 139		
G 134	grain size distribution	Korngrößenverteilung f	répartition f granulométrique	гранулометрический состав
	granular size	s. P 61		
G 135	granulated catalyst	gekörnter Katalysator m	catalyseur m granulé	гранулированный катализатор
G 136	granulating crusher	Granuliermühle f	moulin m à granuler, broyeur-granulateur m	грануляционная мельница
G 137	granulating machine	Granulierapparat m	appareil m granulateur	гранулятор
G 138	granulating worm	Granulierschnecke f	vis f granulatrice	грануляционный шнек
G 139	granulation, graining	Körnung f, Körnigkeit f, Granulierung f, Kornbildung f	grainure f, granulation f	зернение, гранулирование
G 140	granulator	Granulator m, Granulieranlage f	granulateur m	гранулятор

graph 92

	graph	s. A 555		
G 141	graph chart	Schautafel f	tableau m synoptique	диаграмма, график
G 142	Grashof number	Grashof-Zahl f	nombre m de Grashof	число Грасгофа
G 143	gravel filter, pebble filter	Kiesfilter n, Kiespackungsfilter n	filtre m à gravier	набивной гравийный фильтр
G 144	gravimetric analysis	Gewichtsanalyse f	gravimétrie f, analyse f pondérale (gravimétrique)	весовой анализ, гравиметрия
G 145	gravitation, gravity	Gravitation f, Schwerkraft f, Massenanziehung f	gravitation f	тяготение, сила тяжести
G 146	gravitational acceleration	Erdbeschleunigung f	gravité (accélération) f de la pesanteur	ускорение силы тяжести
	gravity	s. G 145		
G 147	grease/to	abschmieren	graisser	смазывать, сменять масло
G 148	grease	Schmierfett n, Schmiere f	graisse f	смазочное масло, смазка
G 149	grid gas	Ferngas n	gaz m amené à longue distance	газ дальнего газоснабжения
	grid operation	s. I 271		
	grind/to	s. C 869		
G 150	grindability	Vermahlungsfähigkeit f, Mahlbarkeit f	broyabilité f	размалываемость, измельчаемость
G 151	grinding, milling, pulverizing	Mahlen n, Mahlung f, Vermahlung f, Ausmahlen n	broyage m, mouture f, pulvérisation f	размалывание, помол, размол, мелкое дробление
G 152	grinding cone	Mahlkegel m	cône m broyeur cannelé	конус мельницы
	grinding drum	s. C 872		
	grinding mill	s. M 218		
	grinding product	s. M 219		
G 153	grinding roll	Mahlwalze f, Brechwalze f	rouleau m de mouture, tambour m de concassage	размалывающий вал
G 154	grinding stock, grain	Mahlgut n	mouture f, matériau m à moudre	размалываемый (измельчаемый) материал
G 155	gross heating value	Verbrennungswert m	pouvoir m calorifique, valeur f calorifique brute (maximum)	высшая теплотворная способность
G 156	gross reaction	Bruttoreaktion f	réaction f brute	брутто-реакция, валовая реакция
	ground coat	s. P 482		
	ground plot	s. P 201		
G 157	ground pump, underground pump	Tiefpumpe f, Tauchpumpe f	pompe f noyée (immergée)	глубинный насос
G 158	ground tank	Tiefbehälter m	réservoir m surbaissé	подземный резервуар
G 159	group redundancy	Gruppenredundanz f	redondance f de groupe	групповое резервирование
G 160	guaranty of reliability	Gewährleistung f der Zuverlässigkeit	garantie f de la fiabilité	обеспечение надежности
G 161	gumming	Verharzung f, Verharzen n	résinification f	осмоление
G 162	gum test	Abdampfprobe f	essai m d'évaporation	проба на отпаривание
G 163	gutter	Abflußrinne f	rigole f d'écoulement, gouttière f	сточный желоб
	gutter of affination	s. A 246		
G 164	gyratory crusher	Kegelbrecher m, Kreiselbrecher m	concasseur m à cônes, broyeur m (concasseur) giratoire	конусная дробилка
G 165	gyratory primary crusher	Rundbrecher m zur Zerkleinerung	préconcasseur m à cônes	вращающаяся дробилка для первичного дробления
G 166	gyratory screen	Drehsieb n, Trommelsieb n	tamis m à tambour, crible m rotatif	вращающийся грохот

H

H 1	Haber process	Haber-Bosch-Verfahren n	procédé m de Haber-Bosch	способ Габера-Боша
H 2	half change value method	Halbwertszeitmethode f, Methode f der Halbwertszeit	méthode f de demi-vie	метод полураспада
H 3	half-life, half-value period, period of decay	Halbwertszeit f	demi-vie f, durée f à mi-hauteur	период полураспада
H 4	hammer crusher	Hammerbrecher m	concasseur m à marteaux	молотковая дробилка
H 5	hammer mill, swinging hammer	Hammermühle f	broyeur m à marteaux	молотковая мельница
H 6	hand control (drive), manual operation	Handbedienung f, Handbetrieb m	commande (exploitation) f manuelle, manœuvre f à la main	ручное производство (управление), ручная эксплуатация (работа)
	handle/to	s. S 226		
H 7	handling device	Fördermittel n, Transportmittel n	moyen m (installation f) de transport	транспортное средство
H 8	hand-operated	handbedient	commandé à la main	ручной
H 9	hand pump	Handpumpe f	pompe f à main	ручная помпа
	happen/to	s. O 8		
H 10	hardener	Härter m	durcisseur m	отвердитель
	hardly soluble	s. S 456		
	hardness	s. D 100		
	haulage, hauling	s. D 149		

H 11	hazard analysis	Gefährdungsanalyse f	analyse f des dangers	аварийный анализ, анализ возможных аварий
H 12	head of column	Kolonnenkopf m	tête f de colonne	верх колонны
H 13	head pressure	Extrusionsdruck m	pression f d'extrusion	давление выдавливания (экструзии)
H 14	head product	Kopfprodukt n	fraction f de tête, produit m en tête	головной продукт
H 15	head tank	Vorlaufbehälter m, Kopfproduktbehälter m	récipient m du produit léger	бак головной фракции
	heaping	s. A 76		
H 16	heart cut	Hauptlauf m, Mittelfraktion f	fraction f de cœur	основная фракция
H 17	heat absorption	Wärmeabsorption f, Wärmeaufnahme f	absorption f thermique (thermale, de chaleur)	поглощение теплоты
H 18	heat accumulation	Wärmestauung f, Wärmestau m	accumulation f de chaleur	накопление (аккумуляция) тепла
	heat accumulator	s. S 591		
H 19	heat addition rate	Wärmezufuhrgeschwindigkeit f	vitesse f d'adduction de chaleur	скорость подвода тепла
H 20	heat balance, thermal balance	Wärmebilanz f	bilan m thermique, balance f de chaleur	тепловой баланс
H 21	heat bath	Heizbad n	bain m de chauffe	нагревательная ванна
H 22	heat capacity	Wärmekapazität f, Wärmeaufnahmevermögen n	capacité f calorifique	теплоемкость
H 23	heat carrier	Wärmeträger m, Wärmetransportmedium n	caloporaustauscher m	échangeur m thermique (de chaleur)
теплообменник				
H 26	heat conductivity, caloric conductibility	Wärmeleitfähigkeit f	conductibilité f calorifique (thermique)	теплопроводность
H 27	heat consumption	Wärmeverbrauch m	dépense (consommation) f de chaleur	потребление (расход) тепла
H 28	heat curve	Wärmekurve f, Erwärmungskurve f	courbe f d'échauffement	кривая нагрева
	heat dissipation	s. H 101		
H 29	heat distribution	Wärmeverteilung f	distribution f de chaleur	распределение тепла
H 30	heat effect	Wärmewirkungsgrad m, thermischer Wirkungsgrad m	rendement m thermique	тепловой коэффициент полезного действия
	heat effects	s. H 111		
H 31	heat energy	Wärmeenergie f	énergie f calorifique	тепловая энергия
H 32	heat engineering	Wärmetechnik f, Heizungstechnik f	technique f thermique (du chauffage)	теплотехника
H 33	heat equivalent	Wärmeäquivalent n, mechanisches Wärmeäquivalent	équivalent m mécanique de la chaleur	эквивалент тепла
H 34	heater	Erhitzer m, Heizapparat m	réchauffeur m, appareil m de chauffage	нагреватель, подогреватель
H 35	heater drum	Heiztrommel f	tambour m de chauffage	сушильный цилиндр (барабан)
	heater flue	s. F 170		
H 36	heat exchange	Wärmeübertragung f, Wärme[aus]tausch m	échange (transfert) m de chaleur	теплообмен, обмен теплоты
H 37	heat exchange equipment	Wärmeübertragungsausrüstung f, Wärmetauschausrüstung f	équipement m d'échange de chaleur, équipement de transfert calorifique	оборудование для теплообмена
H 38	heat exchange medium	Wärmeübertragungsmedium n, Wärmeaustauschmedium n	milieu m d'échange de chaleur, agent m de transfert de chaleur	теплообменная среда
	heat exchanger	s. H 25		
H 39	heat exchange surface	Wärmeübertragungsfläche f, Wärmeaustauschfläche f	surface f d'échange de chaleur	теплопередающая поверхность
H 40	heat expansion joint	Wärmekompensator m, Dehnungsausgleicher m	compensateur m de dilatation thermique	тепловой компенсатор
H 41	heat extraction	Wärmeentzug m, Wärmeabfuhr f	élimination f (soutirage m) de chaleur	отвод тепла, охлаждение, поглощение тепла
H 42	heat factor	Wärmefaktor m	facteur m de chaleur	тепловой фактор
H 43	heat flow	Wärmefluß m, Wärmestrom m	flux m de chaleur	тепловой поток
H 44	heat flow rate	Wärmeflußrate f	taux m (vitesse f) de flux de chaleur	скорость теплового потока
H 45	heat generation	Wärmeerzeugung f	génération f (dégagement m) de chaleur, thermogénèse f	теплообразование, производство тепла
H 46	heat gradient	Wärmegefälle n	chute f du potentiel thermique (de chaleur)	тепловой напор, перепад тепла, разность теплосодержания
H 47	heating chamber	Heizkammer f	régénérateur m, chambre f de chauffe	огневое пространство, топочная камера
H 48	heating coil	Heizspirale f, Heiz[rohr]schlange f	serpentin m chauffant	обогревательная спираль, обогревательный змеевик

H 49	heating current	Heizstrom m	courant m de chauffage	тепловой поток, ток нагрева, ток накала
	heating duct	s. H 53		
H 50	heating effect	Heizwirkung f	effet m de chauffage	теплопроизводительность
H 51	heating element	Heizelement n	élément m thermique (de chauffage), m chaufferette f	нагревательный элемент
H 52	heating fan	Heizgebläse n, Heizlüfter m	thermoventilateur m	вентилятор для отопления
H 53	heating flue, heating duct	Heizkanal m	carneau m de chauffage	нагревательный канал, огневой ход
H 54	heating jacket	Heizmantel m	chemise f (manteau m) de chauffage	паровая (обогревательная) рубашка
H 55	heating load	Wärmebelastung f	charge f thermique	теплонапряженность, тепловая нагрузка
	heating material	s. F 384		
H 56	heating medium	Heizmedium n	milieu (agent) m de chauffe	обогревающая среда, теплоноситель
H 57	heating plant	Heizanlage f	installation f de chauffage	отопительная установка
H 58	heating steam	Heizdampf m	vapeur f de chauff[ag]e	греющий пар
H 59	heating surface	Heiz[ober]fläche f	surface f chauffante (de chauffe)	поверхность нагрева
H 60	heating system	Heizsystem n	système m de chauffage	система нагрева
	heating tube	s. B 185		
H 61	heating tube bundle	Heizrohrbündel n	faisceau m de tubes de chauffage	пучок нагревательных труб
H 62	heating under pressure	Druckerhitzung f	chauffage m sous pression	нагревание под давлением
H 63	heating-up furnace	Aufheizofen m, Vorwärmofen m	four m de préchauffage	печь для предварительного нагрева
H 64	heating zone	Heizzone f	zone f de chauffage	зона нагрева
	heat input	s. H 109		
H 65	heat insulation	Wärmeisolation f	isolation f calorifuge (thermique)	тепловая изоляция, теплоизоляция
H 66	heat insulator	Wärmeisolierstoff m	isolateur m, masses fpl calorifuges	теплоизолирующий материал
H 67	heat load	Wärmelast f, Heizlast f	charge f de chauffage	отопительная (тепловая) нагрузка
H 68/9	heat loss	Wärmeverlust m, Verlustwärme f	perte f thermique, déperdition f de chaleur	потеря тепла, тепловая потеря
	heat of adsorption	s. A 223		
H 70	heat of association	Assoziationswärme f, Assoziationsenthalpie f	chaleur f d'association	теплота ассоциации
H 71	heat of combustion	Verbrennungswärme f, Verbrennungsenthalpie f	chaleur (enthalpie) f de combustion	теплота сгорания (горения), энтальпия сгорания
H 72	heat of compression	Verdichtungswärme f, Kompressionswärme f, Verdichtungsenthalpie f, Kompressionsenthalpie f	chaleur f de compression	теплота сжатия
H 73	heat of condensation	Kondensationswärme f, Kondensationsenthalpie f	chaleur f de condensation	теплота конденсации
H 74	heat of cooling	Abkühlwärme f, Kühlungswärme f, Abkühlenthalpie f, Kühlungsenthalpie f	chaleur f de refroidissement	теплота охлаждения
H 75	heat of decomposition	Zersetzungswärme f, Zerfallswärme f, Zersetzungsenthalpie f, Zerfallsenthalpie f	chaleur f de décomposition	теплота разложения
H 76	heat of desorption	Desorptionswärme f, Desorptionsenthalpie f	chaleur f de désorption	теплота десорбции
H 77	heat of dissociation	Dissoziationswärme f, Dissoziationsenthalpie f	chaleur f de dissociation	теплота диссоциации
H 78	heat of dissolution	Auflösungswärme f, Lösungswärme f, Auflösungsenthalpie f, Lösungsenthalpie f	chaleur f de dissolution	теплота растворения
H 79	heat of evaporation	Verdampfungswärme f, Verdampfungsenthalpie f	chaleur f d'évaporation	теплота парообразования (испарения, конденсации)
H 80	heat of formation	Bildungswärme f, Bildungsenthalpie f	chaleur f de formation	теплота образования
H 81	heat of fusion	Schmelzwärme f, Schmelzenthalpie f	chaleur (enthalpie) f de fusion	энтальпия (теплота) плавления
H 82	heat of hydration	Hydratationswärme f, Hydratationsenthalpie f	chaleur f d'hydratation	теплота гидратации
H 83	heat of liquefaction	Verflüssigungswärme f, Verflüssigungsenthalpie f	chaleur f de liquéfaction	теплота ожижения
H 84	heat of mixing	Mischungswärme f, Mischungsenthalpie f	chaleur f de mixtion	теплота смешения
H 85	heat of mixture	Gemischwärme f, Gemischenthalpie f	chaleur f de mélange	теплота смеси
H 86	heat of neutralization	Neutralisationswärme f, Neutralisationsenthalpie f	chaleur f de neutralisation	теплота нейтрализации
H 87	heat of polymerization	Polymerisationswärme f, Polymerisationsenthalpie f	chaleur f de polymérisation	теплота полимеризации

H 88	**heat of reaction**	Reaktionswärme f, Reaktionsenthalpie f	chaleur f de réaction	теплота (энтальпия) реакции
	heat of reaction	s. a. H 111		
H 89	**heat of separation**	Trennungswärme f, Trennungsenthalpie f	chaleur f de séparation	теплота разделения
H 90	**heat of setting**	Abbindewärme f, Abbindeenthalpie f	chaleur f de prise *(ciment)*, chaleur de durcissement	тепло, выделяемое при схватывании
H 91	**heat of solidification**	Erstarrungswärme f, Erstarrungsenthalpie f	chaleur f de solidification	теплота затвердевания
H 92	**heat of subcooling**	Nachkühlwärme f, Nachkühlenthalpie f	chaleur f de sous-refroidissement	теплота переохлаждения
H 93	**heat of sublimation**	Sublimationswärme f, Sublimationsenthalpie f	chaleur f de la sublimation	теплота сублимации (возгонки)
H 94	**heat of transformation (transition)**	Umwandlungswärme f, Umwandlungsenthalpie f	chaleur f de transformation	теплота превращения
H 95	**heat of transition**	Übergangswärme f, Übergangsenthalpie f	chaleur f de transition	теплота перехода
H 96	**heat of wetting**	Benetzungswärme f, Benetzungsenthalpie f	chaleur f de mouillage	теплота смачивания
H 97	**heat output**	Wärmeleistung f, Wärmeabgabe f	puissance f thermique, débit m calorifique, débit (émission f) de chaleur	теплопроизводительность, теплоотдача, тепловая мощность
H 98	**heat-proof quality**	Wärmebeständigkeit f, Wärmefestigkeit f	résistance f à la chaleur, résistance contre les effets de la chaleur	теплостойкость, термостойкость
H 99	**heat pump**	Wärmepumpe f	thermopompe f, pompe f à (de) chaleur	тепловой насос
H 100	**heat recovery**	Wärmerückgewinnung f	récupération f de la chaleur	утилизация (обратное превращение) тепла
H 101	**heat rejection (removal)**, heat dissipation	Wärmeabfuhr f, Wärmeabgabe f, Wärmedissipation f, Wärmeabführung f	dissipation (émission) f de la chaleur, dissipation à chaud	отвод (отведение, диссипация) тепла, теплоотвод, теплосъем, теплоотдача
H 102	**heat requirement**	Wärmebedarf m, Heizwärmebedarf m	chaleur f nécessaire, besoins mpl en chaleur	требующееся тепло, необходимое количество тепла
H 103	**heat reservoir**	Wärmespeicher m	accumulateur m de chaleur, régénérateur m	тепловой аккумулятор, теплосборник
H 104	**heat resistance**	Hitzebeständigkeit f	résistance f à la chaleur, réfractarité f	жаропрочность, теплостойкость
H 105	**heat source**, source of heat	Wärmequelle f	source f de chaleur	источник тепла, тепловой источник
H 106	**heat stability**, hot strength, high-temperature stability	Wärmefestigkeit f	résistance f à la chaleur	жаростойкость, теплостойкость, нагревостойкость
H 107	**heat storage**	Wärmespeicherung f	accumulation f de chaleur	аккумулирование тепла, тепловое аккумулирование
H 108	**heat supervision**	Wärmekontrolle f	contrôle m de chaleur	тепловой контроль
H 109	**heat supply**, heat input	Wärmezufuhr f	amenée (admission) f de chaleur	приток (затрата, подвод) тепла, подогрев
H 110	**heat surface**, fire surface	Heizfläche f	surface f chauffée (de chauffe)	поверхность нагрева
H 111	**heat tone**, heat change (of reaction, effects)	Wärmetönung f	chaleur f de réaction	тепловой эффект, экзотермия
H 112	**heat transfer**	Wärmeübergang m	transfert m de chaleur	теплопередача, переход тепла
H 113	**heat transfer**	Wärmedurchgang m	passage m (transmission f) de chaleur, écoulement m thermique	теплопередача, передача тепла
	heat transfer	s. a. H 120		
H 114	**heat transfer coefficient**	Wärmeübergangskoeffizient m, Wärmeübergangszahl f	coefficient m de transfert de chaleur	коэффициент теплопередачи
H 115	**heat transfer medium**	Wärmeübertragungsmittel n, Wärmeträger m	moyen m de transfert de chaleur, caloporteur m	теплоноситель, теплопередатчик
H 116	**heat transfer stage**	Wärmeübertragungsstufe f	unité f de transfert de chaleur	ступень теплопередачи
H 117	**heat transfer tray**	Wärmeübertragungsboden m	plateau m de transfert de chaleur	теплообменная тарелка
H 118	**heat transmission**, conduction of heat	Wärmeleitung f	conduction f de chaleur	теплопроводность
H 119	**heat transmission counter flow principle**	Gegenstromwärmeübertragungsprinzip n	principe m de transfert de chaleur à contre-courant	принцип теплопередачи противотоком
H 120	**heat transport**, heat transfer	Wärmetransport m	transport m thermique, transfert m de chaleur	теплопередача, транспорт тепла
	heat value	s. T 137		
H 121	**heavy end**	hochsiedendes Destillat n	queue f de distillation, distillat m à ébullition élevée	тяжелые погоны
H 122	**heavy oil**	Schweröl n, schweres Öl n	huile f lourde	тяжелое масло
H 123	**heavy water**	schweres Wasser n, Deuteriumoxid n	eau f lourde	тяжелая вода
H 124	**height indicator**, level indicator	Inhaltsanzeiger m, Füllstandsanzeiger m	indicateur m de niveau	указатель наличия (уровня), счетчик расхода

H 125	height of lift	Ansaughöhe f	hauteur f d'aspiration	высота всасывания (подачи)
H 126	helical blower, spiral blower	Schraubengebläse n	ventilateur m hélicoïdal, ventilateur â hélice	винтовой (пропеллерный, лопастный) вентилятор
H 127	helical flow	schraubenförmige Strömung f	écoulement m hélicoïde	винтовое течение
H 128	helical ribbon agitator, spiral mixer	Wendelrührer m	agitateur m à bande hélicoidale	спиральная мешалка
H 129	helix angle	Steigungswinkel m (der Extruderschnecke)	angle m d'hélice	угол наклона
H 130	Helmholtz theorem	Helmholtz-Theorem n	théorème m de Helmholtz	теорема Гельмгольца
H 131	Hencky strain measure	Henckysches Dehnungsmaß n	mesure f de dilatation de Hencky, mesure d'extension de Hencky	мера растяжения Генки
H 132	Henry's law	Henry-Daltonsches-Gesetz n	loi f de Henry-Dalton	закон Генри и Дальтона
H 133	hermetic, vacuum-tight	vakuumdicht	hermétique	вакуум-плотный, герметичный
H 134	heterogeneous reactor	Heterogenreaktor m	réacteur m hétérogène	гетерогенный реактор, реактор для гетерогенных реакций
H 135	heterogeneous reactor model	heterogenes Reaktormodell n	modèle m de réacteur hétérogène	гетерогенная модель реактора
H 136	heteropolar	heteropolar	hétéropolaire	гетерополярный, разнополярный
H 137	heteropolar bond	heteropolare Bindung f	liaison f hétéropolaire	гетерополярная связь
H 138	heteropolycondensation, copolymerization	Heteropolykondensation f	polycondensation f hétérogène	гетерополиконденсация, совместная поликонденсация
H 139	heuristic design method	heuristische Entwurfsmethode f	méthode f heuristique de projection	эвристический метод проектирования
H 140	heuristic methods	heuristische Methoden fpl	méthodes fpl heuristiques	эвристические методы
H 141	high boiling	hochsiedend	de point m d'ébullition élevé, à haut point d'ébullition	высококипящий
H 142	high-boiling component	hochsiedende Komponente f	composante f à ébullition élevée	высококипящий компонент
H 143	high-boiling fraction	hochsiedende Fraktion f	fraction f lourde (à ébullition élevée)	фракция с высокой точкой кипения
H 144	high-boiling part	hochsiedender Anteil m	partie f à ébullition élevée	высококипящий компонент
H 145	high-energy	s. E 163		
H 145	high-intensity internal batch mixer	diskontinuierlicher Hochleistungsinnenmischer m	malaxeur (mélangeur) m discontinu à haute intensité	периодический смеситель высокой единичной мощности
H 146	highly disperse material system	hochdisperses Stoffsystem n	système m fortement dispersé de matière	высокодисперсная система веществ
H 147	high polymers	Hochpolymere npl	superpolymères mpl	высокополимерные соединения
H 148	high-pressure air	Hochdruckluft f	air m comprimé	воздух высокого давления
H 149	high-pressure area	Hochdruckbereich m	zone f de haute pression	зона высокого давления
H 150	high-pressure chamber	Hochdruckkammer f	chambre f à haute pression	камера высокого давления
H 151	high-pressure channel	Hochdruckleitung f	conduite f à haute pression	линия высокого давления
H 152	high-pressure circuit	Hochdruckkreislauf m	circuit m (circulation f) à haute pression	схема (контур) высокого давления
H 153	high-pressure distillation plant	Hochdruckdestillationsanlage f	unité f de distillation à haute pression	перегонная установка высокого давления
H 154	high-pressure evaporator	Hochdruckverdampfer m	évaporateur m à haute pression	испаритель высокого давления
H 155	high-pressure facility	Hochdruckanlage f	unité f à haute pression	высоконапорная установка, установка высокого давления
H 156	high-pressure gas, compressed gas	Preßgas n, komprimiertes Gas n	gaz m surpressé (comprimé)	сжатый газ
H 157	high-pressure hot separator	Hochdruckheißabscheider m	séparateur m à chaud à haute pression	горячий сепаратор (высокого давления)
H 158	high-pressure hydrogenation	Hochdruckhydrierung f	hydrogénation f sous haute pression	гидрирование под высоким давлением
H 159	high-pressure process	Hochdruckprozeß m, Hochdruckverfahren m	procédé (processus) m à haute pression	процесс при высоком давлении, метод высокого давления
H 160	high-pressure reactor	Hochdruckreaktor m	réacteur m à haute pression	реактор высокого давления
H 161	high-pressure stage	Hochdruckstufe f	étage m haute pression	ступень высокого давления, высоконапорная ступень
H 162	high-pressure steam	Hochdruckdampf m	vapeur f [à] haute pression	пар высокого давления
H 163	high-pressure synthesis	Hochdrucksynthese f	synthèse f sous haute pression	синтез при высоком давлении
H 164	high-pressure technology	Hochdrucktechnologie f	technologie f haute pression	технология высокого давления
H 165	high-pressure transport	Hochdruckförderung f	transport m sous haute pression	высоконапорный транспорт
H 166	high-pressure tube	Hochdruckrohr n	tuyau m à haute pression	трубопровод высокого давления
H 167	high-pressure vessel	Hochdruckbehälter m	réservoir m à haute pression	сосуд высокого давления
H 168	high-side equipment	druckseitige Ausrüstung f	partie f (équipement m) à haute pression	оборудование на стороне высокого давления

	high-speed evaporator	s. F 198		
	high-temperature carbonization	s. C 364		
H 169	high-temperature distillation	Hochtemperaturdestillation f	distillation f à haute température	перегонка при высокой температуре
H 170	high-temperature reactor	Hochtemperaturreaktor m	réacteur m à haute température	высокотемпературный реактор
H 171	high-temperature short-term evaporation	Hochtemperatur-Kurzzeiteindampfung f	vaporisation f instantanée à haute température	мгновенное упаривание при высокой температуре
	high-temperature stability	s. H 106		
H 172	high vacuum	Hochvakuum n	vide m élevé	высокий вакуум
H 173	high-vacuum condensation	Hochvakuumkondensation f	condensation f à vide élevé	высоко-вакуумная конденсация
H 174	high-vacuum distillation plant	Hochvakuumdestillationsanlage f	unité f de distillation sous vide élevé	установка перегонки под высоким вакуумом
H 175	high-vacuum drying	Hochvakuumtrocknung f	séchage m à vide élevé	сушка под высоким вакуумом
H 176	high-vacuum enclosure	Vakuumbehälter m	enceinte f à vide élevé, bassin m à vide	вакуумная камера (ловушка)
H 177/8	high-vacuum insulated vessel	Behälter m mit Hochvakuumisolierung f	réservoir m isolé à vide élevé	сосуд с высоковакуумной изоляцией
	hoist	s. W 169		
H 179	homogeneous reactor	Homogenreaktor m	réacteur m homogène	гомогенный реактор
H 180	homogeneous temperature	einheitliche Temperatur f	température f uniforme	однородная температура
H 181	homogenization	Homogenisierung f	homogénéisation f	гомогенизирование, гомогенизация
H 182	homogenization time	Homogenisierzeit f	temps m de homogénéisation	время гомогенизации
H 183	homogenize/to	homogenisieren	homogéniser	гомогенизировать
H 184	homopolymerization	Homopolymerisation f	homopolymérisation f	гомополимеризация
H 185	hood, air pipe	Abzug m (für Abgase)	hotte f	вытяжная труба
	hood	s. a. C 808		
H 186	Hookean elasticity	Hookesche Elastizität f	élasticité f de Hooke	Гуковская эластичность (упругость)
H 187/8	Hooke's law	Hookesches Gesetz n	loi f de Hooke	закон Гука
H 189	hopper	Schüttgutbehälter m, Silo n, Vorratsbehälter m, Bunker m, Einfülltrichter m, Fülltrichter m, Schütttrichter m	silo m, réservoir m, trémie f entonnoir (de chargement)	бункер, [загрузочная] воронка
H 190	hopper dryer	Trichtertrockner m	entonnoir m sécheur	воронковая (бункерная) сушилка
H 191	horizontal filter press	Horizontalplattenfilter n	filtre m à plaques horizontal	горизонтальный пластинчатый фильтр, горизонтальный фильтр-пресс
H 192	hot air	Heißluft f, Warmluft f	air m chaud	горячий воздух
H 193	hot-air bath	Heißluftbad n	bain m d'air chaud	баня с нагретым воздухом
	hot-air cabinet	s. D 538		
H 194	hot-air drying	Heißlufttrocknung f	séchage m par l'air chaud	горячевоздушная сушка
	hot blast stove	s. B 123		
H 195	hot-flash drum	Heißabscheider m	séparateur m de détente à chaud	горячий сепаратор со сбросом давления
H 196	hot-gas strainer	Heißgasfilter n	filtre m de gaz chaud	фильтр для горячего газа
H 197	hot polymer	Wärmepolymerisat n	polymère m à chaud	горячий полимер
H 198	hot redundancy	heiße Redundanz f	redondance f chaude	горячее резервирование, горячий редунданс
H 199	hot steam outlet	Heißdampfentnahme f	soutirage m de vapeur surchauffée	отбор перегретого пара
	hot strength	s. H 106		
	hot tube	s. F 170		
H 200	hot water	Heißwasser n	eau f chaude	горячая вода
H 201	hot working	Heißbearbeitung f	traitement m à chaud	горячая обработка
	hourly output	s. P 558		
H 202	housing, casing, shell	Mantel m, Gehäuse n	bâti m, corps m, enveloppe f, chemise f, boîtier m	кожух, корпус
H 203	humid air	feuchte Luft f	air m humide	влажный воздух
H 204	humid gas	feuchtes Gas n	gaz m humide	влажный газ
H 205	humidification	Befeuchtung f, Anfeuchtung f	humidification f, humectation f	увлажнение
H 206	humidification efficiency	Befeuchtungswirkungsgrad m	rendement m d'humectation	коэффициент полезного действия увлажнения
H 207/8	humidification system	Befeuchtungssystem n	système m d'humidification, système d'humectation	система увлажнения
H 209	humidifying capacity	Befeuchtungsleistung f	capacité f d'humidification	производительность по увлажнению
H 210	humidifying chamber	Befeuchtungskammer f	chambre f d'humidification	увлажнительная камера
H 211	humidifying effect	Befeuchtungseffekt m	effet m d'humidification	эффект увлажнения
H 212	humidifying machinery	Befeuchtungsausrüstung f	machinerie f d'humidification, installation f d'humectation	оборудование для увлажнения
H 213	humidity difference	Feuchtigkeitsdifferenz f	différence f d'humidité	разность влагосодержания
H 214	humidity range	Feuchtigkeitsbereich m	plage m d'humidité	диапазон влажности
H 215	humidity variation	Feuchtigkeitsänderung f	variation f d'humidité	изменение влажности

H 216	hybrid computer	Hybridrechner *m*	calculateur *m* hybride	гибридная вычислительная машина
H 217	hybrid models	hybride Modelle *npl*	modèles *mpl* hybrides	гибридные модели
H 218	hydrate formation	Hydratbildung *f*	formation *f* de l'hydrate	образование гидрата
H 219	hydration	Hydratation *f*, Hydratisierung *f*	hydratation *f*	гидратация
H 220	hydration plant	Hydratationsanlage *f*	unité *f* d'hydratation	гидратационная установка
H 221	hydration process	Hydratationsvorgang *m*	procédé *m* d'hydratation	гидратация, процесс гидратации (рафинации масел путем гидратации)
H 222	hydration water	Hydratwasser *n*	eau *f* d'hydratation	гидратная вода
H 223	hydraulic continuous conveyor	hydraulischer Stetigförderer *m*	convoyeur *m* continu hydraulique	гидравлический конвейер непрерывного действия
H 224	hydraulic drive	Druckwasserantrieb *m*, Hydraulikantrieb *m*	commande *f* hydraulique	гидравлический привод
H 225	hydraulic efficiency	hydraulischer Wirkungsgrad *m*	rendement *m* hydraulique	гидравлический коэффициент полезного действия
	hydraulic resistance	*s.* F 262		
H 226	hydraulic water	Druckwasser *n*	eau *f* comprimée (sous pression)	вода под напором (давлением)
H 227	hydrocarbon	Kohlenwasserstoff *m*	hydrocarbure *m*, carbure *m* d'hydrogène	углеводород
H 228	hydrochlorination	Hydrochlorierung *f*	hydrochlorination *f*	гидрохлорирование
H 229	hydrocracker	Hydrokracker *m*, Hydrokrackinganlage *f*	unité *f* de cracking hydrogénant	установка гидрокрекинга
H 230	hydrocracking	Hydrospalten *n*, Hydrokrakken *n*	cracking *m* hydrogénant	гидрокрекинг
H 231	hydrodesulphurization	Wasserstoffentschwefelung *f*	hydrodésulfuration *f*	сероочистка в присутствии водорода
H 232	hydrofiner	Hydrofiner *m*, Hydroraffinator *m*	unité *f* d'hydrofining	установка гидроочистки
H 233	hydrofining	Wasserstoffraffination *f*	hydrofining *m*	гидроочистка
H 234	hydrofinishing plant	Hydrofinishanlage *f*	unité *f* d'hydrofinissage	установка гидроочистки
H 235	hydroforming plant	Hydroformer *m*	unité *f* d'hydroforming	установка гидроформинга
H 236	hydrogenation	Hydrieren *n*, Hydrierung *f*	hydrogénation *f*	гидрирование
H 237	hydrogenation capacity	Hydrierungskapazität *f*, Hydrierkapazität *f*	capacité *f* d'hydrogénation	производственная мощность гидрирования
H 238	hydrogenation catalyst	Hydrierkatalysator *m*	catalyseur *m* d'hydrogénation	катализатор гидрирования
H 239	hydrogenation cracking	spaltende Hydrierung *f*	hydrogénation *f* destructive	деструктивная гидрогенизация
H 239a	hydrogenation plant	Hydrieranlage *f*	unité *f* d'hydrogénation	установка гидрогенизации
H 240	hydrogenation pressure	Hydrierungsdruck *m*	pression *f* d'hydrogénation	давление при гидрировании
H 241	hydrogenation process	Hydrierprozeß *m*	procédé *m* d'hydrogénation	процесс гидрирования
H 242	hydrogenation product	Hydrierprodukt *n*	produit *m* d'hydrogénation	продукт гидрирования
H 243	hydrogenation reactor	Hydrierreaktor *m*	réacteur *m* d'hydrogénation	реактор гидрирования
H 244	hydrogenation selection	selektive Hydrierung *f*	hydrogénation *f* sélective	селективная гидрогенизация
H 245	hydrogenation temperature	Hydriertemperatur *f*	température *f* d'hydrogénation	температура гидрирования
H 246	hydrogenation yield	Hydrierungsausbeute *f*	rendement *m* d'hydrogénation	выход при гидрировании
H 247	hydrogen compressor	Wasserstoffkompressor *m*	compresseur *m* d'hydrogène	компрессор для водорода
H 248	hydrogen consumption	Wasserstoffverbrauch *m*	consommation *f* d'hydrogène	расход водорода
H 249	hydrogen electrode	Wasserstoffelektrode *f*	électrode *f* à hydrogène	водородный электрод
H 250	hydrogen extracting plant	Wasserstoffgewinnungsanlage *f*	unité *f* de séparation de l'hydrogène	установка выделения водорода
H 251	hydrogen ion concentration	Wasserstoffionenkonzentration *f*	concentration *f* des ions d'hydrogène	концентрация ионов водорода
H 252	hydrogen plant	Wasserstoffanlage *f*	installation *f* de fabrication d'hydrogène	установка получения (производства) водорода
H 253	hydrogen production	Wasserstoffherstellung *f*	fabrication (production) *f* d'hydrogène	производство водорода
H 254	hydrogen recovery cost	Wasserstoffgewinnungskosten *pl*	frais *mpl* de production d'hydrogène	стоимость производства водорода
H 255	hydrogen recycle	Wasserstoffkreislauf *m*	circulation *f* de l'hydrogène	циркуляция водорода
H 256	hydrogen recycle stream	Wasserstoffkreisgasstrom *m*	courant *m* d'hydrogène de circulation	поток циркуляционного водорода
H 257	hydrogen upgrading	Wasserstoffanreicherung *f*	enrichissement *m* de l'hydrogène	обогащение водорода
H 258	hydrorefining plant	Hydroraffinationsanlage *f*	unité *f* d'hydroraffinage, unité d'hydrofining	установка гидроочистки
	hygrometer	*s.* M 305		
H 259	hygrometry	Feuchtigkeitsmessung *f*	hygrométrie *f*, psychrométrie *f*	гигрометрия

I

| I 1 | ice plugging (*of a distillation tower*) | Eispfropfenbildung *f* | formation *f* de chevilles en glace | закупорка льдом |

I 2	**ideal borderline cases**	ideale Grenzfälle *mpl*	cas *mpl* limites idéals	идеальные предельные (граничные) случаи
I 3	**ideal Carnot's principle**	idealer Carnot-Prozeß *m*	cycle *m* de Carnot idéal	идеальный цикл Карно
I 4	**ideal compression**	ideale Verdichtung *f*	compression *f* idéale	идеальное сжатие
I 5	**ideal cycle efficiency**	idealer Prozeßwirkungsgrad *m*	rendement *m* de cycle idéal	коэффициент полезного действия идеального цикла
I 6	**ideal elastic body,** ideal elastic body	idealer elastischer Körper *m*, ideal-elastischer Hooke-Körper *m*	solide (corps) *m* parfaitement élastique	идеально-эластичное тело [Гука], идеально-упругое тело [Гука]
I 7	**ideal gas**	ideales Gas *n*	gaz *m* idéal	идеальный газ
I 8	**ideal mixing**	ideale Durchmischung *f*	mélange *m* idéal	идеальное перемешивание
I 9	**ideal models**	ideale Modelle *npl*	modèles *mpl* idéals	идеальные модели
I 10	**ideal plate**	theoretischer Boden *m*	plateau *m* théorique	теоретическая тарелка
I 11	**ideal plate number**	theoretische Bodenzahl *f*	nombre *f* des plateaux théoriques	теоретическое число тарелок
I 12	**ideal refrigerating process**	idealer Kälteprozeß *m*	processus *m* frigorifique idéal	идеальный холодильный цикл (процесс)
I 13	**ideal tray**	idealer Boden *m*	plateau *m* idéal	идеальная тарелка
I 14	**identification,** decision	Identifizierung *f*, Bestimmung *f*	fixation *f*, définition *f*, dosage *m*, analyse *f*	идентификация, определение
I 15	**identification algorithm**	Erkennungsalgorithmus *m*	algorithme *m* d'identification	алгоритм идентификации
I 16	**idle time**	Verlustzeit *f*	temps *m* improductif	потерянное время, время потери
I 17	**idling,** idling speed	Leerlaufdrehzahl *f*	nombre *m* de tours à marche à vide, vitesse *f* à vide	число оборотов холостого хода
	idling	*s. a.* N 39		
I 18	**idling appliance**	Leerlaufeinrichtung *f*	dispositif *m* de ralenti (marche à vide)	система холостого хода
I 19	**idling conditions**	Leerlaufzustand *m*	conditions *fpl* d'inactivité, état *m* de marche à vide	режим холостого хода
	idling speed	*s.* I 17		
I 20	**ignitability**	Zündempfindlichkeit *f*	inflammabilité *f*	воспламеняемость
I 21	**ignition aid,** ignition impulse	Zündinitial *n*	initiale *f* d'amorçage	инициальный (взрывной) импульс
I 22	**ignition groups**	Zündgruppen *fpl*	groupes *mpl* d'inflammabilité	взрывные группы
	ignition impulse	*s.* I 21		
I 23	**ignition mixture**	Zündmischung *f*, Entzündungsgemisch *n*	mélange *m* d'inflammation	воспламенительная смесь
I 24	**ignition point**	Zündpunkt *m*	point *m* d'allumage	точка воспламенения
I 25	**ignition quality**	Zündeigenschaft *f*	aptitude *f* à l'inflammation	способность к воспламенению
I 26	**ignition range**	Zündbereich *m*	domaine *m* d'explosibilité	пределы воспламеняемости, область воспламеняемости
I 27	**ignition source**	Zündquelle *f*	source *f* d'inflammation	очаг зажигания, источник зажигания
I 28	**ignition temperature,** inflammation temperature	Zündtemperatur *f*, Entzündungstemperatur *f*	température *f* d'allumage, température d'ignition, température d'inflammation	температура воспламенения (вспышки)
I 29	**ignition velocity**	Zündgeschwindigkeit *f*	vitesse *f* d'allumage, vitesse de propagation d'une flamme	скорость воспламенения
I 30	**illustration**	Veranschaulichung *f*	illustration *f*	иллюстрация, изображение
I 31	**image equation**	Abbildungsgleichung *f*	équation *f* d'image	уравнение проекции, уравнение изображения
I 32	**imaging unit**	Abbildungsgerät *n*	dispositif *m* de représentation	прибор для изображения
	imitation	*s.* C 736		
I 33	**immersed batterie**	Tauchapparat *m*	appareil *m* de trempage	погружной аппарат, погружная батарея
I 34	**immersed condenser**	Tauchkondensator *m*	condenseur *m* à immersion	погружной конденсатор
I 35	**immersion pipe**	Eintauchrohr *n*	tuyau *m* d'immersion	погружная труба
I 36	**immersion refrigeration**	Tauchkühlung *f*, Eintauchkühlung *f*	réfrigération *f* par immersion	охлаждение погружением
	immersion tube	*s.* P 226		
I 37	**immersion-type cooler**	Tauchkühler *m*	refroidisseur *m* plongeant, refroidisseur plongeur	погружной холодильник, аппарат для охлаждения погружением в жидкость
I 38	**immiscible**	un[ver]mischbar, nicht mischbar	non-miscible, non mélangeable	несмешиваемый
	impact	*s.* C 403		
I 39	**impact factor,** frequency factor	Stoßfaktor *m*, Häufigkeitsfaktor *m*	facteur *m* de collision (fréquence), facteur *m* d'impact	частотный фактор, коэффициент соударений
I 40	**impactor nozzle**	Pralldüse *f*	buse *f* de contre-coup	ударноотражательное сопло
I 41	**impact pressure**	Stoßdruck *m*	pression *f* dynamique	ударное давление
I 42	**impact resistance**	Kerbfestigkeit *f*	résistance *f* à l'entaille	сопротивление на удар
	impact resistance	*s. a.* I 46		

I 43	impact resistant	schlagzäh	résilient	ударопрочно
I 44	impact screen	Prallsieb n, Vibrationssieb n	tamis m à impact, crible m vibreur	грохот, сотрясательный грохот
I 45	impact strength	Kerbschlagzähigkeit f	sensibilité f à l'entaille	прочность на удар бруска с надпилом
I 46	impact strength, impact resistance	Schlagfestigkeit f	résistance f au choc	ударная прочность
I 47	impact theory	Stoßtheorie f	théorie f d'impact	теория столкновений
I 48	impact velocity	Aufprallgeschwindigkeit f	vitesse f d'impact	скорость удара, скорость столкновения
I 49	impeller	drehende Rührstange f, Schnellrührer m	roue f agitatrice	скоростная мешалка, вращающийся вал мешалки
I 50	impermeable	undurchlässig	imperméable, impénétrable	непроницаемый, непропускающий
I 51	impetus, impulsion, power	Antrieb m	commande f, mouvement m, impulsion f	привод
I 52	implementation phase	Durchführungsphase f	phase f de réalisation	фаза реализации, этап осуществления
I 53	implementation time	Ausführungszeit f	temps m d'exécution	время исполнения (работы)
I 54	importance	Bedeutung f	sens m, signification f	значение, важность, значительность
I 55	impoundage	Stau m	retenue f	подпор
I 56	impregnability	Tränkungsvermögen n	imprégnabilité f	впитывающая способность, насыщаемость
I 57	impregnating machine	Tränkgerät n	dispositif m d'imprégnation	пропиточный прибор
I 58	impregnation	Tränken n	imprégnation f	пропитывание, импрегнирование
I 59	impregnation method	Tränkverfahren n	procédé m d'imprégnation	способ пропитки, пропитывание
I 60	imprint, stamp	Aufdruck m	empreinte f, impression f	надпечатка
I 61	improve/to, to refine, to upgrade	veredeln, raffinieren	améliorer, purifier, élever	рафинировать, облагораживать
I 62	improvement	Weiterentwicklung f, Verbesserung f, Entwicklung f	développement m ultérieur	развитие, улучшение
I 63	impulse	Impuls m	impulsion f, quantité f de mouvement	импульс
I 64	impulse balances	Impulsbilanzen fpl	bilans mpl de quantité de mouvement	балансы импульса
	impulsion	s. I 51		
I 65	impure	unrein, verunreinigt	impur, sale	нечистый, загрязненный
	impurity	s. C 627		
I 66	inaccuracy	Ungenauigkeit f	imprécision f, inexactitude f	неточность
	inactive gas	s. R 39		
I 67	inactivity	Inaktivität f	inactivité f	неактивность, дезактивация
I 68	incidence list	Inzidenzliste f	liste f d'incidence	список смежностей
I 69	incidence matrix	Inzidenzmatrix f	matrice f d'incidence	матрица смежностей
I 70	incineration	Veraschung f	incinération f	сжигание
I 71	incinerator	Müllverbrennungsanlage f	usine f d'incinération des ordures ménagères	мусоросжигательная установка
I 72	inclination	Neigung f	déclivité f, inclinaison f, pente f	наклон
I 73	inclined flat surface	schräge flache Fläche f	surface f plate inclinée	наклонная плоская поверхность
I 74	inclined grate	Schrägrost m	grille f inclinée	наклонная колосниковая решетка
I 75	inclined tube, vertical tube, forced-circulation generator	Zwangsumlauferzeuger m, Zwangsdurchlauferzeuger m	générateur m à circulation forcée	наклонная труба, генератор принудительной циркуляции
I 76	incoming gas	eintretender Dampf m	vapeur f d'admission, vapeur d'entrée	поступающий пар
I 77	incoming-material control	Eingangskontrolle f, Eingangswarenkontrolle f	contrôle m des articles achetés, contrôle d'entrée	контроль на входе, входной контроль
I 78	incoming stream, input stream	Eintrittsstrom m, eintretender Strom m	courant m d'entrée, flux m entrant	входящий (поступающий) поток
I 79	incompressibility	Inkompressibilität f	incompressibilité f	несжимаемость
I 80	incompressible body	inkompressibler Körper m	corps m incompressible	несжимаемое тело
I 81	incompressible flow	inkompressible Strömung f	écoulement m incompressible	несжимаемое течение
I 82	inconstancy	Unbeständigkeit f, Instabilität f	instabilité f	непостоянство, нестабильность
I 83	incrustation	Verkrustung f, Bekrustung f	incrustation f	отложение корки (настыли, накипи, окалины, осадка), крустификация, накипь, корка
	indeterminacy	s. U 4		
I 84	indicated work	induzierte (innere) Arbeit f	travail m induit (interne)	внутренняя (индикаторная) работа
I 85	indirect air cooling	indirekte Luftkühlung f	refroidissement m d'air indirect	охлаждение воздуха с промежуточным холодоносителем, непрямое воздушное охлаждение

	English	German	French	Russian
I 86	indirect cooling	indirekte Kühlung f	refroidissement m indirect	непрямое охлаждение
I 87	indirect heat exchange (transfer)	indirekter Wärmeaustausch m, indirekte Wärmeübertragung f	échange (transfert) m de chaleur indirect	непрямая (косвенная) теплопередача
I 88	individual absorption	Eigenabsorption f	autoabsorption f, absorption f propre	собственная абсорбция
	individual parts	s. P 144		
I 89	indoor battery	Raumkühler m	refroidisseur m des locaux	охладитель помещения
I 90	indoor blower	Raumlüfter m	soufflante f intérieure	вентилятор для помещения, камерный вентилятор
I 91	indoor temperature	Raumtemperatur f	température f intérieure	температура помещения
I 92	induced air draft	Saugluftstrom m	courant m d'air d'aspiration	всасывающий воздушный поток, поток всасываемого воздуха
I 93	induced air stream	Ansaugluftstrom m	courant m d'air aspiré	всасываемый поток воздуха
I 94	inducing current	Primärstrom m	courant m primaire, courant inducteur	первичный ток
I 95	induction generator, asynchronous alternator	Asynchrongenerator m	génératrice f asynchrone	асинхронный генератор
I 96	induction motor	Asynchronmotor m	moteur m asynchrone	асинхронный двигатель
I 97	induction period, reaction time	Einwirkungszeit f	temps m d'influence, temps de réaction	время воздействия
I 98	industrial air conditioner	Industrieklimaanlage f	conditionneur m d'air industriel	промышленный кондиционер
I 99	industrial air cooler	Industrieluftkühler m	refroidisseur m d'air industriel	промышленный воздухоохладитель
I 100	industrial building	Industriebau m	bâtiment m industriel	промышленное строительство
I 101	industrial cooler	Industriekühler m	refroidisseur m industriel	промышленный охладитель
I 102	industrial cooling tower	Industriekühlturm m	tour f réfrigérante industrielle	промышленная градирня (охладительная башня)
I 103	industrial ecology	Industrieökologie f	écologie f industrielle	промышленная экология
I 104	industrial furnace	Industrieofen m	four m industriel (pour l'industrie)	промышленная печь
I 105	industrial plant construction	Anlagenbau m	construction f des installations industrielles	капитальное строительство, конструкция и сооружение установок
I 106	industrial plant construction process	Anlagenbauprozeß m	procédé m de construction des installations industrielles	процесс конструкции и сооружения установок, процедура капитального строительства
I 107	industrial product	Industrieerzeugnis n	produit m industriel	промышленное изделие, промышленный продукт
I 108	industrial production	Industrieproduktion f	production f industrielle	промышленное производство, промышленная продукция
I 109	industrial reactor	Industriereaktor m	réacteur m technique (industriel)	промышленный реактор
I 110	industrial refrigerating plant	Industriekälteanlage f	installation f frigorifique industrielle	промышленная холодильная установка
I 111	industrial refrigeration	Industriekühlung f	refroidissement m industriel	промышленное охлаждение
I 112	industrial research	Betriebsforschung f, Industrieforschung f	recherche f industrielle	заводское (промышленное) исследование
I 113	industrial safety	Arbeitssicherheit f	sécurité f des travailleurs	безопасность труда
I 114	industrial safety [regulations]	Arbeitsschutz m	protection f des travailleurs	охрана труда, техника безопасности
I 115	industrial sewage	Industrieabwässer npl	eaux fpl résiduaires	промышленные сточные воды
I 116	industrial turbine	Industrieturbine f	turbine f industrielle	промышленная турбина, турбина для промышленных целей
I 117	industrial waste	Industriemüll m, Industrieabfall m	déchets mpl industriels (de production)	промышленный мусор, производственные отходы
I 118	industrial waste gas	Industrieabgas n	gaz m brûlé (d'échappement) industriel	промышленный выхлопной газ, промышленный газовый выброс
I 119	industrial waste sludge	industrieller Abfallschlamm m	boue f industrielle de déchet	промышленный сточный (отработанный) ил
I 120	industrial water	Industriewasser n	eau f industrielle	промышленная вода, вода для промышленных целей
I 121	inert atmosphere	Atmosphäre f aus Inertgas, inerte Atmosphäre	atmosphère f inerte	инертная газовая среда
I 122	inert gas, protective (shielding, safety) gas	Schutzgas n, Inertgas n	gaz m protectif (protecteur, de protection, inerte)	защитный (инертный) газ, предохраняющий от оксиления газ
I 123	inert gas circuit	Inertgaskreislauf m	circuit m de gaz inerte	схема циркуляции инертного газа
I 124	inert gas plant	Inertgasanlage f	unité f des gaz inertes	установка по производству инертного газа

I 125	inert gas purification plant	Inertgasreinigungsanlage f	unité f de purification des gaz inertes	установка для очистки инертных газов
I 126	inertia	Beharrungsvermögen n, Trägheit f	inertie f, conservation f des forces vives	инерционность, инерция
I 127	inertia, inertness	Trägheitskraft f	force f d'inertie, inertie f	сила инерции
I 128	inertial cooling	Innenkühlung f	refroidissement m interne	внутреннее охлаждение
I 129	inerting blanketing systems	Inertisierungsanlage f, Inertgasanlage f	unité f de gaz inertes	установка для создания инертной атмосферы
I 130	inert material	Ballastmaterial n	matériu m de ballast	инертный (балластный) материал
	inertness	s. I 127		
I 131	infeed, dosage, charging, feeding	Beschickung f, Einspeisung f, Beschicken n	charge f, chargement m, enfournement m, alimentation f, action f de chargement	загрузка, заправка
I 132	infiltration, percolation	Versickerung f	infiltration f	инфильтрация, просачивание
I 133	inflammability	Entflammbarkeit f, Brennbarkeit f	inflammabilité f	воспламеняемость
I 134	inflammation	Entzündung f, Entflammen n	inflammation f	воспламенение
	inflammation temperature	s. I 28		
I 135	inflow	Anströmung f	soufflage m	натекание, приток, обтекание, обдув
	inflow	s. a. 1. I 139; 2. I 189		
I 136	influence	Beeinflussung f, Einfluß m	influence f, intervention f	влияние, воздействие
I 137	influencing function	Eingriffsfunktion f	fonction f d'influence	функция воздействия
I 138	influent, influx	Einlauf m	affluant m, entrée f	вход, приток, впуск
I 139	influx, inflow	Zufluß m, Zustrom m, Zuflußmenge f	afflux m, affluence f	приток, притекание, количество притока
I 140	information flow	Informationsfluß m	flux m d'informations	поток информации
I 141	information processing	Informationsverarbeitung f	traitement m des informations	обработка информации
I 142	ingredient	Bestandteil m, Ingrediens n	partie f intégrante, élément m, ingrédient m, constituant m, composant m	составная часть, компонент, ингредиент
	inhibit/to	s. P 449		
I 143	initial air temperature	Anfangslufttemperatur f	température f d'air initiale	начальная температура воздуха
I 144	initial amount of water	Ursprungswassermenge f, Anfangswassermenge f	quantité f d'eau initiale, quantité d'eau de début	начальное количество воды, количество исходной воды
I 145	initial boiling point	Anfangssiedepunkt m, Siedebeginn m	point m d'ébullition initial	начальная точка кипения, начало кипения, температура начала кипения
I 146	initial boundary element	Anfangsrandelement f	élément m marginal initial	исходный (начальный) граничный элемент
	initial charge	s. I 157		
I 147	initial chilling	Anfangskühlung f	refroidissement m initial	первичное (начальное) охлаждение
I 148	initial concentration	Anfangskonzentration f	concentration f initiale (de départ)	начальная (исходная) концентрация
I 149	initial condensation	Anfangskondensation f	condensation f initiale	первичная (начальная) конденсация
I 150	initial cost	Anlagekosten pl	frais mpl d'établissement	капиталовложения, стоимость установки (оборудования)
I 151	initial crusher	Vorbrecher m, Grobbrecher m	premier concasseur m, préconcasseur m	дробилка для первичного дробления
I 152	initial date	Anfangstermin m	terme m de départ	начальный срок
I 153	initial dimension of the system	Ausgangsgröße f des Systems, Ursprungsgröße f des Systems	grandeur f initiale du système	исходный параметр системы, исходная величина системы
I 154	initial dry content	Anfangstrockengehalt m	teneur m initial en matière sèche	начальное содержание сухого вещества
I 155	initial element	Anfangselement n	élément m initial	начальный элемент
I 156	initial height	Ausgangshöhe f, Ursprungshöhe f	hauteur f initiale	исходная высота
I 157	initial load, initial charge	Anfangsbeschickung f	charge f initiale	начальная нагрузка
I 158	initial loading, starting stress	Anfangsbeanspruchung f	charge (sollicitation) f initiale, effort m initial	исходное напряжение, начальная нагрузка
	initial loading	s. a. S 563		
I 159	initial moisture content	Anfangsfeuchte f	humidité f initiale	начальная влажность
I 160	initial plant	Ausgangsanlage f, Ursprungsanlage f	installation f initiale	исходная установка, исходное положение
I 161	initial point	Anfangspunkt m, Ursprung m	point m initial	исходная (начальная) точка
I 162	inital power	Anzugskraft f	effort m d'entrainement, effort de démarrage	начальная сила, тяговое усилие
I 163	initial pressure	Anfangsdruck m	pression f initiale	начальное давление
	initial product	s. R 63		
I 164	initial production	Anfangsproduktion f	production f initiale	начальное производство, начальная добыча

I 165	**initial rate method**	Methode der Anfangsge-schwindigkeit f	méthode f de la vitesse initiale	метод начальных скоростей
	initial speed	*s.* I 170		
I 166	**initial stability**	Anfangsstabilität f	stabilité f initiale	начальная устойчивость
I 167	**initial state,** original state	Ausgangszustand m	état m initial	исходное (начальное) состояние
I 168	**initial temperature**	Ausgangstemperatur f, Anfangstemperatur f	température f initiale	начальная (исходная) температура
I 169	**initial value**	Anfangswert m	valeur f initiale	начальное значение
I 170	**initial velocity,** initial speed	Anfangsgeschwindigkeit f, Ausgangsgeschwindigkeit f	vitesse f initiale	начальная (исходная) скорость
I 171	**initial water temperature**	Anfangswassertemperatur f	température f d'eau initiale	начальная температура воды
	initiating reaction	*s.* S 566		
I 172	**initiator decay**	Initiatorzerfall m	décomposition f d'excitant	распад инициатора
I 173	**injecting**	Einpressen n, Einspritzen n	injection f	впрыскивание, инъекция, вдавливание, нагнетание
I 174	**injection**	Einspritzung f, Injektion f, Einblasen n	injection f, insufflation f	впрыск, впрыскивание, инъекция, вдувание, продувка
	injection duration	*s.* I 182		
I 175	**injection input,** storage, accumulation	Einspeicherung f, Speicherung f	stockage m, accumulation f, mise f en mémoire (calculatrice)	хранение, накопление
	injection mould	*s.* M 119		
I 176	**injection moulding**	Spritzgußverfahren n, Spritzen n, Spritzguß m	moulage m sous pression, moulage m par injection	литье (метод литья) под давлением
I 177	**injection nozzle**	Einspritzdüse f	gicleur m, injecteur m, pulvérisateur m	форсунка
I 178	**injection pipe**	Einspritzrohr n	tube f d'injection	труба для впрыска
I 179	**injection pressure**	Einspritzdruck m, Einblasedruck m	pression f d'injection, pression f d'insufflation	давление впрыска (впрыскивания, вдувания)
I 180	**injection pump**	Injektorpumpe f, Einspritzpumpe f	pompe f à jet, pompe d'injection	инжекторный насос
I 181	**injection refrigeration**	Einspritzkühlung f	refroidissement m par injection	охлаждение впрыском холодильного агента
I 182	**injection time,** injection duration	Einspritzdauer f	durée f d'injection	продолжительность впрыска, длительность впрыска
I 183	**injection unit,** injector	Injektor m, Dampfstrahlanlage f	injecteur m, installation f à jet de vapeur	инжекционный узел
I 184	**injector,** spraying nozzle	Spritzdüse f	pulvérisateur m, gicleur m	разбрызгивающее сопло, форсунка
I 185	**injector**	Einspritzer m, Injektor m	injecteur m, pulvérisateur, pompe f à jet	инжектор
	injector	*s. a.* S 638		
I 186	**injector nozzle**	Injektordüse f	buse f d'injecteur	сопло инжектора
	injury	*s.* D 2		
I 187	**inleakage**	Eindringung f	affluence f	проникновение внутрь
I 188	**inlet,** admission, occurrence	Einlaß m, Eintritt m	entrée f, admission f	впуск, подвод, подача, вход
I 189	**inlet,** inflow, efflux, water feed	Zulauf m, Einspeisung f, Zuführung f	arrivée f, entrée f, amenée f, alimentation f	приток
	inlet	*s. a.* E 212		
I 190	**inlet air**	eintretende Luft f, Eintrittsluft f	air m d'entrée, air entrant	воздух на входе, поступающий воздух
I 191	**inlet air temperature**	Lufteintrittstemperatur f	température f d'air à l'entrée	температура воздуха на входе
I 192	**inlet area**	Eintrittsquerschnitt m, Eintrittsfläche f	section f d'admission, section d'entrée	входное сечение
I 193	**inlet chamber** (of a pump)	Saugraum m	chambre f d'aspiration	камера всасывания
I 194	**inlet chamber**	Eintrittskammer f	chambre f d'admission	камера впуска
	inlet chamber	*s. a.* F 26		
I 195	**inlet concentration**	Einlaufkonzentration f	concentration f à l'entrée, concentration d'alimentation	входная концентрация, концентрация на входе
I 196	**inlet conduit**	Eintrittsleitung f, Einlaßleitung f	conduite f d'admission, conduite d'entrée	впускной (впусковой) трубопровод
I 197	**inlet connection**	Einlaufstutzen m, Beschickungsstutzen m	tuyau m de charge, tubulure f d'introduction	приемный патрубок
I 198	**inlet file,** input data set	Eingabedatei f	fichier m d'entrée	входной файл, файл ввода
I 199	**inlet filter**	Einlaßfilter n	filtre m d'entrée	входной фильтр
I 200	**inlet flow rate**	Einlaufmenge f	quantité f d'alimentation	входной поток, расход на входе
I 201	**inlet gas**	Eingangsgas n	gaz m d'entrée	исходный газ
I 202	**inlet nozzle**	Einlaufdüse f	buse f d'entrée	впускное (загрузочное) сопло
I 203	**inlet opening**	Eintrittsöffnung f, Einlaßöffnung f	orifice m d'entrée, lumière f d'admission	входное отверстие
I 204	**inlet passage**	Einlaßkanal m	passage (canal) m d'admission	входной канал

I 205	inlet pipe	Einfüllstutzen m, Einströmungsrohr n	tubulure f de remplissage, tube m d'admission, tubulure d'entrée	приемная (впускная) труба
I 206	inlet pressure, running-in pressure	Eintrittsdruck m, Einlaßdruck m, Einlaufdruck m, Eingangsdruck m, Einspeisungsdruck m	pression f d'entrée, pression à l'entrée, pression d'alimentation	давление на входе, давление при впуске, входное давление
I 207	inlet pressure	Ansaugdruck m	pression f d'aspiration	давление всасывания
I 208	inlet state	Einlaufzustand m	état m à l'entrée, état d'entrée	входное состояние, состояние на входе
I 209	inlet system	Einlaßsystem n	système m d'admission	приемная (впускная) система
I 210	inlet temperature	Eintrittstemperatur f, Einlauftemperatur f, Eingangstemperatur f	température f d'entrée, température d'admission	температура на входе, входная температура
I 211	inlet temperature control	Eintrittstemperaturregelung f	réglage m de température d'entrée	регулирование температуры на входе
I 212	inlet tube	Eintrittsrohr n, Einlaßrohr n, Ansaugrohr n	tube f d'admission, tuyau m d'entrée, tuyau d'aspiration	впускная труба
I 213	inlet valve	Einlaßventil n	vanne f d'admission	клапан впуска, пусковой вентиль
I 214	inlet vapour	Eintrittsdampf m	vapeur f d'entrée	пар на входе
I 215	in-line blender	Mischer m in der Pumpleitung, „in-line"-Mischer m	mélangeur m intérieur	труба-смеситель
I 216	in-line blending	Mischen (Vermischen) n in der Pumpleitung, „in-line"-Mischung f	mélange m dans la conduite de refoulement	смешение в трубе
I 217	in-line filter	Einbaufilter n	filtre m incorporé	встроенный фильтр
I 218	inner dimension	Innenmaß n	largeur f intérieure, dimension f intérieure	внутренний размер
	inner energy	s. I 297		
I 219	inorganic compounds	anorganische Verbindung f	composé m minéral	неорганическое соединение
I 220	inorganic refrigerant	anorganisches Kältemittel n	réfrigérant m minéral, agent m frigorifique inorganique	неорганический холодильный агент
I 221	input	Input m, Eingangssignal n, Eingang m, Eingabe f	signal m d'attaque, signal d'entrée, entrée f	вход, ввод, поступление, подача, входной сигнал, сигнал на входе
I 222	input capacitance	Eingangskapazität f	capacité f d'entrée	входная емкость
I 223	input control unit	Eingabesteuereinheit f	unité f de commande d'entrée	устройство управления вводом
I 224	input data	Eingangsdaten pl	données fpl d'entrée	входные данные
	input data set	s. I 198		
I 225	input function	Eingangsfunktion f	fonction f d'entrée	входная функция
I 226	input losses	Eingangsverluste mpl	pertes fpl à l'entrée	потери на входе
I 227	input-output computer	Eingabe-Ausgabe-Rechner m	calculateur m d'entrée et de sortie	вычислительная машина ввода-вывода
I 228	input-output device	Eingabe-Ausgabe-Gerät n	dispositif m entrée-sortie	устройство ввода-вывода
I 229	input-output unit	Eingabe-Ausgabe-Einheit f	ensemble m d'entrée-sortie, unité f entrée-sortie	устройство ввода-вывода
I 230	input perturbation	Eingangsstörung f	trouble m d'entrée	входное возмущение
I 231	input power	Eingangsleistung f	puissance f d'entrée	входная мощность
I 232	input pulse	Eingangsimpuls m	impulsion f d'entrée	входной импульс
I 233	input quantity, input value	Eingangsgröße f	grandeur f d'entrée	входная величина, величина на входе
I 234	input quantity of the system	Eingangsgröße f des Systems	grandeur f d'entrée du système	входной параметр системы, параметр на входе в систему
I 235	input sensitivity	Eingangsempfindlichkeit f	sensibilité f à l'entrée	чувствительность на входе
	input stream	s. I 78		
	input value	s. I 233		
I 236	input variable	Eingangsvariable f	variable m d'entrée	входная переменная
I 237	input voltage	Eingangsspannung f	tension f d'entrée	входное напряжение
I 238	inquiry	Untersuchung f, Anfrage f	exploration f, étude f, examen m, inspection f, demande f	исследование, изучение, запрос
I 239	insertion, switching on	Einschaltung f, Einschalten n	mise f en marche, mise en circuit	включение, пуск
I 240	insertion condenser	Einsatzkühler m	refroidisseur m intérieur	вставной холодильник
I 241	inside diameter	Innendurchmesser m	diamètre m intérieur	внутренний диаметр
I 242	inside surface	Innenfläche f	surface f intérieure, paroi f intérieure	внутренняя поверхность
I 243	insoluble	nicht löslich, unlöslich	insoluble	нерастворимый
I 244	insoluble residue	unlöslicher Rückstand m	résidu m insoluble	нерастворимый остаток
I 245	inspection	Befahrung f, Besichtigung f, Befahren n	inspection f, examen m, examination f, visite f	осмотр, инспекция, проверка, ходовые испытания
I 246	inspection glass, oil-flow indicator	Schauglas n	verre m indicateur, voyant m	очко, глазок для наблюдения
	inspection test	s. C 207		

I 247	**instable,** unstable, deficient	instabil, labil	instable	нестабильный, нестойкий, неустойчивый
	installation	s. 1. A 566; 2. M 349; 3. P 188		
I 248	**installation optimization**	Aufstellungsoptimierung f	optimisation f de disposition	оптимизация размещения (оборудования)
I 249	**installation schedule**	Aufstellungsplan m	plan m de disposition	схема размещения, план расстановки
I 250	**instantaneous reaction**	Momentreaktion f	réaction f instantanée	мгновенная реакция
I 251	**instantaneous yield**	Augenblicksausbeute f	sélectivité f instantanée	мгновенный выход
I 252	**instruction,** direction, regulation, specification	Vorschrift f, Anweisung f, Instruktion f	prescription f, instruction f	инструкция, предписание, регламент
I 253	**instruction manual**	Bedienungsanweisung f	instruction f pour la conduction, règlement m de service	регламент, инструкция
I 254	**instrument leads,** test line, testing wire	Meßleitung f	conduite f de mesure, ligne f de mesure	измерительная линия
I 255	**instrument manufacture**	Gerätebau m	production f d'instruments	приборостроение
	instrument reading	s. T 106		
I 256	**insufficient temperature**	Untertemperatur f	température f trop basse, sous-température f	пониженная температура
I 257	**insulant, insulating material**	Isolierungsmaterial n	isolant m, matière f isolante	изоляционный материал
I 258	**insulation,** lagging, cleading	Dämmung f, Sperrung f, Isolierung f, Wärmedämmung f	isolement m, calorifugeage m	изоляция, изолирование
I 259	**insulator**	Isolator m	isolateur m, isoloir m	изолятор
I 260	**intake duct**	Saugkanal m	tunnel f (canal m) d'aspiration	всасывающий канал, впускной канал
I 261	**intake losses**	Saugverluste mpl	pertes fpl d'aspiration	потери на всасывании
I 262	**intake opening**	Ansaugöffnung f	orifice m d'aspiration	всасывающее отверстие
I 263	**intake volume**	Saugvolumen n	volume m aspiré (d'aspiration)	всасываемый объём
I 264	**integral formula**	Integralformel f	formule f intégrale	интегральная формула
I 265	**integral method**	Integralverfahren n, Integralmethode f	méthode f intégrale	интегральный метод
I 266	**integral representation**	Integraldarstellung f	représentation f intégrale	интегральное представление
I 267	**integral throttle expansion**	Drosseleffekt m	effet m d'étranglement	дроссельный эффект
I 268	**integrated structure**	Vereinigungsstruktur f	structure f intégrée	интегрально-гипотетическая структура
	intelligible	s. C 477		
	intensify/to	s. R 259		
I 269	**interbed cooling**	Zwischenbettkühlung f	refroidissement m aux lits intermédiaires	охлаждение между слоями
I 270	**interchangeability**	Austauschbarkeit f, Vertauschbarkeit f	interchangeabilité f	перестановочность, коммутативность
	interconnected system	s. C 476		
I 271	**interconnection,** grid operation	Verbundbetrieb m	marche f interconnectée	параллельная работа, работа в системе
I 272	**interconversion**	Umwandlung f, Wandlung f	transformation f, conversion f	превращение
I 273	**intercooler**	Zwischenkühler m	refroidisseur m intermédiaire	промежуточный холодильник
I 274	**interface**	Trennungsfläche f	surface f de séparation, interface m	плоскость разделения, перегородка
	interface	s. a. B 228		
I 275	**interfacial**	Zwischenphasen ..., Grenzflächen ...	interfaciale	межфазный
I 276	**interfacial area**	Grenzflächenoberfläche f, Phasengrenzfläche f	aire f interfaciale	межфазная поверхность, граничная поверхность
I 277	**interfacial area**	Austauschfläche f	surface f d'échange	межфазная поверхность, поверхность обмена (тепла, вещества)
I 278	**interfacial tension**	Grenzflächenspannung f	tension f superficielle (interfaciale)	поверхностное натяжение
I 279	**interfacial turbulence**	Zwischenphasenturbulenz f, Phasengrenzflächenturbulenz f	turbulence f interfaciale	межфазная турбулентность
I 280	**interior diameter**	Innendurchmesser m	diamètre m intérieur	внутренний диаметр
I 281	**intermediate,** intermediate product	Zwischenprodukt n, Halbfabrikat n	produit m intermédiaire, demi-produit m, produit semi-fini	полуфабрикат, промежуточное вещество, полупродукт
I 282	**intermediate bin**	Zwischenbunker m	silo m intermédiaire	промежуточный бункер
I 283	**intermediate cooling**	Zwischenkühlung f	réfrigération f intermédiaire	промежуточное охлаждение
I 284	**intermediate overheating**	Zwischenüberhitzung f	surchauffage m intermédiaire, resurchauffe f	промежуточный нагрев
I 285	**intermediate pressure**	Zwischendruck m	pression f intermédiaire	промежуточное давление
	intermediate product	s. I 281		
I 286	**intermediate reaction**	Zwischenreaktion f	réaction f intermédiaire	промежуточная реакция
I 287	**intermediate reactor**	Zwischenreaktor m	réacteur m intermédiaire	промежуточный реактор

	English	German	French	Russian
I 288	intermediate reflux	Zwischenrückfluß m	reflux m interne	промежуточное обратное течение, промежуточная циркуляция
I 289	intermediate storage tank	Zwischentanklager n	réservoir m [de stockage] intermédiaire	промежуточный резервуар
I 290	intermediate store	Zwischenlager n	entreposage m, stockage m intermédiaire	промежуточный склад
I 291	intermeshed circuit	vermaschte Schaltung f	montage m maillé	многоконтурная схема
I 292	intermittent refrigeration	diskontinuierliche Kühlung f	réfrigération f intermittente, refroidissement m discontinu	периодическое (прерывистое) охлаждение
I 293	intermolecular bond	zwischenmolekulare Bindung f	liaison f intermoléculaire	межмолекулярная связь
I 294	internal consumption	Eigenverbrauch m	consommation f propre	собственное потребление
I 295	internal cooling surface	innere Kühlfläche f	surface f intérieure de refroidissement	внутренняя поверхность охлаждения
	internal diameter	s. N 50		
I 296	internal element of the system	inneres Element des Systems n	élément m intérieur du système	внутренний элемент системы
I 297	internal energy, inner energy	innere Energie f	énergie f interne	внутренняя энергия
I 298	internal flow	Innenströmung f	écoulement m intérieur	внутреннее течение
I 299	internal friction	innere Reibung f	frottement m interne, friction f interne	внутреннее трение
I 300	internal furnace	Innenfeuerung f	chauffage m intérieur	внутренняя топка
I 301	internal grid	Innenberohrung f	grille f intérieure	внутренний змеевик
I 302	internal heat	innere Wärme f	chaleur f interne	внутреннее тепло
I 303	internal heat exchange	innerer Wärmeaustausch m, innere Wärmeübertragung f	échange m de chaleur interne	внутренний теплообмен
I 304	internal losses	Eigenverluste mpl	pertes fpl propres	собственные потери
I 305	internal plasticization	innere (interne) Plastifizierung f	plastification f interne	внутренняя пластификация
I 306	internal pressure	Binnendruck m, Innendruck m	pression f interne (intérieure), pression à l'intérieur	внутреннее (когезионное) давление
I 307	internal pressure equalization	innerer Druckausgleich m	compensation f de pression intérieure	внутреннее выравнивание давления
I 308	internal temperature	Innentemperatur f	température f à l'intérieur, température intérieure	внутренняя температура, температура внутри
I 309	interphase	Zwischenphase f	interphase m	граница (поверхность) раздела фаз, промежуточная фаза
I 310	interphase energy transport	Zwischenphasenenergietransport m	transfert m interphase de l'énergie	межфазный транспорт энергии
I 311	interphase momentum transport	Zwischenphasenimpulstransport m	transfert m interphase de la quantité de mouvement	межфазный транспорт импульса
I 312	interphase transport	Zwischenphasentransport m	transport (transfert) m interphase	межфазный транспорт
	intersection	s. C 122		
	intersection point	s. C 856		
I 313	interstage	Zwischenstufe f, Zwischenzustand m	état m intermédiaire	промежуточная ступень (стадия)
I 314	interstage drum	Zwischenabscheider m	séparateur m intermédiaire	промежуточный напорный отделитель
I 315	intimate mixture	inniges Gemisch n	mélange m intime	плотная смесь
I 316	intoxicate/to, to poison, to contaminate	vergiften	empoisonner, intoxiquer, contaminer	отравлять, загрязнять
	intoxicating	s. P 240		
I 317	in-transit temperature	Transporttemperatur f	température f de transport	температура транспорта (при перевозке)
I 318	intraparticle	intrapartikulär	à l'intérieur de la particule	внутричастичный, внутри частицы
I 319	intrinsic kinetics	innere Kinetik f, Eigenkinetik f	cinétique f propre (intrinsèque)	внутренняя (собственная) кинетика
I 320	intrinsic viscosity	Strukturviskosität f, Eigenviskosität f	viscosité f propre	структурная (собственная) вязкость
I 321	introduction	Einleitung f, Zuführung f	introduction f	введение, начало, вступление
I 322	invariable	unveränderlich, invariabel	constant, invariable	неизменный, постоянный
I 323	invariance	Invarianz f	invariabilité f	инвариантность
I 324	invariant	Invariante f	constante f	инвариант
I 325	invariant	invariant	invariable	инвариантный, неизменяющийся
I 326	inventary	Stückliste f, Verzeichnis n, Inventar n, Inventur f	nomenclature f, liste f de pièces, relevé m	инвентаризационная опись, инвентаризация, переучет
I 327	investment expenditure, capital expenditure	Investitionsaufwand m	dépense f d'investissement	объем капиталовложения, затраты на капиталовложение
I 328	investment measure	Investitionsmaßnahme f	mesure f d'investissement	мероприятие капитального строительства
I 329	investment project	Investvorhaben n	projet m d'investissement	капитальное строительство, проект

I 330	**inviscid flow,** non-viscous flow	reibungsfreie Strömung f	écoulement m sans frotte- ment, écoulement non vis- queux	течение без внутреннего трения, течение без учета вязкости
I 331	**ion absorption**	Ionenabsorption f	absorption f d'ions	абсорбция ионов
I 332	**ion exchange**	Ionenaustausch m	échange m d'ions	ионный обмен
I 333	**ion exchange chamber**	Ionenaustauschkammer f	chambre f d'échangeur d'ions	ионообменная камера
I 334	**ion exchanger**	Ionenaustauscher m	échangeur m d'ions	ионообменное вещество, ионообменник, иони- товый фильтр
I 335	**ionic extraction**	Ionenextraktion f	extraction f d'ions	вытягивание (экстракция) ионов
I 336	**ionization of air**	Luftionisation f	ionisation f d'air	ионизация воздуха
I 337	**iron removal**	Enteisenung f	déferrisation f	очистка от железа
I 338	**irradiation**	Bestrahlung f	irradiation f	облучение
I 339	**irrecoverable deformation**	irreversible (nicht rückver- formbare) Deformation f, bleibende Verformung f	déformation f permanente	необратимая деформация
I 340	**irrigation cooler**	Berieselungskühler m	réfrigérant (refroidisseur) m à ruissellement	оросительный холодильник
I 341	**irrigation plant**	Berieselungsanlage f	installation f d'arrosage	оросительная установка (система)
I 342	**irrigation surface**	Berieselungsfläche f	surface f d'arrosage	поверхность орошения
I 343	**irrotational flow**	wirbelfreie Strömung f	écoulement m sans tourbil- lonnement	безвихревое течение
I 344	**isenthalp**	Isenthalpe f	isenthalpe f	изэнтальпа
I 345	**isentrope**	Isentrope f	isentrope f	изэнтропа
I 346	**isentropic change**	adiabate (isentrope) Ände- rung f	changement m isentropique (adiabatique)	адиабатическое (изэнтропи- ческое) изменение
	isentropic compression	s. A 160		
I 347	**isentropic compression work**	isentrope Verdichtungsarbeit f	travail m isentropique de compression	адиабатическая работа сжатия
I 348	**isentropic expansion**	isentrope Expansion f, isen- trope (adiabate) Entspan- nung f	détente f isentropique (adia- batique)	изэнтропическое (адиабати- ческое) расширение
	isentropic exponent	s. A 171		
I 349	**isentropic refrigeration**	adiabate (isentrope) Kühlung f	réfrigération f adiabatique (isentropique)	адиабатическое (изэнтропи- ческое) охлаждение
I 350	**isochore,** isoplere	Isochore f	isochore f	изохора
I 351	**isomeric**	isomer	isomère, métamère	изомерный
I 352	**isometric pipeline plan**	isometrischer Rohrleitungs- plan m	plan m de tuyautage isomé- trique	изометрическая схема тру- бопроводов
	isoplere	s. I 350		
I 353	**isotherm**	Isotherme f	isotherme f	изотерма
I 354	**isothermal addition of heat**	isotherme Wärmezufuhr f	amenée (adduction, admis- sion) f de chaleur isother- mique	изотермический подвод тепла
I 355	**isothermal change**	isotherme Änderung f	changement m isothermique	изотермическое изменение
I 356	**isothermal change**	isothermische Zustandsände- rung f	changement m d'état iso- therme	изотермическое превраще- ние, изотермическое из- менение состояния
I 357	**isothermal compressibility**	isotherme Verdichtbarkeit f	compressibilité f isothermi- que	изотермическая сжима- емость
I 358	**isothermal compression**	isotherme Verdichtung f	compression f isothermique	изотермическое сжатие
I 359	**isothermal efficiency**	isothermischer Wirkungsgrad m	rendement m isotherme	изотермический коэффици- ент полезного действия
I 360	**isothermal expansion**	isotherme Expansion f	détente f isothermique	изотермическое расшире- ние
I 361	**isothermal flow**	isothermische Strömung f	écoulement (flux) m iso- therme	изотермическое течение
I 362	**isothermal measuring data**	isotherme Meßdaten pl	données fpl de mesure iso- thermes	изотермические опытные данные
I 363	**isothermal power consump- tion**	isothermer Energiebedarf m	demande f d'énergie isother- mique	изотермический расход энергии
I 364	**isothermal refrigerated trans- port**	isothermer Kühltransport m	transport m frigorifique iso- thermique	изотермический холо- дильный транспорт
I 365	**isothermal turbulent flow**	isotherme turbulente Strö- mung f	écoulement m isothermique turbulent	изотермическое турбулент- ное течение
I 366	**isotope separation**	Isotopentrennung f	séparation f des isotopes	разделение изотопов
I 367	**isotropic turbulence**	isotrope Turbulenz f	turbulence f isotrope	изотропная турбулентность
I 368	**iteration block**	Iterationsblock m	unité f d'itération	итерационный блок
I 369	**iteration method,** iterative method	Iterationsmethode f	méthode f d'itération	метод итерации
I 370	**iteration procedure**	Iterationsprozeß m	procédé m (méthode f) d'ité- ration	итерационный процесс
I 371	**iteration variable**	Iterationsvariable f	variable f itérative	итерационная переменная
I 372	**iterative computation**	iterative Berechnung f	calcul m itératif	итеративный расчет
	iterative method	s. I 369		

J

	jack	s. W 169		
J 1	jacket	Ummantelung f	enveloppe f, revêtement m, chemise f	покрытие, облицовка
J 2	jacket cooling	Mantelkühlung f	refroidissement m par l'enveloppe, refroidissement de la chemise	охлаждение водяной рубашкой
J 3	jacketed evaporator	Mantelverdampfer m	évaporateur m à double paroi	охлаждающая рубашка с испаряющимся холодильным агентом
J 4	jacketed low-temperature storage tank	Doppelwand-Tieftemperatur-Lagertank m	réservoir m cryogénique à doubles enveloppes	резервуар с рубашкой для низкотемпературного хранения
J 5	jacketed vessel	ummanteltes Gefäß n, Heizmantelgefäß n	récipient m à enveloppe double, cuve f à chemise de chauffage	сосуд с рубашкой, котел с рубашкой
	jacketed wall	s. D 474		
J 6	jacketed-wall vessel	Doppelwandbehälter m	réservoir m à doubles-parois	сосуд с [двойной] рубашкой
J 7	jacket heating	Mantelheizung f	chauffage m de la chemise	обогрев рубашки
J 8	jaw crusher	Backenbrecher m	broyeur (concasseur) m à mâchoires	щековая дробилка, камено-дробилка
J 9	jet, nozzle, ejector, blast pipe	Düse f	buse f, tuyau m, porte-vent m	сопло, насадка
J 10	jet	Verteilerdüse f, Spritzdüse f	pulvérisateur m, gicleur m	распределительное сопло
J 11	jet condenser	Einspritzkondensator m	jet m condenseur, condenseur m par injection	конденсатор смешивающего типа с охлаждением впрыском
	jet cooler	s. F 200		
J 12	jet moulding	Spritzgußverfahren n	moulage m sous pression, coulage m par injection	способ литья под давлением
J 13	jet pipe	Ausflußrohr n	tuyau m de décharge, tube m de déversement	сточная труба
J 14	jet pump, aspirator	Strahlpumpe f	pompe f à jet, injecteur m	струйный насос
J 15	jet widening	Strahlerweiterung f	élargissement m du jet	расширение струи
J 16	jig, fixture, apparatus, appliance, device, facility	Vorrichtung f, Apparatur f	disposition f, dispositif m, appareil m	устройство, приспособление
J 17	jigging screen	Schüttelsieb n, Schwingsieb n	tamis m oscillant, crible m à secousses	вибросито, виброгрохот
J 18	Johnson valve	Johnson-Ventil n	vanne f Johnson	клапан Джонсона
	joining	s. C 470		
J 19	joining element	Verbindungselement n	élément m de jonction, élément de connexion, raccordement m	соединяющий (соединительный) элемент
J 20	joint	Glied n	membre m, bielle f, élément m	звено, элемент, орган
J 21	jolting	Rütteln n	vibration f	встряхивание
J 22	jolting machine	Rüttelmaschine f	machine f à secousses	вибрационная (встряхивающая) машина
J 23	jordan mill, conical mill	Kegelmühle f	concasseur (broyeur) m à cônes	конусная мельница
J 24	Joule heat	Joulesche Wärme f	effet m Joule, chaleur f de Joule	джоулево тепло
J 25	Joule's constant	Joulesche Konstante f	constante f de Joule	постоянная Джоуля
J 26	Joule's equivalent	Joulesches Äquivalent n	équivalent m de Joule	эквивалент Джоуля
J 27	Joule-Thomson cooler	Joule-Thomson-Kühler m	refroidisseur m Joule-Thomson	охладитель, работающий по циклу Джоуля—Томсона
J 28	Joule-Thomson effect	Joule-Thomson-Effekt m	effet m Joule-Thomson	эффект Джоуля-Томсона
J 29	Joule-Thomson expansion process, Joule-Thomson process	Joule-Thomson-Entspannungsprozeß m, Joule-Thomson-Prozeß m	processus m de Joule-Thomson	процесс по Джоулю-Томсону
J 30	Joule-Thomson heat exchanger	Joule-Thomson-Wärmeübertrager m	échangeur m thermique de Joule-Thomson	теплообменник Джоуля-Томсона
	Joule-Thomson process	s. J 29		
	junction	s. C 122		
J 31	just-in-time	Produktion f auf Abruf	production f à convenance	продукция на заказ, производство по короткосрочному заказу

K

K 1	Kargin-Malinsky rule	Kargin-Malinskysche Regel f	règle m de Kargin-Malinsky	правило Каргина-Малинского
K 2	Kargin-Slonimski-Rouse model, KSR model	Kargin-Slonimsky-Rouse'sches Modell n	modèle m de Kargin-Slonimski-Rouse	модель Каргина-Слонимского-Рузе
K 3	Kelvin-Voigt body	Kelvin-Voigtscher Körper m	corps m de Kelvin-Voigt	тело Фогта-Кельвина
K 4	Kelvin-Voigt model	Kelvin-Voigtsches Modell n	modèle m de Kelvin-Voigt	модель Фогта-Кельвина
K 5	key component	Schlüsselkomponente f	composante f clé	ключевой компонент
	kiln	s. O 139		

K 6	kiln drying	Ofentrocknung f	séchage m au four	печная сушка, сушка в печи
K 7	kind of stress	Beanspruchungsart f	mode m de sollicitation	вид нагрузки (напряжения)
K 8	kinematic state	kinematischer Zustand m	état m cinématique	кинематическое состояние
K 9	kinematic viscosity	kinematische Viskosität f	viscosité f cinématique (cinétique)	кинематическая вязкость
K 10	kinetic energy	Bewegungsenergie f, kinetische Energie f	énergie f cinétique	кинетическая энергия, энергия движения
K 11	kinetic measuring data	kinetische Meßdaten pl	valeurs fpl mesurées cinétiques	кинетические опытные данные
K 12	kinetic parameter	kinetischer Parameter m	paramètre cinétique	кинетический параметр
K 13	kinetic theory of gases	kinetische Gastheorie f	théorie f cinétique des gaz	кинетическая теория газа
K 14	Kirkwood-Riseman-Zimm model, KRZ model	Kirkwood-Riseman-Zimm-sches Modell n	modèle m de Kirkwood-Riseman-Zimm	модель Кирквуда-Риземана-Цимма
	K.M. function	s. K 21		
K 15	knead/to	kneten	pétrir	месить, пластицировать
K 16	kneader, kneading machine	Knetmaschine f, Kneter m	pétrisseuse f, machine f à pétrir, pétrin m mécanique	месильная машина, смеситель, пластикатор
K 17	kneading	Knetung f	pétrissage m	пластикация
	kneading machine	s. K 16		
K 18	knockout drum	Ausdrücktrommel f	tambour m d'éjection	барабан-выталкиватель
K 19	knockout vessel	Saugleitungsabscheider m	réservoir m knockout	отделитель жидкости на трубопроводе всасывания
K 20	know-how	Know-how n, Betriebserfahrung f, Fachkenntnisse f, technisches Wissen n	savoir-faire m, know how m	технология, знание дела, «ноу-хау», специальная информация для применения метода
	KRZ model	s. K 14		
K 21	Kuhn-Mark function, K.M. function	Kuhn-Mark-Funktion f	fonction f de Kuhn-Mark	функция Куна-Марка

L

L 1	lability	Labilität f	instabilité f, labilité f	лабильность, неустойчивость
L 2	laboratory equipment (installation)	Laboreinrichtung f	équipement m de laboratoire	лабораторное оборудование
L 3	laboratory measuring equipment	Laboratoriumsmeßgeräte npl	instruments mpl de mesure de laboratoire	лабораторные измерительные приборы
L 4	laboratory reactor	Laborreaktor m	réacteur m de laboratoire	лабораторный реактор
L 5	laboratory scale	Labormaßstab m	échelle f de laboratoire	лабораторный масштаб
L 6	laboratory test	Laborversuch m	essai m de laboratoire	лабораторное испытание
L 7	labour cost	Lohnkosten pl	prix m de revient de la main-d'œuvre	расходы на заработную плату
L 8	labour-saving	Arbeitseinsparung f, Arbeitsersparnis f, Arbeitsvereinfachung f	simplification f du travail	экономия в работе
L 9	lading space	Ladefläche f	plate-forme f de chargement, surface f pour le chargement	площадь нагрузки
L 10	laevorotatory process	Linksprozeß m	procédé m à gauche	левый замкнутый процесс
	lagging	s. I 258		
L 11	Lagrangian function	Lagrangesche Funktion f	fonction f lagrangienne	функция Лагранжа
L 12	Lagrangian multiplier	Lagrangescher Multiplikator m	paramètre m de Lagrange	множитель Лагранжа
L 13	Lagrangian multiplier method	Lagrangesche Multiplikatorenmethode f	méthode f des paramètres de Lagrange	метод множителей Лагранжа
L 14	lamellar heater, cellular-type radiator	Lamellenheizkörper m	radiateur m à lames	пластинчатый нагревательный прибор
L 15	lamellar structure	lamellare Struktur f	structure f lamellée (lamelleuse)	слоистое строение, слоистая структура
L 16	laminar flow	laminare Strömung f, Laminarströmung f	écoulement (courant) m laminaire	ламинарное течение
L 17	laminar flow layer	laminare Strömungsschicht f	couche f d'écoulement laminaire	слой ламинарного потока
L 18	laminar layer	Laminarschicht f	couche f laminaire	ламинарный слой
L 19	laminated flow	Schichtenströmung f	écoulement m en couches	слоистое течение
	lamination	s. L 46		
L 20	Laplace transform[ation]	Laplace-Transformation f, Laplacesche Transformation f	transformation f de Laplace	трансформация Лапласа
L 21	lap roller	Abwickelwalze f	rouleau m débiteur, tambour m déroulant	скатывающий холстовой валик
L 22	large capacity, large-scale process, large-sited production	Großtonnage f	grand tonnage m	валовая вместимость, большое (многотоннажное) производство
L 23	large-capacity shipment	Großraumtransport m	transport m de gros tonnages	большегрузный транспорт
L 24	large-scale operation	Großbetrieb m	établissement m important, grande entreprise f	крупное предприятие
	large-scale process	s. L 22		
L 25	large-scale test	Großversuch m	essai m en grand, essai à grande échelle	опыт большого масштаба

L 26	large-scale unit	Großanlage f, Anlage f industriellen Maßstabs	unité f à échelle industrielle	установка большого масштаба, промышленная установка, установка большой единичной мощности
	large-sited production	s. L 22		
L 27	large-space container	Großraumbehälter m, Großraumcontainer m	conteneur m à grande capacité	контейнер большой емкости
L 28	lasting, permanent, constant	beständig	permanent, résistant, durable	постоянный, неизменный, долговременный
	last runnings	s. B 1		
L 29	late failure	Spätausfall m	défaillance f tardive	поздний отказ
L 30	latency	Latenz f	latence f	латентность, скрытое состояние
L 31	latent cooling effect	latenter Kühleffekt m	effet m de refroidissement latent	скрытая охлаждающая способность
L 32	latent dehumidifying effect	latenter Trocknungseffekt m	effet m de déshumidification latent	скрытое осушающее действие
L 33	latent heat	latente (gebundene) Wärme f	chaleur f latente	скрытая теплота
L 34	latent heat value	latenter Wärmewert m	valeur f de chaleur latente	скрытая теплотворная способность
L 35	latent refrigerating capacity	latente Kälteleistung f	capacité f frigorifique latente	скрытая холодопроизводительность
L 36	latent steam	indirekter Dampf m	vapeur f latente	глухой пар
L 37	lateral branch	Nebenzweig m, Seitenzweig m	branche f latérale	побочная ветвь
	lateral chain	s. S 300		
L 38	lateral strain	Querdehnung f	dilatation f transversale	поперечное растяжение
L 39	lattice constant	Gitterkonstante f	constante f de réseau	постоянная решетки
L 40	lattice energy	Gitterenergie f	énergie f de réseau	энергия решетки
L 41	lattice structure	Gitterstruktur f, Kristallgitterstruktur f	structure f cristalline	структура решетки
L 42	law of constant proportions	Gesetz n der konstanten Proportionen	loi f des proportions constantes	закон постоянных отношений
L 43	law of mass action	Massenwirkungsgesetz n	loi f d'action de masse	закон действия масс
L 44/5	law of similarity (similitude)	Ähnlichkeitsregel f, Ähnlichkeitstheorie f	principe m (règle f) de similitude	теория (принцип, правило) подобия
L 46	layer, couch, bed, ply, coat, lamination	Schicht f, Anstrich m	couche f, enduit m	слой, шихта, отслоение
	layer	s. a. C 327		
	layout	s. 1. A 555; 2. S 70		
L 47	layout of plant	Anlagenlageplan m, Anlagengrundriß m	plan m de l'unité	генеральный план завода
L 48	layout plan	Lageplan m	plan m, plan de situation, tracé m	план расположения
L 49	layout plan, general plan	Übersichtsschema n, Gesamtplan m	plan m [de disposition] d'ensemble	обзорная схема (диаграмма), общий план
L 50	leaching vat	Laugenbehälter m, Laugenbottich m	réservoir m à lessive, cuvier m	выщелачиватель, чан для выщелачивания, емкость для выщелачивания
L 51	lead, feeder	Leitung f, Zuleitung f	conduit m, tuyau m d'alimentation	подводящий питательный провод
L 52	lead	Ganghöhe f, Steigung f (einer Extruderschnecke)	pas m	угол наклона шнека экструдера, глубина нарезки шнека
L 53	lead coating plant	Verbleiungsanlage f	installation f de plombage	установка для освинцовывания
L 54	leading decision	Grundsatzentscheidung f	décision f principale (de principe)	фундаментальное (принципиальное) решение
L 55	lead lining	Bleiauskleidung f	revêtement m en plomb	свинцовая футеровка
L 56	lead reactor	Bleireaktor m, bleiverkleideter Reaktor	réacteur m à revêtement en plomb	свинцовый (освинцованный) реактор
L 57	leaf filter	Plattenfilter n	filtre m à plaques	пластинчатый фильтр
L 58	leak, discharge, drain, outlet	Auslauf m	écouloir m, sortie f, décharge f	выпуск, выход
L 59	leakage, dispersion, variation, scattering, stray	Streuung f	fuite f, dispersion f, variation f	рассеивание, рассеяние
	leakage	s. a. 1. D 488; 2. E 256; 3. L 235		
L 60	leakage losses	Ausströmungsverluste mpl, Leckageverluste mpl	pertes fpl par défaut d'étanchéité	потери от утечки
L 61	leakage of electricity	Elektrizitätsverlust m	perte f d'électricité	потеря электричества
	leak off/to	s. F 248		
	lean gas	s. P 267		
L 62	lean oil	Mageröl n, mageres Absorptionsöl n, armes (frisches, regeneriertes) Waschöl n (für die Absorptionskolonne)	huile m de lavage	десорбированное масло
L 63	leaving air	austretende Luft f	air m de sortie	воздух на выходе, выходящий воздух
L 64	leaving gas	Austrittsdampf m	gaz m de sortie	отходящий газ

L 65	leaving water	Austrittswasser n	eau f sortante	вода на выходе
L 66	length of operation	Funktionsdauer f	durée f de service, temps m de fonctionnement	длительность (время) функционирования, интервал действия
	lenzine	s. B 93		
L 67	less volatile component	schwerere (weniger flüchtige) Komponente f	constituant m peu volatil, composante f plus lourde	малолетучий компонент
L 68	let-down gas	Entspannungsgas n	gaz m détendu, gaz de détente	остаточный газ
L 69	let-down vessel	Entspannungsgefäß n	vase m de détente	расширительная камера
L 70	let-off arrangement	Abwickelvorrichtung f	dérouleuse f, dispositif m de déroulement	разматывающее устройство
L 71	level controller	Flüssigkeitsstandregler m	régulateur m de niveau	регулятор уровня жидкости
	level indicator	s. H 124		
L 72	level line	Niveaulinie f	courbe f de niveau	линия уровня
L 73	level of capacity	Leistungsniveau n	niveau m de puissance	уровень мощности
L 74	level of significance	Signifikanzniveau n	seuil m de probabilité	уровень значимости
L 75	level plane	Niveauebene f, Niveaufläche f	surface f équipotentielle (de niveau)	уровень плоскости
L 76	level structure	Termstruktur f	structure f de terme	структура уровня
L 77	level vessel	Ausgleichsgefäß n	récipient m de détente	уравнительный сосуд
L 78	liberation of energy	Energiefreisetzung f	libération f d'énergie	выделение энергии
	licence pressure	s. D 196		
	lid	s. C 808		
L 79	Liebig condenser	Liebig-Kühler m	refroidisseur m à tube rectiligne	холодильник Либиха
	life	s. D 562		
L 80/1	life expectancy	Lebenserwartung f	attente (durée) f de vie	ожидаемая продолжительность жизни
	life time	s. D 562		
	lift force	s. B 314		
L 82	lifting force	Hubvermögen n, Hubkraft f	force f de levage	подъемная сила, подъемная способность
	lifting height	s. D 152		
L 83	lifting method	Hebeverfahren n, Hubverfahren n	méthode f (procédé m) de levage	метод подъема
L 84	lift line (pipe)	Liftleitung f, Heberleitung f (zum Heben des Katalysators beim katalytischen Kracken), Steigrohr n	tuyau m à siphon	подъемная труба, аэролифт
	lift pump	s. S 802		
	lift tube	s. V 159		
L 85	lift water	Auftriebswasser n	eau f ascencionnelle	нагонная вода
L 86	light end	Destillationsvorlauf m	tête f de distillation, fraction f légère	легкие фракции
L 87	light ends absorption	Vorlaufabsorption f, Absorption f der leichten Anteile	absorption f des fractions légères	абсорбция легких фракций
L 88	light end tower	Kolonne f für leichte Destillate	colonne f pour distillats légers	колонна для легких дистиллятов
	light industry	s. C 609		
L 89	lighting installation	Lichtanlage f, Beleuchtungsanlage f	installation f d'éclairage	осветительная установка
L 90	lightning protection	Blitzschutz n, Blitzableiter m	parafoudre m, paratonnerre m	молниеотвод, громоотвод, грозовой разрядник
L 91	lightning protective system	Blitzschutzanlage f	installation f parafoudre	грозозащитная установка
L 92	light resistance	Lichtbeständigkeit f	résistance (stabilité) f à la lumière	светостойкость
L 93	lightweight construction method	Leichtbauweise f	méthode f de construction légère	сроительство легких конструкций, облегченная конструкция
L 94	lime kiln	Kalkofen m	chaufour m, four m à chaux	печь для обжига известняка
	limit	s. B 225		
L 95	limit case, limiting case	Extremfall m, Grenzfall m	cas m limite, cas extrême	предельный случай
L 96	limited matrix	begrenzte Matrix f	matrice f limitée	ограниченная матрица
L 97	limited power	begrenzte Leistung f	puissance f limitée	ограниченная мощность
L 98	limited value	begrenzter Wert m	valeur f bornée	ограниченная величина
L 99	limited variable	begrenzte Variable f	variable f bornée	ограниченная переменная
	limiting case	s. L 95		
L 100	limiting condition	Grenzbedingung f, Randbedingung f	condition f limite	краевое (граничное) условие
L 101	limiting current area	Grenzstromgebiet n	régime m de courant limite	область предельного [по]тока
L 102	limiting current density	Grenzstromdichte f	densité f de courant limite	плотность предельного [по]тока
L 103	limiting pressure	Grenzdruck m	pression f limite	предельное давление
L 104	limiting steam flow	Grenzdampfmenge f	débit m de vapeur limite	предельное количество пара
L 105	limiting temperature	Grenztemperatur f	température f limite	предельная температура
L 106	limiting value, critical value	Grenzwert m	valeur f limite, limite f	предельное значение
L 107	limit law	Grenzgesetz n	loi f limite	предельный закон
L 108	limit of capacity	Leistungsgrenze f	limite f de puissance	предел мощности, предел производительности

L 109	limit of error	Fehlergrenze f	limite f d'erreur	предел ошибок, граница погрешности
L 110	limit of measurement	Meßgrenze f	limite f de mesure	предел измерения, граница области измерения
L 111	limit of resistance	Festigkeitsgrenze f	résistance f à la rupture, limite de rupture	предел прочности
L 112	limit of scrapping	Verschrottungsgrenze f	limite f de démolition	граница списывания в скрап
L 113	limit of stability	Stabilitätsgrenze f	limite f de stabilité	предел устойчивости
L 114	limits of integration	Integrationsbereich m, Integrationsgrenzen fpl	domaine m d'intégration, limites fpl d'intégration	пределы интегрирования
L 115	line/to, to plate	auskleiden, umkleiden	enrober	облицовывать, обшивать
L 116	linear equation	lineare Gleichung f	équation f linéaire	линейное уравнение
L 117	linearization	Linearisierung f	linéarisation f	линеаризация
L 118	linear optimization	Linearoptimierung f	optimisation f linéaire	линейная оптимизация
L 119	linear polymer	lineares Polymer n	polymère m linéaire	линейный полимер
L 120	line current	Netzstrom m	courant m de ligne, courant du secteur	линейный ток, ток в сети
	line drop	s. P 276		
L 121	line of attack	Anströmlinie f	ligne f de soufflage	линия натекания жидкости на тело
L 122	line of force	Kraftlinie f	ligne f de force	силовая линия
L 123	line pressure drop	Leitungsdruckverlust m	chute f de pression de conduite	падение давления в трубопроводе
L 124	line variation	Netzspannungsregler m	régulateur m de tension de secteur	регулятор напряжения сети
L 125	line voltage	Netzspannung f	tension f du secteur, tension de réseau	линейное напряжение, напряжение сети
L 126	lining	Auskleidung f	revêtement m	облицовка, обшивка, футеровка
L 127	link, bond, binding	Bindung f	liaison f	связь, соединение, скрепление
	link	s. a. C 580/1		
L 128	linkage forces	Bindungskräfte fpl	forces fpl de liaison	силы сцепления (связи), связующие силы
L 129	linked graph	zusammenhängender Graph m	graphe m cohérent	связанный граф
	linking energy	s. B 101		
	liquefaction	s. L 164		
L 130	liquefaction method	Verflüssigungsverfahren n	méthode f (procédé m) de liquéfaction	способ ожижения
L 131	liquefaction of air	Luftverflüssigung f	liquéfaction f d'air	ожижение воздуха
L 132	liquefaction of gases	Gasverflüssigung f	liquéfaction f du gaz	ожижение газа, сжижение газа
L 133	liquefaction pressure	Verflüssigungsdruck m	pression f de liquéfaction	давление сжижения
L 134	liquefaction process	Verflüssigungsprozeß m	processus m de liquéfaction	процесс ожижения
L 135	liquefaction step	Verflüssigungsstufe f	stade m de liquéfaction	ступень сжижения
L 136	liquefaction temperature	Verflüssigungstemperatur f	température f de liquéfaction	температура разжижения
L 137	liquefaction yield	Verflüssigungsleistung f	rendement m de liquéfaction	производительность по ожижению
L 138	liquefiable	verflüssigbar	liquéfiable, condensable	разжижаемый, разжижающийся
L 139	liquefied gas	Flüssiggas n	gaz m liquéfié	сжиженный (жидкий) газ
L 140	liquefied gas container	Flüssiggasbehälter m	citerne f pour les gaz liquéfiés	резервуар для сжиженного газа
L 141	liquefied natural gas	verflüssigtes Erdgas n	gaz m naturel liquéfié	сжиженный природный газ
L 142	liquefier	Verflüssigungsapparat m	appareil m à liquéfier	разжижитель, аппаратура для разжижения
	liquefier	s. a. C 548		
L 143	liquefy/to, to condensate, to fluidize	verflüssigen, kondensieren, flüssig machen, schmelzen, zerlassen	fluidifier, liquéfier, condenser, fondre	сжижать, разжижать, превращать в жидкость, расплавлять
L 144	liquefying agent	Verflüssigungsmittel n, Verflüssiger n	liquéfacteur m	разжижающее средство
L 145	liquefying plant	Verflüssigungsanlage f	installation f de liquéfaction	установка для сжижения
L 146	liquid air	flüssige Luft f	air m liquide	жидкий воздух
L 147	liquid bypassing	Flüssigkeitsbypass m	by-passage m de liquide	байпас жидкости, перепуск жидкости
L 148	liquid circulation	Flüssigkeitsumlauf m	circulation f (cycle m) de liquide	циркуляция жидкости
L 149	liquid cooler	Flüssigkeitskühler m	refroidisseur m à liquide	холодильник для жидкости
L 150	liquid cooling	Flüssigkeitskühlung f	refroidissement m par liquide	жидкостное охлаждение
L 151	liquid density	Flüssigkeitsdichte f	densité f de liquide (fluide)	плотность жидкости
L 152	liquid-drop model	Tröpfchenmodell n	modèle m de gouttelettes	капельная модель
L 153	liquid film model	Flüssigkeitsfilmmodell n	modèle m du film liquide	модель жидкой пленки
L 154	liquid firing	Flüssigkeitsfeuerung f, Ölfeuerung f	foyer m aux combustibles liquides, chauffe f au mazout	сжигание жидкого топлива
L 155	liquid fuel	flüssiger Brennstoff m	combustible m, carburant m	жидкое топливо
L 156	liquid-fuel reactor	Flüssigkeitsreaktor m	réacteur m pour liquides	реактор с жидким топливом
L 157	liquid gas analysis	Flüssiggasanalyse f	analyse f de gaz liquéfié	анализ сжиженного газа

L 158	liquid gas interface	Flüssigkeit-Gas-Grenzfläche, Grenzfläche f flüssig-gas-förmig, Phasengrenzfläche f	interface f gaz-liquide	поверхность раздела [фаз] жидкость-газ
L 159	liquid gas precooler	Flüssiggasvorkühler m	précondenseur m de gaz liquide	предохладитель сжиженного газа
L 160	liquid gas spray column	Gas-Flüssigkeits-Sprühkolonne f, Gas-Flüssigkeits-Sprühturm m	colonne f à pulvérisation gaz-liquide	газожидкостная распылительная колонна
L 161	liquid gas vaporizer	Flüssiggasverdampfer m	vaporisateur m de gaz liquide	испаритель жидкого газа
L 162	liquid head	Flüssigkeitshöhe f	niveau m du liquide	гидростатическая высота, высота уровня жидкости
L 163	liquid heat exchanger	Flüssigkeitswärmeaustauscher m, Flüssigkeitswärmeübertrager m	échangeur m de chaleur pour liquides	жидкостный теплообменник
L 164	liquidification, liquefaction	Verflüssigen n, Verflüssigung f	liquéfaction f, fluidification f	ожижение, разжижение, расплавление
L 165	liquid jet	Flüssigkeitsstrahl m	jet m liquide	струя жидкости
L 166	liquid level	Flüssigkeitsstand m	hauteur (niveau) m du liquide	уровень жидкости
L 167	liquid-liquid extraction	Flüssig-Flüssig-Extraktion f	extraction f liquide-liquide	жидкостно-жидкостная экстракция
L 168	liquid-liquid interface	Flüssigkeit-Flüssigkeit-Grenzfläche f, Grenzfläche f flüssig-flüssig	interface f liquide-liquide	поверхность раздела фаз жидкость-жидкость
L 169	liquid manure pump	Jauchepumpe f	pompe f à purin	насос для навозной жижи
L 170	liquid manure tank	Jauchebehälter m	réservoir m à purin	яма для навозной жижи
L 171	liquid mist removal	Nebelabscheidung f	séparation f de brume	каплеулавливание, каплеотделение, отделение тумана
L 172	liquid mixture	Flüssigkeitsgemisch n	mélange m liquide	жидкая смесь
L 173	liquid motion	Flüssigkeitsbewegung f	mouvement m de liquide	движение жидкости
L 174	liquid phase	Flüssigphase f, Flüssigkeitsphase f	phase f liquide	жидкая фаза
L 175	liquid-phase alkylation process	Verfahren zur Flüssigphasenalkylierung f	procédé m d'alkylation en phase liquide	процесс алкилирования в жидкой фазе
L 176	liquid recirculating system	Flüssigkeitsumlaufsystem n	cycle m de liquide, système m de recirculation de liquide	система рециркуляции жидкости
L 177	liquid recirculation pump	Umlaufpumpe f	pompe f à circulation, pompe d'accélération	циркуляционный насос
L 178	liquid surface	Flüssigkeitsoberfläche f	surface f de liquide	поверхность жидкости
L 179	liquid turbulence	Flüssigkeitsturbulenz f	turbulence f de liquide	турбулентность потока жидкости
L 180	liquid-vapour mixture	Dampf-Flüssigkeits-Gemisch n	mélange m de liquide-vapeur	парожидкостная смесь, смесь жидкости и пара
L 181	liquid-vapour stream	Flüssigkeitsdampfstrom m	courant m de liquide-vapeur, flux m de liquide-vapeur	поток парожидкостной смеси
	liquor	s. L 306		
L 182	list presentation	Listendarstellung f	représentation f en forme de listes	представление в виде списков
	live load	s. W 190		
L 183	live steam	Frischdampf m, direkter Dampf m	vapeur f vive (fraîche, directe)	острый (свежий, прямой) пар
L 184	lixiviation	Auslaugen n, Auslaugung f	lixiviation f	выщелачивание
L 185	load/to, to ship	verladen	charger, expédier	грузить, нагружать
L 186	load, weight, strain	Gewicht n, Beschickung f, Last f, Belastung f	poids m, charge f	вес, груз, нагрузка, загрузка
	load	s. a. C 181		
L 187	load area	Belastungsfläche f, Ladefläche f	surface f de charge	площадь нагрузки, нагруженная площадь
L 188	load-bearing means	Tragmittel n	moyen m de support	подъемное средство, приспособление для подъема, грузоподъемное приспособление
L 189	load behaviour	Lastverhalten n	comportement m de charge	поведение груза, поведение при нагрузке
L 190	load-carrying capacity	Belastbarkeit f, Tragfähigkeit f	portée f, capacité f de charge, limite f de charge	грузоподъемность, допускаемая нагрузка, подъемная способность
L 191	load-carrying capacity range	Tragfähigkeitsbereich m, Belastungsbereich m	domaine m de charge	область нагрузки
	load coefficient	s. D 104		
L 192	load compensation	Belastungsausgleich m	compensation (égalisation) f de charge	компенсация нагрузки
L 193	load curve	Belastungskurve f	ligne f de charge	кривая воздействия нагрузки
L 194	load distribution	Lastaufteilung f	répartition f des charges	распределение нагрузки
	loaded polymer	s. F 69		
L 195	loader	Aufladevorrichtung f	dispositif m de chargement	загрузочное приспособление

L 196	load factor	Beschickungsfaktor m, Belastungsfaktor m, Belastungsgrad m	facteur m de charge, facteur m d'utilisation	коэффициент (степень) нагрузки
L 197	loading, charging, strain	Beladen n, Beladung f, Verladung f, Aufladen n, Ladebetrieb m	chargement m, opération f de chargement	загрузка, нагрузка, погрузка, погрузочные работы
L 198	loading capacity	Belastungsvermögen n, zulässige Belastung f	capacité f admissible de charge	предельная допустимая нагрузка
L 199	loading crane	Verladekran m	grue f de chargement	погрузочно-разгрузочный кран, погрузочный кран
L 200	loading density	Füllungsdichte f, Fülldichte f	densité f au remplissage, densité de chargement	плотность наполнения
L 201	loading device	Ladeeinrichtung f	installation f de chargement	зарядная установка
L 202	loading material	Fördergut n	matières à transporter	транспортируемый материал
L 203	loading pattern	Beschickungsschema n, Beladungsplan m	modèle m de chargement	схема загрузки
L 204	loading point	Staugrenze f	capacité f de charge	предел напора
L 205	load limit	Belastungsgrenze f, Lastgrenze f	limite f de charge, charge f limite	предел нагрузки
L 206	load moment	Lastmoment n	couple m résistant	момент сил (представляющих нагрузку)
L 207	load peak	Belastungsspitze f, Maximallast f	pointe f de charge, charge f maximum	максимум нагрузки
L 208	load per surface	Flächenbelastung f	charge f par unité de surface, charge au mètre carré	нагрузка на единицу поверхности
L 209	load phase	Belastungsphase f	phase f de charge	фаза нагрузки
L 210	load shift	Lastverlagerung f	déplacement m de charge	смещение (сдвиг) груза
L 211	load take-up	Lastaufnahmemittel n	accessoire m de levage, porte-charge m	средство для приема нагрузки
L 212	local climate zone	lokale Klimazone f	zone f climatique locale	местная климатическая зона
L 213	local heat transfer	lokale Wärmeübertragung f	transfert m de chaleur local	местная теплопередача
L 214	local iteration block	lokaler Iterationsblock	bloc m d'itération local	локальный итерационный блок
L 215	localized corrosion, selective corrosion	Lokalkorrosion f	corrosion f locale (localisée)	местная коррозия
L 216	local thermostat	lokaler Thermostat m	thermostat m local	местный термостат, местное термореле
L 217	local velocity gradient	lokaler Geschwindigkeitsgradient m	gradient m local de vitesse	местный градиент скорости
L 218	local velocity of current	Anströmgeschwindigkeit f	vitesse f de soufflage	скорость занесения (обтекания)
L 219	location, position, place, station	Standort m, Aufstellungsort, Position f	station f, emplacement m, position f	местоположение, место установки (монтажа, стоянки), стоянка
L 220	location approval	Standortgenehmigung f	autorisation f de l'emplacement	разрешение на строительство на определенном месте
L 221	location of tearing	Trennstelle f, Schnittstelle f	position f de la coupe, point m de séparation	место разрыва (разделения, среза)
L 222	logarithmic additivity	logarithmische Additivität f	additivité f logarithmique	логарифмическая аддитивность
L 223	logarithmic mean temperature difference	logarithmisches Temperaturmittel n	différence f moyenne logarithmique de température	средняя логарифмическая разность температур
	logical	s. C 477		
L 224	long-chain	langkettig	à chaîne longue	длинноцепный, содержащий длинную цепь
L 225	long-chain branching	langkettige Verzweigung f	ramification f à chaînes longues	длинноцепное разветвление
L 226	longevity	Langlebigkeit f	longévité f	долговечность, выживаемость
L 227	longitudinal elongation	Längsdehnung f	allongement m longitudinal	продольное растяжение
L 228	long-term continuous duty (operation)	kontinuierliche Fahrweise f	opération f continue	непрерывный режим работы
L 229	long term effect	Langzeitwirkung f	effet m à long terme	длительный эффект, долгосрочное воздействие
L 230	loop	Schleife f, Kreislauf m, [geschlossener] Regelkreis m	boucle f, circuit m fermé	петля, контур, контур регулирования, цикл
	loop	s. a. C 254		
L 231	loop dryer	Schleifentrockner m	sécheur m à boucle	петлевая сушилка
L 232	loop rule	Schleifenregel f	règle m de boucle	правило для разрыва циклов
L 233	loose goods	Schüttgut n	matière f en vrac (tas)	сыпучий материал
L 234	Loschmidt's number	Loschmidtsche Zahl f	nombre m de Loschmidt	число Лошмидта
L 235	loss, leakage, dissipation, waste	Verlust m	perte f, déperdition f, déchet m	потеря, утечка
L 236	loss angle	Verlustwinkel m	angle m de pertes	угол потери
L 237	loss by diffusion	Diffusionsverlust m	perte f par diffusion	потеря на диффузии, диффузионная потеря

	loss by evaporation	s. E 286		
L 238	loss by leakage	Ableitungsverlust m, Lecka- geverlust m	perte f de dérivation, cou- lage m, fuite f	потеря отвода
L 239	loss due to dressing	Aufbereitungsverlust m	perte de traitement	потеря при обогащении (приготовлении)
L 240	loss factor	Verlustfaktor m	facteur m de pertes	коэффициент потери
	loss in weight	s. S 292		
L 241	loss modulus	Verlustmodul m	module m de perte	модуль потери
L 242/3	loss of charge	Ladeverlust m	perte f de charge	потеря заряда
L 244	loss of work	Arbeitsverlust m, Energiever- lust m	perte f d'énergie	потеря труда (в работе), производственная потеря
L 245	lost current	Verluststrom m	courant m perdu	ток потери, поток потери
L 246	lost heat, waste heat	Abhitze f, Abwärme f	chaleur f perdue, chaleur d'échappement	отводимое (отходящее, от- работанное) тепло
L 247	low-boiling	leicht siedend, tiefsiedend, niedrigsiedend	à bas point d'ébullition, bouillant à basse tempéra- ture, de point d'ébullition inférieur	кипящий при низкой темпе- ратуре, легкокипящий, низкокипящий
L 248	low-boiling component	niedrigsiedende Komponente f	composante f à bas point d'ébullition	низкокипящий компонент
L 249	low-boiling mixture	niedrigsiedendes Gemisch n	mélange m à bas point d'ébullition	низкокипящая смесь
L 250	low-consuming technology	ressourcensparende Techno- logie f	technologie f à consomma- tion réduite, technologie économique	ресурсосберегающая тех- нология
L 251	lowering of the melting point	Schmelzpunkterniedrigung f	abaissement m cryoscopique	понижение точки плавления
L 252	lowering of vapour pressure	Dampfdruckerniedrigung f	abaissement m de la pression de vapeur	понижение упругости (дав- ления) пара
L 253	lowest pressure, minimum pressure	Minimaldruck m, Mindest- druck m	pression f minimum	минимальное давление
L 254	low limiting curve	untere Grenzkurve f	courbe f limite inférieure	нижняя граничная (предель- ная) кривая
L 255	low pressure	Niederdruck m	basse pression f, pression ré- duite	низкое давление
L 256	low pressure, depression, vacuum	Unterdruck m, Vakuum n	vide m partiel, dépression f	пониженное (нижнее) дав- ление, вакуум, разреже- ние
L 257	low-pressure air	Niederdruckluft f	air m de basse pression	воздух низкого давления
L 258	low-pressure blower, low- pressure fan	Niederdruckgebläse n	soufflerie f à basse pression	воздуходувка (вентилятор) низкого давления
L 259	low-pressure boiler	Niederdruckkessel m	chaudière f à basse pression	котел низкого давления
L 260	low-pressure chamber	Niederdruckkammer f	chambre f à basse pression	камера низкого давления
L 261	low-pressure channel	Niederdruckkanal m, Nieder- druckleitung f	conduit m de basse pression	линия низкого давления, ка- нал (трубопровод) низ- кого давления
L 262	low-pressure circuit	Niederdruckkreislauf m	circuit m de basse pression	схема низкого давления
L 263	low-pressure evaporator	Niederdruckverdampfer m	évaporateur m à basse pres- sion	испаритель низкого давле- ния
	low-pressure fan	s. L 258		
L 264	low-pressure part, low-pres- sure subsystem	Niederdruckteil n	partie f [à] basse pression	часть (подсистема) низкого давления
L 265	low-pressure preheater	Niederdruckvorwärmer m	préchauffeur m à basse pres- sion	подогреватель низкого дав- ления
L 266	low-pressure process	Niederdruckprozeß m	processus m à basse pression	процесс низкого давления
L 267	low-pressure range	Niederdruckbereich m	domaine m de basse pression	диапазон низких давлений
L 268	low-pressure side	Niederdruckseite f	côté m de basse pression	сторона низкого давления
L 269	low-pressure stage	Niederdruckstufe f	étage m [à] basse pression	ступень низкого давления
L 270	low-pressure steam	Niederdruckdampf m	vapeur f à basse pression	пар низкого давления
L 271	low-pressure steam engine, low-pressure turbine	Niederdruckturbine f	turbine f [à] basse pression	турбина низкого давления
L 272	low-pressure steam-heating	Niederdruckdampfheizung f	chauffage m à basse pression	паровое отопление низкого давления
	low-pressure subsystem	s. L 264		
	low-pressure tank	s. L 273		
	low-pressure turbine	s. L 271		
L 273	low-pressure vessel, low pressure tank	Niederdruckbehälter m	réservoir m à basse pression	резервуар (емкость) низ- кого давления
	low product	s. A 259		
L 274	low-stage compressor	Niederdruckkompressor m	compresseur m à basse pres- sion	компрессор ступени низ- кого давления
L 275	low-stage cycle	Niederdruckstufenprozeß m	procédé m à basse pression à plusieurs étages	цикл ступени низкого дав- ления
L 276	low temperature	Tieftemperatur f	température f basse	глубокая (низкая) темпера- тура
L 277	low-temperature bath	Tieftemperaturbad n	bain m à basse température	низкотемпературная ванна (баня)
L 278	low-temperature carboniza- tion	Tieftemperaturverkokung f, Schwelen n	cokéfaction (carbonisation) f à basse température	коксование при низких тем- пературах, полукоксова- ние
L 279	low-temperature carboniza- tion gas	Schwelgas n	gaz m de distillation à basse température	газ, выделяющийся при по- лукоксовании

L 280	low-temperature carbonization plant	Schwelanlage f, Schwelerei f	installation f pour la lente distillation, installation de carbonisation à basse température	установка для полукоксования
L 281	low-temperature carbonizing furnace	Schwelofen m	four m à lente distillation, four de distillation à basse température	печь для коксования при низкой температуре
L 282	low-temperature cooler	Tieftemperaturkühler m	refroidisseur m à basse température	низкотемпературный охладитель
L 283	low-temperature cooling	Tieftemperaturkühlung f	refroidissement m à basse température	охлаждение глубоким холодом, низкотемпературное охлаждение
L 284	low-temperature dehydration	Tieftemperaturtrocknung f	déshydratation (séchage) f à basse température	низкотемпературная сушка
L 285	low temperature dryer	Kaltlufttrockner m	sécheur m à l'air froid	сушилка, работающая с помощью холодного воздуха, низкотемпературная сушилка
L 286	low-temperature evaporation	Tieftemperaturverdampfung f	évaporation f à basse température	низкотемпературное испарение, низкотемпературная выпарка
L 287	low-temperature field	Tieftemperaturbereich m	champ (domaine) m de basse température	низкотемпературный диапазон
L 288	low-temperature gas scrubbing	Tieftemperaturgasreinigung f	épuration f de gaz à basse température	низкотемпературная очистка газа
L 289	low-temperature insulation	Kälteschutz m	isolation f calorifuge, calorifugeage m	низкотемпературная изоляция
L 290	low-temperature load	Tieftemperaturwärmelast m	charge f calorifique à basse température	тепловая нагрузка при низкой температуре
L 291	low-temperature operation	Tieftemperaturbetrieb m	opération f à basse température	низкотемпературный режим
L 292	low-temperature plant	Tieftemperaturanlage f	installation f frigorifique	установка глубокого холода, установка низких температур
L 293	low-temperature polymerization	Tieftemperaturpolymerisation f	polymérisation f à basse température	низкотемпературная полимеризация
L 294	low-temperature processing	Tieftemperaturbehandlung f	traitement m à basse température	низкотемпературная обработка
L 295	low-temperature reactor	Niedertemperaturreaktor m	réacteur m à basse température	низкотемпературный реактор
L 296	low-temperature rectification	Tieftemperaturrektifikation f	rectification f à basse température	низкотемпературная ректификация
L 297	low tension	Niederspannung f	basse tension f	низкое напряжение
L 298	lubricant	Schmierstoff m, Schmiermittel n, Schmierflüssigkeit f, Gleitmittel n	lubrifiant m, graisse f	смазочный материал, смазка, смазочное масло (вещество)
L 299	lubricating	Abschmieren n	graissage m	смазывание
L 300	lubricating oil	Schmieröl n	huile m de lubrification	смазочное масло
L 301	lubricating oil cooler	Schmierölkühler m	réfrigérant m à huile lubrifiante	охладитель смазочного масла
L 302	lubricating power, lubricity, oilness	Schmierfähigkeit f	pouvoir m lubrifiant, onctuosité f	смазывающая способность
L 303	lubrication	Schmierung f	lubrification f, graissage m	смазка, смазывание
L 304	lubrication system lubricity	Schmiersystem n s. L 302	système m de graissage	система смазки
L 305	lug, shoulder, rim, set-off, rest	Ansatz m	épaulement m, bout m	исходная смесь
L 306	lye, liquor	Lauge f	lessive f alcaline, lessive f	щелок, щелочный раствор
L 307	lye bath	Laugenbad n	bain m à lessive, lessive f	щелочная ванна
L 308	lye container	Laugenbehälter m	réservoir m à lessive	сборник для щелока
L 309	lyosorption	Benetzen n (durch Lösungsmittel)	mouillage m	смачивание

M

M 1	machine attendant	Maschinenwärter m	mécanicien m, machiniste m, surveillant m	машинист
M 2	machine drive	Maschinenantrieb m	commande f par machine (moteur)	привод машины
M 3	machinery diagram	konstruktives Fließbild n	flow-sheet m constructif	конструктивная диаграмма, конструктивная схема
M 4	machining step	Bearbeitungsstufe f	degré m d'usinage, phase m d'usinage	ступень обработки, технологическая операция
M 5	Mach number	Mach-Zahl f, Machsche Zahl f	nombre m de Mach, Mach m	число Маха
M 6	macrocorrosion	Makrokorrosion f	corrosion f macroscopique	макрокоррозия
M 7	macromixture	Makromischung f	macromélange m	макросмешение
M 8	macromolecular branching	Makromolekularverzweigung f	ramification f macromoléculaire	макромолекулярная разветвленность
M 9	macromolecular coil	makromolekulare Spirale f	spirale f macromoléculaire	макромолекулярная спираль

M 10	macropore	Makropore f	macropore m	макропора
M 11	magnetic field	Magnetfeld n, Kraftfeld n	champ m magnétique	силовое (магнитное) поле
M 12	magnetic flux	Kraftlinienfluß m	flux m magnétique	поток [магнитных] силовых линий
M 13	magnetic gasket	Magnetabdichtung f	garniture f magnétique	магнитная прокладка
M 14	magnetic separation	magnetische Aufbereitung f	préparation f (séparation f, triage m) magnétique	магнитное обогащение
M 15	magnetohydrodynamics	Magnetohydrodynamik f	magnétodynamique f des fluides, magnétohydrody-namique f	магнитогидродинамика
M 16	magnitude	Größe f, Wert m, Betrag m, Menge f, Größenklasse f	grandeur f, taille f, étendue f, ampleur f	величина, размер, значе-ние, количество
M 17	main absorber	Hauptabsorber m	absorbeur m principal	главный абсорбер
M 18	main absorption	Hauptabsorption f	absorption f principale	основная абсорбция
M 19	main absorption stage	Hauptabsortionsstufe f	phase f d'absorption princi-pale	основная стадия абсорбции
M 20	main column	Hauptkolonne f	colonne f principale	главная колонна
M 21	main component	Hauptkomponente f, Haupt-bestandteil m	constituant m principal	основной (главный) компо-нент, главная составная часть
M 22	main condenser	Hauptkondensator m	condenseur m principal	главный конденсатор
M 23	main constituent	Hauptanteil m	constituant m principal, com-posante f prédominante	главная доля
M 24	main contractor	Generalauftragnehmer m, Hauptauftragnehmer m	mandataire m principal	генеральный (главный) под-рядчик
M 25	main discharge, main out-flow	Hauptabfluß m	déchargeoir m principal	главный сток
M 26	main equipment	Hauptausrüstung f	équipement m principal	главное оборудование (оснащение)
M 27	main flow, basic flow	Grundstrom m, Hauptstrom m, Hauptströmung f	courant (écoulement) m prin-cipal	основной (главный) поток, главное течение
M 28	main fraction	Hauptfraktion f	fraction f principale	главная фракция
M 29	main gear	Hauptantrieb m	commande f principale	главный привод
M 30	main girder	Hauptträger m	poutre f principale, longeron m	главная опора (несущая балка)
M 31	main group	Hauptgruppe f	groupe m principal	главная группа, главный узел
M 32	main line	Hauptleitung f	conduite f principale	магистральная (главная) ли-ния
M 33	main outflow main plant	s. M 25 Hauptanlage f	installation f principale	главная (ведущая, основ-ная) установка
M 34	main reaction main reactor	s. C 234 Hauptreaktor m	réacteur m principal	основной (главный) реактор
M 35	main refrigerating system	Hauptkälteanlage f	installation f frigorifique prin-cipale	главная холодильная уста-новка
M 36	mains	Rohrnetz n	tuyautage m	трубопроводная сеть, сеть трубопроводов
M 37	main steam piping	Hauptdampfleitung f	conduite f de vapeur princi-pale	главный паропровод
M 38	main stream	Hauptstrom m	courant (flux) m principal	главный ток, магистральный (главный) поток
M 39	main stream column configu-ration	Hauptkolonnensystem n, Hauptstromkolonnenschal-tung f	système m de colonnes principales	система, состоящая из главных колонн
M 40	maintain/to	aufrechterhalten	maintenir	поддерживать
M 40a	maintained temperature	aufrechterhaltene Tempera-tur f	température f maintenue	поддерживаемая темпера-тура
M 41	maintenance, servicing, at-tendance	Instandhaltung f, Pflege f, Unterhaltung f, Wartung f	entretien m, maintien m en bon état, surveillance f	уход, надзор, ремонт, тех-ническое обслуживание, содержание в исправно-сти
M 42	maintenance costs	Instandhaltungskosten pl, In-standhaltungsaufwand m	frais mpl d'entretien	стоимость ремонта, сто-имость технического об-служивания, затраты на ремонт
M 43	maintenance documentation	Instandhaltungsdokumenta-tion f	documentation f d'entretien	ремонтная документация, документация по техни-ческому обслуживанию
M 44	maintenance instruction	Instandhaltungsanweisung f, Betriebsanweisung f	instructions fpl d'entretien, instructions de service	инструкция по уходу и об-служиванию, инструкция по ремонту, эксплуата-ционная инструкция
M 45	maintenance part	Ersatzteil n	pièce f de rechange, élément m de réserve	запасная часть
M 46	maintenance period	Instandhaltungsperiode f	période f d'entretien	период технического об-служивания, ремонтный период (цикл)
M 47	maintenance process	Instandhaltungsprozeß n	processus m d'entretien	процесс ремонта, процесс технического обслужива-ния

M 48	maintenance routine work	Instandhaltungsarbeiten *fpl*	travaux *mpl* d'entretien	ремонтные работы, работы по уходу и обслуживанию
M 49	maintenance strategy	Reparaturstrategie *f*	stratégie *f* de remise en état, politique *f* d'entretien	стратегия ремонта
M 50	maintenance theory	Instandhaltungstheorie *f*	théorie *f* d'entretien	теория технического обслуживания, теория ремонта
M 51/2	maintenance unit	Instandhaltungseinheit *f*	unité *f* d'entretien	ремонтная единица, ремонтный узел
M 53	main trunk	Hauptstrang *m*	artère *f* principale	магистраль, магистральная линия
M 54	main vaporizer	Hauptverdampfer *m*	èvaporateur *m* principal	главный испаритель, главный корпус выпарной установки
M 55	main wind direction	Hauptwindrichtung *f*	direction *f* principale du vent	преобладающее направление ветра, главное направление ветра
M 56	maker's serial number	Chargenzahl *f*	nombre *f* de charge	количество загрузок (партий)
M 57	make-time	Einschaltverzug *m*	retard *m* à l'enclenchement	запаздывание при включении
M 58	make-up	Ausgleich *m*, Auffüllung *f*	compensation *f*	компенсация, выравнивание, наполнение
M 59	make-up gas	Zusatzgas *n*	gaz *m* d'appoint	свежий газ
M 60	make-up solvent	Zusatzlösungsmittel *n*	solvant *m* d'appoint, solvant additionnel	свежий (дополнительный, добавочный) растворитель
M 61	make-up water	Zusatzwasser *n*, Frischwasser *n*	eau *f* fraîche	свежая (дополнительная, добавочная) вода
	malleability	*s.* E 37		
M 62	manhole	Mannloch *n*	trou *m* d'homme à nettoyage, orifice *m* de nettoiement	горловина, лаз, люк
	manhole	*s. a.* A 574		
M 63	manifold	Verteiler *m*, Verteilerrohr *n*	tuyau *m* distributeur, rèseau *m* d'alimentation	коллектор, разветвленный трубопровод, патрубок, распределитель, распределительный трубопровод
	manipulated variable	*s.* C 658		
	manipulation	*s.* P 515		
M 64	man-machine-system	Mensch-Maschine-System *n*	système *m* homme-machine	человеко-машинная система
M 65	manner of production	Herstellungsweise *f*	mode *m* de production	способ изготовления, метод производства
M 66	manometer	Manometer *n*, Druckanzeiger *m*	manomètre *m*, indicateur *m* de pression	манометр
M 67	manual adjustment	manuelle Regelung *f*	réglage *m* manuel, régulation *f* manuelle	ручное регулирование
	manual operation	*s.* H 6		
M 68	manufactured gas	Industriegas *n*	gaz *m* industriel	промышленный газ
M 69	manufacturing chemical equipment	chemischer Apparatebau *m*	construction *f* des appareils pour l'industrie chimique	химическое аппаратостроение
M 70	manufacturing engineering	Fertigungstechnik *f*	technique *f* de fabrication	технология изготовления
M 71	manufacturing method	Fertigungsmethode *f*	méthode *f* de production (fabrication)	метод производства (изготовления), технология
M 72	manufacturing method	Bearbeitungsmethode *f*	méthode *f* (procédé *m*) d'usinage	метод обработки
	manufacturing method	*s.a.* P 562		
M 73	manufacturing organization	Fertigungsorganisation *f*	organisation *f* de fabrication (production)	организация изготовления (производства)
M 74	manufacturing process	Fertigungsprozeß *m*, Verarbeitungsprozeß *m*, Verarbeitungsverfahren *n*	procédè *m* de fabrication (manufacture), processus *m* de traitement (transformation)	технология, технологический процесс, процесс изготовления, процесс (технология) переработки
	manufacturing schedule	*s.* P 563		
	manufacturing technique	*s.* P 564		
M 75	marginal service life	Grenznutzungsdauer *f*	vie *f* utile marginale (limite)	предельная длительность использования (применения)
M 76	marker generator	Zeitmarkengeber *m*, Markengeber *m*	marqueur *m* de temps	маркер
M 77	Mark-Kuhn-Honwink-Sakurada equation	Mark-Kuhn-Honwink-Sakuradasche Gleichung *f*	équation *f* de Mark-Kuhn-Honwink-Sakurada	уравнение Марка-Куна-Хонвинка-Закурады
M 78	Martin constant	Martin-Konstante *f*	constante *f* de Martin	постоянная Мартина
M 79	mass balance	Massenbilanz *f*, Materialbilanz *f*, Mengenbilanz *f*, Stoffbilanz *f*	bilan *m* massique, bilan de matière	массовый (материальный) баланс
M 80	mass coefficient of absorption	Massenabsorptionskoeffizient *m*	coefficient *m* d'absorption massique	массовый коэффициент поглощения
	mass concentration	*s.* C 520		

M 81	mass consumption	Massenverbrauch m, Mengenverbrauch m	consommation f de matière	расход массы
M 82	mass conversion	Massenumwandlung f	conversion f de masse	превращение массы
M 83	mass density	Massendichte f	densité f massique	плотность массы
M 84	mass exchange	Masseaustausch m, Stoffaustausch m	transfert m de matière (masse), échange m de matière	массообмен
M 85	mass flow rate	Massendurchsatz m, Massenstromgeschwindigkeit f	débit m massique	массовая скорость потока, массовый расход
M 86	mass flow-sheet	Mengenfließschema n	flow-sheet m quantitatif	технологическая схема материальных потоков
M 87	mass flux	Durchfluß m [in Masseneinheiten], Durchflußmenge f [je Zeiteinheit], Massenfluß m, Massenstrom m	débit m massique	массовый поток
M 88	mass fraction (ratio)	Massenbruch m	fraction f massique	массовая доля
M 89	mass transfer, mass transport	Stoffübergang m, Stoffübertragung f, Stofftransport m, Stoffaustausch m	transfert m de matière, transport m de matière	массопередача, массоперенос, перенос вещества, транспорт веществ, массообмен
M 90	mass-transfer coefficient	Stoffübergangszahl f, Stoffübergangskoeffizient m	coefficient m de transfert de matière	коэффициент (число) массопередачи
	mass-transfer controlled	s. M 91/2		
	mass transport	s. M 89		
M 91/2	mass-transport controlled, mass-transfer controlled	stofftransportlimitiert	limité par le transfert de matière	лимитировано массообменом, ограничено вследствие массообмена
	mass unit	s. Q 10		
M 93	masterbatch	Masterbatch m, Vormischung f, Grundmischung f (Plast)	mélange m maître	предварительная (основная) смесь
M 94	master controller	Hauptregler m	régulateur m principal	главный регулятор
M 95	material	Stoff m, Material n, Werkstoff m	matière f, substance f, matériel m, matériau m	вещество, материал
M 96	material balance	Stoffbilanz f, Materialbilanz f	bilan m matière	материальный баланс
M 97	material constant	Werkstoffkonstante f	caractéristique m du matériel	коэффициент (постоянная) материала
M 98	material consumption	Materialverbrauch m	consommation f des matériaux	расход материала
M 99	material consumption norm (standard)	Materialverbrauchsnorm f	consommation f normalisée de matériaux	норма расхода материала
M 100	material conversion models	Stoffumwandlungsmodell n	modèle m de transformation de matière	модель превращения вещества
M 101	material cost	Materialkosten pl	coût m des matières premières	стоимость материала
M 102	material data	Stoffdaten pl, Stoffwerte mpl	paramètres mpl du matériau, valeurs fpl caractéristiques des matériaux	термодинамические данные, термодинамические свойства веществ
M 103	material defect	Werkstoffehler m, Materialfehler m	défaut m de matériel (matière), vice m de matière	дефект материала, материал с дефектом
	material economy	s. F 52		
M 104	material flow	Materialfluß m	écoulement (flux) m de matière	поток материала
M 105	material intensity	Materialintensität f	intenité f en matériaux	материалоемкость
M 106	material model	Stoffmodell n	modèle m de matière	материальная модель
M 107	material models	materielle Modelle npl	modèles mpl matériels, maquettes fpl	материальные модели
M 108/9	material parameter, data	Stoffwert m	constante f propre à la substance	физико-химический коэффициент вещества, физико-химическая постоянная вещества, свойство вещества
	material property	s. P 589		
M 110	material regeneration	stoffliche Regeneration f	régénération f matérielle	регенерация вещества
M 111	materials handling	Materialtransport m	transport m de la matière, transport des matériaux	транспортирование материалов
M 112	material stream	Stoffstrom m, Stofffluß m	écoulement m de matière, transport m de matière, flux m de matière	массовый (материальный) поток, поток веществ
M 113	material testing	Werkstoffprüfung f	essai m (épreuve f) des matériaux	испытание материала
M 114	material to be dried	Trockengut n	matériau m à sécher	материал для сушки
M 115	material to be ground	Mahlgut n	matériau m à moudre	измельчаемый материал
M 116	material to be mixed	Mischgut n	matériaux mpl à mélanger	смешиваемый материал
M 117	material transport models	Stofftransportmodelle npl	modèles mpl de transport de matière	модели переноса вещества
	material utilization	s. F 52		
M 118	mathematical model	mathematisches Modell n	modèle m mathématique	математическая модель
M 119	matrix [form], injection mould	Matrize f, Form f, Spritzmundstück n, Düse f	buse f, matrice f	матрица, форма, мундштук

M 120/1	maximal value, maximum, maximum value	Maximalwert m, größter Wert m	valeur f maximale	максимальное значение, максимальная (наибольшая) величина, максимум
M 122	maximum charging	Maximalbelastung f	charge f maximale	максимальная нагрузка
M 123	maximum demand indicator	Maximumzähler m	compteur m à indicateur de maximum	максимальный счетчик
M 124	maximum efficiency	Höchstwirkungsgrad m	rendement m maximal	максимальная эффективность, максимальный коэффициент полезного действия
M 125	maximum evaporative capacity	Höchstverdampfungsfähigkeit f	débit m évaporatif maximal, puissance f d'évaporation maximum	максимальная паропроизводительность, высокая способность к испарению
	maximum inflow	s. P 77		
M 126	maximum pressure	Höchstdruck m	pression f maximale, très haute pression	максимальное давление
M 127	maximum principle	Maximumprinzip n	principe m maximum	принцип максимума
M 128	maximum temperature	Höchsttemperatur f	température f maximale	максимальная температура
M 129	maximum temperature difference	maximales Temperaturgefälle n	gradient m maximal de température	максимальный перепад температур
	maximum value	s. M 120/1		
M 130	Maxwell body	Maxwellscher Körper m	corps m maxwellien (de Maxwell)	тело Максвелла
M 131	Maxwell-Boltzmann law	Maxwell-Boltzmann-Gesetz m	loi f de Maxwell-Boltzmann	закон Максвелла-Больцмана
M 132	Maxwell element	Maxwellsches Element n, Maxwell-Körper m	élément m de Maxwell	элемент Максвелла
M 133	mean free path	mittlere freie Weglänge f	longueur f de chemin libre, libre parcours m moyen	свободная средняя длина пробега
M 134	mean of attach	Aufschlußmittel n	agent m d'attaque, agent de dissolution	расщепляющее средство, растворяющее средство
M 135	mean of conveying	Transportmittel n	moyen (agent) m de transport	транспортное средство
M 136	mean of transportation	Beförderungsmittel n	moyen m de transport (locomotion)	транспортное средство
M 137	mean pressure	Mitteldruck m	moyenne pression f	среднее давление
M 138	mean temperature difference	mittlere Temperaturdifferenz f	différence f moyenne de température	средняя разность температур
M 139	mean value	Mittelwert m	valeur f moyenne	среднее значение
M 140	measured quantity	Meßgröße f	grandeur f mesurée (à mesurer)	измеряемая величина
M 141	measuring apparatus	Meßvorrichtung f, Meßapparat m	appareil m mesureur (de mesure)	измерительный прибор, измерительное устройство (приспособление)
M 142	measuring arrangements	Meßanordnung f	montage m de mesure	измерительное устройство, измерительная схема
M 143	measuring bridge	Meßbrücke f	pont m Wheatstone, ohmmètre m à pont	измерительный мостик
M 144	measuring channel	Meßkanal m	voie f de mesure	измерительный канал
M 145	measuring circuit	Meßschaltung f	montage m de mesure	измерительная схема
M 146	measuring device	Meßeinrichtung f	dispositif m de mesure	измерительное устройство
M 147	measuring error	Meßfehler m	erreur f de mesure	ошибка измерения
	measuring flask	s. G 131		
	measuring glass	s. G 130		
M 148	measuring impulse	Meßimpuls m	impulsion f de mesure	измерительный импульс
M 149	measuring range	Meßbereich m	amplitude f (domaine m) de mesure, étendue f de mesurage	диапазон измерения, пределы измерения, область измерения
M 150	measuring rod	Meßlatte f	jalon m, latte f de mesure, règle f divisée	мерный жезл, измерительная рейка
M 151	measuring system	Maßsystem n	système m de mesure	система мер (размерности)
M 152	measuring vessel, graduated jar	Meßbehälter m	réservoir m de mesure, bac m jaugeur	мерный бак
M 153	mechanical dust collection	mechanische Entstaubung f	dépoussiérage m mécanique	механическое обеспыливание
M 154	mechanical efficiency	mechanischer Wirkungsgrad m	rendement m mécanique	механический коэффициент полезного действия
M 155	mechanical energy balance	mechanische Energiebilanz f	bilan m de l'énergie mécanique	баланс механической энергии
M 156	mechanical property	mechanische Eigenschaft f	propriété f mécanique	механическое свойство
M 157	mechanical refrigeration	maschinelle (mechanische) Kühlung f	réfrigération f mécanique	машинное охлаждение
M 158	mechanical treatment	mechanische Aufbereitung f	traitement m, séparation f, préparation f mécanique	механическое обогащение
M 159	mechanochemical effect	mechanochemische Einwirkung f, mechanochemischer Effekt m	effet m mécano-chimique	механохимическое воздействие
M 160	medium pressure hydrogenation	Mitteldruckhydrierung f	hydrogénation f à pression moyenne	гидрирование при среднем давлении
M 161	medium screening	Siebung f mit mittlerer Korngröße, mittlere Siebung	criblage m moyen, tamisage m moyen	средний рассев

M 162	melt/to, to fuse	verschmelzen	fondre ensemble, fondre, fu-ser	смешивать, плавить, рас-плавлять
M 163	melt/to	[zer]schmelzen, flüssig wer-den, sich auflösen	fondre, fuser, porter à fusion	расплавить, разжижать, растворяться
M 164	melt, fusion, molten mass, melting flux	Schmelze f, Schmelzfluß m	fonte f, masse f fondue, fu-sion f, matière f fusée, fu-sion complète	плавка, плав, расплав, по-ток расплава, флюс, пла-вень, расплавленная ко-лоша
M 165	melt fracture	Schmelzebruch m	rupture f de fusion	дробление расплава
M 166	melting	Schmelzen f, Aufschmelzen n	dégel m, fusion f, fondage m	плавление, расплавление
M 167	melting bath, molten bath	Schmelzbad n	bain m de fusion	жидкая ванна, ванна для расплавления, сварочная ванна
M 168	melting chamber	Schmelzkammer f	chambre f de fusion	камера таяния (для расплавления)
M 169	melting curve	Schmelzverlauf m	courbe f de fusion	кривая таяния (расплавле-ния)
	melting-down	s. F 416		
	melting flux	s. M 164		
M 170	melting furnace	Schmelzofen m	four (fourneau) m de fusion	плавильная (варочная) печь
M 171	melting losses	Schmelzverluste f	pertes fpl de fusion	потери при расплавлении, потери от таяния
M 172	melting point, fusion point	Schmelzpunkt m	point m de fusion	точка (температура) плавле-ния
M 173	melting point diagram	Schmelzdiagramm n	diagramme m des points de fusion	диаграмма плавления
M 174	melting pot	Schmelzkessel m	chaudière f à fusion	плавильник, плавильный ко-тель
M 175	melting surface	Schmelzoberfläche f	surface f de fusion	поверхность плавления (та-яния)
M 176	melting time	Schmelzdauer f	durée f de fusion	продолжительность таяния (плавления)
M 177	melting zone	Aufschmelzzone f	zone f de fusion	зона расплавления
M 178	membrane filter	Membranfilter n	filtre m microporeux	мембранный фильтр
M 179	membrane valve	Membranventil n	soupape f à membrane (dia-phragme)	мембранный вентиль
M 180	memory equation	Gedächtnisgleichung f	équation f de mémoire	функция (уравнение) па-мяти
M 181	mercury rotating pump	Kapselpumpe f	pompe f blindée	капсюльный насос
M 182	mesh	Siebmasche f	maille f [de crible]	отверстие сита
M 183	mesh screen, screen	Sieb n, Siebvorrichtung f, Fil-ter n, Gitter n, Schirm m	crible m, tamis m, passoire f, dispositif m de criblage, filtre m	сито, решето, решетка, экран, фильтр
M 184	metal bound, metallic link-age	metallische Bindung f	liaison f métallique	металлическая связь
M 185	metal deposition	Metallabscheider m	séparateur m de métal	металлоотделитель
M 186	metallic conductance	metallische Leitung f	conductibilité f métallique	металлическая проводи-мость
	metallic linkage	s. M 184		
M 187	metathesis	chemische Umsetzung f, Austauschreaktion f	transformation f chimique, réaction f, réaction chimi-que, réaction de double décomposition	химическое превращение, обменная реакция, реак-ция обменного разложе-ния
M 188	metering	Beschickung f	alimentation f	загрузка, засыпка
M 189	metering zone	Dosierzone f	zone f de dosage	зона дозирования
M 190	methanol synthesis	Methanol-Synthese-Verfah-ren n	procédé f de synthèse de méthanol	способ синтеза метанола
	method	s. P 494		
M 191	method of approach	Annäherungsmethode f	méthode f d'approximation	метод аппроксимации (при-ближения)
M 192	method of charging	Beschickungsweise f	mode m de chargement, mode d'alimentation, mé-thode f de charge	метод (способ) загрузки
M 193	method of construction	Bauweise f	forme (méthode) f de construction	способ стройки (выемки, строительства)
M 194	method of decomposition	Dekompositionsmethode f	méthode f de décomposition	декомпозиционный метод, метод декомпозиции
M 195	method of determination	Bestimmungsmethode f	méthode f de détermination, méthode d'analyse, ana-lyse f	метод определения
M 196	method of least squares	Methode f der kleinsten Qua-drate	méthode f des moindres car-rés	метод наименьших квадра-тов
M 197	method of preparation	Vorbereitungsmethode f	méthode f de préparation	метод приготовления (под-готовки)
M 198	method of processing	Verarbeitungsmethode f	méthode f de traitement (fa-brication)	метод переработки (обра-ботки), способ перера-ботки
M 199	method of proof	Beweismethode f	méthode f de démonstration	метод доказательства
M 200	method of separation	Abscheidungsverfahren n	méthode f de séparation	способ разделения (отделе-ния)

M 201	method of treatment procedure	Behandlungsverfahren n	procédé m de traitement	метод обработки, способ обращения
M 202	microbicide	Konservierungsmittel n	préservatif m antiseptique	консервирующее средство
M 203	microburner	Mikrobrenner m	microbruleur m	микрогорелка
M 204	microcomputer	Mikrorechner m	microcalculatrice f	микропроцессорная вычислительная машина, микрокалькулятор, микрокомпьютер
M 205	microcorrosion	Mikrokorrosion f	corrosion f microscopique	• микрокоррозия
M 206	microcrack	Mikroriß m	microfente f	микротрещина
M 207	microcrystalline	mikrokristallin, feinkristallin	microcristallin	тонкокристаллический, микрокристаллический
M 208	micromixture	Mikromischung f	micromélange m	микросмешение, перемешивание в микромасштабе
M 209	microparticle	Mikroteilchen n	particule f microscopique, microparticule f	микрочастица
M 210	microrheology	Mikrorheologie f	microrhéologie f	микрореология
M 211	microstructure	Feinstruktur f, Mikrostruktur f	structure f fine	микроструктура, тонкая структура
M 212	microturbulence	Mikroturbulenz f	microturbulence f	микротурбулентность
M 213	microviscoelasticity	Mikroviskoelastizität f	microviscoélasticité f	микровязкоупругость
M 214	migration velocity, velocity of migration	Wanderungsgeschwindigkeit f	vitesse f de migration	скорость передвижения (перемещения, миграции)
	mill/to	s. C 869		
M 215	mill, pulverizer	Mühle f	moulin m, broyeur m	мельница, дробилка
M 216	milling	Mahlarbeit f	broyage m	затраты на размол
	milling	s. a. G 151		
M 217	milling cycle	Mahlgang m, Mühlgang m	tournant m (moulin)	мельничный постав
M 218	milling plant, grinding mill	Mahlanlage f, Mahlmaschine f	installation f de concassage, installation de broyage, appareil m à moudre	измельчающая установка, мельница, установка для измельчения, размольный аппарат, размольная машина
M 219	milling product, grinding product	Mahlprodukt n	produit m moulu	продукт помола, размалываемый материал
M 220	mineral oil processing	Mineralölverarbeitung f, Erdölverarbeitung f	traitement m du pétrole	переработка минерального масла, переработка нефти
M 221	mineral oil technology	Mineralöltechnologie f	technologie f pétrolière	технология минерального масла
	mingle/to	s. B 134		
M 222	minimization of tear variables	Schnittzahlminimierung f	minimalisation f des coupes	минимизация разрывающего множества
M 223	minimum, minimum value	Minimum n, Minimalwert m, kleinster Wert m	minimum m, valeur f minimale	минимум, наименьшее значение
M 224	minimum capacity (charge)	Mindestbelastung f	charge f minimum	минимальная нагрузка
M 225	minimum efficiency	Mindestleistung f	puissance f minimum	минимальная мощность
	minimum extent	s. M 229		
M 226	minimum height	Mindesthöhe f	hauteur f minimum	минимальная высота
M 227/8	minimum ignition energy	Mindestzündenergie f	énergie f minimum d'inflammation	минимальная энергия воспламенения
	minimum pressure	s. L 253		
M 229	minimum size, minimum extent	Mindestmaß n	minimum m, mesure f minimum	минимальный размер
M 230	minimum speed	Minimalgeschwindigkeit f	vitesse f minimale	минимальная скорость
M 231	minimum stability	Mindeststabilität f	stabilité f minimum	минимальная устойчивость
M 232	minimum temperature	minimale Temperatur f	température f minimale	минимальная температура
M 233	minimum temperature difference	minimale Temperaturdifferenz f	différence f de température minimale	минимальная разность температур
	minimum value	s. M 223		
	mining loss	s. E 380		
M 234	miscibility	Mischbarkeit f	miscibilité f	смешиваемость
M 235	miscibility gap	Mischungslücke f	lacune f de miscibilité, non-miscibilité f	диапазон несмешиваемости, интервал несмешиваемости
M 236	miscible, mixable	mischbar, vermischbar	miscible	смешивающийся
M 237	mist flow	Nebelströmung f	écoulement m brumeux	поток тумана, фонтанирование в виде распыления
	mix/to	s. B 134		
M 238	mix	Mischgut n	matériau m mixte	смешиваемый материал
	mixable	s. M 236		
M 239	mixed catalyst	Mischkatalysator m	catalyseur m mixte	смешанный катализатор
	mixed condensation	s. C 332		
M 240	mixed gas	Mischgas n	gaz m mixte	смешанный газ
M 241	mixer, mixing apparatus, blender	Mischer m, Mischvorrichtung f, Mischapparat m	mélangeur m, malaxeur m, appareil m à mélanger	смеситель, смесительное устройство, смесительная аппаратура, мешалка
M 242	mixer head	Mischkopf m	tête f de mélange	смесительная головка
M 243	mixer platform	Mischerbühne f	plate-forme f de mélange	миксерная площадка
M 244	mixer-settler tower	Turmextraktor m	colonne f d'extraction	башенный экстрактор

M 245	mixing	Mischen n, Verrühren n, Vermischung f	mélangeage m, agitation f, malaxage m, malaxation f	смешение, смешивание, перемешивание
	mixing apparatus	s. M 241		
M 246	mixing behaviour	Durchmischungsverhalten n	mélangeage m, comportement m de mélange	поведение при смешении
M 247	mixing chamber	Mischkammer f	chambre f de mélange	смесительная камера
M 248	mixing condenser	Mischkondensator m	condenseur m mixte	конденсатор смешивающего типа
M 249	mixing drum	Mischtrommel f	tambour-malaxeur m, tambour m à mélanger	смесительный барабан
M 250	mixing heat	Mischungswärme f, Mischungsenthalpie f	chaleur f de mélange	теплота смешения
M 251	mixing index	Mischindex m	index m de mélange	индекс смешения
M 252	mixing jet	Mischdüse f	tuyère f mélangeuse, buse f mélangeuse	смесительное сопло, смесительная форсунка
	mixing kettle	s. A 288		
M 253	mixing machine	Mischmaschine f	machine f à mélanger, mélangeur m, malaxeur m	смесительная машина, смеситель
M 254	mixing mill	Mischmühle f	moulin-mélangeur m	смешивающая мельница, мельница-смеситель
M 255	mixing of gases and liquid, gas-liquid mixing	Begasen n, Gaseinleitung f	introduction f de gaz	газирование, пропускание газа через жидкость, аэрация
M 256	mixing place	Mischstelle f	point m de mélange	место (точка) смешения
M 257	mixing plant	Mischanlage f	installation f de mélange	смешивающая установка, установка для смешения
M 258	mixing rolls	Mischwalzen fpl	rouleaux mpl mélangeurs	смесительные вальцы
M 259	mixing screw	Mischschnecke f, Schneckenrührer m	vis f sans fin de mélange	червячный смеситель, червячная мешалка, смесительный червяк
M 260	mixing state	Mischungszustand m	état m de mélange	состояние смешения (смеси)
	mixing tap	s. M 262		
M 261	mixing tube	Mischrohr n	tuyau m de mélange	смешивающая труба, труба смешения
M 262	mixing valve, mixing tap	Mischventil n, Mischhahn m	vanne f de mélange, robinet m mélangeur	смесительный кран, смешивающий клапан (вентиль)
M 263	mixing vessel	Mischbehälter m	cuve-mélangeur f, mélangeoir m	смесительный резервуар, смесительная емкость, бак для смешения
M 264	mixture, blend, compound	Mischung f, Gemisch n, Stoffgemisch n, Mischprodukt n	mélange m, produit m mixte	смесь, смешивание, смешение, смесь веществ, продукт смешения
M 265	mixture composition	Gemischzusammensetzung f	composition f de mélange	состав смеси
M 266	mixture concentration	Gemischkonzentration f	concentration f de mélange	концентрация смеси
M 267	mixture contamination	Gemischverunreinigung f	contamination f du mélange	загрязнение смеси
M 268	mixture cooling	Gemischkühlung f	refroidissement m de mélange	охлаждение смеси
M 269	mixture line	Mischungslinie f	ligne f de miscibilité	кривая состава смеси
M 270	mixture of vapours	Dampfgemisch n	mélange m de vapeurs	смесь паров
M 271	mixture ratio	Mischungsverhältnis n	rapport m (proportion f) de mélange, dosage m	соотношение компонентов смеси
M 272	mixture rule, rule of mixtures	Mischungsregel f	règle f des mélanges	правило смешения
	MMD	s. M 326		
M 273	mobile lifting gear	mobiles Hebezeug n	organes mpl de levage mobiles	мобильное подъемное приспособление
M 274	model application	Modellanwendung f, Modellnutzung f	application (utilisation) f de modèle	применение модели
M 275	model design	Modellprojektierung f	projection f par modèle	модельное проектирование
M 276	model development	Modellentwicklung f	développement de modèle	разработка модели
M 277	model equation	Modellgleichung f	équation f de modèle	модельное уравнение
M 278	model experiment, model test	Modellexperiment n, Modellversuch m	expérience f sur maquette, essai m sur modèle	эксперимент на модели, модельный эксперимент, испытание при помощи модели, моделирование
M 279	modelling, simulation	Modellierung f	modélisation f	[имитационное] моделирование
M 280	model of component parts	Bauelementemodell n	modèle m des composants	модель конструктивных элементов
M 281	model parameter	Modellparameter m	paramètre m de modèle	модельный параметр, параметр модели
M 282	model scale	Modellmaßstab m	échelle f de modèle	модельный масштаб
M 283	model simplification	Modellvereinfachung f	simplification f de modèle	упрощение модели
M 284	model term	Modellterm m	terme m de modèle	член модели
	model test	s. M 278		
M 285	model theory	Modelltheorie f	théorie f de modèle	модельная теория
M 286	mode of application	Anwendungsweise f	mode m d'application	способ применения, инструкция по применению
M 287	mode of driving	Antriebsart f	genre (mode) m de commande	вид тяги, вид привода

M 288	mode of operation, operating mode, function	Betriebsweise f, Wirkungsweise f	mode m opératoire, manière f d'opérer	способ производства, производственный способ (метод), принцип работы (действия), рабочий режим
M 289	modification	Modifikation f, Modifizierung f	variété f, modification f	модификация
M 290/1	modular principle	Modulprinzip n, Baukastenprinzip n	conception f bloc-éléments, principe m modulaire	модульный принцип
	modulus of cross elasticity	s. P 241		
M 292	modulus of elasticity, Young's modulus	Elastizitätsmodul m, E-Modul m	coefficient (module) m d'élasticité	модуль упругости (эластичности)
M 293	moistening	Befeuchten n, Befeuchtung f	humidification f, humectation f	смачивание, увлажнение
M 294	moistening machine	Befeuchtungsmaschine f	humidificateur m, scrubber m	увлажнительная машина
M 295	moistening plant	Befeuchtungsanlage f	humidificateur m, installation f de l'humectation	увлажнитель, увлажнительная установка
M 296	moistening tower	Befeuchtungsturm m	tour f d'humidification	увлажнительная башня
M 297	moisture absorption	Feuchtigkeitsabsorption f	absorption f d'humidité	абсорбция влаги
M 298	moisture accumulation	Feuchtigkeitsansammlung f	accumulation f d'humidité	аккумуляция (накопление) влаги
M 299	moisture adsorption	Feuchtigkeitsadsorption f	adsorption f d'humidité	адсорбция влаги
M 300	moisture agent, wetting agent	Befeuchtungsmittel n	agent m de l'humectation, agent humidificateur	увлажнительное средство, вещество для увлажнения
M 301	moisture capacity	Wasseraufnahmefähigkeit f	propriété f hygroscopique	водопоглощающая способность
M 302	moisture condensation	Feuchtigkeitskondensation f	condensation f d'humidité	конденсация влаги
M 303	moisture content, amount of moisture	Feuchtigkeitsgehalt m	quantité f (teneur m, taux m) d'humidité, degré m d'humidité	влажность, содержание влаги, влагосодержание
	moisture content	s. a. F 277		
M 304	moisture content difference	Feuchtedifferenz f	différence f de teneur en humidité	разность влагосодержаний
M 305	moisture content meter, hygrometer	Feuchtigkeitsmesser f	hygromètre m, psychromètre m	прибор для измерения влажности, гигрометр
M 306	moisture contents	Naßgehalt m, Feuchtegehalt m	humidité f	влагосодержание
M 307	moisture diffusion	Feuchtigkeitsdiffusion f	diffusion f d'humidité	диффузия влаги
M 308	moisture equilibrium	Feuchtegleichgewicht n	équilibre m d'humidité	равновесная влажность, равновесное влагосодержание
M 309	moisture exchange	Feuchteaustausch m	échange m d'humidité	влагообмен
M 310	moisture intake	Ansaugen n von Feuchtigkeit, Feuchtigkeitseintritt m	admission f d'humidité	всасывание (проникновение) влаги
M 311	moisture of vapour	Dampffeuchtigkeit f	humidité f de vapeur	влажность пара, влагосодержание пара
M 312	moisture precipitation	Feuchtigkeitsausscheidung f	précipitation f d'humidité	влаговыпадение, влаговыделение
M 313	moisture resistance	Feuchtigkeitsbeständigkeit f	résistance f à l'humidité	влагостойкость
M 314	moisture separator	Feuchtigkeitsabscheider m	séparateur m d'humidité	отделитель влаги
M 315	moisture transfer	Feuchtigkeitsaustausch m, Feuchteübertragung f	transport m d'humidité	влагопередача, влагообмен
M 316	moisture withdrawal	Feuchtigkeitsabführung f, Feuchtigkeitssenkung f	élimination f d'humidité	отвод влаги, отвод влажности, понижение влажности
M 317	molar balance	Molbilanz f	bilan m moléculaire	мольный баланс
M 318	molar concentration	Molkonzentration f	concentration f moléculaire	мольная концентрация
M 319	molar flow rate	Moldurchsatz m	débit m moléculaire	мольный расход
	molar fraction	s. M 331		
M 320	molar heat, molecular heat	Mol[ekular]wärme f	chaleur f moléculaire	молокулярная теплоемкость
	molar volume	s. M 329		
M 321	molar weight	Molmasse f	poids m (masse f) moléculaire	молекулярный вес, молекулярная масса
M 322	molecular coil	Molekülspirale f	spirale f moléculaire	молекулярная спираль
M 323/4	molecular distillation	Moleculardestillation f	distillation f moléculaire	молекулярная дистилляция
	molecular heat	s. M 320		
M 325	molecular lattice	Molekülgitter n	réseau m moléculaire	молекулярная решетка
M 326	molecular mass distribution, MMD	Molekülmassenverteilung f	distribution f de masse moléculaire	молекулярно-массовое распределение, распределение молекулярной массы
M 327	molecular pressure	Molekulardruck m	pression f moléculaire	молекулярное давление
M 328	molecular structure	Molekularstruktur f	structure f moléculaire	молекулярная структура
M 329	molecular volume, molar volume	Molvolumen n, Molekularvolumen n	volume m molaire (moléculaire)	мол[екул]ярный объем
M 330	molecular weight	relative Molekülmasse f, Molekulargewicht n, Molmasse f	poids m moléculaire	относительная молекулярная масса, молекулярный вес
M 331	mole fraction, molar fraction	Molenbruch m	fraction f de mole, fraction (titre m) molaire	молярная доля, мольная доля
M 332	Mollier diagram	Mollier-Diagramm n, s-Diagramm n	diagramme m Mollier	энтальпийная диаграмма, s-Диаграмма Мольера

	molten bath	*s.* M 167		
	molten mass	*s.* M 164		
M 333	momentary balance	Zeitpunktbilanz f, Augen-blicksbilanz f	bilan m momentané	баланс к определенному моменту времени
M 334	moment of inertia	Trägheitsmoment n	moment m d'inertie	момент инерции
M 335	momentum balance	Impulsbilanz f	bilan m de quantité de mouvement	баланс импульса (количества движения)
M 336	momentum flux	Impulsfluß m, Impulsstrom m	flux m de quantité de mouvement	поток импульса (количества движения)
M 337	momentum transport	Impulstransport m	transport m de quantité de mouvement	перенос импульса (количества движения)
	monitor	*s.* M 204		
M 338	monomer transfer	Monomerübertragung f	transfert m de monomère	перенос мономера
M 339	monophase system	Einphasensystem n	système m à phase unique	однофазная (гомогенная) система
M 340	**Monte-Carlo method**	Monte-Carlo-Methode f	méthode f Monte-Carlo	метод Монте-Карло
M 341	**Mooney-Rivlin elastic body**	Mooney-Rivlinscher elastischer Körper m	corps m élastique de Mooney-Rivlin	упругое тело Муни-Ривлина
M 342	**Mooney-Rivlin function**, MR function	Mooney-Rivlinsche Funktion f	fonction f de Mooney-Rivlin	функция Муни-Ривлина
	mordant	*s.* 1. C 111; 2. D 39		
	mordanting	*s.* C 286		
M 343	**motive power**	Triebkraft f	force f motrice	движущая сила
	motor fuel	*s.* F 385		
M 344	**mould,** die	Form f, Werkzeug n, Preßform f (Plastverarb.)	moule m, matrice f de compression	форма, формующий инструмент, пресс-форма
M 345	**mould filling**	Formfüllung f	remplissage m de moule	наполнение (заполнение) формы
M 346	**moulding [compound]**	Preßmasse f	matière f à mouler par compression	прессовочная масса, пресс-масса
M 347	**moulding factor**	Formfaktor m	facteur m de forme	фактор формы
M 348	**moulding powder**	pulvrige Preßmasse f, Preßpulver n, Pulver n (Plastverarbeitung)	poudre f à mouler	порошковая прессмасса, пресс-порошок, порошок
M 349	**mounting,** installation	Einbau m	installation f, montage m	встройка, установка
M 350	**mount opening**	Einbauhöhe f	hauteur f de montage	размер установки (встройки)
M 351	**moving bed**	Wanderbett n	lit m mobile	подвижный (движущийся) слой
M 352	**moving-bed cracking plant**	Wanderbettkrackanlage f	unité f de cracking au lit mobile	крекинг-установка с подвижным слоем
M 353	**moving-bed process**	Wanderbettprozeß m	procédé m à lit mobile	процесс с подвижным слоем
M 354	**moving-bed reactor**	Wanderbettreaktor m	réacteur m à lit mobile	реактор с подвижным (движущимся) слоем (катализатора)
	mud	*s.* S 379		
M 355	mud filter	Schlammfilter n	filtre m de boue	шламовый фильтр
M 356	**mud pump,** slurry pump	Dickstoffpumpe f	pompe f à matières épaisses	шламовый насос, насос для откачки сгустка
M 357	**mud water,** sludge water	Schlammwasser n	eau f boueuse	смывная вода, вода с илом
M 358	multicomponent	Mehrkomponenten ..., Mehrstoff ..., Vielstoff ...	à plusieurs constituants	многокомпонентный, многовещественный
M 359	multicomponent mixture	Mehrkomponentengemisch n	mélange m à plusieurs constituants	многокомпонентная смесь
M 360	multidimensional	mehrdimensional	à plus de trois dimensions	многомерный
M 361	multidimensional evaluation	mehrdimensionale Bewertung f	évaluation f à plusieures dimensions	многомерная (многокритериальная) оценка, многокритериальное оценивание
M 362	multidimensional evaluation system	mehrdimensionales Bewertungssystem n	système m d'évaluation à plusieures dimensions	система многокритериальной оценки, многомерная система оценки
M 363	multilayer condenser	Schichtenkondensator m	condenseur m à plusieurs couches	слойный конденсатор
M 364	multilayered film	mehrschichtige Folie f, Mehrschichtfolie f	feuille f à plusieurs couches	многослойная пленка
M 365	multilevel optimization	Mehrebenenoptimierung f	optimisation f à plusieurs niveaux	многоуровневая оптимизация
M 366	multiphase contactor	Mehrphasentrenneinrichtung f, Destillationsanlage f	installation f de fractionnement	установка для фазового разделения, дистилляционная установка
M 367	multiple belt dryer	Mehrbandtrockner m	sécheur m à plusieurs bandes	многоленточная сушилка
M 368	multiple bound	Mehrfachbindung f	liaison f multiple	множественная (многократная) связь
M 369	multiple-crane construction	Mehrkranmontage f	montage m à l'aide de plusieurs grues	монтаж при помощи нескольких кранов, многокрановый монтаж
M 370	multiple effect evaporator	Vielkörperverdampfanlage f, Mehrfachverdampfer m	évaporateur m à effet multiple	многокорпусный выпарной аппарат, многокорпусная выпарная установка

M 371	multiple evaporator refriger-ating plant	Mehrverdampferkälteanlage f	installation f frigorifique à évaporateurs multiples	многоиспарительная холо-дильная установка
M 372	multiple feed charging	Mehrfachbeschickung f	alimentation f multiple	одновременная загрузка нескольких заготовок, многопоточная загрузка
M 373	multiple-hearth dryer	Tellertrockner m	sécheur m vertical à plateaux	тарельчатая сушилка
M 374	multiple-hearth roaster	Etagenofen m	four m aux étages	этажная (многоподовая) печь
M 375	multiple heater	Mehretagenofen m	four m à plusieurs étages	многоярусная печь
M 376	multiple-purpose plant	Mehrzweckanlage f	installation f à plusieurs usages	универсальная установка, мобильная схема много-ассортиментной химико-технологической уста-новки
M 377	multiple rotary screen	Trommelsieb n mit mehreren Sieben	tambour m cribleur à plu-sieurs cribles	вращающийся грохот с не-сколькими ситами
M 378	multiple screw extruder	Mehrschneckenextruder m	extrudeuse f à plusieurs vis	многошнековый экструдер
M 379	multiple staff crushing	mehrstufige Zerkleinerung f	concassage m à plusieurs étages	многоступенчатое измель-чение
M 380	multiple stage compressor	mehrstufiger Verdichter m	compresseur m à plusieurs étages	многоступенчатый ком-прессор
M 381	multiple story cooler	Tellerkühler m	refroidisseur m à plateaux	тарельчатый (дисковый) хо-лодильник
M 382	multiple story dryer	Mehretagentrockner m	sécheur m à plusieurs étages	многоярусная сушилка (су-шильная часть)
M 383	multiple story machine	Mehretagenmaschine f	machine f à plusieurs étages	многоярусная машина
M 384	multiproduct plant	Mehrproduktenanlage f	unité f à plusieurs produits	совмещенная технологиче-ская схема
M 385	multipurpose plant	Mehrzweckanlage f	unité f à plusieurs fonctions	мобильная технологическая схема
M 386	multistage	vielstufig	aux multiples étages, à plu-sieurs étages	многоступенчатый
M 387	multistage compression	mehrstufige Kompression f	compression f à plusieurs étages	многоступенчатое сжатие
M 388	multistage evaporation	mehrstufige Verdampfung f, Mehrstufenverdampfung f	évaporation à plusieurs étages	многоступенчатое испаре-ние, многокорпусная (многоступенчатая) выпарка
M 389	multistage expansion	mehrstufige Entspannung f	détente f à plusieurs étages	многоступенчатое расши-рение
M 390	multistage fractionation	vielstufige Fraktionierung f	fractionnement m à plusieurs étages	многоступенчатое фрак-ционирование
M 391	multistage process	Vielstufenverfahren n	procédé m à plusieurs étages	многоступенчатый процесс (способ)
M 392	multistage throttling	mehrstufige Drosselung f	étranglement m à plusieurs étages	многоступенчатое дроссе-лирование
M 393	multistart	mehrgängig	aux passages multiples	многоходовой
M 394	multistop quenching	gestufte Abschreckung f	refroidissement m brusque à plusieurs étages	ступенчатая закалка
M 395	multitube revolving dryer	Röhrentrockner m	séchoir m tubulaire	трубчатая (барабанная) су-шилка

N

N 1	narrow distributed polymer	engverteiltes Polymer n	polymère m à distribution serrée	полимер с узкой фракцией распределения
N 2	natural gas	Naturgas n, Erdgas n	gaz m naturel	природный газ
N 3	natural gas liquefaction	Erdgasverflüssigung f	liquéfaction f de gaz naturel	ожижение природного газа
N 4	natural resin	Naturharz n	résine f naturelle, poix-résine f	природная смола
N 5	natural size	natürlicher Maßstab m	grandeur f naturelle	естественный масштаб
N 6	nature	Beschaffenheit f, Natur f, Zu-stand m	nature f, qualité f intrinsè-que, état m	свойство, качество, струк-тура, консистенция
N 7	Navier-Stokes equations	Navier-Stokessche Gleichun-gen fpl	équations f de Navier-Stokes	уравнения Навье-Стокса
N 8	necessary optimality condi-tions	notwendige Optimalitätsbe-dingungen f	conditions fpl nécessaires d'optimalité	необходимые условия опти-мальности
N 9	neglect/to, to disregard, to omit	vernachlässigen	négliger	пренебрегать
N 10	neglect, omission	Vernachlässigung f	négligence f	пренебрежение
N 11	neighbour element	Nachbarelement n	élément m voisin (adjacent)	соседный (смежный) эле-мент
	nephelometer	s. T 333		
N 12	Nernst heat theorem	Nernstsches Wärmetheorem n	théorème m de la chaleur, théorème de Nernst	тепловая теорема Нернста, принцип Нернста
	net	s. N 19		
N 13	net calorific value	unterer Heizwert m	valeur f calorifique nette (mi-nimum)	рабочая теплотворность, нижняя теплотворная способность
N 14	net energy	Nutzenergie f	énergie f utile	полезная (эффективная) энергия

N 15	net refrigerating capacity	Nutzkälteleistung f	capacité f frigorifique utile	полезная холодопроизводительность (нетто)
N 16	net tonnage	Nettotonnage f	tonnage m net, port m net	чистый тоннаж, тоннаж нетто
N 17	net volume flow	Gesamtdurchfluß m	débit m total	общий (суммарный) расход
N 18	net weight	Nettogewicht n, Füllgewicht n	poids m net, poids de remplissage	вес нетто (наполнения, заполнения)
	net weight	s. a. D 551		
N 19	network, gauze, net	Netz n, Netzwerk n	réseau m, système m du réseau	сеть, сетка, схема
N 20	network density	Netzdichte f	densité f de réseau	плотность сетки (сети)
N 21	network junction	Netzknoten m	nœud m de réseau	узел сетки
N 22	network technique	Netzplantechnik	technique f de réseau	сетевая техника
N 23	neutralization	Neutralisieren n, Neutralisation f	neutralisation f	нейтрализование, нейтрализация
N 24	neutralization value	Neutralisationszahl f	indice m de neutralisation	кислотное число
N 25	neutral point	Neutralpunkt m	point m neutre	нейтральная точка
N 26	new construction, revised design	Neuausführung f, Neukonstruktion f	construction f nouvelle	новая конструкция, новое исполнение
N 27	Newtonian fluid mechanics	Newtonsche Strömungsmechanik f, Mechanik f Newtonscher fluider Medien	mécanique f des fluides newtoniens	ньютоновская гидромеханика, механика ньютоновских жидких сред
N 28	Newtonian liquid	Newtonsche Flüssigkeit f	liquide m newtonien	ньютоновская жидкость
	Newton method	s. N 31		
N 29	Newton-Raphson method	Newton-Raphson-Verfahren n	méthode f de Newton-Raphson	метод Ньютона-Рафсона
N 30	Newton's approximation method	Newtonsches Näherungsverfahren n	méthode f d'approximation newtonienne	метод приближений (итераций) Ньютона, приближенный способ (метод) Ньютона
N 31	Newton's method, Newton method	Newton-Methode f	méthode f newtonienne	метод Ньютона
N 32	nitrogen fertilizer	Stickstoffdüngemittel n	engrais m azotique (azoté)	азотнотуковое удобрение
	noble gas	s. R 39		
N 33	nodal adjacency matrix	Knotenadjazenzmatrix f	matrice f associée	матрица смежности
N 34	nodal network	Knotennetz n	réseau m de nœuds	сеть (схема) узлов
	node	s. C 122		
	noise absorption	s. S 438		
N 35	noise emission	Lärmemission f	émission (propagation) f du bruit	эмиссия (распространение) шума
N 36	noise measurement	Lärmmessung f	mesure f de bruit	измерение шума
N 37	noise protection	Lärmschutz m	protection f contre les bruits	защита от шума
N 38	noise suppression	Lärmdämpfung f	amortissement m de bruit	шумоглушение
N 39	no-load, idling	Leerlauf m	marche f à vide	холостой ход, ход без нагрузки
N 40	no-load control	Leerlaufkontrolle f	contrôle m de marche à vide	контроль (управление) холостого хода
	nominal capacity	s. N 49		
N 41	nominal diameter	Nenndurchmesser m	diamètre m nominal	условный диаметр
N 42	nominal output	Solleistung f	production f théorique (exigée), rendement m prévu	проектная (заданная, необходимая) мощность
N 43	nominal pressure, desired pressure	Solldruck m	pression f exigée (de consigne)	заданное (необходимое, проектное) давление
N 44	nominal pressure	Nenndruck m	pression f nominale	номинальное (условное) давление
N 45	nominal pressure stage	Nenndruckstufe f	degré m de pression nominal	номинальная ступень давления
N 46	nominal size	Nennmaß n	cote f nominale	номинальный размер
N 47	nominal temperature	Nenntemperatur f	température f nominale	номинальная температура
N 48	nominal temperature, desired temperature	Solltemperatur f	température f de consigne	заданная температура
N 49	nominal value, required output, nominal capacity	Soll n, Nennwert m	grandeur f théorique (de consigne), valeur f nominale (préscrite)	номинальное значение, дебит, плановое задание, норма
N 50	nominal width, internal diameter	Nennweite f	portée f nominale, diamètre m nominal	условный (номинальный) размер
N 51	nomogram	Funktionsnetz n, Nomogramm n	nomogramme m, abaque m	функциональная сеть, номограмма
N 52	non-blocking	verstopfungsfrei	sans engorgement	свободно от блокировки (забивки)
N 53	non-catalytic gas-solid reaction	nichtkatalytische Gas-Feststoff-Reaktion f	réaction f gaz-solide non catalytique	некаталитическая реакция между газом и твердой фазой
N 54	non-condensable component	nichtkondensierende Komponente f	composante f non condensable	неконденсирующийся компонент
N 55	non-condensable gases	nichtkondensierende Gase f	gaz mpl non condensables	неконденсирующиеся газы
N 56	non-condensable vapour	nichtkondensierbarer Dampf m	vapeur f non condensable	неконденсирующийся пар
N 57	non-condensed air	nichtkondensierte Luft f	air m non condensé	неконденсированный воздух

N 58	non-corroding	korrosionsfrei	noncorrosif, résistant à la corrosion	свободно от коррозии
N 59	non-destructive	zerstörungsfrei	non-destructif, sans destruction	бездеструктивный
N 60	non-dimensional variable	dimensionslose Variable f	variable f sans dimension	безразмерная переменная
N 61	non-directed graph	ungerichteter Graph m	graphe m non dirigé	неориентованный граф
N 62	non-inflammability	Unbrennbarkeit f	incombustibilité f	невоспламеняемость
N 63	non-isothermal measuring data	nichtisotherme Meßdaten pl	données fpl de mesure non isothermes	неизотермические данные измерений
N 64	non-Newtonian fluid	nicht-Newtonsches fluides (flüssiges) Medium n	fluide m non-newtonien	неньютоновская жидкая среда
N 65	non-Newtonian liquid	nicht-Newtonsche Flüssigkeit f	liquide m non-newtonien	неньютоновская жидкость
N 66	non-Newtonian medium	nicht-Newtonsches Medium n	milieu m non-newtonien	неньютоновская среда
N 67	non-operating period	Leerlaufperiode f	période f de marche à vide	холостой период
N 68	non-polar	nichtpolar, unpolar, apolar	non-polaire, homéopolaire	неполярный
N 69	non-recirculating cooling	Durchlaufkühlung f	refroidissement m par passage simple	однократное прямоточное охлаждение
N 70	non-stationary flow	instationäre Strömung f	courant (écoulement) m non stationnaire	нестационарное течение
N 71	non-stationary process	instationärer Prozeß m	procédé m non-stationnaire	нестационарный процесс
N 72	non-steady running conditions	nichtstationärer Betriebszustand m	état m de marche non-stationnaire	неустановившийся рабочий режим
N 73	non-steady temperature	nichtstationäre (instationäre) Temperatur f	température f non-stationnaire	нестационарная температура
	non-viscous flow	s. I 330		
N 74	non-zero elements	Nicht-Null-Elemente npl	éléments mpl non nuls	ненулевые элементы
N 75	no-return valve	Rückschlagventil n	soupape f de retenue	обратный вентиль
N 76	normal ambient temperature	normale Umgebungstemperatur f	température f d'ambiance normale	нормальная температура окружающей среды
N 77	normal boiling point	normaler Siedepunkt m	point m d'ébullition normal	нормальная точка кипения
N 78	normal conditions	Normalbedingungen fpl	conditions fpl normales	нормальные условия
N 79	normal distribution, Gaussian distribution	Normalverteilung f, Gauß'sche Verteilung f	distribution f normale (de Gauss)	нормальное распределение, распределение Гаусса
	normal element	s. S 536		
N 80	normal force	Normalkraft f, Normalkomponente f der Kraft	force f normale	нормальная сила, нормальная составляющая силы, составляющая силы по нормали
N 81	normalization	Normierung f, Normung f	normalisation f, standardisation f	нормирование
N 82	normal pressure	Normaldruck m	pression f atmosphérique standardisée	нормальное давление
	normal solution	s. S 547		
N 83	normal state	Normalzustand m	état m normal	нормальный уровень, нормальное состояние
N 84	normal stress	Normalspannung f	tension f normale	нормальное напряжение
N 85	nozzle, orifice	Ausflußdüse f	buse f de décharge	сопло, отверстие истечения
	nozzle	s. a. 1. D 246; 2. J 9		
N 86	nuclear charge	Kernladung f	charge f nucléaire	заряд атомного ядра
N 87	nuclear energetics	Kernenergetik f	économie f de l'énergie nucléaire	ядерная энергетика
N 88	nuclear energy, atomic energy	Kernenergie f	énergie f nucléaire	ядерная (атомная) энергия
N 89	nuclear evaporation	Kernverdampfung f	évaporation f nucléaire	испарение ядерных частиц
N 90	nuclear fission	Atomspaltung f, Atomzertrümmerung f, Kernspaltung f	destruction f des atomes, fission f nucléaire	расщепление атомного ядра, деление ядра
N 91	nuclear fuel	Kernbrennstoff m	combustible m nucléaire	ядерное горючее (топливо)
	nuclear pile	s. A 593		
N 92	nuclear power plant	Kernkraftwerk n, Kernenergieanlage f	installation f énergétique nucléaire, centrale f nucléaire (atomique)	ядерная энергетическая установка, атомная [электро]станция
N 93	nuclear reaction	Kernprozeß m	réaction f nucléaire	ядерный процесс
	nuclear reactor	s. A 593		
N 94	nucleate boiling, bubble boiling	Blasenverdampfung f, Blasensieden n	ébullition f à bulles, vaporisation f par bulles	пузырьковое кипение
N 95	nucleation	Keimbildung f	germination f	образование зародышей, зародышеобразование
N 96	nucleation site	Dampfbildungszentrum n	site m de formation de bulles	центр парообразования
N 97	number of bubble trays	Glockenbodenzahl f	nombre m des plateaux à calotte	число колпачковых тарелок
N 98/9	number of dimensions	Dimensionszahl f	nombre m de dimensions	число размерности, размерность
N 100	number of moles	Molzahl f	nombre m de moles	число молей
N 101	number of reactors	Reaktoranzahl f	nombre m des réacteurs	количество (число) реакторов
	number of revolutions	s. S 475		
N 102	number of similarity	Ähnlichkeitskennzahl f	nombre m de similitude	критерий подобия

	English	German	French	Russian
N 103	number of stage	Stufenzahl f	nombre m d'étages	число ступеней (секции)
N 104	number of tear variable	Schnittzahl f	nombre m de coupes	разрываемое множество
N 105	number of theoretical plates	Zahl f der theoretischen Böden	nombre m de plateaux théoriques	число теоретических тарелок
N 106	number of transfer units	Zahl f der Austauscheinheiten	nombre m d'unités de transfert	число обменных единиц, число единиц обмена
N 107	numerical control	Datensteuerung f, numerische Steuerung f	contrôle m numérique	численное (цифровое) управление
N 108	numerical value	Zahlenwert m	valeur f numérique	численное значение
N 109	Nusselt number	Nusselt-Zahl f	nombre m de Nusselt	число (критерий) Нуссельта

O

	English	German	French	Russian
	object function	s. O 2		
O 1	objective, criterion	Zielgröße f	grandeur f visée	критерий (параметр) оптимизации, целевая функция
O 2	objective function, object function	Zielfunktion f	fonction f économique, fonction-objectif m	целевая функция, функция цели
O 3	objective function characteristic	Zielfunktionscharakteristik f	caractéristique f de la fonction économique	характеристика целевой функции
O 4	objective function curve	Zielfunktionskurve f	courbe f de la fonction économique	кривая (ход) целевой функции
O 5	observability	Beobachtbarkeit f	observabilité f	наблюдаемость
O 6	obsolescence	Veralten n, Alterung f	vieillissement m	устарелость, старение
O 7	obstruction, blockage, plugging, stoppage	Verstopfung f, Verstopfen n, Verlegen n, Zusetzen n	engorgement m, colmatage m, bouchage m	затыкание, закупорка, засорение, закупоривание
O 8	occur/to, to appear, to happen	vorkommen, auftreten, erscheinen	apparaître	происходить, бывать
	occurrence	s. 1. I 188; 2. P 387		
O 9	off-cycle	Aussetzperiode f	période f d'arrêt	нерабочая часть цикла
O 10/1	off-gas treater	Abgaswaschkolonne f	tour f de lavage des gaz résiduels	промывная колонна для отходящего газа
O 12	off-gas treating	Abgasbehandlung f, Restgasbehandlung f	traitement m des gaz résiduaires	обработка отходящих газов
O 13	off-peak power	Normalleistung f, Leistung f außerhalb der Spitzenzeit	puissance f en déhors de pointe	энергия, вырабатываемая во внепиковые периоды
O 14	off-sites	Nebenanlagen fpl	installations fpl annexes	вспомогательные установки
	off-take tank	s. W 177		
O 15	off-take tower	Wasserentnahmeturm m	tour f de soutirage d'eau	водозаборная башня
O 16	oil bath	Ölbad n	bain m d'huile	масляная баня (ванна)
O 17	oil cooler	Ölkühler m	réfrigérant m à huile	масляный холодильник
O 18	oil extraction	Ölextraktion f	extraction f d'huile	масляная экстракция
O 19	oil feed pump	Öldruckpumpe f	pompe f de pression à huile	масляный нагнетательный насос
O 20	oil film	Ölfilm m	film m d'huile	масляный фильм, масляная пленка
O 21	oil filter	Ölfilter n	filtre m d'huile	масляный фильтр, фильтр для масла
O 22	oil firing, oil furnace	Ölfeuerung f	four (foyer) m à l'huile	нефтяное отопление, отопление жидким топливом
	oil-flow indicator	s. I 246		
	oil furnace	s. O 22		
O 23	oil immersed capacitor	Ölkondensator m	condensateur m dans l'huile	масляный конденсатор
	oilness	s. L 302		
O 24	oil pressure	Öldruck m	pression f d'huile	масляное давление, давление масла
O 25	oil pressure meter	Öldruckmesser m	manomètre m de pression d'huile	масляный манометр
O 26	oil pressure pipe	Öldruckleitung f	conduite f d'huile sous pression	маслонапорный трубопровод
O 27	oil product	Erdölprodukt n	produit m petrolier	нефтепродукт
O 28	oil pump	Ölpumpe f	pompe f à huile	масляный насос
O 29	oil refinery	Erdölraffinerie f	raffinerie f de pétrole	нефтеперегонная установка
O 30	oil refining	Ölraffination f	raffinage m d'huile	рафинация (очистка) масла, очистка нефти
O 31	oil resistant	ölfest, ölbeständig	résistant à l'huile	маслостойкий
O 32	oil separation	Ölabtrennung f	séparation d'huile	отделение (сепарация) масла
O 33	oil separator, trap	Ölabscheider m, Entöler m	séparateur m d'huile, déshuileur m	маслоотделитель, сепаратор [для] масла
O 34	oil smoke	Öldampf m	gaz m d'huile	масляный пар
O 35	oil sump	Ölsumpf m	carter m à l'huile, auget m à l'huile	масляный зумпф, картер масляной ванны
O 36	oil transport	Ölförderung f, Öltransport m	extraction f (transport m) de pétrole	подача (транспорт) масла
O 37	oil-wetted air cleaner	Naßluftfilter n	filtre m à air humide	контактный (мокрый) воздушный фильтр
O 38	Oldroyd operator	Oldroydscher Operator m	opérateur m d'Oldroyd	оператор Ольдройда
	omission	s. N 10		

	omit/to	s. N 9		
O 39	one-component system	Einstoffsystem n, Einkomponentensystem n	système m à un composant	однофазная система
O 40	one-dimensional evaluation	eindimensionale Bewertung f	évaluation f unidimensionnelle	одномерная оценка
O 41	one-dimensional evaluation system	eindimensionales Bewertungssystem n	système m d'évaluation à une dimension	одномерная система оценки (качества)
O 42	one-dimensional flow	eindimensionale Strömung f	écoulement m unidimensionnel	одномерное течение
O 43	one-layer filter	Einfachfilter n	filtre m simple	простой (одностадийный) фильтр
O 44	one-stage compression cycle	einstufiger Verdichtungsprozeß m	cycle m de compression à un étage	одноступенчатый процесс сжатия (уплотнения)
O 45	on-process temperature	Arbeitstemperatur f, Prozeßtemperatur f	température f de régime (travail)	рабочая температура, температура процесса
O 46	opacity, opaqueness turbidity, cloudiness	Trübung f, Undurchsichtigkeit f	opacité f, louche m, trouble m	муть, мутнение, непрозрачность
O 47	open-air construction	Freibauweise f	construction f à ciel ouvert	конструкция открытого типа, открытая конструкция
O 48	open-air piping	Außenberohrung f	tuyauterie f extérieure	наружная трассировка трубопроводов
O 49	open-air storage	Freilager n	stockage m à ciel ouvert	открытый склад
O 50	open-circuit crushing	Durchlaufzerkleinerung f	concassage m direct, broyage m à passage direct	измельчение в открытом цикле
O 51	open-hearth furnace	Siemens-Martin-Ofen m	four m Siemens-Martin	мартеновская печь
O 52	open pressure, outlet pressure	Ausflußdruck m, Abflußdruck m	pression f à la sortie du fluide (liquide)	давление на выходе, напор
O 53	open system	offenes System m	système m ouvert	открытая система
O 54	operability	Bedienbarkeit f	maniabilité f	управляемость, маневренность
O 55	operable, workable	betriebsfähig	prêt pour le service, en ordre de fonctionnement	работоспособный, готовый к эксплуатации
O 56	operating characteristic	Betriebskennlinie f	caractéristique f de fonctionnement	рабочая характеристика
O 57/8	operating characteristics	Betriebsverhalten n	caractéristiques mpl de fonctionnement	рабочая характеристика, эксплуатационная (оперативная) характеристика
O 59	operating conditions	Betriebsregime n	mode m de fabrication (fonctionnement)	производственный режим
	operating conditions	s. a. O 68		
O 60	operating costs	Betriebskosten pl	frais mpl d'exploitation	производственные расходы, издержки производства, стоимость эксплуатации, производственная стоимость
O 61	operating engineer	Betriebsingenieur m	ingénieur m de fabrication	инженер по технической эксплуатации, производственный инженер
O 62	operating instruction	Betriebsanleitung f, Betriebsvorschriften fpl	instructions fpl de service, instructions	инструкция по эксплуатации (обслуживанию), регламент
O 63	operating limit	Betriebsgrenze f	limite f de fonctionnement	производственный лимит, лимит производства, предел эксплуатации, рабочий предел
	operating line	s. W 188		
	operating mode	s. M 288		
O 64	operating optimization	Betriebsoptimierung f	optimisation f de fonctionnement	оптимизация производства (режима) (процесса)
O 65	operating parameter	Betriebskennwert m	paramètre m d'exploitation	производственный (рабочий) параметр
O 66	operating power	Betriebsleistung f	puissance f d'opération	рабочая (эксплуатационная, полезная) мощность
O 67	operating pressure	Betriebsdruck m	pression f normale (de service)	рабочее давление
O 68	operating regime, operating conditions	Betriebsbedingungen fpl, Arbeitsbedingungen fpl	régime m d'opération, conditions fpl de fonctionnement (travail)	рабочие условия, режим эксплуатации (работы), условия работы (эксплуатации)
O 69	operating result	Betriebsergebnis n	résultat m d'exploitation	результат предприятия (работы завода)
O 70	operating state (status)	Betriebszustand m	état m en fonctionnement	эксплуатационное состояние, режим работы
O 71	operating temperature, working temperature	Betriebstemperatur f	température f de service (fonctionnement)	рабочая температура
O 72	operating temperature range	Arbeitstemperaturbereich m, Betriebstemperaturbereich m	plage f de température de fonctionnement, domaine m de température de régime	рабочий диапазон температур

O 73	operating time	Herstellungsdauer f	durée f de fabrication	время производства, длительность изготовления
O 74	operating time	Betriebszeit f	temps m de fonctionnement	производственное время, время производства
O 75	operating time, test time	Laufzeit f	durée f de fonctionnement	время пробега, продолжительность хода
	operating trouble	s. B 254		
O 76	operating voltage	Betriebsspannung f	tension f d'emploi	рабочее напряжение
O 77	operation, attendance, control	Bedienung f	maniement m, conduite f, commande f, manipulation f	обслуживание, эксплуатация, управление
O 78	operation, procedure	Arbeitsgang m, Arbeitsprozeß m	opération f, suite f des opérations, phase f de travail	операция, работа, рабочий ход, процесс работы, технологический процесс
O 79	operation, severity level	Fahrweise f	régime f de fonctionnement, mode m opératoire	режим эксплуатации (работы)
O 80	operational analysis	Verfahrensanalyse f	analyse f de procédé	анализ технологии (работы установки, процесса)
O 81	operational procedure	Arbeitsablauf m	succession f des opérations, déroulement m du travail, phases fpl de travail	рабочий (технологический) процесс, рабочий режим
O 82	operational procedure	Betriebsablauf m	fonctionnement m	функционирование, эксплуатация, режим, порядок эксплуатации
O 83	operation control	Betriebskontrolle f	contrôle m de fabrication, service m de contrôle	контроль производства
O 84	operation data	Betriebsdaten fpl	caractéristiques fpl de régime	эксплуатационные характеристики
O 85	operation desk, control desk	Bedienungsfeld n	tableau m de commande	панель управления, контрольная панель
O 86	operation reliability, safety of operation	Betriebssicherheit f	sécurité f de fonctionnement (service)	безопасность эксплуатации (в работе)
O 87	optical-stress law	optisches Spannungsgesetz n	loi f photoélastique	оптический закон напряжения
O 88	optimality criterion	Optimalitätskriterium n	critérium m d'optimalité, critère m d'optimalité	критерий оптимальности (оптимизации)
O 89	optimality demand	Optimalitätsforderung f	demande f d'optimalité	критерий оптимальности, требование к оптимальности
O 90	optimization method	Optimierungsmethode f	méthode f d'optimisation	метод оптимизации
O 91	optimization problem	Optimierungsproblem n	problème m d'optimisation	проблема оптимизации
O 92	optimization strategy	Optimierungsstrategie f	politique f d'optimisation	стратегия оптимизации
O 93	optimization variable	Optimierungsvariable f	variable f d'optimisation	переменная оптимизации
O 94	optimum point	Optimalpunkt m	point m optimum	оптимальная точка, оптимум
O 95	optimum structure	optimale Struktur f	structure f optimum	оптимальная структура
O 96	order of magnitude	Größenordnung f	ordre m de grandeur	порядок величины
O 97	order of reaction	Reaktionsordnung f	ordre m de réaction	порядок реакции
O 98	order parameter	Ordnungsparameter m	paramètre m d'ordre	порядковый параметр, параметр упорядочивания
	ore sorter	s. C 277		
O 99	orientation system	Orientierungssystem n	système m d'orientation	система ориентации
	orifice	s. 1. N 85; 2. O 111		
O 100	orifice equation	Durchflußgleichung f	équation f de débit	уравнение расхода
O 101	original solution	Urlösung f	solution f primitive	эталонный (первичный) раствор
	original state	s. I 167		
O 102	oscillatory motion	Schwing[ungs]bewegung f, schwingende Bewegung f	mouvement m oscillatoire	колебательное (осцилирующее) движение
O 103	osmotic pressure	osmotischer Druck m	pression f osmotique	осмотическое давление
O 104	Ostwald-de Waele model	Ostwald-de Waelesches Modell n	modèle m de Ostwald-de Waele	модель Оствальда-де Веля
O 105	outdoor intake air flow	Außenluftsaugstrom m	courant m d'air extérieur	поток всасываемого наружного воздуха
O 106	outdoor substation	Freiluftschaltanlage f	poste m extérieur	распределительная подстанция открытого типа
O 107	outer casing	Außenmantel m, Außenhülle f	enveloppe f extérieure	наружная рубашка
O 108	outfall ditch	Vorfluter m	canal m de dérivation	водосборная емкость для сточных вод
	outflow	s. O 111		
O 109	outflow curve	Abflußmengenlinie f	courbe f de débit	кривая расходов
O 110	outflow velocity	Ausströmgeschwindigkeit f	vitesse f d'évacuation, vitesse f d'émission	выпускная скорость, скорость выпуска
O 111	outlet, discharge, orifice, drain, outflow	Abfluß m, Ausfluß m, Abzug m, Ausflußöffnung f, Ablauf m	débouché m, déchargoir m, voie f d'écoulement, orifice m d'écoulement, décharge f	утечка, истечение, вытяжка, сливное отверстие, сток, слив
O 112	outlet, conduit	Abzug m, Dunsthaube f	extracteur m, évacuateur m, hotte f	отводный канал, вытяжка
	outlet	s. a. L 58		

O 113	outlet air temperature	Luftaustrittstemperatur f	température f de l'air sortant	температура воздуха на выходе
O 114	outlet chamber	Austrittskammer f	chambre f de sortie	выходная камера
O 115	outlet channel	Ausflußkanal m	canal m de déversement (décharge)	канал истечения, спускной канал
O 116	outlet control	Abflußregelung f	régulation f de débit	регулирование стока
O 117	outlet controller	Abflußregler m	régulateur m de débit	регулятор расхода
O 118	outlet cross section	Ausflußquerschnitt m	section f de la voie d'écoulement	поперечное сечение истечения
	outlet of affinage	s. A 246		
O 119	outlet opening tube	Abflußöffnung f	déversoir m, décharge f	отверстие истечения
	outlet pipe	s. E 348		
	outlet pressure	s. O 52		
O 120	outlet ratio	Abflußverhältnis n	taux m de décharge	коэффициент стока
O 121	outlet slide	Abflußschieber m	soupape f de décharge	спускной клинкет, спускная задвижка
O 122	outlet state	Austrittszustand m	état m à la sortie, état à l'issue	состояние на выходе
	outlet stream	s. O 128		
O 123	outlet tank	Ausflußbehälter m	réservoir m de déversement	резервуар (емкость) истечения
O 124	outlet temperature	Austrittstemperatur f	température f de sortie	выходная температура, температура на выходе
O 125	outlet temperature difference	Austrittstemperaturdifferenz f	différence f de température à la sortie	разность выходных температур
O 126	outlet tube	Austrittsrohr n	tuyau m de sortie, conduite f sortante	выпускная (спускная) труба
O 127	outlet valve, exhaust valve	Auslaßklappe f	clapet m d'échappement	выпускной клапан
O 128	output, output current, outlet stream	Ausstoß m, Ausgangsstrom m	débit m, courant m sortant, courant de sortie, débit sortant	выпуск, выход, выходной ток
	output	s. a. 1. C 32; 2. G 1		
	output capacity	s. P 550		
	output current	s. O 128		
O 129	output element	Ausgangselement n	élément m de sortie	исходный (выходной, краевой) элемент
O 130/1	output filter	Ausgangsfilter n	filtre de sortie	выходной фильтр
	output pressure	s. F 136		
O 132	output stage	Endstufe f	stade (procédé) m final, opération f finale	последняя ступень, конечная ступень
O 133	outside air load	Außenluftlast f	charge f d'air extérieur	нагрузка внешнего (наружного) воздуха
O 134	outside air total heat	Außenluftenthalpie f	enthalpie f de l'air extérieur	теплосодержание наружного воздуха
O 135	outside heating	Außenbeheizung f	chauffage m extérieur	внешний обогрев
O 136	outside radius	äußerer Radius m	rayon m extérieur	внешний радиус
O 137	outside temperature	Außentemperatur f	température f extérieure	наружная (внешняя) температура
O 138	outstream efficiency	Auslastungsgrad m	pourcentage m d'utilisation	степень загрузки установки
O 139	oven, furnace, kiln	Ofen m	four m, fourneau m	печь
	overage	s. V 183		
O 140	overall consumption	Gesamtverbrauch m	consommation f totale	общее потребление, общий расход
O 141	overall efficiency	Gesamtwirkungsgrad m, Gesamtgüte f	rendement m total, qualité f totale	общий коэффициент полезного действия
O 142	overall heat transfer coefficient	Wärmedurchgangskoeffizient m, Gesamtwärmeübertragungskoeffizient m	coefficient m de transmission de chaleur, taux m de conductibilité thermique	общий коэффициент теплопередачи
O 143	overall plant, complete plant	Gesamtanlage f	installation f totale	полная установка, установка в целом, вся установка
O 144	overall process theory	allgemeine Prozeßtheorie f	théorie f générale des processus	общая теория процессов
O 145	overall reaction rate	Bruttoreaktionsgeschwindigkeit f	vitesse f globale de réaction	общая скорость реакции, брутто-скорость реакции
	overall system	s. C 476		
O 146	overcharge, overload	Über[be]lastung f, Überlast f	effort m excessif, surcharge f	перегрузка
O 147	overdesign	Überdimensionierung f	surdimensionnement m	проектирование с запасом
O 148	overdesign factor	Überdimensionsierungsfaktor m	facteur m de surdimensionnement	фактор (коэффициент) запаса
O 149	overdesign range	Überdimensionierungsbereich m	domaine m de surdimensionnement	интервал избыточных размеров, интервал запаса
O 150	overflow	Überlauf m, Überfluß m	trop-plein m	перепуск, перелив, избыток
	overflowing	s. F 226		
O 151	overflow pipe, waste tube	Überlaufrohr n	trop-plein m, tuyau m de trop-plein	переливная (переточная) трубка
O 152	overflow tank	Überlaufbecken n	bassin m à débordement	водосливный бассейн
O 153	overflow tank	Überlaufkammer f	chambre f de trop-plein	камера перелива (с водосливом)
O 154	overhead battery	Deckenwärmeübertrager m, Deckenkühler m	batterie f plafonnière	потолочный теплообменник

O 155	overhead bunker	Hochbunker *m*	trémie *f* surélevée	надземный бункер
	overhead coil	s. C 115		
O 156	overhead evaporator	Deckenverdampfer *m*	évaporateur *m* plafonnier	потолочный испаритель
O 157	overhead irrigation	Beregnung *f*	arrosage *m*	орошение
O 158	overhead plate evaporator	Deckenplattenverdampfer *m*	évaporateur *m* aux plaques de plafond	потолочный плиточный испаритель
O 159	overhead product	Kopfdestillat *n*, Kopfprodukt *m*	produit *m* de tête, distillat *m*	верхний дистиллят, отогнанный (отделенный) продукт
O 160	overhead temperature	Kopftemperatur *f*	température *f* en tête	температура в головной (верхней) части колонны
O 161	overheating, superheating	Überhitzung *f*, Überhitzen *n*, Heißlaufen *n*	surchauffe *f*, échauffement *m*	перегревание, перегрев, горячий ход
	overload	s. O 146		
O 162	overload cut-out	Überlastschalter *m*	déclencheur *m* de surcharge	выключатель при перегрузке, прибор защиты от перегрузки
O 163	overload protection	Überlastungsschutz *m*	protection *f* contre les surcharges	защита от перегрузки
O 164	over point	Destillationsbeginn *m*	point *m* initial de distillation	начало дистилляции (кипения)
O 165	overpressure, additional pressure, pressure burden	Überdruck *m*	excès *m* de pression, surpression *f*	избыточное давление
O 166	overpressure distillation	Überdruckdestillation *f*	distillation *f* sous pression	перегонка при избыточном (сверхвысоком) давлении
	oversaturation	s. S 834		
O 167	oversize, sieve residue	Überkorn *n*, Siebrückstand *m*	gros *m* du crible	верхний продукт (класс), надрешетный продукт
O 168	oversizing	Überdimensionierung *f*	surdimensionnement *m*	применение избыточных размеров, расчет с учетом запаса
O 169	overvoltage	Überspannung *f*	surtension *f*, survoltage *m*	перенапряжение
	oxygen breathing apparatus	s. O 173		
O 170	oxygen content	Sauerstoffgehalt *m*	teneur *m* en oxygène	содержание кислорода
O 171	oxygen cylinder	Sauerstoffflasche *f*	bouteille *f* à oxygène	кислородный баллон
O 172	oxygen demand	Sauerstoffbedarf *m*	demande *f* d'oxygène	потребность в кислороде
O 173	oxygen inhaling apparatus, oxygen breathing apparatus	Sauerstoffatemschutzgerät *n*	appareil *m* de respiration à oxygène, masque *m* respiratoire protecteur à oxygène	кислородный респиратор (горноспасательный аппарат)
O 174	oxyhydrogen gas	Knallgas *n*	gaz *m* explosif, gaz oxyhydrique	гремучий газ

P

P 1	pace length	Schrittweite *f*	longueur *f* de pas	длина шага
P 2	package	Packung *f*, Schüttung *f*, Ballung *f*, Verdichtung *f*	tassement *m*, bourrage *m*, amas *m*, accumulation *f*	упаковка, набивка, уплотнение
P 3/4	packaging machine	Verpackungsmaschine *f*	machine *f* à emballer	упаковочная машина
	packed-bed reactor	s. F 188		
P 5	packed [distillation] column	Füllkörperkolonne *f*, Füllkörpersäule *f*	colonne *f* à garnissage	насадочная колонна, колонна с насадкой
P 6	packed-tower absorber	Füllkörperabsorber *m*	absorbeur *m* à garnissage	насадочный абсорбер, башня с насадкой
P 7	packed trickle column	Füllkörperrieselkolonne *f*	colonne *f* garnie à ruissellement	насадочная орошаемая колонна
P 8	packing, filling, stuffing	Füllung *f*, Verpackung *f*, Packung *f*	remplissage *m*, emballage *m*, garnissage *m*	наполнение, упаковка, паковка
	packing	s. a. S 759		
P 9	packing density	Packungsdichte *f*	degré *m* de tassement	плотность упаковки, толщина набивки
P 10	packing effect	Packungseffekt *m*	effet *m* de tassement	эффект упаковки
P 11	packing fluid, sealing (confining) liquid	Sperrflüssigkeit *f*	liquide *m* obturant	жидкостный затвор
P 12	packing for fractionating columns	Füllkörperfüllung *f* der Fraktionierkolonne	garnissage *m* de la colonne	башенная насадка
P 13	packing material	Packungsmaterial *n*	matériau *m* de garniture	закладочный (набивочный) материал, насадочный материал
	packing material	s. a. C 615		
	packing ring	s. G 55		
P 14	packings	Füllkörper *mpl*	garnissage *m*	башенная насадка, элементы башенной насадки
P 15	paddel mixer	Schaufelrührer *m*	agitateur *m* turbinaire	лопастная мешалка
P 16	panel, switch board	Schalttafel *f*	panneau *m* de distribution, tableau *m* des instruments	распределительный коммутационный щит, доска
P 17	paper pulp	Papierzellstoff *m*, Zellstoff *m* für die Papierindustrie, Papiermasse *f*	pâte *f* à papier	целлюлозная масса, бумажная масса
P 18	parallel-connected	nebeneinandergeschaltet, parallelgeschaltet	couplé en parallèle	параллельно включенный

P 19	parallel connection, connection in parallel	Parallelschaltung f	montage m en parallèle	параллельное соединение (включение)
P 20	paralleled compressors	parallelgeschaltete Kompressoren mpl	compresseurs mpl à montage parallèle	параллельно соединенные компрессоры
P 21	parallel flow, uniflow current	Parallelstrom f, Gleichstrom m	flux mpl parallèles, courant m parallèle	прямоточное течение, параллельный ток, прямоток
	parallel flow	s. a. C 525		
P 22	parallel-flow air cooler	Parallelstromluftkühler m	refroidisseur d'air à flux parallèles	воздухоохладитель с параллельным движением воздуха и холодоносителя
P 23	parallel-flow condensation	Gleichstromkondensation f	condensation f à courant parallèle	конденсация с параллельным током
P 24	parallel-flow cooler	Gleichstromkühler m, Parallelstromkühler m	refroidisseur m à flux parallèle	охладитель с параллельным потоком, прямоточный охладитель
P 25	parallel-flow heat exchanger	Gleichstromwärmeübertrager m, Parallelstromwärmeübertrager m	échangeur thermique à courants parallèles	прямоточный теплообменник
P 26	parallel-flow heat transfer	Gleichstromwärmeübertragung f, Parallelstromwärmeübertragung f	transfert m de chaleur à courants parallèles	теплопередача при параллельном движении сред, прямоточная теплопередача
P 27	parallel reaction	Parallelreaktion f	réaction f parallèle	параллельная реакция
P 28	parallel system	Parallelsystem n	connexion f en parallèle	раздельная система
P 29	parallel working	Parallelbetrieb m	marche f en parallèle	параллельная работа
	parameter	s. C 177		
P 30	parameter determination	Parameterbestimmung f	détermination f de paramètres	определение параметров
P 31	parameter hierarchy	Parameterrangfolge f	suite f hiérarchique des paramètres	последовательность параметров по их значимости
P 32	parametric sensitivity	parametrische Empfindlichkeit f	sensibilité f paramétrique	параметрическая чувствительность
P 33	part	Teil n, Element n	partie f, élément m	часть, доля, элемент, деталь
P 34	partial-capacity operation	Teillastbetrieb m	opération f à charge partielle, fonctionnement m sous charge réduite	работа с частичной производительностью
P 35	partial combustion	Teilverbrennung f	combustion f partielle	частичное сгорание
P 36	partial condensation	Teilkondensation f, partielle Kondensation f	condensation f partielle	частичная (неполная) конденсация, дефлегмация
P 37	partial cooling	partielle Kühlung f	refroidissement m partiel	частичное охлаждение
P 38	partial dehydration	partieller Feuchtigkeitsentzug m	déshydratation f partielle	частичное обезвоживание
P 39	partial drop of pressure	Teildruckgefälle n	perte f de charge partielle	частичный напор, частичный перепад давления
P 40	partial drying	partielle (teilweise) Trocknung f	séchage m partiel	частичная сушка
P 41	partial feeding, partial load, branch current	Teilförderung f, Teilbeschickung f	transport m partiel, alimentation f partielle	частичная подача, неполная нагрузка
P 42	partial flow	Teilströmung f	écoulement m partiel	элементарная струя, частичное течение
P 43	partial liquefaction	partielle Verflüssigung f	liquéfaction f partielle	частичное ожижение
P 44	partial load	Teillast f	charge f partielle	частичная (неполная) нагрузка
	partial load	s. a. P 41		
P 45	partial load behaviour	Teillastverhalten n	comportement m sous charge partielle	поведение при частичной нагрузке
P 46	partial losses	Teilverluste mpl	pertes fpl partielles	частичные потери
P 47	partial miscibility	partielle Mischbarkeit f	miscibilité f partielle	парциальная (частичная) смешиваемость
P 48	partial open-air structure	Teilfreibau m	installation f partielle à l'air libre	конструкция частично открытого типа
P 49	partial pressure	Teildruck m, Partialdruck m	pression f partielle	парциальное давление
P 50	partial purification	Teilreinigung f	épuration f partielle	частичная очистка
P 51	partial reaction	Teilreaktion f	réaction f partielle	частичная реакция
P 52	partial repair	Teilinstandsetzung f	réparation f partielle	частичный (неполный) ремонт
P 53	partial treatment	Teilbehandlung f	traitement m partiel	частичная обработка
P 54	partial vacuum	Luftverdünnung f	raréfaction (dépression) f de l'air	разрежение воздуха
P 55	partial vapour pressure	Partialdampfdruck m	pression f partielle de vapeur	парциальное давление пара
P 56	partial volume	Partialvolumen n, Teilvolumen n	volume m partiel	парциальный объем
P 57	partial water flow	Teilwasserstrom m	courant m d'eau partiel	частичный расход воды
P 58	particle	Teilchen n, Bauteil n	particule f, élément m de construction	частица, часть, элемент
P 59	particle diameter	Teilchendurchmesser m	diamètre m de particule	диаметр частицы
P 60	particle energy	Teilchenenergie f	énergie f de particules	энергия частицы
P 61	particle size, granular size	Teilchengröße f, Korngröße f	grandeur f de particule, grosseur f de grain, calibre m	размер частицы, крупность (величина) зерна

P 62	particle size analysis	Korngrößenanalyse f	granulométrie f	анализ размеров частиц, анализ крупности зерен, гранулометрический анализ
P 63	particle structure	Teilchenstruktur f	structure f de particule	структура частицы
P 64	particulate solid	Feststoffpartikel f, Feststoff-teilchen n	particule f solide	частица твердых веществ, твердая частица
P 65	partition coefficient, distribution coefficient	Verteilungskoeffizient m, Aufteilungskoeffizient m, Teilungsverhältnis n	coefficient m de distribution (partage)	коэффициент распределения
	partition law	s. D 438		
P 66	part-load performance	Teillastleistung f	performance f de charge partielle	частичная производительность
P 67	parts list, specification	Stückliste f	liste f de pièces, nomenclature f	список мест, спецификация
P 68	passage, transmission, traversal	Durchlaß m, Durchgang m, Durchlauf m	passage m, passant m	прохождение, проход, пассаж, ход, канал пропускаемости, истечение, поток, пробег
P 69	passage of gases	Durchblasen n von Gasen	purge f, soufflement m	продувка
P 70	pass-out pipe	Entnahmeleitung f, Abfüll-leitung f	conduite f de soutirage	трубопровод регулируемого отбора пара, водозаборный водовод, трубопровод для отбора проб
P 71	pass-out pressure	Entnahmedruck m	pression f de soutirage	давление отбора
P 72	pass-out steam	Entnahmedampf m	vapeur f soutirée (de soutirage)	отборочный пар, пар из отбора
P 73	pass-out turbine	Entnahmeturbine f	turbine f à prise de vapeur	турбина с отбором пара
P 74	paste	Paste f, Brei m, Teig m, Kleister m, Klebstoff m	pâte f	паста, каша, тесто, клей
P 75	pasty material	pastöses Material n	matière f pâteuse	пастообразный материал
P 76	path of the rays	Strahlengang m	marche f des rayons	ход лучей
P 77	peak effluent, maximum inflow, peak influent	Höchstzuflußmenge f	alimentation f maximum	максимальный (пиковый) расход
P 78	peak demand	Spitzenbedarf m	demande f de pointe	пиковая нагрузка (потребность)
	peak influent	s. P 77		
P 79	peak load	Spitzenbelastung f, Spitzenlast f	charge f maximum (de pointe)	пиковая (максимальная) нагрузка
P 80	pebble bed reactor	Kieselbettreaktor m	réacteur m à zone active en forme de galets	реактор с кремнеземистой насадкой
	pebble filter	s. G 143		
P 81	pebble heater	Wärmeüberträger m mit Kieselfüllung f	réchauffeur m à galets	нагреватель (теплообменник) с галечным теплоносителем
	pebble mill	s. B 30		
P 82	pebbles	körnige Trägersubstanz f, Körner npl, feines Geröll n	grains mpl	зерна
P 83	Péclet's number	Péclet-Zahl f	nombre m de Péclet	число (критерий) Пекле
P 84	peep hole	Schauloch n	trou m de regard, fenêtre f	смотровое отверстие (окно)
P 85	pelleting, tabletting, pilling	Tablettierung f, Tablettieren n, Pelletierung f	transformation f en pastilles, pelletage m	таблетирование, изготовление таблеток
P 86	pelleting press, pelletizing extruder	Tablettiermaschine f, Pelletizer m	machine f à comprimer les tablettes, pastilleuse f	таблетировочная машина, машина для таблетирования, таблеточный пресс
P 87	pellets	Granulat n	granulé m	гранулированный продукт, гранулят, гранула
P 88	Peltier coefficient	Peltier-Koeffizient m	coefficient m Peltier	коэффициент Пельтье
P 89	Peltier cooling device	Peltier-Kühler m	installation f frigorifique Peltier	термоэлектрический прибор охлаждения
P 90	Peltier effect	Peltier-Effekt m	effet m Peltier	эффект Пельтье
	Peltier heat exchanger	s. T 152		
	pendulum-type hydroextractor	s. U 6		
P 91	penetrating power	Durchdringungsvermögen n	pouvoir m de pénétration	проницаемость, способность проникновения
P 92	penetration	Penetration f, Durchdringung f	pénétration f	проницание, пенетрация, проникновение
P 93	penetration theory	Penetrationstheorie f	théorie f de pénétration	теория пенетрации
P 94/5	percentage by volume	Volum[en]prozent n	pour-cent m volumique	объемный процент
	percentage purity	s. D 108		
P 96	percent [by] weight	Gewichtsprozent n	pourcentage m en poids, pour-cent m pondéral	процент по весу, весовой процент
P 97	percolating filter	Tropfkörper m	percolateur m	орошаемый биофильтр
	percolation	s. I 132		
P 98	perfectly mixed reactor	ideal durchmischter Reaktor m, Reaktor mit idealer Vermischung	réacteur m parfaitement mélangé	реактор идеального перемешивания, идеально перемешенный реактор
P 99	perforated area	perforierte Fläche f	surface f perforée	перфорированная поверхность

	perforated bottom	s. S 307		
P 100	perforated tube	perforiertes Rohr n	tube m perforé	перфорированная труба
P 101	perforation	Lochung f, Perforierung f, Durchlöcherung f	perforation f	перфорация
P 102	performance	Kenndaten pl, Betriebsverhalten n, Durchführung f, Güte f, Funktion f, Leistungsfähigkeit f	puissance f, rendement m, capacité f productive	характеристика, эксплуатационные качества, производственность, коэффициент полезного действия, интенсивность труда
	performance	s. a. 1. B 76; 2. P 493		
P 103	performance coefficient	Leistungsfaktor m	facteur m de puissance	коэффициент мощности
P 104	performance factor	Gütefaktor m	coefficient m de qualité	качественный коэффициент, показатель качества
P 105	performance of work	Arbeitsleistung f	rendement m effectif	производительность [труда], мощность, рабочая мощность
P 106	periodic adsorption	satzweise (diskontinuierliche) Adsorption f	adsorption f discontinue (cyclique)	периодическая адсорбция
P 107	period of aging	Alterungsdauer f, Altern n	période f de vieillissement, vieillissement m	продолжительность (длительность) старения
	period of decay	s. H 3		
P 108	period of time balances	Zeitbereichsbilanzen fpl	bilans mpl des intervalles temporels	балансы за определенный промежуток времени
P 109	period of use	Benutzungsdauer f, Nutzungsdauer f	durée f d'utilisation	продолжительность использования
P 110	peripheral speed	Umfangsgeschwindigkeit f	vitesse f périphérique	окружная скорость
	permanent	s. L 28		
P 111	permanent costs	[feste] Kosten pl	frais mpl fixes	капитальные затраты, стоимость амортизации
P 112	permeability	Durchlässigkeit f, Permeabilität f, Wasserdurchlässigkeit f	perméabilité f	проницаемость, пропускаемость
P 113	permeability to gas	Gasdurchlässigkeit f	perméabilité f aux gaz	газопроницаемость
P 114	permeance	Leitvermögen n	conductivité f	проводоспособность, проводимость
P 115	permutation	Permutation f	permutation f	перестановка, перемещение
P 116	perpendicular	senkrecht, normal, lotrecht	vertical, perpendiculaire, à plomb	нормальный, перпендикулярный
P 117	pervious	durchlässig	perméable	проницаемый
P 118	petrol chemistry	Petrolchemie f	pétrochimie f	нефтяная химия, нефтехимия
P 119	petroleum	Erdöl n, Rohöl n, Petroleum n	brut m, pétrole m brut, huile f brute	нефть, керосин
P 120	petroleum plant	Erdölanlage f, Erdölverarbeitungsanlage f	installation f pétrolière	установка переработки нефти
P 121	petroleum product	Erdölerzeugnis n, Erdölprodukt n	produit m pétrolier	нефтепродукт, нефтяной продукт, продукт переработки нефти
P 122	petroleum refinery	Erdölraffinerie f	raffinerie f de pétrole	нефтеперегонный завод, дистилляция нефти, нефтеперерабатывающий завод, нефтеперерабатывающая установка
P 123	petroleum spirit	Leichtbenzin n	essence f minérale, essence légère	лаковый бензин, легкий бензин
P 124	phase angle	Phasenwinkel m	angle m de phase, phase f	угол фаз, наклон фазы
P 125	phase boundary	Phasengrenze f	limite f de phase	предел (граница) фаз
P 126	phase change	Phasenänderung f	changement m de phase	изменение (превращение) фаз
P 127	phase combination	Phasenkombination f	combinaison f de phases	комбинация фаз
P 128	phase diagram	Phasendiagramm n, Zustandsdiagramm n	diagramme m de phases, diagramme d'état	фазовая диаграмма, диаграмма состояния
P 129	phase equilibrium	Phasengleichgewicht n	équilibre m des phases	равновесие фаз, фазовое равновесие
P 130	phase inversion	Phasenumkehrung f	inversion f des phases	инверсия (обращение) фаз
P 131	phase lag	Phasenverzögerung f	retard m de phase	отставание по фазе
P 132	phase of motion	Bewegungsphase f	phase f de mouvement	фаза движения
P 133	phase ratio	Phasenverhältnis n	rapport m de phases	[со]отношение фаз
P 134	phase rule	Phasenregel f, Phasengesetz n	règle f des phases	правило фаз
P 135	phase separation	Phasentrennung f	séparation f des phases	раздел (разделение, сепарация) фаз
P 136	phase transition	Phasenübergang m, Phasenumwandlung f	changement m (transformation f) de phase	фазовые превращения, обращение фаз, фазовый переход
P 137	phasing	Phasenabgleich m (Drehstrom)	réglage m en courant déphasé	уравновешивание по фазе
	phenol removal	s. D 169		
P 138	photochemical reaction	photochemische Reaktion f	réaction f photochimique	фотохимическая реакция
P 139	pH value	pH-Wert m	valeur f [du] pH	значение pH

P 140	physical form	Aggregatform f	état m physique	агрегатное состояние, агрегатная форма
	pickle	s. A 84		
P 141	pickling agent	Abbeizmittel n	mordant m, corrosif m	травильное (разъедающее) средство
P 142	pick-up fraction	Zwischenfraktion f	fraction f intermédiaire	промежуточная фракция
P 143	picture variable	Abbildungsvariable f	variable f des images	переменная (параметр) изображения
P 144	piece goods, individual parts	Stückgut n	marchandise f en colis	штучное изделие, кусковой материал, штучный груз
P 145	piling space	Ladevolumen n, Stauraum m	espace m d'empilage	грузовой объем
	pilling	s. P 85		
P 146	pilot flame	Zündflamme f	flamme f d'allumage, veilleuse f	пламя зажигания, запальное (контрольное) пламя
P 147	pilot model, experimental model	Versuchsmodell n	modèle m d'expérience	опытная модель, испытательный образец
P 148	pilot plant, experimental unit	Pilotanlage f, Modellanlage f, Versuchsstand m, Teststand m, Versuchsanlage f	installation f d'essai, installation pilote, pilote m, banc m d'essai, banc d'épreuve	опытная установка, модельная установка, экспериментальный стенд, пилотная установка, испытательный стенд, экспериментальная установка
P 149	pin beater mill	Schlagstiftmühle f	moulin m à pilons, broyeur m à barres	ударная штифтовая мельница, дезинтегратор
P 150	pinched-plate screen	Siebblech n	tôle f perforée (à tamissage)	металлическая (ситовая, дырчатая) решетка
P 151	pinch point	Knickpunkt m (Punkt des geringsten Abstandes zwischen zwei Kurven)	point m de brisure	точка перегиба (перелома)
P 152	pipe	Rohr n, Rohrschlange f	tube m, conduit m, serpentin m, tuyau m [serpentin]	труба, змеевик
P 153	pipe bend	Krümmer m, Rohrknie n	coude m, coude d'un conduit	колено, отвод, закругленный поворот
P 154	pipe bridge	Rohrbrücke f	pont m à tuyaux	трубопроводная панель, трубопроводный мост
P 155	pipe coil evaporator	Schlangenrohrverdampfer m	évaporateur m de type serpentin	змеевиковый испаритель
P 156	pipe connection	Rohrverbindung f	raccord m de tuyaux	соединение трубопроводов
P 157	pipe diameter	Rohrdurchmesser m	diamètre m de tube	диаметр трубы
P 158	pipe fittings	Rohrarmaturen fpl	robinetterie f	трубчатая арматура, арматура для труб
P 159	pipe friction factor	Rohrreibungsbeiwert m	coefficient m de rugosité	коэффициент трения труб
P 160	pipe joint	Rohrstutzen m	tubulure f	патрубок, штуцер трубы
P 161	pipe layout	Rohrleitungsplan m	plan m de tuyautage	схема трубопроводов, трассирование трубопроводов, план расположения трубопроводов
P 162	pipeline	Rohrleitung f	conduit m de tuyaux	трубопровод
P 163	pipeline system	Rohrleitungssystem n	système m de tuyautage	система трубопроводов
P 164	pipe network	Rohrnetz n	réseau m de tuyaux	трубопроводная сеть, сеть трубопроводов
P 165	pipe still	Röhrenofen m	fourneau m tubulaire	трубчатая печь
P 166	pipe support	Rohrunterstützung f	support m de tuyau	опора для труб, трубчатая стойка, кронштейн для труб
P 167	piping	Rohrleitung f	conduite f, canalisation f, conduit m de tuyaux	трубопровод
P 168	piping arrangement	Rohranordnung f	disposition f des tuyaux	расположение трубопроводов
P 169	piston blower, piston compressor	Kolbengebläse n	soufflet m à piston	поршневая воздуходувная машина
P 170	piston capacity, working volume	Hubraum m	cylindrée f	рабочий объем цилиндра
	piston compressor	s. P 169		
P 171	piston flow, displacement (graft) flow	Verdrängungsströmung f, Kolbenströmung f, Pfropfströmung f	écoulement m piston	вытесняющее (поршневое, выталкивающее) течение
P 172	piston pressure	Kolbendruck m	pression f au piston	давление на поршень, поршневое давление
P 173	piston steam engine	Kolbendampfmaschine f	machine f à vapeur à piston	поршневая паровая машина
P 174	piston stoke	Kolbenbewegung f, Kolbenhub m	course f de piston	движение поршня
P 175	piston type valve	Kolbenventil n	soupape f à piston	клапан поршня
P 176	pit bin	Tiefbunker m	soute f surbaissée	подземный бункер
P 177	Pitot tube	Pitot-Rohr n, Staurohr n, Pitotsche Röhre f	tube m (trombe f) de Pitot	трубка Пито
	place	s. L 219		
P 178	place of reaction	Reaktionsort m	place f de réaction	место реакции
	plan	s. A 555		
P 179	Planck distribution	Planck-Verteilung f	distribution f de Planck	распределение Планка
P 180	Planck's radiation law	Plancksches Strahlungsgesetz n	loi f de Planck, théorème m du rayonnement de Planck	закон излучения Планка

P 181	plane of symmetry	Symmetrieebene f	plan m de symétrie	плоскость симметрии
P 182	plane stress	ebener [zweiachsiger] Spannungszustand	tension f biaxiale	плоское [двухмерное] напряженное состояние
P 183	planetary roll mill	Planetarwalzenmühle f, Planetarwalzenstuhl m	moulin m à cylindres planétaire	планетарная валковая (вальцевая) мельница
	planning	s. P 583		
P 184	planning documents	Planungsgrundlage f, Planungsunterlagen fpl	documents mpl de planification	основа планирования, документация планирования (проектирования)
P 185	planning method	Planungsmethode f	méthode f de planification	метод планирования (проектирования)
P 186	planning models	Planungsmodelle npl	modèles mpl de planification (conception)	модели планирования
P 187	plan of operation sequence	Arbeitsablaufplan m	planification f de la suite des opérations	операционный график
P 188	plant, installation, works, establishment, factory	Anlage f, Werk n, Betrieb m, Betriebsanlage f	installation f, fabrique f, usine f, entreprise f, unité f industrielle	установка, завод, производство, промышленная установка
P 189	plant computer system	Betriebsrechnersystem n	système m de calculateurs de production	вычислительная система для управления производством (АСУП)
P 190	plant control	Anlagensteuerung f, Anlagenüberwachung f	commande f (contrôle m) de l'unité	управление установкой
P 191	plant cooling capacity	Kälteanlagenleistung f	capacité f d'installation frigorifique	холодопроизводительность установки
P 192	plant design	Anlagenauslegung f, anlagentechnischer Entwurf m	projet m de l'installation, projet m	расчет установки, проект технической установки
P 193	plant operation	Betrieb m von Anlagen	exploitation f d'installations	эксплуатация установок, работа установок
P 194	plant preparation	Anlagenvorbereitung f	préparation f d'installations	подготовка установки
P 195	plant realization	Anlagenrealisierung f, Errichtung f von Anlagen	réalisation f de l'installation	реализация строительства установки, создание установки
P 196	plant safety	Anlagensicherheit f	sécurité f de l'installation	безопасность установки
P 197	plant stoppage	Anlagenstillstand m	arrêt m de l'installation	простой установки
P 198	plant structure	Anlagenstruktur f	structure f de l'installation	структура установки
P 199	plant subject control	überwachungspflichtige Anlage f	installation f soumise à la surveillance	установка, подлежащая периодическому надзору, установка, требующая постоянного контроля
P 200	plant units	Anlageneinheiten fpl	unités fpl de l'installation	единицы мощности установки
P 201	plan view, ground plot	Grundrißdarstellung f, Grundriß m	projection f horizontale, tracé m, plan m	изображение вида сверху, план, чертеж
P 202	plasma jet reactor	Plasmastrahlreaktor m	réacteur m à jet de plasma	реактор с плазменной струей
P 203	plasma reaction	Plasmareaktion f	réaction f plasmatique	плазменная реакция, реакция в плазме
P 204	plasma state	Plasmazustand m	état m du plasma	плазменное состояние
P 205	plasticate/to	plasti[fi]zieren, weichmachen	plastifier	пластифицировать
P 206	plastication	Plastizieren n, Plastizierung f	plastication f	пластикация
P 207	plasticizer, emollient, softener	Weichmacher m, Plastifikator m, Erweichungsmittel n	plastifiant m, assouplissant m, émollient m, moyen m de ramollissement	мягчитель, смягчитель, пластификатор
P 208	plastics	Plaste mpl, Kunststoffe mpl, Werkstoff m	matières fpl plastiques (artificielles), matière f synthétique	пластмассы, пластические массы, конструктивный материал
	plastics	s. a. S 912		
P 209	plastics processing	Plastverarbeitung f	transformation f de matière plastique	переработка пластмасс (пластических масс)
	plastic viscosity	s. B 107		
P 210	plastifying, plasticization	Plastifizierung f	plastification f	пластификация, пластифицирование
P 211	plastisol	Plastisol n	plastisol m, pâte f	пластизол
	plate/to	s. L 115		
P 212	plate	Teller m, Platte f, Beschlag m, Schild n, Anode f, Boden m	plateau m, disque m, plaque f, panneau m	пластинка, плита, анод, тарелка, диск
P 213	plate air-cooling unit	Plattenluftkühler m	refroidisseur m d'air aux plaques	пластинчатый воздухоохладитель
P 214	plate and frame filter	Rahmenfilter n	filtre m à cadres et à plateaux	рамный фильтр
P 215	plate column	Bodenkolonne f, Bodensäule f	colonne f à plateaux	тарельчатая колонна
P 216	plate conveyor	Plattenband n	ruban à plaques, transporteur m à palettes	плиточный (пластинчатый) конвейер, плиточная лента
P 217	plate cooler	Plattenkühler m	refroidisseur m à plaques	пластинчатый охладитель

	English	German	French	Russian
P 218	**plate efficiency**	Gesamtbodenwirkungsgrad m	rendement m de plateau total	коэффициент полезного действия тарелки, отношение числа теоретических тарелок к числу действующих тарелок
	plate filter	s. F 116		
P 219	**plate fin evaporator**	Rippenplattenverdampfer m	évaporateur m aux plateaux cannelés	плиточный испаритель с оребренными (ребристыми) трубами
P 220	**plate-fin heat exchanger**	Wärmeübertrager m mit gerippten Platten	échangeur m de chaleur aux plateaux cannelés	теплообменник с ребристыми пластинами
P 221	**plate heat exchanger**	Plattenwärmeübertrager m	échangeur m thermique aux plateaux	пластинчатый теплообменник
P 222	**plate-type condenser**	Plattenverflüssiger m, Plattenkondensator m	condenseur m à plaques	пластинчатый (листотрубный) конденсатор
P 223	**plate-type evaporator**	Plattenverdampfer m	évaporateur m de type plaque	пластинчатый испаритель, испаритель плиточного типа
P 224	**platform**	Bühne f, Podest n	plate-forme m, passerelle f	платформа, полка
	plugging	s. O 7		
	plug up/to	s. B 136		
P 225	**plume air**	Rauchfahne f, Rauchsäule f	colonne f de fumée	столб дыма
P 226	**plunge pipe**, immersion tube	Tauchrohr n	tube m plongeur	трубка для погружения, погружная трубка
	plunger	s. D 293		
P 227	**plunger injection moulding**	Kolbenspritzgußverfahren n	procédé m de moulage par injection à l'aide d'un piston	поршневое литье под давлением
P 228	**plunger pump**	Kolbenpumpe f	pompe f à piston	поршневой насос
	plunger pump	s. a. P 420		
	plunging	s. S 768		
	ply	s. L 46		
P 229	**pneumatic conveying chute**	pneumatische Förderrinne f	rigole f pneumatique de transport	пневматический транспортный желоб
P 230	**pneumatic conveyor**	Druckluftförderer m	transporteur m pneumatique	пневматический транспортер
P 231	**pneumatic conveyor dryer**	Stromtrockner m	séchoir m pneumatique	пневматическая сушилка
P 232	**pneumatic mixer**	pneumatischer Mischer m	mélangeur m pneumatique	пневматический смеситель
P 233	**pneumatic pick**	Abbauhammer m	marteau-pic m, marteau m d'exploitation	отбойный (расклепочный) молоток
	pneumatic pressure	s. A 402		
P 234	**pneumatic temperature controller**	pneumatischer Temperaturregler m	régulateur m pneumatique de température	пневматический регулятор температуры
P 235	**pneumatic transport**	pneumatischer Transport m	transport m pneumatique	пневматический транспорт
	point diagram	s. C 891		
P 236	**point of attachment**	Ansatzpunkt m	point m d'attache	место заложения, начальная (исходная) точка
P 237/8	**point of inflection**	Wendepunkt m	point m d'inflexion	точка поворота (перегиба)
P 239	**point of stop**	Anschlagpunkt m	point m d'arrêt	упорная точка
	poison/to	s. I 316		
P 240	**poisoning**, intoxicating, contamination	Vergiftung f	empoisonnement m, intoxication f	отравление
P 241	**Poisson's ratio,** modulus of cross elasticity	Querdehnungszahl f, Querelastizitätsmodul m	constante f de Poisson	число Пуассона, число поперечного растяжения
P 242	**polarizability**	Polarisierbarkeit f	aptitude f à être polarisé	поляризуемость
P 243	**polarization optical method**	optische Polarisationsmethode f	méthode optique f de polarisation	оптический поляризационный метод
P 244	**pollutant exhaustion**	Schadstoffabsaugung f	exhaustion f de l'élément polluant	отсасывание вредных веществ
P 245	**pollutant spreading**	Schadstoffausbreitung f	propagation f de l'élément polluant	распространение вредных веществ
	pollute/to	s. C 624		
	pollution	s. C 627		
P 246	**pollution load**	Abwasserbelastung f	charge f en eaux d'égouts	нагрузка (норма отведения) сточных вод
P 247	**polyatomic**	mehratomig, vielatomig, polyatomar	polyatomique	многоатомный
P 248	**polycondensation**	Polykondensation f	polycondensation f	поликонденсация
P 249	**polycrystalline**	polykristallin	polycristallin	поликристаллический, многокристаллический
P 250	**polydisperse polymer**	polydisperses Polymer n	polymère m polydispersé	полидисперсный полимер
P 251	**polydispersity**	Polydispersität f	polydispersité f	полидисперсность
P 252	**polyethylene production**	Polyethylen-Herstellungsverfahren n	production f de polyéthylène	способ получения полиэтилена, производство полиэтилена
P 253	**polymer,** polymerizate	Polymerisationsprodukt n, Polymer n	corps m polymère, produit m polymérisé	полимеризат, продукт полимеризации
P 254	**polymer engineering and science**	Polymertechnik f und -wissenschaft f	science f et technique f des polymères	техника и наука о полимерах, полимерная техника и наука
P 255	**polymeric chain**	Polymerkette f	chaîne f polymère	полимерная цепь

P 256	polymeric liquid	polymere Flüssigkeit f	liquide m polymère	полимерная жидкость
P 257	polymeric material	polymeres Material n	matérial m polymère	полимерный материал
P 258	polymeric system	Polymersystem n	système m polymère	полимерная система
	polymerizate	s. P 253		
P 259	polymerization autoclave	Polymerisationsautoklav m	autoclave m de polymérisation	полимеризатор, автоклав для полимеризации
P 260	polimerization reaction	Polymerisationsreaktion f	réaction f de polymérisation	реакция полимеризации
P 261	polymer melt	Polymerschmelze f	polymère m fondu, fonte f polymère	расплав полимера
P 262	polymer melt rheology	Rheologie f der Polymerschmelze	rhéologie f du polymère fondu	реология расплава полимера
P 263	polymer processing	Polymerverarbeitung f	transformation f (traitement m) des polymères	переработка полимеров
P 264	polymer processing machinery	Polymerverarbeitungsmaschinen fpl	machines fpl pour le traitement des polymères	машины для переработки полимеров
P 265	polymer-solvent interaction	Polymer-Lösungsmittel-Wechselwirkung f	interaction f polymère-solvant	взаимодействие полимер-растворитель
P 266	polyoptimization	Polyoptimierung f, mehrkriterielle Optimierung f	optimisation f multiple, polyoptimisation f	полиоптимизация, многокритериальная оптимизация
P 267	poor gas, lean gas, weak gas	Schwachgas n, armes Gas n	gaz m pauvre	колосниковый (бедный, низкокалорийный, слабый) газ
P 268	pore formation	Porenbildung f	formation f de pores	порообразование
P 269	pore model	Porenmodell n	modèle m de pores	модель пор
P 270	pore volume	Porenvolumen n	volume m des pores	объем пор
P 271	porous medium	poröses Medium n	milieu (matériau) m poreux	пористая среда
P 272	portable cooling device	ortsveränderlicher Kühler m	refroidisseur m déplacable	портативное охлаждающее устройство
	position	s. L 219		
P 273	positive displacement pump	Verdrängerpumpe f	pompe f à déplacement	объемный насос, насос вытеснения
P 274	post-reactor processing	Bearbeitung f nach Verlassen des Reaktors	traitement m ultérieur	обработка после выхода из реактора
P 275	potential difference between electrodes	Elektrodenspannung f	tension f aux électrodes	напряжение на электродах, электродный потенциал
P 276	potential drop, voltage (line) drop	Spannungsabfall m	chute f de tension (potentiel)	падение (перепад) напряжения
P 277	potential energy	potentielle Energie f	énergie f potentielle	потенциальная энергия
P 278	potential flow	Potentialströmung f	courant (écoulement) m sans tourbillonnement	потенциальное течение
P 279	potential function	Potentialfunktion f	fonction f de potentiel	потенциальная функция
P 280	potential gradient	Potentialgradient m	gradient m du potentiel	градиент потенциала
P 281/3	potential gradient	Potentialgefälle n	chute f du potentiel	падение потенциала, потенциальный перепад
	potting	s. E 100		
	pour point	s. F 251		
P 284	Powell method	Powell-Verfahren n	méthode f de Powell	метод Поуэля
	power	s. I 51		
P 285	power absorption, power input (consumption)	Leistungsaufnahme f	puissance f absorbée	потребление мощности
P 286	power constant	Leistungskonstante f	constante f de puissance	постоянная мощности
P 287	power consumption, power demand	Kraftbedarf m, Leistungsbedarf m, Kraftverbrauch m, Energiebedarf m	force f nécessaire, puissance f consommée (nécessaire), demande f d'énergie	потребность в энергии, силовая потребность, потребляемая мощность, расход мощности, силовые затраты, потребление энергии
	power consumption	s. a. P 285		
P 288	power conversion	Energieumformung f	transformation f d'énergie	преобразование (трансформация) энергии
P 289	power current	Kraftstrom m	force f motrice, courant m force	электрическая энергия, силовой ток
	power demand	s. P 287		
P 290	power density	Leistungsdichte f	densité f de puissance	плотность мощности
P 291	power drop	Leistungsabfall m	perte f de puissance	падение мощности
P 292	power economy	Energiewirtschaft f	économie f de l'énergie	энергетическое хозяйство, энергетика, энергохозяйство
P 293	power engineering	Energietechnik f	technologie f énergétique	энергетика
P 294	power formula	Leistungsformel f	formule f de puissance	формула мощности
P 295	power fuel gas	Brenngas n	gaz m combustible	горючий (топливный) газ
P 296	power gain	Leistungsgewinn m	gain m de puissance, profit m	приращение мощности
P 297	power gas, propellant	Treibgas n, Kraftgas n	carburant m gazeux, gaz m combustible	сжиженный (силовой, генераторный) газ
P 298	power increase	Leistungsanstieg m	augmentation f de puissance	приращение мощности
	power input	s. P 285		
P 299	power lead, power supply (pack)	Kraftanschluß m	raccordement m de force	силовой ввод
P 300	power loss	Leistungsverlust m	perte f de puissance	потеря мощности

P 301	**power loss,** dissipation	Verlustleistung f	puissance f perdue	мощность потерь
	power network	s. P 308		
P 302	**power of performance curve**	Leistungskennlinie f	courbe f de puissance, caractéristique f	кривая характеристики мощности
P 303	**power output**	Ausgangsleistung f, Leistungsabgabe f	puissance (rendement m) de sortie, puissance f disponible (débitée), rendement m	выходная (излучаемая, начальная) мощность, отдача мощности
	power pack	s. P 299		
P 304	**power pile,** power reactor	Leistungsreaktor m	réacteur m de puissance	энергетический реактор
	power plant	s. P 306		
	power reactor	s. P 304		
P 305	**power requirement,** acceptance power	Abnahmeleistung f	puissance f de réception	отдаваемая мощность
	power source	s. S 448/9		
P 306	**power station,** power plant	Kraftwerk n, Energieerzeugungsanlage f	centrale f de force motrice, centrale (usine f) électrique	силовая станция, электростанция
P 307	**power supply**	Netzanschluß m, Energieversorgung f	alimentation f en énergie, alimentation sur secteur, branchement m au réseau	включение в сеть, энергообеспечение, энергоснабжение, подключение к сети
	power supply	s. a. P 299		
P 308	**power supply system,** power network	Energieversorgungssystem n	système m d'alimentation en énergie	система энергообеспечения
P 309	**power supply unit**	Energieblock m, Energieversorgungseinheit f	unité f d'alimentation en énergie	блок энергоснабжения
P 310	**power transmission**	Kraftübertragung f, Energieübertragung f	transmission f d'énergie, transport m de force	передача энергии (силы), электропередача
P 311	**power unit**	Leistungseinheit f	unité f de puissance	единица мощности
P 312	**practicability test**	Durchführbarkeitsversuch m	essai m de faisabilité	проверка осуществимости
P 313	**practical experience**	Betriebserfahrung f	expérience f d'exploitation	производственный опыт
P 314	**Prandtl-number**	Prandtl-Zahl f	nombre m de Prandtl	число Прандтля
P 315	**preassembly**	Vormontage f	montage m préliminaire	предварительный монтаж, предварительная сборка
P 316	**preassembly site**	Vormontageplatz m	place f du montage préliminaire	площадка для предварительного монтажа
P 317	**preblend**	Vormischung f	action f de prémélange	предварительное смешение
P 318/9	**precalculation,** estimation	Vorausberechnung f	calcul m préalable	калькуляция, предварительная оценка
	prechilling	s. P 333		
P 320	**precipitable**	ausfällbar	précipitable	осаждаемый
P 321	**precipitant**	Fällbad n	précipitant m	осадитель, среда, в которой происходит осаждение
	precipitate	s. D 175		
P 322	**precipitated,** settled	abgesetzt, ausgeschieden, ausgefällt	déposé, précipité	осажденный, осевший
P 323	**precipitation**	Ausfällung f, Niederschlag m, Fällung f	condensation f	конденсирование, выпадение, осадок
	precipitation	s.a. E 85		
P 324	**precipitation sediments**	Ausscheidungssedimente npl, Niederschlag m	sédiment m de précipitation, dépôt m	химический осадок
	precipitator	s. S 216		
P 325	**precleaning**	Vorreinigung f	premier nettoyage m	предварительная очистка
P 326	**precompacting**	Vorverdichtung f	précompression f	предварительное сжатие, наддув, предварительное уплотнение
P 327	**precompactor**	Vorverdichter m	précompresseur m	нагнетатель, наддувной агрегат, предварительный компрессор
P 328	**precompress / to**	vorverdichten	précomprimer	подвергать предварительному сжатию, подвергать предварительному уплотнению
P 329	**precondensate**	Vorkondensat n	précondensat m	предконденсат, форконденсат
P 330	**precondensation**	Vorkondensation f	précondensation f	предварительная конденсация
P 331	**precondition**	Vorbedingung f	précondition f	предварительное условие
	preconditioning	s. P 377		
P 332	**precooler**	Vorkühler m	prérefroidisseur m	предохладитель
P 333	**precooling,** forecooling, prechilling	Vorkühlung f	précondensation f, prérefroidissement m	предварительное охлаждение
P 334	**precooling cycle**	Vorkühlprozeß m	cycle m de préréfrigération	цикл предварительного охлаждения
P 335	**precooling degree**	Vorkühlstufe f	degré m de préréfrigération, étage m de refroidissement préliminaire	степень предварительного охлаждения
P 336	**precooling level**	Vorkühlniveau n	niveau m de prérefroidissement	уровень предварительного охлаждения
P 337	**precooling period**	Vorkühlperiode f	période f de prérefroidissement	период предварительного охлаждения

P 338	precooling temperature	Vorkühltemperatur f	température f de prérefroidissement	температура предварительного охлаждения
P 339	precooling time	Vorkühlzeit f	durée f de prérefroidissement	продолжительность предварительного охлаждения, время предварительного охлаждения
P 340	precooling zone	Vorkühlzone f	zone f de prérefroidissment	зона предварительного охлаждения
P 341	precrushing	Vorzerkleinerung f	désintégration f préliminaire	предварительное дробление (измельчение)
	predetermine / to	s. C 10		
P 342	predewatering	Vorentwässerung f	assèchement m préalable	предварительное осушение
P 343	prediction, prognosis, prognostication	Prognosearbeit f	prévision f, prédiction f	прогнозирование
P 344	predry/to	vortrocknen	présécher	подвергать предварительной сушке
P 345	predryer	Vortrockner m	sécheur m préliminaire, préséchoir m, présécheur m	аппарат для подсушки, предварительная сушилка
P 346	predrying	Vortrocknung f	préséchage m, séchage m préliminaire	подсушка, подсушивание, предварительная сушка
P 347	pre-evaporation	Vorverdampfung f	prévaporisation f	предварительное испарение
P 348	pre-evaporator	Vorverdampfer m	prévaporisateur m	нуль-корпус выпарной установки, предварительная ступень выпарной установки
P 349	prefabricate/to	vorfertigen	préfabriquer	предварительно изготовлять
P 350	prefabrication	Montagebau m	mise f en œuvre des éléments préfabriqués	сборное строительство
P 351	prefabrication	Vorfertigung f	préfabrication f	предварительный монтаж, предварительная сборка, предварительное изготовление
P 352	preflash tower	Vorentspannungskolonne f	tour f de détente incomplète	колонна предварительного испарения
P 353	prefractionation	Vorzerlegung f, Vorfraktionierung f	préfractionnement m	предварительное фракционирование (разделение)
P 354	prefractionation tower	Vorkolonne f, Vorfraktionierer m, Vorzerlegungsapparat m	colonne f de préfractionnement	колонна предварительного фракционирования
P 355/6	preheat/to	vorheizen, vorwärmen	préchauffer, réchauffer	предварительно нагревать, подогревать
P 357	preheating	Vorheizung f, Vorwärmung f	préchauffage m, réchauffage m	подогрев, предварительный нагрев
P 358	preheating zone	Vorwärmzone f	zone f de réchauffage	зона подогрева
P 359	preliminary assembly drawing	Entwurfszeichnung f	dessin m de projet	эскизный чертеж
P 360	preliminary calculation	Vorkalkulation f, Überschlagsrechnung f	calculation f au préalable	предварительная калькуляция (оценка), смета
P 361	preliminary calculation sequence	vorläufige Berechnungsreihenfolge f	suite f de calculs provisoire	предварительная последовательность расчета (процесса, системы)
P 362	preliminary condenser	Vorkondensator m	condenseur m préliminaire	предварительный конденсатор
P 363	preliminary decision on investment	Investvorentscheidung f	décision f préliminaire de l'investissement	предварительное решение о капиталовложении
P 364	preliminary degassing	Vorentgasung f	dégazage m préliminaire	предварительное удаление газа, предварительная дегазация
P 365	preliminary design	Vorprojektierung f	projection f de départ	эскизное проектирование
P 366	preliminary design	Vorentwurf m	avant-projet m	эскизный проект
P 367	preliminary erection	Vormontage f	montage m préliminaire	предварительный монтаж
P 368	preliminary investigation	Voruntersuchung f	examen m préliminaire	предварительное исследование
	preliminary planning	s. P 381		
	preliminary project	s. P 382		
	preliminary selection	s. S 87		
	preliminary step	s. P 445		
P 369	preliminary test	Vorprüfung f, Vorversuch m	essai m préliminaire	предварительный опыт (эксперимент), предварительное испытание
P 370	preliminary treatment	Voraufbereitung f	prétraitement m, traitement m préliminaire	предварительное обогащение
P 371	preliminary washing	Vorwaschen n	lavage m préalable	предварительная промывка
P 372	preliminary washing drum	Vorwaschtrommel f	tambour m laveur préliminaire	барабан предварительной промывки
P 373	preliminary work	Vorarbeit f, Vorbereitung f	travail m préliminaire, préparation f	подготовительная работа, подготовка
P 374	premature	vorzeitig	prématuré	преждевременно, досрочно

P 375	premium fuel (gasoline)	Premium-Kraftstoff *m*, Superkraftstoff *m*, Premium-Benzin *n*, Superbenzin *n*	supercarburant *m*	высокооктановый бензин
P 376	premix burner	Injektorbrenner *m*	brûleur *m* à injection	инжекторная горелка
P 377	preparation, pretreatment, dressing, preconditioning	Vorbereitung *f*, Vorbehandlung *f*	préparation *f*, travaux *mpl* préparatoires, traitement *m* préalable	подготовка, предварительная обработка, первичная обработка
P 378	preparation of the bath	Badbeschickung *f*	chargement *m* de bain	загрузка ванны
P 379	preparation of the sludge	Schlammaufbereitung *f*	traitement *m* des boues	обогащение шлама
P 380	preparatory phase	Vorbereitungsphase *f*	phase *f* préparatoire	подготовительная фаза
P 381	preplanning, preliminary planning	Vorplanung *f*	planning *m* préliminaire, prévision *f*	предварительное (перспективное) планирование
P 382	preproject, preliminary project	Vorprojekt *n*	projet *m* préliminaire	эскизный проект
P 383	prepurify/to	vorreinigen	épurer préalablement	производить предварительную очистку
P 384	prerequisite	Voraussetzung *f*, Vorbedingung *f*	supposition *f*	предположение, предпосылка
P 385	prescrubber	Vorwäscher *m*	laveur *m* préliminaire, tour *f* de prélavage	предварительный скруббер
P 386	preselect/to	vorwählen	présélectionner	производить предварительный выбор
P 387	presence, existence, occurrence, deposit	Vorkommen *n*, Existenz *f*, Auftreten *n*	présence *f*, existence *f*, occurence *f*, apparition *f*	месторождение, залегание, нахождение
	preservation	*s.* C 585		
P 388	preset/to	vorher festlegen	déterminer au préalable	предварительно задавать
P 389	presizing	Vorklassierung *f*	préclassement *m*	предварительный отсев
	press cake	*s.* F 93		
P 390	press cooler	Druckkühler *m*	refroidisseur *m* sous pression	холодильник под давлением
P 391	press filter	Preßfilter *n*	filtre *m* sous pression	фильтр-пресс
P 392	pressing out, squeezing out, extrusion	Auspressen *n*	extrusion *f*	выжимание, отжимание
	pressure adapter (balance)	*s.* P 424		
P 393	pressure boiler	Drucksiedekessel *m*	chaudière *f* à pression	котел для кипячения под давлением
	pressure burden	*s.* O 165		
P 394	pressure column	Druckkolonne *f*	colonne *f* sous pression élevée	колонна, работающая под давлением
P 395	pressure compensation	Druckausgleich *m*	compensation *f* de pression	выравнивание давления
P 396	pressure controlled	druckgesteuert	contrôlé par la pression	управляемый давлением
P 397	pressure curve	Druckkurve *f*	ligne *f* des pressions	кривая давления
	pressure difference	*s.* P 400		
P 398	pressure distillate	Krackdestillat *n*	distillat *m* du cracking	пресс-дистиллят
P 399	pressure distillation	Druckdestillation *f*	distillation *f* sous pression	дистилляция под давлением
	pressure distillation	*s. a.* D 428		
P 400	pressure drop, pressure difference (loss)	Druckabfall *m*, Druckminderung *f*, Druckgefälle *n*, Druckverlust *m*	perte *f* de pression (charge), différence (diminution, chute) *f* de pression	перепад (снижение, потеря) давления, падение напора
P 401	pressure drop calculation	Berechnung *f* des Druckabfalls, Druckverlustberechnung *f*	calcul *m* de la perte de charge	расчет падения давления
P 402	pressure drop rate	Druckverlustwert *m*	valeur *f* de la perte de charge	величина потерь давления
P 403	pressure electrolysis	Druckelektrolyse *f*	électrolyse *f* sous pression	пьезоэлектролиз
P 404	pressure-enthalpy chart	i-Diagramm *n*, Druck-Enthalpie-Diagramm *n*	diagramme *m* de pression-enthalpie	диаграмма энтальпия-давление
P 405	pressure filter	Druckfilter *n*, Filterpresse *f*	filtre-presse *m*	фильтр, работающий под давлением, напорный фильтр
P 406	pressure filtration	Druckfiltration *f*	filtration *f* sous pression	фильтрация под давлением
P 407	pressure fluctuation	Druckschwankung *f*	variation *f* de pression	колебание давления
P 408	pressure gas	Druckgas *n*	gaz *m* comprimé	сжатый газ
P 409	pressure gasification process	Druckvergasungsprozeß *m*	procédé *m* de gazéification sous pression	процесс газификации под давлением
P 410	pressure gas plant	Druckgasanlage *f*	installation *f* à gaz comprimé	установка сжатого газа
P 411	pressure gauge	Dampfmesser *m*	débitmètre *m*, mesureur *m* de débit	паромер
P 412	pressure gauge pipe stop valve	Manometerabsperrventil *n*	robinet *m* de manomètre	запорный вентиль к манометру
	pressure increase	*s.* P 427		
P 413	pressure line, delivery pipe, discharge line (pipe)	Druckleitung *f*	conduite *f* forcée, conduite sous pression, tuyau *m* forcé	напорный трубопровод
	pressure loss	*s.* P 400		
P 414	pressure measurement pipe	Druckmeßstutzen *m*	tubulure *f* pour la prise de pression	патрубок для измерения давления
P 415	pressure moulding	Druckformen *n*	moulage *m* par pression	формование под давлением
P 416	pressure oil	Drucköl *n*	huile *f* sous pression	нагнетаемое масло

P 417	pressure oil wash	Druckölwäsche f	lavage m de l'huile sous pression	промывка маслом под давлением
P 418	pressure polymerization	Druckpolymerisation f	polymérisation f sous pression	полимеризация под давлением
P 419	pressure-proof	druckfest	résistant à la pression	герметичный
P 420	pressure pump, plunger (forcing) pump	Druckpumpe f	pompe f refoulante (de pression)	нагнетательный насос
P 421	pressure range	Druckbereich m	domaine m de pression	диапазон давления
P 422	pressure ratio	Druckverhältnis n	rapport m de pression	режим давления, отношение давления
P 423	pressure reducing valve	Druckminderer m, Druckminderventil n, Reduzierventil n	soupape f réductrice (de réduction)	редукционный вентиль (клапан)
P 424	pressure regulator, pressure adapter (balance)	Druckregler m	régulateur m de pression, manostat m	регулятор давления
P 425	pressure release	Entspannung f	détente f, relaxation f	расширение, понижение давления
P 426	pressure relief technology	Druckentlastungstechnik f	technique f de détention	техника разгрузки давления
P 427	pressure rise, pressure increase	Druckanstieg m	accroissement m (augmentation f) de pression	возрастание давления
P 428	pressure-sealed, pressure-tight	druckdicht	étanche à la pression	герметический
P 429	pressure sensitive element	Druckfühler m	élément m sensible à pression, capteur m de pression	элемент, чувствительный к давлению, датчик давления
P 430	pressure spread	Druckausbreitung f, Druckübertragung f	transmission f de pression	распространение давления
P 431	pressure support	Druckstütze f	tige f de compression	опорная стойка, суппорт
P 432	pressure-temperature-relationship	Druck-Temperatur-Beziehung f	relation f pression-température	соотношение давления и температуры
P 433	pressure testing	Abdrücken n, Druckprüfung f	mise f à l'épreuve de pression	испытание под давлением
	pressure-tight	s. P 428		
P 434	pressure transducer	Druckgeber m, Druckfühler m	capteur m de pression	датчик давления
P 435	pressure type reactor	Druckreaktor m	réacteur m à pression	реактор под давлением
P 436	pressure variation	Druckänderung f	changement m de pression	изменение давления
P 437	pressure vessel	Druckbehälter m, Druckkessel m, Druckgefäß n	autoclave m, récipient m sous pression, récipient de pression	автоклав, аппарат высокого давления, напорный резервуар
P 438	pressure vessel with agitator	Rührwerksdruckkessel m	réservoir m de pression à agitateur	автоклав с мешалкой
P 439	pressure volume	Druckvolumen n	volume m de pression	объем под давлением, нагнетаемый объем
P 440	pressure-volume-temperature law	Druck-Volumen-Temperatur-Beziehung f	relation f pression-volume-température	закон отношения давление-объем-температура
P 441	pressure water wash	Druckwasserwäsche f	lavage m à l'eau comprimée	промывка водой под давлением
P 442	pressure wave	Druckwelle f	onde f de pression	волна давления
P 443	pressurized fluidized bed	Druckwirbelschicht f, Wirbelschicht f unter erhöhtem Druck	couche f fluidisée sous pression	кипящий слой под давлением
P 444	pressurized water reactor	Druckwasserreaktor m	réacteur m à eau pressurisée	реактор, охлаждаемый водой под давлением
P 445	prestage, preliminary step	Vorstufe f	état m préalable, stade m antérieur	предварительная ступень
P 446	pretreat/to	vorbehandeln	préparer, traiter au préalable	предварительно обработать
	pretreatment	s. P 377		
P 447	prevailing, prevalent	vorherrschend	prédominant	преобладающий
P 448	prevent/to	vorbeugen	prévenir	предупреждать, предохранять, предотвращать
P 449	prevent/to, to avoid, to inhibit	verhindern, vermeiden, hindern	empêcher, éviter	избежать, препятствовать
P 450	preventation	Verhütung f, Vorbeugung f	prévention f	предотвращение, предохранение
P 451	prevention	Verhinderung f, Hinderung f, Hemmung f	empêchement m, inhibition f	предупреждение, предотвращение, предохранение, препятствие
P 452	prevention of foaming	Entschäumung f	élimination f de mousse	устранение пены, обеспенивание
P 453	prevention of pollution	Luftreinhaltung f	protection f de l'air contre la pollution	защита воздуха от загрязнений, очистка воздуха
P 454	preventive maintenance	planmäßige vorbeugende Instandhaltung f	entretien m préventif planifié	планово-предупредительный ремонт
P 455	prilling (process)	Sprühkristallisieren n, Sprühkristallisation f, Prillen n	cristallisation f à pulvérisation	распылительная кристаллизация
P 456	primary air	Primärluft f	air m primaire	первичный воздух
P 457	primary airstream	Primärluftstrom m	courant m d'air primaire	поток первичного воздуха
P 458	primary air temperature	Primärlufttemperatur f	température f de l'air primaire	температура первичного воздуха
P 459	primary cell	Primärelement n	pile f primaire	первичный элемент

P 460	**primary cleaned gas**	vorgereinigtes Gas n	gaz m prépurifié	грубоочищенный газ, предварительно очищенный газ
P 461	**primary column**	Vordestillationskolonne f	colonne f de prédistillation	колонна предварительной перегонки
P 462	**primary condenser**	Primärkondensator m	condenseur m principal	основной (первичный) конденсатор
P 463	**primary coolant,** primary refrigerant	Primärkühlmittel n, Primärkältemittel n	agent m de refroidissement primaire, réfrigérant m primaire, agent frigorifique primaire	первичный хлад[о]агент (хладоноситель, холодильный агент)
P 464	**primary cooling water circuit**	Primärkühlkreislauf m	circuit m de refroidissement primaire	первичный контур охлаждающей воды
P 465	**primary crusher,** primary mill	Grobmühle f, Brecher m	concasseur m	мельница для крупного размола
P 466	**primary crushing**	Vorzerkleinerung f	désintégration f préliminaire, broyage m primaire	первичное измельчение
P 467	**primary desulphurization**	Primärentschwefelung f	désulfuration f primaire	первичное обессеривание
P 468/9	**primary distillation**	Vordestillation f	prédistillation f	предварительная дистилляция
P 470	**primary drying**	Primärtrocknung f	séchage m primaire	первичная сушка
	primary effect	s. P 473		
P 471	**primary filter**	Primärfilter n	filtre m primaire	фильтр предварительной очистки, первичный фильтр
P 472	**primary flash column**	Vorentgasungskolonne f	colonne f de séparation flash primaire	колонна предварительной отгонки газов, колонна первичной дегазации
	primary material	s. C 186		
	primary mill	s. P 465		
P 473	**primary process,** primary effect	Primärvorgang m, Primärprozeß m	procédé m primaire	первичный процесс
P 474	**primary product**	Primärprodukt n	produit m primaire	первичный продукт
P 475	**primary reaction**	Primärreaktion f	réaction f primaire	первичная реакция
P 476/7	**primary reactor**	Primärreaktor m	réacteur m primaire	первичный реактор
P 478	**primary refrigerating medium**	Primärkälteträger m	milieu m frigorifique primaire	первичный холодоноситель
P 479	**primary shaping**	Urformen n	formage m primaire	первичное формование
P 480	**primary steam piping,** primary vapour supply pipe	Primärdampfleitung f	conduite f de vapeur primaire	первичный паропровод
P 481	**primary water**	Primärwasser n	eau f primaire	первичная вода
P 482	**prime coat,** ground (first) coat	Grundanstrich m	première couche f, couche de fond	грунтовка
P 483	**prime cost**	Selbstkosten pl	prix m coûtant (de revient)	себестоимость
P 484	**primer**	Initialzündung f	fusée f d'amorçage	первичное зажигание, детонатор
P 485	**principal stress**	Hauptspannung f, Hauptbeanspruchung f	tension (sollicitation) f principale	главное напряжение
P 486	**principle of action**	Aktionsprinzip n	principe m d'action	принцип активности (действия)
P 487	**principle of design**	Entwurfsprinzip n	principe m de projet	принцип проектирования
P 488	**principle of process**	Verfahrensprinzip n	principe m de procédé	технологический принцип
P 489	**principle scheme**	Prinzipschema n	schéma m de principe	принципиальная схема
P 490	**probability of survival**	Überlebenswahrscheinlichkeit f	probabilité f de survie	вероятность долговечности, выживаемость
P 491	**probability theory**	Wahrscheinlichkeitstheorie f	théorie f des probabilités	теория вероятности
P 492	**problem preparation (solution) process**	Problembearbeitungsprozeß m, Problemlösungsprozeß m	processus m de traitement de problèmes, voie f de solution	алгоритм решения проблем
P 493	**procedure,** performance procedure	Arbeitsweise f s. a. O 78	mode m de travail	режим [работы]
P 494	**proceeding, process,** method, technique, approach	Verfahren n, Methode f, Vorgehensweise f, Verfahrensweise f	processus m [opératoire], méthode f, procédé m, mode m opératoire	процесс, метод, способ, прием, подход
P 495	**process analysis**	Prozeßanalyse f	analyse f de procédé	анализ процесса, процессный анализ
P 496	**process automation**	Prozeßautomatisierung f	automatisation f de procédé	автоматизация процесса
P 497	**process computer**	Prozeßrechner m	calculateur m de processus	управляющая вычислительная машина
P 498	**process condition**	Verfahrensbedingung f	condition f de procédé	условие (технологический режим) процесса
P 499	**process control**	Prozeßsteuerung f, Verfahrenssteuerung f, Prozeßregelung f, Verfahrensregelung f, Betriebskontrolle f, Prozeßführung f, Prozeßüberwachung f	contrôle m de procédé, contrôle et commande f des procédés	управление процессом, регулирование (контроль) процесса, заводской контроль, ведение процесса, мониторинг
P 500	**process design**	Verfahrensauslegung f, Verfahrensplanung f, Verfahrensentwurf m	projet m de processus, planification f de procédé, projet de procédé	проектирование (расчет, планирование, эскизный проект) процесса, проектный набросок, эскизное проектирование

P 501	process development	Prozeßentwicklung f, Verfahrensentwicklung f	développement m de procédé, développement technologique	разработка процесса (технологии, технологического процесса)
P 502	process diagram, flow sheet	[schematisches] Fließbild n, Fließschema n, Flußbild n, Verfahrensfließbild n	flow-sheet m [schématique], flow-sheet technologique	технологическая схема, схематическая диаграмма процесса
P 503	process documentation	Verfahrensdokumentation f	documentation f de procédé	документация технологического процесса
	process engineer	s. C 214		
P 504	process engineering, process technology, chemical engineering	Verfahrenstechnik f	génie f chimique	химическая технология, технология, процессы и аппараты химической технологии
P 505	process equipment	Verfahrensausrüstung f	équipement m technologique	оборудование для проведения процесса, аппараты (химической технологии)
P 506	process equipment design	Ausrüstungsentwurf f, Ausrüstungsauslegung f	projet m des installations (équipements technologiques)	расчет (проектирование) оборудования
P 507	process evaluation	Verfahrenseinschätzung f, Verfahrensbewertung f	évaluation f de procédé	оценка процесса
P 508	process flow	Fertigungsfluß m, technologischer Durchlauf m	suite f des opérations	производственный поток, линия изготовления
P 509	process flow diagram	Verfahrensfließdiagramm n, Verfahrensschema n	schéma m de fonctionnement de procédé, schéma technologique	технологическая схема, графическая схема процесса
P 510	process flowsheeting package	Programmpaket n zur Verfahrensberechnung	paquet m de programme pour le calcul de procédés	система программ для расчета химико-технологических схем
P 511	process group	Prozeßgruppe f	groupe m de processus	группа (ступень) процесса, отделение
P 512	process guarantee	Verfahrensgarantie f	garantie f de procédé	гарантийность процесса
P 513	processibility	Verarbeitbarkeit f, Verarbeitungseignung f, Verarbeitungsmöglichkeit f	possibilité f de traitement, aptitude f au traitement	перерабатываемость
P 514	process industry	stoffwandelnde Industrie f	industrie f chimique et parachimique	химическая и смежная с ней промышленность
P 515	processing, manipulation	Bearbeitung f, Behandlung f	traitement m, travail m, formage m, usinage m	обработка, отделка, обращение, уход
	processing	s. a. D 501/2		
P 516	processing direction	Verarbeitungshinweis m	instruction f pour le traitement	инструкция по переработке
P 517	processing instruction	Verarbeitungsrichtlinie f	instruction f de transformation	инструкция по переработке
P 518	processing machine	Verarbeitungsmaschine f	machine f de transformation	машина для переработки
P 519	processing plant	stoffwirtschaftlicher Betrieb m	entreprise f chimique ou parachimique	химическое и смежное с ним предприятие
	processing plant	s. a. P 521		
P 520	processing properties	Verarbeitungseigenschaften fpl	propriétés fpl de traitement	свойства переработки, свойства материала при переработке
P 521	processing system, processing plant	Verarbeitungsanlage f	installation f de traitement, système m de transformation	установка для переработки веществ (материалов)
P 522	processing technology	Verarbeitungstechnik f, Verarbeitungstechnologie f	technologie f de transformation	техника процессов переработки материалов, технология переработки
P 523	processing temperature	Verarbeitungstemperatur f	température f de transformation	температура переработки
	processing time	s. F 267		
P 524	process intensification	Prozeßintensivierung f	intensification f de processus	интенсификация процесса
P 525	process of boiling	Siedeprozeß m	procédé m d'ébullition	процесс кипения
P 526	process of conversion	Umwandlungsprozeß m	procédé m de conversion	процесс превращения (преобразования) материала
P 527	process of disintegration	Auflösungsprozeß m, Zersetzungsvorgang m	procédé m de décomposition (dissolution)	процесс растворения
P 528	process of formation	Bildungsvorgang m, Entstehungsprozeß m	procédé m de formation	процесс образования (формирования)
P 529	process of solidification	Erstarrungsvorgang m, Verfestigung f	processus m de solidification (congélation)	процесс затвердевания
P 530	process optimization	Prozeßoptimierung f	optimisation f de processus	оптимизация процесса
P 531	process period, run period, process time	Belegungszeit f	temps m d'occupation	продолжительность процесса
P 532	process refrigeration	Prozeßkühlung f	réfrigération f de procédé	охлаждение процесса, охлаждение по ходу технологического процесса
P 533	process stabilization	Prozeßstabilisierung f	stabilisation f du processus	стабилизация процесса
P 534	process stage	Verfahrensstufe f	étage m de processus	отделение (химико-технологической системы)

P 535	process steam	Betriebsdampf m	vapeur f de procédé	производственный (технологический) пар, пар применяемый для обогрева технологического оборудования
P 536	process system	Verarbeitungssystem n, verfahrenstechnisches System n	système m technologique (de traitement)	система переработки (обработки), химико-технологическая система
	process technology	s. P 504		
	process time	s. P 531		
P 537	process train	Verfahrenszug m	suite f des processus	совокупность химико-технологических систем, объединенных общим видом сырья, технологическая линия
P 538	process unit	Prozeßeinheit f	unité f de procédé	отдельный процесс, единица (элемент) процесса, ступень, процесс
P 539	process utilization factor	Prozeßausnutzungsfaktor m	facteur m d'utilisation du procédé	коэффициент использования процесса
P 540	process variable	Prozeßvariable f	variable m de processus	переменная процесса
P 541	produce, yield	Ausbringen n, Ausstoß m	produit m net, rendement m	добыча, выход
	produce	s. a. G 1		
P 542	produced quantity of steam	Dampfleistung f	débit m (production f) de vapeur	паропроизводительность
P 543	producer factory	Erzeugerbetrieb m	usine f de producteur	завод-изготовитель
	producer gas	s. G 117		
P 544	producer shaft	Generatorschacht m	cuve f de générateur	шахта генератора
P 545	product checking	Erzeugnisprüfung f	essai m de produit	испытание изделий (продукта)
P 546	product cycle	Produktkreislauf m	cycle m de produit	циркуляция (оборотный цикл) продукта
P 547	product development	Erzeugnisentwicklung f	développement m de produit	разработка изделий
P 548	production break down	Produktionsausfall m	perte f de production	простой производства
P 549	production capability	Produktionsfähigkeit f	productivité f, capacité f productive	мощность (объем) производства
P 550	production capacity, output capacity	Produktionskapazität f	capacité f de production	производительность, объем производства, производственная мощность
P 551	production change-over	Produktionsumstellung f	changement m de production	перестановка производства
P 552	production control	Betriebsüberwachung f	surveillance f d'exploitation	диспетчерский (производственный) контроль, контроль производства, управление производством, производственный надзор
P 553	production costs	Herstellungskosten pl	prix m de revient, prix de fabrique	производственные расходы, себестоимость
P 554	production engineering	Fertigungsplanung f	planification f de production	планирование производства (изготовления)
P 555	production line	Produktionslinie f	ligne f de production	производственная линия, линия производства
P 556	production of steam	Dampfproduktion f	production f de vapeur	производство пара, паропроизводство
P 557	production program	Fertigungsprogramm n, Produktionsprogramm n	programme m de production	производственная программа
P 558	production rate per hour, hourly output	Stundenleistung f	débit m horaire, rendement m par heure	часовая производительность
P 559	production reactor	Produktionsreaktor m	réacteur m de production	реактор для производства, производственный реактор
	production scheduling	s. P 563		
P 560	production stream	Produktionsstrom m	courant m de production	поток производства, производственный поток
P 561	production system	Produktionssystem n	système m de production	система производства, производственная система
P 562	production technique, manufacturing method	Fertigungsart f	mode m de fabrication	способ изготовления
P 563	production technology, production scheduling, manufacturing schedule	Fertigungsablauf m	suite f des opérations technologiques	технология, технологический процесс, последовательность производства
P 564	production technology, manufacturing technique	Herstellungstechnologie f	technologie f de fabrication	технология производства (изготовления)
P 565	production unit	Produktionsanlage f	unité de production	производственная установка
P 566	product of catalysis	Katalysat n	produit m catalytique	продукт каталитической реакции
P 567	product of combustion	Verbrennungsprodukt n	produit m de combustion	продукт сгорания
P 568	product of dressing	Aufbereitungsgut n	produit m traité, produit préparé	продукт обогащения, обогащенный продукт
P 569	product of enrichment	Anreicherungsprodukt n	produit m d'enrichissement	продукт обогащения
	product of fission	s. F 180/1		

P 570/1	product of regeneration	Regenerationsprodukt n	produit m régénéré	продукт регенерации
P 572	product separation	Produktabspaltung f, Produkt-abtrennung f	séparation f de produit	выделение (отделение, отщепление) продукта
P 573	product stability	Erzeugnisstabilität f	stabilité f de produit	стабильность изделия
P 574	profiled filament	Profilfaden m	fil m profilé	профилированное волокно
P 575	profitability	Rentabilität f	rentabilité f, rendement m	рентабельность
	prognosis	s. P 343		
	prognostication	s. P 343		
P 576	program control	Programmsteuerung f	commande f par programme	программное управление
P 577	program controlled	programmgesteuert	à réglage programmé	с программным управлением
P 578	programming language	Programmsprache f, Programmiersprache f	langage m de programmation	язык программирования
P 579	program system	Programmsystem n	système m de programmes	система программ
P 580	progress, behaviour, course	Verlauf m, Ablauf m, Gang m	allure f, marche f	ход, течение, протекание (процесса)
P 581	project costs	Objektkosten pl	frais mpl de projet	расходы на объект, объектная стоимость
P 582	project documents	Projektunterlagen fpl	documents mpl de projet	проектная документация
P 583	projecting, planning, designing	Projektierung f	projection f	проектирование, проект
P 584	projecting model	Projektierungsmodell n	modèle m de projet	модель проектирования, проектная модель
P 585	project preparation performance	Projektierungsleistung f	travaux mpl de projection (projet)	объем проектирования, затраты на проектирование
P 586	propagation reaction	Wachstumsreaktion f	réaction f de développement (croissance)	реакция роста
P 587	propagation velocity	Ausbreitungsgeschwindigkeit f	célérité f [de propagation]	скорость распространения
P 588	propellant	Treibmittel n	agent m moteur	вещество для двигателя (вспенивания)
	propellant	s. a. P 297		
P 589	property of material, material property	Materialeigenschaft f, Stoffeigenschaft f, Werkstoffeigenschaft f	propriété f de matériau (matière)	свойство материала (вещества)
P 590	proper value	Eigenwert m	valeur f propre	собственное значение
P 591	proportionality factor	Verhältniszahl f	nombre m proportionnel	передаточное число, фактор пропорциональности
P 592	proportioning	Dimensionierung f	dimensionnement m	определение геометрических размеров, расчет
P 593	proportioning procedure	Dimensionierungsverfahren n	procédé m de dimensionnement	метод расчета, процедура расчета размеров
	proportioning pump	s. B 135		
	protection coat	s. A 512		
P 594	protective coating	Schutzüberzug m	gaine (enveloppe) f protectrice	защитное покрытие
	protective gas	s. I 122		
	provision	S 686		
P 595	pseudoplastic	quasiplastisch, pseudoplastisch	quasi-plastique	квазипластичный, псевдопластичный
P 596	pseudosolution	kolloidale Lösung f	solution f colloidale	коллоидный раствор
P 597	psychrometry	Luftfeuchtigkeitsmessung f	psychrométrie f, hygrométrie f, hygroscopie f	измерение влажности воздуха, психрометрия
P 598	pulldown operation	Temperaturabsenkungsprozeß m	opération f d'abaissement de température	процесс понижения температуры
P 599	pulldown period	Temperaturabsenkungsperiode f, Abkühlperiode f	période f de refroidissement	период понижения температуры, период охлаждения
	puller	s. E 332		
	pulling force	s. T 87		
P 600	pulse column	pulsierende Kolonne f, Pulskolonne f	colonne f pulsée (à pulsations)	колонна с пульсирующим потоком
P 601	pulsed supply	schubweise Versorgung f	alimentation f pulsée (par pulsation)	пульсирующее питание, пульсирующее снабжение
P 602	pulverization	Pulverisierung f	pulvérisation f	порошкообразование
	pulverize/to	s. C 869		
	pulverized coal burner	s. C 310		
P 603	pulverized coal firing	Kohlenstaubfeuerung f	chauffage m au poussier de charbon	пылеугольная топка
	pulverizer	s. M 215		
	pulverizing	s. G 151		
P 604	pumpback reflux	zirkulierender Rücklauf m	reflux m circulant (repompé)	рециркуляционная флегма, рециркуляционный обратный поток
P 605	pump delivery	Pumpenförderung f	transport m par pompes	насосная подача
	pumped-storage hydropower plant	s. P 610		
	pumping head	s. D 152		
P 606	pumping plant	Pumpenanlage f	installation f de pompes, station f de pompage	насосная установка
P 607	pumping speed	Sauggeschwindigkeit f	vitesse f d'aspiration	скорость всасывания

P 608	pump out/to, to evacuate	abpumpen, leerpumpen	pomper, vider par pompage, évacuer	откачивать, откачать
P 609	pump speed	Pumpendrehzahl *f*	vitesse *f* de pompe	число оборотов насоса
P 610	pump storage station, pumped-storage hydro-power plant	Pumpspeicherwerk *n*	usine *f* d'accumulation par pompage	насосноаккумулирующая электростанция
P 611	puncture strength	Durchschlagfestigkeit *f*	rigidité *f* diélectrique	прочность на пробой, пробивная прочность
P 612	pure gas	Reingas *n*	gaz *m* purifié	очищенный (чистый) газ
P 613	pure shear	reine Scherung *f*	cisaillement *m* pur	чистый сдвиг
P 614	purge system	Reinigungssystem *n*, Spülsystem *n*	système *m* de purge (rinçage)	система очистки, продувная (промывная) система
P 615	purging	Austreiben *n*	expulsion *f*	вытеснение, продувка, выталкивание
P 616	purification	Reinigung *f*, Purifikation *f*, Raffination *f*, Reindarstellung *f*	purification *f*, préparation *f* à l'état pur, nettoyage *m*	очистка, ректификация, рафинация, приготовление чистого продукта
	purification	*s. a.* W 16		
P 617	purity test	Reinheitsprüfung *f*	test *m* de pureté	испытание на чистоту
P 618	pushing conveyance	Schubförderung *f*	transport *m* par poussée	порционная (периодическая) транспортировка, толкающий (толчковый) транспорт
	pushing force	*s.* T 202		
	push out/to	*s.* D 381		
	put in/to	*s.* B 75		
P 619	putting into operation	Inbetriebnahme *f*	mise *f* en marche (exploitation)	пуск в ход, ввод в эксплуатацию
	putting out of action	*s.* S 694		

Q

Q 1	qualitative analysis	qualitative Analyse *f*	analyse *f* qualitative	качественный анализ
	quality inspection	*s.* A 67		
Q 2	quality loss	Güteverlust *m*, Qualitätsminderung *f*	perte *f* de qualité	потеря качества
Q 3	quantitative analysis	quantitative Analyse *f*	analyse *f* quantitative, dosage *m*	количественный анализ
Q 4	quantity flow-sheet	Mengenfließbild *n*	flow-sheet *m* quantitatif	схема количественных потоков (процесса)
Q 5	quantity of cooling water required	Kühlwasserbedarf *m*	débit *m* d'eau réfrigérante nécessaire	необходимое количество охлаждающей воды, потребность в охлаждающей воде
Q 6	quantity of electricity	Elektrizitätsmenge *f*	quantité *f* d'électricité	количество электричества
Q 7	quantity of gas	Gasmenge *f*	quantité *f* de gaz	количество газа
Q 8	quantity of heat	Wärmemenge *f*	quantité *f* de chaleur	количество тепла
Q 9	quantity of lubricant	Schmiermittelmenge *f*	quantité *f* de lubrifiant	количество смазочного материала
Q 10	quantum unity, mass unit	Masseneinheit *f*	unité *f* de masse	единица массы
Q 11	quantum yield	Quantenausbeute *f*	rendement *m* quantique	квантовый выход
Q 12	quasi-continuous operation	quasikontinuierlicher Betrieb *m*	exploitation *f* quasi-continue	квазинепрерывное производство
Q 13	quasi-homogeneous	quasihomogen	quasi-homogène	квазигомогенный
Q 14	quasi-homogeneous reactor model	quasihomogenes Reaktormodell *n*	modèle *m* de réacteur quasi-homogène	квазигомогенная модель реактора
Q 15	quasi-stable	quasistabil	métastable	квазиустойчивый
Q 16	quasi-static	quasistatisch	quasi-statique	квазистатический
Q 17	quasi-stationary	quasistationär	quasi-stationnaire	квазистационарный
Q 18	quasi-stationary behaviour	quasistationäres Verhalten *n*	comportement *m* quasi-stationnaire	квазистационарное поведение
Q 19	quench/to	löschen, abschrecken, kühlen	tremper, refroidir	закаливать, охлаждать, тушить, гасить
Q 20	quench column	Abschreckturm *m*, Abschreckkolonne *f*	tour (colonne) *f* de refroidissement brusque	квенчинг-башня, закалочная колонна
Q 21	quenched spark gap	Löscheinrichtung *f*	installation *f* d'extinction	искрогасительное (затушительное) устройство
	quencher	*s.* Q 24		
Q 22	quenching, cooling	Abschreckung *f*	refroidissement *m* brusque	закалка, быстрое охлаждение
Q 23	quenching	Dämpfung *f* (Unterdrückung einer Reaktion)	amortissement *m*	подавление, затухание
Q 24	quenching medium, quencher	Abkühlmittel *n*	agent *m* réfrigérant	охладитель, закалочная среда
Q 25	quenching oil	Abschrecköl *n*	huile *f* de trempe	закалочное масло
Q 26	quenching temperature	Abschrecktemperatur *f*	température *f* de refroidissement brusque	температура закалки
Q 27	quickness, velocity of detonation	Detonationsgeschwindigkeit *f*	vitesse *f* de détonation	скорость детонации (взрыва)

R

R 1	rabble/to	durchrühren	brasser, agiter	перемешивать
R 2	raceway coil	Parallelrohrkühler m	refroidisseur m aux tuyaux parallèles	змеевик из параллельных труб
R 3	racking	Abfüllen n (in ein Faß)	décuvage m, mise f en tonneau	разлив, расфасовка
R 4	racking plant	Abfüllanlage f	installation f de soutirage	разливочная машина (установка)
R 5	radial bag filter	Rundfilter n	filtre m circulaire	круглый фильтр
R 6	radial flow	Radialströmung f	écoulement (courant) m radial	радиальное течение
R 7	radiant energy	Strahlungsenergie f	énergie f rayonnante (rayonnée)	энергия излучения, радиационная энергия
R 8	radiant energy transfer	Strahlungsenergieübertragung f	transfert m énergétique par rayonnement	перенос энергии излучением, радиационный энергоперенос, радиационная энергопередача
R 9/10	radiant heat transfer	Strahlungswärmeübertragung f	transfert m de chaleur rayonnante	радиационная теплопередача (тепловая передача)
R 11	radiation shield	Strahlungsabschirmung f, Strahlungsschutz m	protection f contre les rayonnements	радиационный щит, защита от радиации
R 12	radioactive waste	radioaktiver Abfall m	déchet m radioactif	радиоактивные отбросы
R 13	raffinate	Raffinat n	produit m raffiné	рафинированный продукт, рафинат
R 14	railroad spur, siding	Gleisanschluß m	ligne f de raccordement	соединение путей, стык, ветка
R 15	raising of the boiling point	Siedepunkterhöhung f	élévation f du point d'ébullition	повышение точки кипения
R 16	ram air	aufprallende Luft f	air m d'impact	набегающий воздух
R 17	ram extruder	Kolbenstrangpresse f	extrudeuse f à piston	поршневой экструдер
R 18	ram extrusion	Kolbenstrangpressen n	extrusion f à piston	поршневая экструзия
R 19	ram-feed injection moulding machine	Kolbenspritzgießmaschine f	presse à injecter à piston	поршневая машина литья под давлением, поршневая литьевая машина под давлением
R 20	ram pressure	Staudruck m	pression f dynamique	напор встречного потока
R 21	random	stochastisch, regellos, zufällig	aléatoire, irrégulier, par hasard	случайный, стохастический, беспорядочный
R 22	random-dumped tower packing	zufällig angeordnete Kolonnenschüttung f	garnissage m de colonne à disposition accidentelle	насадка со случайным распределением
R 23	random failure	Zufallsausfall m	défaillance f accidentelle	случайный отказ
R 24	random filling	regellose Füllung f	garnissage m désordonné	беспорядочная (случайная) загрузка, беспорядочная насадка
R 25	randomization	zufällige Anordnung f	disposition f par hasard	рандомизация
R 26	randomize/to	zufällig zuordnen	disposer au hasard	рандомизировать
R 27	random sample	Stichprobe f	échantillon m pris au hasard	выборка
R 28	random variable	Zufallsgröße f	grandeur f aléatoire	случайная величина
R 29	range of loading	Belastungsbereich m, Beanspruchungsbereich m	domaine m de sollicitation	область нагрузки, диапазон эксплуатации
R 30	range of sensitivity	Empfindlichkeitsbereich m	domaine m de sensibilité	область чувствительности
R 31	range of temperature measurement	Temperaturmeßbereich m	domaine m de mesure de température, champ m de thermométrie	интервал измерения температуры
R 32	Rankine cycle	Rankine-Prozeß m	cycle m de Rankine	цикл Ранкина
R 33	Rankine temperature scale	Rankine-Temperaturskale f	échelle f de température Rankine	шкала температур по Ранкину
R 34	Ranque effect	Ranque-Effekt m	effet m Ranque	эффект Ранка
R 35	Raoult's law	Raoultsches Gesetz n	loi f de Raoult	закон Рауля
R 36	rapid filter	Schnellfilter n	filtre m rapide	быстродействующий фильтр
R 37	rarefaction of gases	Verdünnen n von Gasen	raréfaction f des gaz	разрежение газов
R 38	rarefied gas	verdünntes Gas n	gaz m raréfié	разреженный газ
R 39	rare gas, noble (inactive) gas	Edelgas n	gaz m rare (noble)	благородный (инертный) газ
R 40	Raschig ring	Raschig-Ring m	anneau m Raschig	кольцо Рашига
R 41	rated capacity	Nennkapazität f	capacité f nominale	номинальная мощность
R 42	rated current	Nennstrom m	intensité f nominale	номинальный ток
R 43	rated load	Nennbelastung f	charge f nominale	номинальная нагрузка
R 44	rated output (power)	Nennleistung f	débit m nominal, puissance f normale	условная мощность
R 45	rated speed	Nenngeschwindigkeit f, Nenndrehzahl f	vitesse f nominale	условное число оборотов, номинальная скорость, номинальное число оборотов в минуту
R 46	rated voltage	Nennspannung f	tension f nominale	номинальное напряжение
R 47	rate of combustion, burning rate, combustion rate (velocity)	Verbrennungsgeschwindigkeit f	vitesse f de combustion	скорость сгорания (горения)
R 48	rate of condensation	Kondensationsgeschwindigkeit f	vitesse f de condensation	скорость конденсации

R 49	rate of descent	Sinkgeschwindigkeit f, Absetzgeschwindigkeit f	vitesse f de dépôt, vitesse de décantage	скорость осаждения
	rate of flow	s. F 255		
R 50/1	rate of formation	Bildungsgeschwindigkeit f	vitesse f de formation	скорость образования
R 52	rate of material change	Stoffänderungsgeschwindigkeit f, Umsetzungsrate f	vitesse f de changement de matière, taux m de conversion	скорость превращения веществ, скорость химической реакции
R 53	rate of melting	Schmelzgeschwindigkeit f	taux m (vitesse f) de fusion	скорость таяния (плавления)
R 54	rate of reaction	Reaktionsgeschwindigkeit f, Umsatz m	vitesse f de réaction, taux m de conversion	скорость реакции (превращения)
R 55	rate of runoff	Abflußmenge f	débit m de l'écoulement	расход, количество стока
R 56	rate of solution	Lösungsgeschwindigkeit f	vitesse f de dissolution	скорость растворения
R 57	rate of transformation	Umwandlungsgeschwindigkeit f	vitesse f de transformation	скорость превращения (конверсии)
R 58	rate of vaporization	Verdampfungsgeschwindigkeit f, Verdampfungsrate f	vitesse f d'évaporation, vitesse (taux m) de vaporisation	скорость испарения
R 59	rate of vertical descent	Fallgeschwindigkeit f	vitesse f de chute	скорость падения
	rating conditions	s. D 187		
R 60/1	rating curve	Abflußkurve f, Durchsatzkurve f	courbe f de débit	кривая стока
	ratio of compression	s. C 492		
R 62	raw energy	Primärenergie f	énergie f brute (primaire)	первичная энергия
R 63	raw material, feed[stock], base (initial) product, basic (starting) material	Ausgangsprodukt n, Ausgangsstoff m, Rohmaterial n, Rohstoff m, Grundmaterial n, Einsatzgut n, Einsatzprodukt n	produit m de départ (base), produit initial, matière brute (première), matériau m de base, charge f	исходный продукт, сырье, исходное вещество, основной материал, основное сырье
R 64	raw material distribution	Rohstoffverteilung f	distribution f de matière brute	распределение сырья
R 65	raw product	Rohprodukt n	produit m brut	сырой продукт, сырье
R 66	reactant, reaction component	Reaktant m, Reaktionspartner m, Reaktionskomponente f, Reagens n	composante f de réaction, réactif m	реагент, участник (компонент) реакции
R 67	reaction	chemische Reaktion f	réaction f chimique	реакция
R 68	reaction	Reaktion f (Rückstoß)	coup m en arrière, réaction f	отдача, реактивный удар
R 69	reaction	s. a. 1. B 76; 2. T 349		
	reaction chain	Reaktionskette f	chaîne f de réactions	цепь реагирующих веществ, цепная реакция
R 70	reaction chamber	Reaktionskammer f	chambre f de réaction	реакционная камера
	reaction component	s. R 66		
R 71	reaction condition	Reaktionsbedingung f	condition f de réaction, condition réactionnelle	условие реакции
R 72	reaction constant	Reaktionskonstante f	constante f de réaction	константа реакции
R 73	reaction cycle	Reaktionszyklus m	cycle m réactionnel (de réaction)	цикл реакций, реакционный цикл
R 74	reaction effect	Reaktionswirkung f	effet m de réaction	эффект реакции
R 75	reaction energy	Reaktionsenergie f	énergie f de réaction	энергия реакции
R 76	reaction enthalpy	Reaktionsenthalpie f	enthalpie f de réaction	энтальпия реакции
R 77	reaction entropy	Reaktionsentropie f	entropie f de réaction	энтропия реакции
R 78	reaction equation	Reaktionsgleichung f	équation f réactionnelle	уравнение реакции
R 79	reaction force	Reaktionskraft f, Rückstoß m	force f de réaction	сила реакции, сила отдачи
R 80	reaction forming	Reaktionsformen n	conformation f par réaction	реакционное формирование (формование)
R 81	reaction front	Reaktionsfront f	front m de réaction	фронт реакции, реакционный фронт
R 82	reaction kinetics	Reaktionskinetik f	cinétique f réactionnelle (de réaction)	кинетика реакции
R 83	reaction liquid	Reaktionsflüssigkeit f	liquide m de réaction	жидкий реактив
R 84	reaction mass	Reaktionsmasse f	masse f provenant d'une réaction, masse réactionnelle	реакционная масса
R 85	reaction mechanism	Reaktionsmechanismus m	mécanisme m de réaction	механизм реакции
R 86	reaction medium	Reaktionsmedium n	milieu m réactionnel	реакционная среда, среда реакции
R 87	reaction of formation	Bildungsreaktion f	réaction f de formation	реакция образования
R 88	reaction pressure	Reaktionsdruck m	pression f de réaction	давление реакции
	reaction pressure	s. a. C 799		
R 89	reaction product	Reaktionsprodukt n	produit m réactionnel (de réaction)	продукт реакции
R 90	reaction rate equation	Reaktionsgeschwindigkeitsgleichung f	équation f de vitesse de réaction	уравнение скорости химической реакции
R 91	reaction scheme	Reaktionsschema n	schéma m réactionnel	схема реакций
R 92	reaction space	Reaktionsraum m	chambre f de réaction	реакционное пространство
R 93	reaction stage	Reaktionsstufe f	degré m de réaction	ступень реакции
R 94	reaction time	Reaktionsdauer f, Reaktionszeit f	temps m de réaction	время (длительность, продолжительность) реакции
	reaction time	s. a. I 97		
R 95	reaction tower	Reaktionsturm m, Reaktionskolonne f	colonne f à réaction	реакционная башня (колонка)
R 96	reaction type	Reaktionstyp m	type m réactionnel (de réaction)	тип реакции

R 97	reaction vessel	Reaktionsgefäß n	récipient m de réaction	реактор, реакционный сосуд
R 98	reaction volume	Reaktionsvolumen n	volume m de réaction	объем реакции
R 99	reaction zone	Reaktionszone f	zone f de réaction	зона реакции
R 100	reactivation temperature	Regenerationstemperatur f, Reaktivierungstemperatur f	température f de réactivation	температура регенерации (реактивации)
R 101	reactivity, activity	Reaktionsfähigkeit f, Reaktionsvermögen n	pouvoir m de réaction, réactivité f	реакционная способность, реактивность
R 102	reactor	chemischer Reaktor m	réacteur m [chimique]	реактор
R 103	reactor, catalytic reactor	Kontaktapparat m	appareil m de contact, réacteur m catalytique	контактный аппарат
R 104	reactor calculation, reactor computation	Reaktorberechnung f	calcul m de réacteur	расчет реактора
R 105	reactor cell	Reaktorzelle f	cellule f de réacteur	ячейка реактора
	reactor computation	s. R 104		
R 106	reactor coolant	Reaktorkühlmittel n	réfrigérant m de réacteur	хлад[о]агент реактора
R 107	reactor cooling	Reaktorkühlung f	réfrigération f de réacteur	охлаждение реактора
R 108	reactor cooling system	Reaktorkühlsystem n	système m réfrigérant de réacteur	система охлаждения реактора
R 109	reactor design	Reaktorkonstruktion f	construction f de réacteur	конструкция реактора
R 110	reactor dynamics	Reaktordynamik f, Dynamik f von Reaktoren	dynamique f de réacteur	динамика реактора
R 111	reactor element	Reaktorelement n	élément m de réacteur	элемент реактора, тепловыделяющий элемент
R 112	reactor kettle	Reaktorkessel m	cuve f réactionnelle	бак реактора
R 113	reactor loading	Reaktorbeschickung f	alimentation f de réacteur	загрузка реактора
R 114	reactor modelling	Reaktormodellierung f	modélisation f de réacteur	моделирование реактора
R 115	reactor shell	Reaktormantel m	enveloppe f de réacteur	рубашка реактора
R 116	reactor sizing	Reaktordimensionierung f	dimensionnement m de réacteur	расчет реактора
R 117	reactor theory	Reaktortheorie f	théorie f des réacteurs	теория реакторов
R 118	reactor volume	Reaktorvolumen n	volume m de réacteur	объем реактора
R 119	reactor wall	Reaktorwand f	paroi f de réacteur	стена реактора
R 120	reactor with outerloop	Reaktor m mit äußerem Kreislauf	réacteur m à recirculation extérieure	реактор со внешней циркуляцией
R 121	readiness for operation	Betriebsbereitschaft f	état m de service, ordre m de marche	готовность производства
R 122	reading error	Ablesefehler m	erreur f de lecture	ошибка отсчета
R 123	readjustability	Nachstellbarkeit f	réglabilité f, ajustabilité f	регулируемость
R 124	readjustment	Nachregelung f	[r]ajustage m	дополнительная регулировка
R 125/6	reagent	Reaktionsmittel n	agent m de réaction	реактив, реагент
	real gas	s. A 127		
	realization	s. C 603		
R 127	realization extent	Realisierungsumfang m	étendue f de réalisation	объем реализации (строительства)
R 128	realization time	Realisierungsdauer f	temps m (durée f) de réalisation	длительность реализации, продолжительность строительства
R 129	real plate	wirklicher Boden m	plateau m réel	реальная тарелка
R 130	real time	Echtzeit f	temps m réel	реальное истинное время
R 131	real volume stream	tatsächlicher Volumenstrom m	débit m volumique réel	действительный объемный поток (расход)
R 132	rearrangement	Umlagerung f	arrangement m nouveau	перегруппировка
R 133	reboiler, evaporator, vaporizer	Destillationsgefäß n, Destillierblase f, Blase f, Verdampfungsofen m, Aufkocher m, Aufkochofen m, Rückverdampfer m, Wiederaufkocher m, Reboiler m, Aufwärmer m	vaporiseur m, vaporisateur m, rebouilleur m, appareil chauffant	испаритель, кипятильник, рибойлер, подогреватель
R 134	rebound, recoil	Rückprall m, Zurückprallen n	rebondissement m	отскок, отдача
	recalculation	s. C 674		
R 135	receiver	Aufnehmer m	absorbeur m	приемник, абсорбер
R 136	receiver	Vorlage f	récipient m	приемник, сборник, бак
	receiver	s. a. 1. B 28; 2. B 77; 3. S 305		
R 137	receiving bin	Auffanggefäß n	bac m collecteur (de réception)	приемник, приемный сосуд
R 138	receiving container	Aufnahmebehälter m	container m réceptif	приемник, приемный резервуар
R 139	receiving hopper	Auffangtrichter m (für Schüttgut)	trémie f réceptive	сборная (приемная) воронка
R 140	reciprocating compressor	Kolbenverdichter m, Kolbenkompressor m	compresseur m à piston	поршневой компрессор
R 141	recirculate/to	rückpumpen, zurückführen, im Kreislauf fahren, umwälzen	recirculer	рециркулировать, циркулировать в замкнутом контуре
R 142	recirculated air	Umluft f	air m circulé (de circulation)	циркуляционный воздух
R 143	recirculated air cooler	Umluftkühler m	refroidisseur m d'air circulé	охладитель рециркулирующего воздуха
R 144	recirculation evaporator	Umlaufverdampfer m	évaporateur m à recirculation	циркуляционный выпарной аппарат

R 145	recirculation ratio	Umlaufverhältnis n	taux m de récirculation	коэффициент рециркуляции
R 146	recirculation type	Rezirkulationstyp m	type m de recirculation	тип [ре]циркуляции
R 147	reclaim/to	zurückgewinnen, regenerieren	récupérer, régénérer	регенерировать
R 148	reclaim	Regenerat n, regenerierter Kautschuk m, Regenerativgummi m	matière f régénérée, caoutchouc m régénéré, régénéré m	регенерат
R 149	reclaim cycle	Regenerationsprozeß m, Reaktivierung f	cycle m de réactivation (régénération)	цикл регенерации
	reclaiming	s. R 244		
R 150	recognition method	Erkennungsverfahren n	méthode f de reconnaissance	метод распознавания
	recoil	s. R 134		
R 151	reconstruction measures	Rekonstruktionsmaßnahmen fpl	mesures fpl de reconstruction	мероприятия по реконструкции
R 152	recooler	Sekundärkühler m	refroidisseur m secondaire	вторичный охладитель
R 153	recooling	Rückkühlen n	refroidissement m de retour	охлаждение циркулирующей водой
R 154	record grouping, block formation	Blockbildung f	groupement m d'enregistrements	образование блоков, запись блоками
R 155	recoverable deformation	reversible Deformation f	déformation f réversible	обратимая деформация
	recovery	s. 1. D 501/2; 2. R 166		
R 156	recovery economy	Aufbereitungswirtschaft f, Rückgewinnungswirtschaft f	économie f de récupération	рекуперационное (обогатительное) хозяйство
R 157	recovery factor	Wiedergewinnungsrate f	taux m de récupération	коэффициент регенерирования
R 158	recovery of solvent	Lösungsmittelrückgewinnung f	récupération f du solvant	рекуперация растворителей
R 159	recrystallization	Umkristallisieren n	recristallisation f	перекристаллизование
R 160	recrystallize/to	auskristallisieren	séparer par cristallisation	выкристаллизовывать
R 161	rectification	Rektifikation f, Gegenstromdestillation f	rectification f	ректификация
R 162	rectifying apparatus	Rektifizierapparat m, Rektifikationsapparat m	appareil m de rectification	ректификационный аппарат
R 163	rectifying column	Rektifikationskolonne f	colonne f à rectifier, colonne de rectification	ректификационная колонна
R 164	rectilinear flow	geradlinige Strömung f	écoulement (courant) m rectiligne	прямолинейное течение
R 165	rectilinear shear flow	geradlinige Scherströmung f	écoulement m rectiligne de cisaillement	прямолинейное сдвиговое течение
R 166	recuperation, recovery	Wiedergewinnung f, Rückgewinnung f	récupération f	рекуперация
	recuperator	s. R 247		
R 167	recyclable waste material	Sekundärrohstoffe mpl, Altrohstoffe mpl	matériaux mpl secondaires, vieux materiaux mpl	вторичное (отработанное) сырье
R 168	recycle/to	umpumpen, zurückführen, rückführen, wiedereinspeisen	recycler	[ре]циркулировать
R 169	recycle blower	Gasumlaufgebläse n	ventilateur m de recyclage	нагнетатель циркуляционного газа
R 170	recycle pump	Umlaufpumpe f, Umwälzpumpe f	pompe f de recirculation	циркуляционный насос
R 171	recycle reactor	Schlaufenreaktor m	réacteur m à circulation	циркуляционный реактор
R 172	recycling	Rückführung f, Wiederverwendung f, Kreislauffahrweise f	recirculation f	рецикл, рисайклинг
R 173	recycling	Abfallverwertung f	utilisation f des déchets, recyclage m des déchets	утилизация отбросов
	redissolution	s. R 319		
	redistillation	s. C 353		
	reduced capacity	s. D 291		
R 174	reduced crude cracking process	Verfahren n zum Kracken von atmosphärischem Rückstand	processus m de craquage du résidu atmosphérique	крекинг-процесс остатка от атмосферной дистилляции
R 175	reduced scale	verkleinerter Maßstab m	échelle f réduite	уменьшенный масштаб
R 176	reducible	reduzierbar	réductible	восстанавливаемый, восстановляющийся, редуцируемый, сводимый
R 177	reduction	Abbau m, Reduktion f, Zerkleinerung f, Verengung f, Reduzierung f, Einschränkung f	séparation f, dissolution f, dégradation f, décomposition f, réduction f	редукция, восстановление, приведение, понижение, измельчение
	reduction	s. a. D 44		
R 178	reduction ratio	Abbaugrad m, Zersetzungsrate f	taux m de décomposition	степень разложения (деструкции)
R 179	redundancy degree	Redundanzgrad m	degré m de redondance	масштаб резервирования
R 180	reference electrode	Bezugselektrode f	électrode f de comparaison	электрод сравнения
R 181	reference element	Bezugselement m	élément m de référence	условный элемент
R 182	reference level	Bezugsniveau n	niveau m de référence	условный горизонт, уровень отсчета
R 183	reference magnitude	Bezugsmaß n	mesure f de référence	исходный размер

R 184	reference point	Bezugspunkt *m*	point *m* de référence, référence *f*	репер, условная точка
R 185	reference pressure	Bezugsdruck *m*	pression *f* de référence	исходное давление
R 186	reference system	Bezugssystem *n*	système *m* de référence	основная система
R 187	reference temperature	Bezugstemperatur *f*	température *f* de référence	исходная температура
R 188	refine/to	abtreiben, läutern	coupeller, ressuer	отгонять, перегонять
	refine/to	s. a. I 61		
R 189	refinery	Raffinerie *f*, Raffinationsanlage *f*	raffinerie *f*, installation *f* de raffinerie	рафинировочная установка
R 190	refinery gas	Raffineriegas *n*	gaz *m* de raffinerie	нефтезаводской газ
R 191	refinery gas decomposition process	Raffineriegaszerlegungsverfahren *n*	processus *m* de décomposition du gaz de raffinage	способ фракционирования [нефтяного] газа
R 192	refinery scheduling	Raffinerieproduktionsplan *m*	programme *m* de production de raffinerie	производственный план нефтеперерабатывающего завода
R 193	refining	Raffination *f*, Raffinieren *m*, Läuterung *f*	raffinage *m*, affinage *m*, épurement *m*	рафинация, очистка, рафинирование, очищение, купеляция
R 194	refining furnace	Abtreibeofen *m*	fourneau *m* d'affinerie	купеляционная печь
R 195	refining loss	Verarbeitungsverlust *m*, Raffinerieverlust *m*	perte *f* de raffinage	потери при нефтепереработке
R 196	refining plant	Raffinationsbetrieb *m*	installation *f* de raffinage	рафинационная установка
R 197	refining process	Raffinationsprozeß *m*	procédé *m* de raffinage	процесс рафинирования
R 198	reflux	Rückfluß *m*, Rücklauf *m*	reflux *m*	флегма, орошение, обратный поток
	reflux condenser	s. D 171		
R 199	reflux ratio	Rücklaufverhältnis *n*	taux *m* de reflux	коэффициент возврата (дефлегмации), флегмовое число
R 200	reflux valve	Rückflußventil *n*	soupape (valve) *f* de reflux	обратный клапан, вентиль орошения
R 201	reformer, reforming plant	Reformer *m*, Reformieranlage *f*	unité *f* de reforming	риформинг, установка риформинга
R 202	refractive index	Brechungszahl *f*, Brechungsindex *m*	indice *m* de refraction	индекс преломления
	refractory	s. F 167		
R 203/4	refractory brick	feuerfester Stein *m*, Schamottestein *m*, Schamotteziegel *m*	brique *f* réfractaire	огнеупорный камень
	refrigerant	s. C 701		
R 205	refrigerant absorber	Kältemittelabsorber *m*	absorbeur *m* de réfrigérant	абсорбер холодильного агента
R 206	refrigerant absorption	Kältemittelabsorption *f*	absorbtion *f* de réfrigérant	абсорбция холодильного агента
R 207	refrigerant bath	Kältemittelbad *n*	bain *m* de fluide frigorigène	холодильная баня, баня холодильного агента
R 208	refrigerant bleed	Kältemittelabführung *f*	soutirage *m* de fluide frigorigène	спуск холодильного агента
R 209	refrigerant demand	Kältemittelbedarf *m*	demande *f* de fluide frigorigène	потребность в холодильном агенте
R 210	refrigerant effluent	Kältemittelausströmung *f*	effluent *m* de fluide frigorigène	вытекающий поток холодильного агента
R 211	refrigerant escape	Kältemittelaustritt *m*	échappement *m* de fluide frigorigène	утечка холодильного агента
R 212	refrigerant head	Kältemitteldruck *m*	pression *f* de fluide frigorigène	напор холодильного агента
R 213	refrigerant quantity	Kältemittelmenge *f*	quantité *f* de fluide frigorigène	количество холодильного агента
R 214	refrigerant vapour condensation	Kältemitteldampfkondensation *f*	condensation *f* de vapeur de fluide frigorigène	конденсация пара холодильного агента
R 215	refrigerated processing	Kältebearbeitung *f*	traitement *m* à froid	холодильная обработка
R 216	refrigerated transport	Kühltransport *m*	transport *m* frigorifique	холодильный транспорт
R 217	refrigerating apparatus	Kälteerzeuger *m*	appareil *m* de réfrigération, producteur *m* de froid	холодильный аппарат
R 218	refrigerating arrangement	Kühlanlage *f*	installation *f* frigorifique, installation de réfrigération, réfrigérant *m*	холодильная установка
R 219	refrigerating capacity	Kälteleistung *f*	capacité *f* frigorifique	холодопроизводительность
R 220	refrigerating influence	Kälteeinfluß *m*	influence *f* du froid	влияние холода
R 221	refrigerating machine	Kältemaschine *f*	machine *f* frigorifique (à froid)	холодильная машина
R 222	refrigerating plant	Kälteanlage *f*	installation *f* frigorifique	холодильная установка
R 223	refrigerating plant load	Kälteanlagenwärmelast *f*	charge *f* calorifique de l'installation frigorifique	тепловая нагрузка холодильной установки
R 224	refrigerating system efficiency	Kälteanlagenwirkungsgrad *m*	rendement *m* de l'installation frigorifique	коэффициент полезного действия холодильной установки
R 225	refrigerating temperature	Kühltemperatur *f*, Abkühltemperatur *f*	température *f* de réfrigération, température *f* de refroidissement	температура охлаждения
R 226	refrigeration, cold production	Kühlung *f*, Kälteerzeugung *f*	réfrigération *f*, refroidissement *m*, production *f* du froid	охлаждение, замораживание, производство [искусственного] холода

	refrigeration		*s. a.* C 700	
R 227	refrigeration compressor	Kälteverdichter *m*, Kältekompressor *m*	compresseur *m* frigorifique	холодильный компрессор
R 228	refrigeration cycle	Kälteprozeß *m*	cycle *m* frigorifique	холодильный цикл
R 229	refrigeration demand	Kühlbedarf *m*	demande *f* de refroidissement	потребность в холоде
R 230	refrigeration distribution	Kälteverteilung *f*	distribution *f* du froid	распределение холода
R 231	refrigeration engineering	Kältetechnik *f*	technique *f* de froid	холодильная техника
R 232	refrigeration industry	Kälteindustrie *f*	industrie *f* frigorifique	холодильная промышленность
R 233	refrigeration losses	Kälteverluste *mpl*	pertes *fpl* du froid	потери холода
R 234	refrigeration range	Kühltemperaturbereich *m*	plage *f* de réfrigération	диапазон температур охлаждения
R 235	refrigeration source	Kältequelle *f*	source *f* du froid	холодный источник
R 236	refrigeration technique	Kühltechnik *f*	technique *f* de refroidissement	технология охлаждения
	refrigerator		*s.* C 698	
R 237	refuse, waste	Müll *m*, Abfall *m*	immondices *fpl*, ordures *fpl* ménagères, déchets *mpl* industriels	отброс, отход, мусор
R 238	refuse collecting plant	Abfallsammelanlage *f*	installation *f* de collecte pour les déchets	установка для сбора отходов
	refuse combustion		*s.* R 241	
R 239	refuse disposal	Abfallvernichtung *f*	élimination *f* des déchets	утилизация отходов
R 240	refuse disposal plant	Abfallvernichtungsanlage *f*	installation *f* pour l'élimination des déchets	установка по утилизации отходов
R 241	refuse incineration, refuse combustion	Müllverbrennung *f*	combustion *f* des déchets	сжигание мусора
R 242	refuse processing	Müllverwertung *f*, Müllbehandlung *f*	traitement *m* des ordures	переработка отходов (мусора)
R 243	regasification	Wiedervergasung *f*	regazéification *f*	регазификация
R 244	regeneration, reclaiming	Regenerierung *f*, Reaktivierung *f*	régénération *f*, épuration *f*	регенерация
R 245	regeneration device	Regenerationsgerät *n*	dispositif *m* de régénération	регенерация, установка для регенерации
R 246	regenerative furnace	Speicherofen *m*	four *m* à régénération	регенеративная печь
R 247	regenerator, recuperator	Winderhitzer *m*, Abhitzeverwerter *m*	réchauffeur *m* d'air, régénérateur *m*	котел-утилизатор
R 248	regression coefficient	Regressionskoeffizient *m*	coefficient *m* de régression	коэффициент регрессии
R 249	regular gasoline *(motor spirit)*	Normalbenzin *n*; [normales] Fahrbenzin *n*	essence *f* ordinaire, essence standarde	моторный бензин
R 250	regulating unit	Stelleinrichtung *f*	dispositif *m* (installation *f*) de réglage	исполнительный орган, исполнительный механизм
	regulation		*s.* I 252	
R 251	regulation speed	Regelgeschwindigkeit *f*	vitesse *f* de réglage	скорость регулировки
R 252	regulation systems	Regelanlagen *fpl*	installations *fpl* de régulation	регулирующие механизмы
R 253	regulator	Einsteller *m*	ajusteur *m*	установочное звено, регулятор
	regulator mixture		*s.* B 291	
R 254	regulator system	Puffersystem *n*	système *m* de tampon	буферная система
R 255	Reiner elastic body	Reinerscher elastischer Körper *m*	corps *m* élastique de Reiner	эластичное тело Рейнера
R 256	Reiner function	Reinersche Funktion *f*	fonction *f* de Reiner	функция Рейнера
R 257	Reiner-Philippoff model	Reiner-Philippoffsches Modell *n*	modèle *m* de Reiner-Philippoff	модель Рейнера-Филиппова
R 258	Reiner-Riwlin equation	Reiner-Riwlinsche Gleichung *f*	équation *f* de Reiner-Riwlin	уравнение Рейнера-Ривлина
R 259	reinforce/to, to concentrate, to intensify	verstärken	renforcer, armer, concentrer, amplifier	концентрировать, усилить, армировать
	reinforced		*s.* C 511	
R 260	reinforced concrete	Stahlbeton *m*	béton *m* armé	железобетон, сталебетон
	reinforcement		*s.* A 479	
R 261	relative density	Dichteverhältnis *n*	densité *f* relative	относительная плотность
R 262	relaxation function	Relaxationsfunktion *f*	fonction *f* de relaxation	функция релаксации
R 263	relaxation time spectrum	Relaxationszeitspektrum *n*	spectre *m* du temps de relâchement	спектр времени релаксации
R 264	reliability	Zuverlässigkeit *f*	fiabilité *f*	надежность
R 265	reliability analysis	Zuverlässigkeitsanalyse *f*	analyse *f* de fiabilité	анализ надежности
R 266	reliability function	Zuverlässigkeitsfunktion *f*	fonction *f* de fiabilité	функция надежности
R 267	reliability in service	Betriebszuverlässigkeit *f*	fiabilité *f* de service	надежность установки [цеха]
R 268	reliability-logical diagram	zuverlässigkeitslogisches Schema *n*	schéma *m* de fiabilité	надежностная схема
R 269	reliability parameter	Zuverlässigkeitskenngröße *f*	caractéristique *m* de fiabilité	показатель (параметр) надежности
R 270	reliability work	Zuverlässigkeitsarbeit *f*	travail *m* pour l'augmentation de la fiabilité	работа (исследование) по надежности
R 271	relief valve, safety valve	Entlastungsventil *n*, Überdruckventil *n*, Sicherheitsventil *n*	soupape *f* de sûreté	предохранительный клапан
R 272	remote control	Fernsteuerung *f*	commande *f* à distance, télécommande *f*	дистанционное управление

R 273	remote indicator	Fernanzeiger m	téléindicateur m	дистанционный указатель
R 274	remote pressure controller	Ferndruckregler m	régulateur m de pression à distance	дистанционный регулятор давления
R 275	remote temperature control	Ferntemperaturregelung f	régulation f de température à distance	дистанционное регулирование температуры
R 276	removal	Verdrängung f	déplacement m, refoulement m	вытеснение, водоизмещение
R 277	rendering inert	Inertisierung f	inertisation f	инертизация
R 278	renewal interval	Erneuerungsintervall m	intervalle m de renouvellement	интервал восстановления
R 279	renewal theory	Erneuerungstheorie f	théorie f de renouvellement	теория восстановления
R 280	renovation	Erneuerung f	remise f en état, renouvellement m	восстановление, реконструкция
R 281	repair conditions	Instandsetzungsbedingungen fpl	conditions fpl de réparation	условия ремонта, предпосылки технического обслуживания
R 282	repair cycle	Reparaturzyklus m	cycle m de réparation	ремонтный цикл
R 283	repair probability	Instandsetzungswahrscheinlichkeit f	probabilité f de réparation	вероятность ремонта
R 284	repair rate	Instandsetzungsrate f	taux m de réparation	интенсивность ремонта
R 285	repair rate	Instandsetzungszeit f	temps m de réparation	время ремонта
R 286	repair strategy	Reparaturstrategie f	politique f de réparation	ремонтная стратегия, стратегия по техническому обслуживанию
R 287	replaceable filter	Austauschfilter n	filtre m remplaçable (d'échange)	сменный фильтр
R 288	reprocessing	Wiederverwendung f, Wiederaufbereitung f	réutilisation f, retraitement m	регенерация
R 289	repulsive force	abstoßende Kraft f, Abstoßungskraft f	force f répulsive	отталкивающая сила
R 290	request, demand	Anforderung f	demande f, exigence f	запрос, требование
R 291	request button	Abfragetaste f	touche f de requête	клавиша запроса
	required output	s. N 49		
R 292	rerun	Redestillat n	redistillat m	вторичный перегон
R 293	research and development	Forschung f und Entwicklung f	recherche f et développement m	исследование и развитие
R 294	research centre	Forschungszentrum n	centre m de recherches	исследовательский центр
R 295	research method	Forschungsmethode f	méthode f de recherche	экспериментальный метод
R 296	research reactor	Forschungsreaktor m	réacteur m de recherche	исследовательский реактор
R 297	research work	Forschungsarbeit f	travail m de recherche	исследовательская работа
R 298	reserve element	Reserveelement n	élément m de réserve	резервный элемент
R 299	reserve of stability	Stabilitätsreserve f	réserve f de stabilité	запас устойчивости
R 300	reservoir, basin, tank	Tank m, Behälter m, Bassin n	réservoir m, citerne f, tank m	резервуар, емкость, танк, бак
	reservoir	s. a. 1. C 401; 2. S 710		
R 301	residence time	Verweilzeit f, Aufenthaltszeit f, Retentionszeit f, Durchgangszeit f	temps m de séjour	время пребывания (прохождения)
R 302	residence time spectrum	Verweilzeitspektrum n	spectre m (distribution f) de temps de séjour	спектр времени пребывания
R 303	residual air	Restluft f	air m résiduel	остаточный воздух
R 304	residual gas	Restgas n, Endgas n	gaz m résiduaire (final, résiduel, de queue)	остаточный газ
R 305	residual heat	Restwärme f	chaleur f restante	остаточное тепло
	residual product	s. B 332		
R 306	residual stress	Restspannung f	effort m résiduel, tension f résiduaire	остаточное напряжение
R 307	residual water	Restwasser n	eau f résiduelle	остаточная вода
R 308	residue after evaporating	Eindampfrückstand m	résidu m de vaporisation	остаток после выпарки
R 309	residue of rock oil	Erdölrückstand m	résidu m de pétrole	остаток нефти, нефтяной остаток
R 310	residue on evaporation	Trockenrückstand m	résidu m d'évaporation, résidu sec	сухой остаток
R 311	residuum	Rückstand m	résidu m	остаток
R 312	resilience	Nachgiebigkeit f, Anpassungsfähigkeit f	résilience f	эластичность, работоспособность, упругая деформация, вязкость
R 313	resin-injection moulding	Spritzgießen n, Einspritzverfahren n	moulage m sous pression, procédé m de moulage par injection	литье под давлением
R 314/5	resistance capacity	Widerstandskapazität f (Leitfähigkeitsgefäß)	constante f de la pile	емкость сопротивления
	resistance to flow	s. F 261		
R 316	resistance to tearing	Reißfestigkeit f	résistance f à la rupture, résistance à déchirure	сопротивление разрыву
R 317	resistance to wear	Verschleißfestigkeit f	résistance f à l'usure	сопротивление износу, прочность на износ, износостойкость
	resistant to adhesion	s. A 150		
	resistivity	s. S 471		
R 318	resizing	Nachklassierung f	reclassement m	дополнительная сортировка (классификация)

R 319	**resolution,** redissolution	Wiederauflösung *f*	redissolution *f*	обратное (повторное) растворение
R 320	**resolving power**	Auflösungsvermögen *n*	pouvoir *m* résolvant	разрешающая (растворяющая) способность
R 321	**respirator filter**	Atemschutzfilter *n*, Atemfilter *n*	filtre *m* respirateur	фильтрующий респиратор, фильтр респиратора
R 322	**respiratory protection**	Atemschutz *m*	protection *f* respiratoire	защита дыхательных путей
	response	*s.* B 76		
	rest	*s.* L 305		
R 323	**resultant product**	Folgeprodukt *n*	produit *m* résultant	продукт реакции
R 324	**resulting damage**	Folgeschaden *m*	dommage *m* de suite	результаты (последствия) повреждений
R 325	**result of research**	Forschungsergebnis *n*	résultats *mpl* de recherche	результат исследования
R 326	**retardation time spectrum**	Verzögerungszeitspektrum *n*	spectre *m* du temps de retard	спектр запаздывания
	retort	*s.* C 362		
	retort furnace	*s.* C 155		
R 327	**return connection**	Rückführschaltung *f*	rétrocouplage *m*, montage *m* rétroactif	схема с рециклом
R 328	**return piping**	Rückleitung *f*	conduite *f* de retour	обратный трубопровод
R 329	**return screen**	Rücklaufsieb *n*	crible *m* de retour	обратное сито
R 330	**return shock**	Rückschlag *m*	choc *m* en arrière, contre-coup *m*	отскок
R 331	**reverberatory furnace**	Flammenofen *m*	four *m* à réverbère	пламенная печь
R 332	**reverse reaction**	Rückreaktion *f*	réaction *f* inverse (d'inversion)	обратная реакция
R 333	**reversibility**	Reversibilität *f*, Umkehrbarkeit *f*	réversibilité *f*	обратимость
R 334	**reversibility degree**	Reversibilitätsgrad *m*	degré *m* de réversibilité	степень обратимости
R 335	**reversible expansion**	umkehrbare Entspannung *f*	détente *f* réversible	обратимое расширение
R 336	**reversible reaction**	Gleichgewichtsreaktion *f*, reversible Umsetzung *f*	réaction *f* équilibrée (d'équilibre)	равновесная реакция
	revised design	*s.* N 26		
	revision	*s.* C 201		
R 337	**revolving disk feeder**	Tellerspeiser *m*	alimentateur *m* à disques tournants	дисковый (тарельчатый) питатель
R 338	**revolving drum screen**	Trommelsiebmaschine *f*, Trommelsieb *n*	tambour *m* cribleur (classeur)	барабанный вращающийся грохот
R 339	**revolving dryer**	Röhrentrockner *m*	sécheur *m* à tubes	трубчатая сушилка
R 340	**revolving filter**	Zellenfilter *n*, Vakuumzellenfilter *n*	filtre *m* cellulaire aspirateur	секционный (вакуумный) фильтр
R 341	**revolving furnace**	Drehofen *m*	four *m* tournant	вращающаяся печь
	revolving screen	*s.* C 905		
R 342	**Reynolds equation**	Reynoldssche Gleichung *f*	équation *f* de Reynolds	уравнение Рейнольдса
R 343	**Reynolds number**	Reynoldssche Zahl *f*	nombre *f* de Reynolds	число Рейнольдса
	rheogoniometer	*s.* C 566		
R 344	**rheological equation of state**	rheologische Zustandsgleichung *f*	équation *f* d'état rhéologique	реологическое уравнение состояния
R 345	**rheological homogeneous liquid**	rheologisch homogene Flüssigkeit *f*	liquide *m* rhéologiquement homogène	реологическая гомогенная жидкость
R 346	**ribbed radiator,** gilled cooler	Rippenkühler *m*	réfrigérant *m* à ailettes	ребристый охладитель
R 347	**ribbon-blade agitator**	Bandrührer *m*	agitateur *m* à vis	ленточная мешалка
R 348	**rich gas**	Starkgas *n*	gaz *m* riche	высококалорийный газ
R 349	**riddle,** riddling equipment	Siebsortierer *m*	crible *m* de triage	ситовая сортировка
	riddle	*s. a.* S 260		
	riddling equipment	*s.* R 349		
R 350	**rigid body**	starrer Körper *m*	corps *m* rigide	жесткое тело
R 351	**rigid-chain polymer**	starrkettiges Polymer *n*	polymère *m* à chaîne rigide	жесткоцепной полимер
R 352	**rigid macromolecule**	starres Makromolekül *n*	macromolécule *f* rigide	жесткая макромолекула
	rim	*s.* L 305		
R 353	**rim condition**	Randbedingung *f*	condition *f* marginale	граничное условие
R 354	**ring-roll crusher**	Ringwalzenbrecher *m*	concasseur *m* à cylindres annulaires	кольцевая валковая дробилка
	rinsing	*s.* W 17		
R 355	**rinsing water,** wash water	Spülwasser *n*	eau *f* d'écurage, rinçure *f*	смывная вода, вода для промывки
R 356	**ripple tray**	Wellsiebboden *m*	plateau *m* à ondulations	волнистая тарелка
R 357	**rise in temperature**	Temperatursteigerung *f*	élévation *f* de température	повышение температуры
	rising pipe	*s.* V 159		
R 358	**rising velocity**	Steiggeschwindigkeit *f*	vitesse *f* de montée	скорость подъема
R 359	**risk of fracture**	Bruchgefahr *f*	danger *m* de rupture	предельное удлинение, опасность разрыва
R 360/2	**rocket fuel**	Raketenbrennstoff *m*	carburant *m* à fusée	ракетное топливо
	rollcrown	*s.* C 19		
	roll crusher	*s.* R 367		
	roller	*s.* D 518		
R 363	**roller mill**	Walzenmühle *f*	moulin *m* à cylindres	вальцовая (валковая) мельница
R 364	**roller screen**	Rollenrost *m*, Walzenrost *m*	grille *f* (crible-classeur *m*) à rouleaux	роликовый грохот
R 365	**rolling**	Walzen *n*	laminage *m*	вальцевание, прокатка, размол, дробление

R 366	rolling ability	Walzbarkeit f	aptitude f.au laminage	вальцуемость
R 367	rolling crusher, roll crusher	Walzenbrecher m	broyeur m à cylindre, broyeur giratoire	валковая (вальцовая) дробилка
R 368	rolling friction	Rollreibung f, rollende Reibung f	friction f roulante, frottement m de roulement	трение качения
R 369	roll-jaw crusher	Walzenbackenbrecher m	broyeur m à mâchoires cylindriques	валково-щековая дробилка
R 370	roll mill	Walzenstuhl m, Walzwerk n	laminoir m mélangeur	смесительные вальцы
R 371	roll screen	Walzenrost m, Rollenklassierer m	classeur m à rouleaux	валковый грохот
R 372	room temperature	Zimmertemperatur f, Raumtemperatur f	température f ambiante, température intérieure	комнатная температура, температура помещения
R 373	Rosenbrock method	Rosenbrock-Verfahren n	méthode f de Rosenbrock	метод Розенброка
	rotary blower	s. C 127		
R 374	rotary compressor	Rotationskompressor m	compresseur m rotatif	ротационный компрессор
R 375	rotary crusher	Rundbrecher m, Kegelbrecher m, Kreiselbrecher m	concasseur m à cônes, concasseur giratoire	конусная дробилка
	rotary disk filter	s. D 370		
R 376	rotary distillation column	Destillationsdrehkolonne f	colonne f à disques rotatifs	колонка с ротационными тарелками
R 377	rotary drum filter	Trommeldrehfilter n	filtre m à tambour rotatif	вращающийся барабанный фильтр
R 378	rotary dryer	Drehtrockner m, Rührtrockner m	sécheur m rotatif	барабанная (вращающаяся) сушилка
R 379	rotary dryer	Trockentrommel f	séchoir m rotatif, tambour m de séchage	сушильный барабан
	rotary dryer	s. a. C 904		
R 380	rotary filter	Drehfilter n, Trommelfilter n	filtre m à tambour	вращающийся фильтр
R 381	rotary gear pump	Zahnradpumpe f	pompe f à engrenages	шестеренчатый насос
R 382	rotary grate gas producer	Drehrostgenerator m	générateur de gaz à grille tournante	газогенератор с вращающейся колосниковой решеткой
R 383	rotary hearth furnace	Tellerofen m	four m à sole tournante	печь с вращающимся подом
R 384	rotary impact mill	Trommelschlagmühle f, Schlägermühle f	broyeur m à barres cylindrique	ротационная (барабанная) ударная мельница
R 385	rotary impulse	Drehimpuls m	quantité f de mouvement angulaire	импульс вращения
R 386	rotary kiln	Drehrohrofen m	four m rotatif (tubulaire tournant)	вращающаяся трубчатая печь
R 387	rotary mixer, tumbling mixer	Rührtrommel f, Mischtrommel f, Kreiselmischer m	tambour m à mélanger, mélangeur m centrifuge	барабан-мешалка, волчковый смеситель, ротационный смеситель
R 388	rotary pump, drum pump	Drehkolbenpumpe f, Rotationspumpe f, Umlaufkolbenpumpe f	pompe f à piston rotatif	ротационный (роторный) насос
R 389	rotary screen	Drehsieb n, Trommelsieb n	tambour-classeur m, tamis m rotatif	вращающийся грохот
R 390	rotary slide valve vacuum pump	Drehschiebervakuumpumpe f	pompe f rotative à palettes	ротационный пластинчатый вакуумный насос
R 391	rotary vacuum dryer	Vakuumschaufeltrockner m	séchoir m rotatif à vide	вращающаяся вакуумная сушилка
	rotary vacuum filter	s. V 18		
R 392	rotary vacuum pump	Drehkolbenvakuumpumpe f	pompe f rotative à piston excentré	ротационный вакуумный насос
R 393	rotating auger	Förderschnecke f	hélice f transporteuse, transporteur m à vis sans fin	вращающийся (транспортный) шнек
R 394	rotating disk contactor	Drehscheibenextraktor m	contacteur (extracteur) m rotatif à disques	колонка с вращающимися дисками, дисковый экстрактор
R 395	rotating drum	Drehtrommel f	tambour m rotatif	вращающийся барабан
R 396	rotating strip column	rotierende Kolonne f	colonne f rotative de stripping	вращающаяся колонна
	rotational flow	s. E 13		
R 397	rotational speed	Umlaufgeschwindigkeit f	vitesse f périphérique, vitesse de rotation (circulation)	скорость вращения
R 398	rotation cooling	Umlaufkühlung f	réfrigération f en circuit fermé	циркуляционное охлаждение
R 399	rotor type dust collector	Rotationsabscheider m	séparateur m rotatif	ротационный сепаратор
R 400	rough-crushing	Vorbrechen n	préconcassage m	предварительное (крупное) дробление
R 401	roughness factor	Rauheitsgrad m	facteur (degré) m de rugosité	коэффициент шероховатости
R 402	rough outline	Grobübersicht f	plan m d'ensemble approximatif	грубый обзор
R 403	round capillary	Rundkapillare f	tube m capillaire circulaire	капилляр с круглым поперечным сечением
R 404	route of travel	Beförderungsweg m	parcours m, chemin m de transport	маршрут перевозок
R 405	row nucleation	Reihenkristallisationskeimbildung f	germination f en rangée	цепное образование центров кристаллизации

R 406	rubber elasticity	Gummielastizität f	élasticité f de caoutchouc	резиноэластичность
R 407	rubber industry	Gummiindustrie f	industrie f du caoutchouc	резиновая промышленность
R 408	rubberizing	Gummierung f	revêtement m au caoutchouc	прорезинивание, гуммирование
R 409	rubber-like state	gummiartiger Zustand m	état m gommeux	резинообразное состояние
R 410	rubber mixer	Gummimischer m	malaxeur m de caoutchouc	резиносмеситель
R 411	rubber mixture	Gummimischung f	mélange m de caoutchouc	резиновая смесь
R 412	rubber packing	Gummidichtung f	garniture f en caoutchouc	резиновое уплотнение
R 413	rubber processing	Gummiverarbeitung f	transformation f de caoutchouc	переработка резины
	rule of mixtures	s. M 272		
R 414	rule of thumb	Faustregel f	règle f empirique	эмпирическое правило
	rules	s. S 545		
R 415	run	Fahrperiode f, Betriebsperiode f, Versuchsperiode f	cycle m de marche, période f d'essai	цикл, работа, ход, рабочая партия, рабочий цикл
R 416	runaway surface ignition	weglaufende Oberflächenzündung f	allumage m sur surface chaude à avance croissante	разносное поверхностное воспламенение
R 417	run cycle, working cycle	Arbeitsprozeß m	cycle m de marche, opération f	рабочий цикл
	run-down tank	s. A 79		
R 418	run-in period	Einlaufzeit f	temps m de mise en œuvre	время пуска (откатки)
R 419	run-in speed	Einlaufgeschwindigkeit f	vitesse f à l'entrée	скорость входа, скорость впуска
R 420	runner	Einspritzkanal m	passage m d'injection	литник, инжектор
	running-in pressure	s. I 206		
R 421	running point	Destillationstemperatur f, Destillationspunkt m	température f de la distillation, point m de distillation	температура перегонки, пределы кипения, температурные пределы выкипания
R 422	run-off coefficient	Abflußkoeffizient m	coefficient m d'écoulement, coefficient de décharge	коэффициент стока
R 423	run parallel/to	parallel laufen	marcher en parallèle	параллельно работать
	run period	s. P 531		
R 424	rupture	Reißen n, Spaltung f, Bruch m	rupture f, déchirement m	разрыв, расщепление, разрушение
R 425	rupture diaphragm	Berstscheibe f	disque m d'explosion, disque d'éclatement	разрывная мембрана
R 426	rupture-disk, shear-disk	Berstscheibe f	disque m d'éclatement	разрывающая диафрагма
R 427	rupture pressure	Zerreißdruck m, Berstdruck m	pression f d'éclatement	давление разрыва (продавливания)
	rusting	s. C 743		
R 428	rust-preventative paint	Rostschutzanstrich m	enduit m antirouille	антикоррозионная окраска
R 429	rust prevention	Rostschutz m	protection f contre la rouille	защита от ржавления (коррозии)

S

S 1	sack/to	absacken, einsacken	ensacher, mettre en sac	засыпать в мешки
S 2	sack-filling tunnel, bagging-off facility	Absacktrichter m	trémie f d'ensachage	выбойный закром
S 3	safe area	Sicherheitszone f, Sicherheitsbereich m	marge (zone) f de sécurité	опасная зона
S 4	safe design	Sicherheitsentwurf m, Sicherheitskonstruktion f	projet m sûr, construction f sûre	безопасный проект, безопасная конструкция
	safe distance	s. S 8		
S 5	safety and fire protection regulation	Arbeitsschutz- und Brandschutz[an]ordnung f	règlements mpl concernant la sécurité de travailleurs et la protection contre le feu	инструкция по технике безопасности и противопожарному делу
S 6	safety appliance	Sicherheitseinrichtung f	mécanisme m de sûreté, dispositif m de sécurité	предохранительное устройство
S 7	safety container	Sicherheitsbehälter m	conteneur m de sûreté	предохранительный контейнер
S 8	safety distance, safe distance	Sicherheitsabstand m	distance f de sécurité	безопасное расстояние
S 9	safety factor	Sicherheitsfaktor m	coefficient m de sécurité	коэффициент надежности (запаса)
	safety gas	s. I 122		
S 10	safety hazard	Sicherheitsrisiko n	risque m de sécurité	риск опасности
S 11	safety measures	Schutzmaßnahmen fpl, Sicherheitsmaßnahmen fpl	mesures fpl de sécurité (précaution)	мероприятия по обеспечению безопасности
	safety of operation	s. O 86		
S 12	safety requirement	Sicherheits[an]forderung f, Sicherheitserfordernis n	exigence f de sécurité	требование техники безопасности
S 13	safety switch	Sicherheitsschalter m	interrupteur m de sécurité	предохранительный выключатель
S 14	safety technology	Sicherheitstechnik f	technique f de sécurité	техника безопасности
S 15	safety technology factors	sicherheitstechnische Kennzahlen fpl	coefficients mpl de la technique de sécurité, facteurs mpl de sécurité.	показатели техники безопасности

S 16/7	safety technology means	sicherheitstechnische Mittel npl	moyens mpl de la technique de sécurité	средства техники безопасности
	safety valve	s. R 271		
	sagging point	s. S 394		
S 18	salable product	verkaufsfähiges Produkt n	produit m marchand	продажный продукт
S 19	salification, salifying, salt formation	Salzbildung f	salification f	солеобразование
S 20	salt bath	Salzbad n	bain m salin (de sel)	солевая баня
S 21	salt concentration	Salzkonzentration f	concentration f de sel	концентрация соли
S 22	salt content	Salzgehalt m	teneur m en sel, salinité f	содержание соли
	salt formation	s. S 19		
S 23	salt solution	Salzlösung f, Salzlake f	solution f saline, saumure f	рассол
S 24	salvage value	Schrottwert m, Bergungswert m (Schiff)	valeur f de sauvetage	плата (стоимость) скрапа (лома)
S 25	same direction/in the	gleichsinnig, gleichgerichtet	de même sens	синфазно (синхронно) работающий
S 26	sample	Probe f	épreuve f, échantillon m	испытание, проба, образец
S 27	sampling	Probenahme f	prélèvement m des échantillons	отбор проб
S 28	sand bath	Sandbad n	bain-marie m	песочная баня
S 29	sand filter	Sandfilter n	filtre m de sable	песочный фильтр
S 30	sanitary equipment	sanitäre Einrichtungen fpl	appareils mpl sanitaires	санитарное оборудование
S 31	Sankey diagram	Sankey-Diagramm n	diagramme m thermique	Сенкей-диаграмма, количественная диаграмма
S 32	saponifiable constituent	verseifbarer Anteil m	constituant m saponifiable	омыляемый компонент
S 33	saponification	Verseifung f	saponification f	омыление
S 34	saponification number	Verseifungszahl f	indice (coefficient) m de saponification	число омыления
S 35	saponification of esters	Esterverseifung f	saponification f d'esters	омыление сложного эфира
S 36	saponification test	Verseifungstest m	essai m de saponification	проба на омыляемость
S 37	Sargent-Westerberg algorithm	Algorithmus m von Sargent und Westerberg	algorithme m de Sargent-Westerberg	алгоритм Саржента и Вестерберга
S 38	saturate/to	sättigen, absättigen, saturieren	saturer	насыщать, пропитывать, сатурировать
S 39	saturated condition	Sättigungszustand m	état m saturé (de saturation)	состояние насыщения
S 40	saturated solution	gesättigte Lösung f	solution f saturée	насыщенный раствор
S 41	saturated steam	gesättigter Dampf m, Sattdampf m, Naßdampf m, satter Dampf	vapeur f saturée (humide)	насыщенный (влажный) пар
S 42	saturated temperature	Sättigungstemperatur f	température f de saturation	температура насыщения
S 43	saturated vapour enthalpie	Enthalpie f des gesättigten Dampfes	enthalpie f de vapeur saturée	энтальпия насыщенного пара
S 44	saturation capacity	Sättigungskapazität f	capacité f de saturation	насыщаемость, способность к насыщению, поглотительная способность
	saturation coefficient	s. S 47		
S 45	saturation concentration	Sättigungskonzentration f	concentration f de saturation	концентрация насыщения
S 46	saturation curve	Sättigungskurve f	courbe f de saturation	кривая насыщения
S 47	saturation degree (factor), saturation coefficient	Sättigungsfaktor m, Sättigungskoeffizient m, Sättigungsgrad m	coefficient m de saturation	коэффициент насыщения
S 48	saturation index	Sättigungsindex m	indice m de saturation	индекс (коэффициент) насыщения
S 49	saturation limit	Sättigungsgrenze f	limite f de saturation	коэффициент (предел) насыщения
S 50	saturation point	Sättigungspunkt m	point m de saturation	точка насыщения
S 51	saturation pressure	Sättigungsdruck m	pression f de saturation	давление насыщения
S 52	saturation process	Sättigungsprozeß m	processus m de saturation	процесс насыщения
S 53	saturation value	Sättigungswert m	valeur f de saturation	величина насыщения
S 54	saturator	Sättiger m, Saturateur m	saturateur m	сатуратор
S 55	scaffold[ing]	Gerüst n	charpente f, tréteau m, échafaud m	леса, подмости, клеть, станина
S 56	scale, graduation	Skale f, Graduierung f	gamme f, cadran m, graduation f	шкала, масштаб
S 57	scale	Kesselstein	tartre m, incrustations fpl	накипь в котле, котельная накипь
S 58	scale factor	Maßstabsfaktor m	facteur m d'échelle	масштабный фактор (коэффициент)
S 59	scale formation	Niederschlagbildung f	formation f de précipité	нарастание накипи
S 60	scale of turbulence	Turbulenzlänge f, Turbulenzgrad m, Turbulenzmaßstab m	degré m de turbulence	масштаб (степень) турбулентности
S 61	scale up factor	Vergrößerungsfaktor m	facteur m d'échelle	масштабный коэффициент
S 62	scaling furnace	Beizofen m	four m de décapage	травильная печь
S 63	scanning system	Abtastsystem n	système m de balayage	система развертки
S 64	scarcity value	Defizitwert m	valeur f de déficit	величина дефицита
S 65	scatter diagram	Korrelationsdiagramm n	nuage m de corrélation	диаграмма рассеяния
S 66	scattering	Zerstäuben n (Feststoffe)	pulvérisation f	пульверизация
	scattering	s. a. L 59		
S 67	scattering factor	Streufaktor m	coefficient m de dispersion	коэффициент рассеяния

S 68	scavenging gas, flush (circulation) gas	Spülgas n	gaz m de balayage	циркулирующий (промывной) газ, обогревающий газ при внутреннем обогреве
S 69	scheduling	Planung f	projet m, planification f, plan m, conception f	планирование, технологическая проработка
S 70	scheme, layout	Bauplan m	dessin m du génie civil, projet m de construction	проект строительства
	scheme	s. a. C 200		
S 71	Schmidt number	Schmidtsche Zahl f	nombre m de Schmidt	критерий Шмидта
	scoria	s. S 364		
S 72	scorification, scorifying	Verschlackung f	scorification f	шлакование, ошлакование
S 73	scrap/to	verschrotten	démolir, mettre à la ferraille	разделывать скрап
S 74	scrap	Schrott m	ferrailles fpl	скрап
S 75	scraper agitator	Kratzrührer m	agitateur m à racloires	скребковая мешалка
S 76	scrapping	Verschrottung f	démolition f	разделка скрапа
S 77	scrap plastics	Regeneratplast m	plastique m de rebut	полимерный регенерат
S 78/9	screen/to	sieben, klassieren, sichten, filtern	trier, tamiser, cribler	просеивать, отсеивать, сортировать, фильтровать
	screen	s. 1. M 183; 2. S 91		
	screen aperture	s. S 92		
	screen classification	s. S 95		
S 80/1	screen cloth filter	Flächenfilter n, Planfilter n	filtre m plan	плоский фильтр
S 82	screen dryer	Siebtrockner m	sécheur m cribleur	устройство для сушки на ситах
S 83	screen fabric	Siebgewebe n	grillage m pour crible, toile f de tamissage	ситоткань, ткань для сит
S 84	screen-faced ball mill, screening mill	Siebkugelmühle f	broyeur m à boulets avec tamis périphérique	шаровая мельница с периферической разгрузкой
S 85	screen hammer mill	Siebhammermühle f	broyeur à marteaux avec tamis	молотковая мельница с ситами
S 86	screening, sifting	Sieben n, Siebung f	criblage m, tamisage m, triage m	просеивание, просевание, грохочение
S 87	screening, preliminary selection	Vorauswahl f	choix m préalable	предварительный отбор
S 88	screening analysis	Siebanalyse f	analyse f granulométrique	гранулометрический анализ
S 89	screening device	Scheideanlage f	affinerie f, installation f de séparation, unité f de triage	сортировочная установка, сортировка
S 90	screening drum, sizing (sieve) drum	Siebtrommel f, Klassiertrommel f, Sortiertrommel f	tambour m, crible m à tambour, tambour classeur, tambour cribleur, trommel m classeur	барабанный грохот, барабанное сито, сортировочный (сетчатый) барабан
S 91	screening machine, screen	Siebapparat m	appareil m tamisseur, crible m, tamis m	грохот, решето
	screening mill	s. S 84		
S 92	screen opening, screen aperture	Siebweite f, Siebgröße f	numéro m de tamis	размер отверстия решета
S 93	screen overflow	Siebrückstand m, Siebüberlauf m	refus m de criblage, refus	отсев, надрешетный продукт
S 94	screen oversize	Siebgrobes n, Siebübergang m	refus m d'un crible, produit m traversé	отсев, остаток на грохоте, надрешетный продукт
	screen riddle	s. S 260		
S 95	screen sizing, screen classification	Siebklassierung f, Kornklassierung f	triage m à tamis, classement m granulométrique, granulométrie f	грохочение, классификация по величение (размеру) зерен, классификация рассевом
S 96	screen structure	Siebstruktur f	structure f de crible	ситовая структура
S 97	screen underflow	Siebdurchgang m, Siebdurchlauf m	passant m du crible, produit m passé, passée f	просев, подрешетный продукт
S 98	screen underframe	Siebboden m	fond m perforé, fond à tamis	дно ситовой рамы, рабочее полотно сита
S 99	screen undersize	Siebfeines n, Siebdurchfall m	produit m passé, menus mpl du crible	просев
S 100	screw	Schnecke f	vis f sans fin	червяк, винт, шнек
S 101	screw-bed injection moulding machine	Schneckenspritzgießmaschine f	machine (presse) f d'injection, machine de moulage par extrusion	шнековая литьевая машина под давлением, шнековая машина литья под давлением
S 102	screw conveyor	Vollförderschnecke f	hélice f transporteuse pleine	закрытый шнековый транспортер
S 103	screw conveyor, worm conveyor	Förderschnecke f, Schneckenförderer m, Transportschnecke f	hélice f transporteuse, vis f [sans fin] transporteuse	винтовой транспортер, транспортный червяк (шнек), шнековый конвейер
S 104	screw-conveyor drum dryer	Schneckentrockner m	sécheur-convoyeur m à vis	шнековая сушилка
S 105	screw-conveyor extruder, screw extruder	Schneckenextruder m, Schneckenstrangpresse f	boudineuse f [à vis], extrudeuse f [à vis]	червячный пресс, экструдер, шнекпресс, шнековый экструдер
S 106	screw extrusion	Schneckenpressen n	extrusion f	шнековое экструдирование, червячное прессование

S 107	screw feeder	Schneckenspeiser *m*	alimentateur *m* à vis sans fin	шнековый питатель, винтовой (червячный) питатель
S 108	screw machine	Schneckenmaschine *f*	machine *f* à vis	червячная машина
S 109	screw mixer	Schneckenmischer *m*	mélangeur *m* à vis	шнековый смеситель
S 110	screw pump	Schneckenpumpe *f*	pompe *f* à vis sans fin	винтовой насос
S 111	scrubber	Gasreiniger *m*, Gaswäscher *m*	laveur *m* de gaz, épurateur *m*, scrubber *m*	скруббер, газоочиститель, газопромыватель
	scrubber	*s. a.* 1. S 506; 2. W 20		
S 112	scrubbing tower	Rieselturm *m*, Berieselungsturm *m*, Waschturm *m*	tour *f* de lavage	скруббер, газоочиститель
S 113	scrub solution	Waschlösung *f*	solution *f* de lavage	промывной раствор
S 114	seal/to	abdichten, verstopfen	étouper, calfeutrer, boucher	уплотнять, герметизировать, конопатить, закупоривать
S 115	sealant	Dichtungsmaterial *n*, Abdichtungsmaterial *n*	matériel *m* d'étoupage, matériaux *mpl* d'étanchéité	материал для уплотнения
S 116	sealing	Abdichtung *f*	étoupage *m*, étoupement *m*, bourrage *m*	герметизация, уплотнение
S 117	sealing area, sealing face	Abdichtfläche *f*, Dichtfläche *f*	face *f* de joint, surface *f* de contact	поверхность уплотнения
S 118	sealing compound	Dichtungsmittel *n*, Dichtung *f*	garniture *f*, étoupe *f*, matériel *m* d'étoupage	уплотнительный материал, уплотнение
	sealing face	*s.* S 117		
	sealing liquid	*s.* 1. P 11; 2. S 122		
S 119	sealing roller	Abdichtwalze *f*	rouleau *m* d'étanchéité	уплотнительный валик
S 120	sealing strength	Haftfestigkeit *f*	adhésivité *f*, adhérence *f*, force *f* d'adhérence	прочность прилипания
S 121	sealing surface	Dichtungssitzfläche *f*	surface *f* d'assise du joint	поверхность уплотняющего устройства
S 122	sealing water, sealing liquid	Absperrwasser *n*	eau *f* obturante, liquide *m* obturant	запруженная (ограждённая) вода
S 123	seal leg	Dichtungsstück *n*	pièce *f* de garniture, pièce *f* d'étanchéité	уплотняющее колено, жидкостный затвор
S 124	seal pressure	Sperrdruck *m*	pression *f* de barrage	давление при закрытой задвижке
S 125	seamless pipe	nahtloses Rohr *n*	tube *m* sans soudure	бесшовная труба
S 126	searching method	Suchverfahren *n*	méthode *f* de recherche	метод поиска
S 127	searching point	Suchpunkt *m*	point *m* de recherche	точка на траектории поиска
S 128	searching strategy	Suchstrategie *f*	politique *f* de recherche	стратегия поиска
	secondary action	*s.* S 136		
S 129	secondary air	Sekundärluft *f*	air *m* secondaire	вторичный воздух
S 130	secondary airstream	Sekundärluftströmung *f*	courant *m* d'air secondaire	поток вторичного воздуха
	secondary circuit	*s.* S 139		
S 131	secondary condenser	Sekundärkondensator *m*	condenseur *m* secondaire	вторичный конденсатор
S 132	secondary crushing	Nachzerkleinerung *f*	broyage *m* secondaire	вторичное (повторное) измельчение
S 133	secondary cyclone	Sekundärzyklon *m*	cyclone *m* secondaire	вторичный циклон
S 134	secondary drum	Sekundärtrommel *f*	tambour *m* secondaire	вторичный барабан
S 135	secondary drying	Sekundärtrocknung *f*	séchage *m* secondaire	вторичное обезвоживание
S 136	secondary effect, secondary action	Nebenwirkung *f*, Nebeneffekt *m*	effet *m* (action *f*, réaction *f*) secondaire	побочный эффект
S 137	secondary flow	Nebenströmung *f*	écoulement (courant) *m* secondaire	вторичное течение, вторичный поток
S 138	secondary function	Nebenfunktion *f*	fonction *f* secondaire	вторичная функция
S 139	secondary method, secondary circuit (process)	Sekundärverfahren *n*	procédé *m* secondaire	вторичный способ
S 140	secondary pressure	Sekundärdruck *m*	pression *f* secondaire	вторичное давление
S 141	secondary process	Nebenbetrieb *m*, Hilfsanlage *f*	service *f* annexe (secondaire)	вспомогательное хозяйство
	secondary process	*s. a.* S 139		
S 142	secondary quality	Nebeneigenschaft *f*	qualité (propriété) *f* secondaire	побочное свойство
S 143	secondary reaction	Sekundärreaktion *f*, Nebenreaktion *f*	réaction *f* secondaire	вторичная реакция
S 144	secondary shaping	Nachformen *n*	postformage *m*	вторичное формование
S 145	secondary valence	Nebenvalenz *f*, Nebenwertigkeit *f*	valence *f* secondaire	побочная валентность
S 146	secondary vapour	Sekundärdampf *m*	vapeur *f* secondaire	вторичный пар
S 147	second crusher	Nachbrecher *m*	reconcasseur *m*, broyeuse *f* secondaire	вторичная дробилка
S 148	second drying process	Nachtrocknen *n*	séchage *m* final	дополнительная сушка
S 149	second filter	Nachfilter *n*	filtre *m* secondaire (en aval)	дополнительный фильтр
	sectional plant	*s.* S 779		
S 150	sectioning	Sektionierung *f*	sectionnement *m*	секционирование, подразделение по секциям
S 151	section under construction, stage of construction	Bauabschnitt *m*	tranche *f* de construction	выемочное поле, выемочный участок
S 152	sediment	Bodensatz *m*, Sediment *n*	dépôt *m*, sédiment *m*, résidu *m*, couche *f*	отстой, осадок
S 153	sedimentation	Sedimentieren *n*, Ausfällen *n*, Fällen *n*, Sedimentation *f*	sédimentation *f*, précipitation *f*	седиментирование, оседание, осаждение

	sedimentation basin	s. S 241		
S 154	sedimentation centrifuge	Absetzzentrifuge f	centrifuge f de sédimentation (décantation)	обезвоживающая центрифуга
S 155	sedimentation potential	Sedimentationspotential n	potentiel m de sedimentation	потенциал седиментации (оседания, осаждения)
S 156	sedimentation velocity	Absetzgeschwindigkeit f	vitesse f de décantation	скорость осаждения (седиментации)
S 157	sedimentation zone	Absetzzone f	zone f de décantation	зона седиментации (осаждения)
S 158	sediment out/to	absedimentieren	déposer, déposer des sédiments	оседать, седиментировать
S 159	segregation	Segregation f, Seigerung f, Trennung f, Aussonderung f, Sichten n, Klassieren n	ségrégation f, triage m	сегрегация, разделение, выделение, сортировка
	segregation	s. a. D 177		
S 160	segregation limit	Aussonderungsgrenze f, Klassengrenze f	limite f de triage	предел отбора, граница сегрегации
	selection	s. C 249		
S 161	selection principle	Auswahlprinzip n	principe m de sélection	принцип селекции (выбора, отбора)
S 162	selection rule	Auswahlregel f	règle f de sélection	правило селекции (выбора, отбора)
S 163	selective adjustment	Teilregelung f, Selektivregelung f, selektive Regelung f, lokale Regelung	réglage m sélectif, réglage local	избирательное регулирование
S 164	selective cooling	selektive Kühlung f	refroidissement m sélectif	избирательное охлаждение
	selective corrosion	s. L 215		
S 165	selective procedure	Auswahlverfahren n	procédure f de sélection	метод селекции (выбора, выборки)
S 166	selectivity, separation effect (sharpness)	Trennschärfe f, Trennungsgenauigkeit f, Selektivität f, Trennungsvermögen n	sélectivité f, pouvoir m séparateur, précision f de séparation	селективность, избирательность, острота настройки
S 167	self-cleaning underfeed stoker	selbstreinigende Unterschubfeuerung f	foyer m à propulsion inférieure autonettoyant	самоочищающаяся топка с нижней подачей
S 168	self-cooling	Selbstkühlung f	refroidissement m naturel	естественное остывание
S 169	self-deashing underfeed stoker	selbstentaschende Unterschubfeuerung f	foyer m à propulsion inférience autodécendrant	самообеззоливающая топка с нижней подачей
S 170	self-diffusion	Selbstdiffusion f, Eigendiffusion f	autodiffusion f	собственная диффузия
S 171	self-evaporation	Eigenverdampfung f, Selbstverdampfung f	auto-évaporation f	самоиспарение
	self-igniting	s. S 493		
S 172	self-induction	Selbstinduktion f	induction f propre, self-induction f	самоиндукция
S 173	self-purification	Selbstreinigung f	auto-épuration f, autonettoyage m	самоочищение
S 174	semifinished product	Halbfabrikat m, Halbzeug n, Halbfertigprodukt n	semi-produit m, demi-produit m	полупродукт, полуфабрикат
S 175	semigas	Halbgas n	gaz mpl étranglés	средний (генераторный) газ
S 176	semirefrigeration	Halbkühlung f	semi-refroidissement m	полуохлаждение
S 177	semivolatile	halbflüchtig	semi-volatil	полулетучий
S 178	sensibility, sensitivity	Empfindlichkeit f, Sensibilität f	sensibilité f	чувствительность, светочувствительность, сенсибилизация
S 179	sensibility quantity	Empfindlichkeitswert m	valeur f de sensibilité	величина чувствительности
S 180	sensible cooling effect	fühlbare Kälteleistung f	capacité f frigorifique sensible	холодопроизводительность по ощутимому теплу
S 181	sensible heat	fühlbare Wärme f	chaleur f sensible	ощутимое тепло
S 182	sensible heat capacity	fühlbare Wärmeleistung f	rendement m thermique sensible	производительность по ощутимой теплоте
S 183	sensible heat factor	fühlbarer Wärmeanteil m	quantité f de chaleur sensible	доля ощутимого тепла
S 184	sensible heat load	fühlbare Wärmelast f	charge f thermique sensible	тепловая нагрузка по ощутимому теплу
S 185	sensible heat removal	Abführung f fühlbarer Wärme	élimination f de chaleur sensible	отвод ощутимой теплоты
S 186	sensing contact	Abfühlkontakt m	contact m d'exploration	считывающий контакт
S 187	sensing element	Fühlelement n, Fühler m	élément m sensible, palpeur m	чувствительный элемент
	sensitivity	s. S 178		
S 188	sensitivity analysis	Empfindlichkeitsanalyse f	analyse f de sensibilité	анализ чувствительности
S 189	sensitivity criterion	Empfindlichkeitskriterium n	critère m de sensibilité	критерий чувствительности
S 190	sensitivity curve	Empfindlichkeitskurve f	courbe f de sensibilité, caractéristique f	кривая чувствительности, тарировочная характеристика датчика
S 191	sensitivity function	Empfindlichkeitsfunktion f	fonction f de sensibilité	функция чувствительности
S 192	sensitivity limit	Empfindlichkeitsgrenze f	limite f de sensibilité	порог чувствительности
S 193	sensitivity of measurement	Meßempfindlichkeit f	sensibilité f de mesure	чувствительность (точность) измерения
S 194	separating	Zerteilen n, Zersetzen n, Trennen n	décomposition f	раздробление, разделение
S 195	separating agent	Ausscheidungsmittel n	réactif m, agent m de séparation	осадитель

	English	German	French	Russian
S 196	**separating column**	Abtrennkolonne f	colonne f de séparation	ректификационная колонна
S 197	**separating condenser**	Trennkondensator m	condenseur m séparateur	разделительный конденсатор
S 198	**separating funnel,** separatory	Scheidetrichter m	entonnoir m à séparation (décantation)	делительная воронка
S 199	**separating into components**	Entmischen n, Inhomogenisierung f	déshomogénéisation f	сегрегация, сепарация, разделение, расслоение
S 200	**separating nozzle**	Trenndüse f	buse f de séparation	разделительное сопло
S 201	**separating plant**	Sortieranlage f	installation f de triage, trieur m	сортировочная установка, сортировка
S 202	**separating screen**	Sortiersieb n	crible m de triage	сито для сортировки
S 203	**separating tube,** separation tube	Trennrohr n	tube m séparateur	разделительная труба
S 204	**separation,** classifying	Lösen n, Trennen n, Trennung f; Entmischung f, Scheidung f; Sortierung f, Klassierung f	séparation f, décomposition f, classement m, triage m	отделение, сепарация, разделение, сортировка, классификация, тонкий рассев, отсадка
	separation	s. a. 1. D 177; 2. E 85		
S 205	**separation column**	Trennsäule f, Trennkolonne f	colonne f à fractionner, colonne de rectification	колонна разделения, фракционнирующая колонна
S 206	**separation effect**	Trenneffekt m	séparation f, effet m de séparation	эффект разделения
	separation effect	s. a. S 166		
S 207	**separation efficiency**	Trennleistung f	rendement m séparateur (de séparation)	эффективность разделения
S 208	**separation factor**	Trennfaktor m	facteur m de séparation, coefficient m séparateur	фактор разделения
S 209	**separation method**	Trennungsmethode f	méthode f de séparation	метод разделения (отделения)
S 210	**separation plant**	Trennanlage f	installation f séparatrice, unité f de séparation	разделительная установка
S 211	**separation pressure**	Trenndruck m	pression f de séparation	давление отбора
S 212	**separation process**	Trennungsvorgang m	processus m de séparation	процесс разделения, сепарационный процесс
	separation sharpness	s. S 166		
S 213	**separation stage (step)**	Trennstufe f	étage m de séparation	ступень разделения
S 214	**separation system**	Trennsystem n, Entmischungssystem n	système m de séparation (décantation)	система разделения (удаления)
	separation tube	s. S 203		
S 215	**separation unit**	Stofftrenneinheit f	unité f de séparation de matière	сепаратор, узел разделения, элемент сепарации
S 216	**separator,** precipitator, trap settler	Abscheider m, Scheider m, Separator m, Trenner m, Trennapparat m	séparateur m, décanteur m, trieur m	разделитель, ловушка, отделитель, отстойник, сепаратор
	separator	s. a. 1. D 171; 2. S 308		
	separatory	s. S 198		
S 217	**sequence of charging**	Beschickungsfolge f, Chargierfolge f	suite f des opérations de chargement	последовательность загрузки
S 218	**sequence of operations**	Bearbeitungsfolge f	suite f des opérations	порядок обработки, технологический процесс
S 219	**sequence of processes**	Prozeßfolge f	suite f de procédés	последовательность процессов
S 220	**sequential calculation**	sequentielle Berechnung f	calcul m séquentiel	последовательное вычисление, последовательный расчет
S 221	**sequential procedure**	sequentielle Vorgehensweise f, Folgeprozeß m	procédé m séquentiel, processus m séquentiel	последовательный подход (расчета системы)
S 222	**series connection**	Reihenschaltung f	montage m en série	последовательное соединение
S 223	**series connection with store**	Reihenschaltung f mit Speicher	montage m en série avec accumulateur	последовательное соединение со сборником
S 224	**series of measurements**	Meßreihe f	série f de mesures	ряд (серия) измерений
S 225	**series-parallel circuit (connection)**	Serienparallelschaltung f	couplage m série-parallèle	последовательно-параллельное включение
S 226	**serve/to,** to handle	bedienen, betreiben	conduire, manœuvrer, manier	обслуживать
S 227	**service condition**	Betriebsverhältnisse npl	conditions fpl de service, régime m de fonctionnement	условия эксплуатации, положения производства, режимы работы
S 228	**service life**	Betriebsdauer f, Nutzungsperiode f, Benutzungsdauer f, Nutzungsdauer f	vie f utile, durée f de service	продолжительность [ис-] пользования
S 229	**service strategy**	Reparaturstrategie f	politique f de réparation	ремонтная стратегия, стратегия технического обслуживания
S 230	**service technology**	Reparaturtechnologie f, Wartungstechnologie f	technologie f d'entretien, technologie f de remise en état	ремонтная технология, технология технического обслуживания
S 231	**service time**	Betriebszeit f	temps m de service	время обслуживания
S 232	**service water**	Brauchwasser n	eau f industrielle	вода для технических целей, хозяйственная (техническая) вода

	servicing	s. M 41		
S 233	servicing schedule	Wartungsvorschrift f	instruction f d'entretien	правила по уходу
S 234	servicing staff	Instandhaltungspersonal n	personnel m d'entretien	ремонтный персонал, персонал по техническому обслуживанию
S 235	servo system	Folgesystem n	système m de poursuite	следящая система
	set-off	s. L 305		
S 236	set point	Sollwert m	valeur f préscrite, valeur de consigne	заданное значение, заданная величина
	setting point	s. S 403		
S 237	setting quality	Abbindefähigkeit f	capacité f de durcissement, faculté f de prise (ciment)	схватываемость, способность к схватыванию
S 238	setting time	Erstarrungszeit f	temps m de solidification	время затвердения (застывания)
	settled	s. P 322		
S 239	settler	Absetzbecken n, Absetzbehälter m, Absetztank m, Absetzer m, Klärbecken n, Klärbehälter m, Klärtank m; Klärwanne f, Klärgrube f; Scheidebehälter m, Abscheider m	bassin m de décantation, séparateur m, cuve f de décantation	отстойник, сепаратор
	settler	s. a. S 216		
S 240	settling	Setzen n, Absetzen n	déposition f, formation f de dépôt, sédimentation f	отсадка, осаждение, оседание, седиментирование
S 241	settling basin, sedimentation basin	Absetzbecken n	bassin m sédimentaire, bassin f de sédimentation	отстойник, седиментатор
	settling pond	s. S 372		
S 242	settling tank	Absetzbehälter m	réservoir m de décantation, cuve f de sédimentation	отстойник, отстойный бак
S 243	settling tower	Klärturm m, Klärkolonne f	colonne f de clarification	вертикальный отстойник
S 244	set value (of the staff)	Soll-Bestand m	effectif m prescrit, stock m exigé	штатный состав
	severity level	s. O 79		
	sewage	s. D 340		
S 245	sewage disposal	Abwasserabführung f, Abwasseraustritt m	sortie f (déchargement m) des eaux usées	спуск сточных вод
	sewage disposal	s. a. W 63		
S 246	sewage disposal plant	Abwasserreinigungsanlage f, Abwasseranlage f	unité f d'épuration des eaux usées, installation f des eaux résiduaires	установка для очистки сточных вод
S 247	sewage flow	Abwasserzuführung f	adduction f des eaux d'égouts	выпуск сточных вод
S 248	sewage plant	Kläranlage f	installation f de décantation, station f d'épuration	очистная установка
S 249	sewage sludge	Abwasserschlamm m	vase f d'eaux d'égout	шлам сточных вод
S 250	sewer	Abflußrohr n, Abwasserkanal m, Ablaßkanal m, Schleuse f	tube m d'écoulement, tuyau m (rigole f) de décharge	канализационная (сточная) труба, коллектор
	sewerage	s. D 489		
S 251	sewer gas	Faulgas n, Klärgas n	gaz m de curage (digestion)	биологический газ, септик-тэнк-газ
S 252	sewer system	Abwassersystem n, Abwassersammelsystem n	système m de collecte des eaux usées	система канализации для сточных вод
S 253	shaft cooler	Schachtkühler m	refroidisseur m à cuve	шахтный воздухоохладитель
S 254	shaft furnace (kiln)	Schachtofen m	four[neau] m à cuve	шахтная печь
S 255	shaker	Rüttler m	secoueur m, vibreur m	вибратор, качающийся грохот
S 256	shaking, agitation	Schütteln n, Erschüttern n	mouvement m de secousses, agitation f, battement m (de liquides)	качание, встряхивание, взбалтывание
S 257	shaking apparatus	Schüttelapparat m	appareil m trembleur, dispositif m de secouage	аппарат для встряхивания, трясучка
S 258	shaking feeder	Schüttelrinne f	gouttière f à secousses, distributeur m vibrant	качающийся желоб (транспортер)
S 259	shaking grate	Schüttelrost m, Rüttelrost m	grille f à secousses, tamis f vibrant, tamis oscillant	встряхивающая (качающаяся колосниковая) решетка
S 260	shaking screen (sieve), [screen] riddle, vibrating screen, swing sieve	Schüttelsieb n, Rüttelsieb n, Vibrationssieb n, Schwingsieb n, Schwingsiebmaschine f, Vibrationssortierer m	tamis m oscillant, crible m secoueur (vibrateur) à secousses, crible vibrant, tamis à vibration (secousses)	качающийся (сотрясательный, вибрационный) грохот, качающееся (сотрясательное, грубое, крупнопетлистое, вибрационное) сито, отсадочная машина, вибросито, вибрационная сортировка
S 261	shaking table	Rütteltisch m	table f vibrante	вибрационный стол (стенд)
	shape/to	s. D 64		
S 262	shaping	Verformung f, spanlose Verformung f, Formgebung f	formage m, façonnage m	изменение формы, переформование, формоизменение

S 263	shear axis	Scherachse f	axe m de cisaillement	ось сдвига
	shear disk	s. R 426		
S 264	shear gradient	Schergefälle n	gradient m de cisaillement	градиент сдвига
S 265	shearing displacement	Scherverschiebung f	déplacement m de cissaillement	сдвиговое перемещение
S 266	shearing flow	Scherströmung f	écoulement m de cisaillement	сдвиговое течение
S 267	shearing force	Scherkraft f	force f de cisaillement	напряжение при сдвиге
S 268/9	shearing plane	Gleitebene f, Scherebene f	plan m de glissement	плоскость сдвига
	shearing strain (stress)	s. S 272		
S 270	shearing surface	Gleitfläche f, Scherfläche f	surface f de glissement	плоскость сдвига
S 271	shear modulus	Scherelastizitätsmodul m	module m d'élasticité au cisaillement	сдвиговый модуль упругости
S 272	shear strain, shear stress, shearing strain (stress)	Scherdehnung f, Scherspannung f, Schubspannung f	déformation f de (par) cisaillement, tension f (effort m) de cisaillement	касательное (сдвигающее) напряжение
S 273	shear strain	Scherverformung f	déformation f de cisaillement	формование сдвига
S 274	shear strength	Scherfestigkeit f	résistance f au cisaillement	сопротивление к сдвигу, предел прочности на сдвиг
S 275	shear stress	Scherspannung f, Scherbeanspruchung f	effort m (tension f) de cisaillement	напряжение сдвига
S 276	shear viscosity	Scherungsviskosität f	viscosité f au cisaillement	сдвиговая вязкость
S 277	sheet breaking	Aufbersten n der Schichten	éclatement m des couches	разрыв пластов
	sheeting	s. F 75		
S 278	sheet production	Folienherstellung f	fabrication f de feuilles	изготовление пленок
S 279	shelf time	Lagerzeit f, Lagerungsbeständigkeit f, Lagerstabilität f, Lagerfähigkeit f, Lagerungseignung f	durée f de stockage, aptitude f au stockage	время хранения, стабильность при хранении
	shell	s. H 202		
S 280	shell-and-tube cooler	Rohrbündelkühler m	refroidisseur m à faisceau tubulaire	кожухотрубный охладитель
S 281	shell-and-tube evaporator	Rohrbündelverdampfer m	évaporateur m multitubulaire, évaporateur à faisceau tubulaire	кожухотрубный испаритель
S 282	shell-and-tube heat exchanger	Rohrbündelwärmeübertrager m	échangeur m thermique à faisceau tubulaire	кожухотрубный теплообменник
S 283	shell-and-tube liquefier	Rohrbündelgasverflüssiger m	liquéfacteur m multitubulaire, condenseur m de gaz à faisceau tubulaire	кожухотрубный ожижитель
S 284	shell-and-tube reactor	Rohrbündelreaktor m	réacteur m à faisceau de tubes, réacteur à faisceau tubulaire	кожухотрубчатый реактор
S 285	shell-and-tube vessel	Rohrbündelapparat m	appareil m à faisceau de tubes, appareil multitubulaire, appareil à faisceau tubulaire	кожухотрубный сосуд, кожухотрубный аппарат
S 286	shell insulation	Mantelisolation f	isolement m extérieur (de manteau)	изоляция кожуха
S 287	shell still	Destillationsblase f	matras m à distillation, alambic m, vase m distillatoire	перегонный куб
S 288	shell-type condenser	Mantelkondensator m	condenseur m multitubulaire à enveloppe	кожуховый конденсатор
	shielding gas	s. I 122		
	ship/to	s. L 185		
S 289	shock wave	Stoßwelle f	onde f de choc	ударная волна
S 290	shop drawing	Konstruktionszeichnung f	plan m de construction, dessin m d'exécution	рабочий чертеж, конструкционный проект
	shortage	s. S 292		
S 291	short-circuit current	Kurzschlußstrom m, Bypass m	écoulement m en bypass	поток короткого замыкания, байпас
S 292	short weight, loss in weight; shortage	Gewichtsverlust m	perte f de poids	потеря в весе, потеря веса
	shoulder	s. L 305		
S 293	shower cooler, vertical cooler	Rieselkühler m	refroidisseur m à ruissellement	оросительный холодильник, градирня
	shrinkage stress	s. C 645		
S 294	shrinkage value	Masseverlust m, Schwundverlust m	perte f massique, perte f de retrait	величина усушки, потеря веса
S 295	shutdown	Betriebsstillegung f	interruption f de travail, arrêt m du travail	остановка, консервация [завода]
S 296	shut-off element	Absperrorgan n	organe m d'arrêt, robinet m d'arrêt	запорная арматура, стопор, запор
S 297	shut-off valve, gate valve	Absperrschieber m	vanne f, vanne d'arrêt	отсечная заслонка, запорная задвижка
S 298	shuttle conveyor	Pendelförderer m, Pendelbecherwerk n	élévateur m à godets oscillants, convoyeur m oscillant	транспортер по челночной схеме
S 299	siccative, drying agent	Trocknungsmittel n, Trockenmittel n	desséchant m, déshydratant m, agent m dessiccateur	сушильный агент, осушитель, высушивающее вещество

S 300	**side chain,** lateral chain	Seitenkette f	chaîne f latérale	боковая цепь
S 301	**sidestream,** circulating flow	Nebenstrom m, Kreislaufstrom m	by-pass m, écoulement m secondaire	байпас
S 302	**side wall burner**	Wandbrenner m	brûleur m latéral	настенная горелка
	siding	s. R 14		
S 303	**sieve bend**	Bogensieb n	crible m courbé	криволинейный грохот
	sieve drum	s. S 90		
S 304	**sieve grate**	Siebrost m	grille f de crible	ситообразная колосниковая решетка, колосниковый грохот
S 305	**sieve liquid,** receiver	Siebbehälter m	réservoir m de criblage	сборник для подсеточной воды
S 306	**sieve mesh**	Siebmasche f	maille f d'un tamis	ячейка в сите, меш
S 307	**sieve plate,** perforated bottom	Siebboden m	fond m perforé	ситчатая тарелка, решетчатое днище
	sieve residue	s. O 167		
S 308	**sifter,** separator	Sichter m	séparateur m	просеватель
	sifting	s. S 86		
S 309	**sigma blade mixer**	Z-förmiger Schaufelmischer m	malaxeur m à pales sigma	z-образный лопастной смеситель
S 310	**significance criterion**	Signifikanzkriterium n	critère m de signification	критерий значимости
S 311	**significance point**	Signifikanzgrenze f	limite f de signification	граница значимости
S 312	**significance study**	Signifikanzuntersuchung f	étude f de signification	исследование значимости
S 313	**sign of wear**	Abnutzungserscheinungen fpl	signes mpl d'usure, indices mpl de détérioration	признаки износа
	silo	s. B 94		
S 314	**similarity theory,** theory of similitude	Ähnlichkeitstheorie f	théorie f de la similitude	теория подобия
S 315	**simple distillation**	einfache Destillation f	distillation f simple	простая перегонка
S 316	**simple shear**	einfache Scherung f	cisaillement m simple	простой сдвиг
S 317	**simplex method**	Simplexverfahren n	méthode f simplexe	симплексный метод
	simulation	s. M 279		
S 318	**simulation computation**	Simulationsrechnung f	calcul m de simulation	расчет математической модели
S 319	**single-effect evaporator**	Einstufenverdampfer m	évaporateur m à simple effet	однокорпусной выпарной аппарат
S 320	**single-evaporator refrigerating machine**	Einzelverdampferkälteanlage f	installation f frigorifique à un seul évaporateur	одноиспарительная холодильная машина
S 321	**single-line process**	Einstrangverfahren n	procédé m en une ligne	технология (процесс) с одной линией (нитью)
S 322	**single nozzle**	Einzeldüse f	buse f unique, tuyau m individuel	[отдельное] сопло
S 323	**single-pass evaporator**	Einwegverdampfer m	évaporateur m à passage unique	одноходовой кожухотрубный испаритель
S 324	**single-phase fan**	Einphasenlüfter m	ventilateur m monophasé	однофазный вентилятор
S 325	**single-phase reactor**	Einphasenreaktor m	réacteur pour un état physique	реактор с одной фазой, реактор с охладителем неизменного агрегатного состояния
S 326	**single-screw extruder (injection machine)**	Einschneckenspritzgußmaschine f	machine f de moulage à vis unique	одношнековая литьевая машина, одношнековый экструдер
S 327	**single-stage absorber**	einstufiger Absorber m	absorbeur m à un étage	одноступенчатый абсорбер
S 328	**single-stage absorption**	Einstufenabsorption f	absorption f à simple effet	одноступенчатая абсорбция
S 329	**single-stage compression**	einstufige Kompression (Verdichtung) f	compression f à un étage	одноступенчатое сжатие
S 330	**single-stage compressor**	einstufiger Verdichter m	compresseur m à simple effet	одноступенчатый компрессор
S 331	**single-stage design**	einstufige Ausführung f	type m à simple effet	одноступенчатая конструкция
S 332	**single-stage distillation**	Einstufendestillation f	distillation f à un étage	одноступенчатая дистилляция
S 333	**single-stage drive**	einstufiger Antrieb m	commande f monoétagée	одноступенчатый привод
S 334	**single-stage process**	Einstufenprozeß m	procédé m à simple effet	одноступенчатый процесс
S 335	**single-step method**	Einzelschrittverfahren n	méthode pas à pas	одношаговый метод
S 336	**single-strand large-scale plant**	einsträngige Großanlage f	installation f de grande capacité en une ligne	однониточная крупнотоннажная установка
S 337	**single-strand unit**	Einstranganlage f	installation f en une ligne	единичная (однониточная) установка
S 338	**single-zone reactor**	Einzonenreaktor m	réacteur m à zone unique	однозонный реактор
	sinker	s. D 293		
S 339	**sintering**	Sintern n, Sinterung f	frittage m	спекание, агломерация, получение синтера
S 340	**sintering apparatus**	Sinterapparat m	appareil m de frittage	спекательный аппарат
S 341	**sintering furnace**	Sinterofen m	four m de frittage	агломерационная печь
S 342	**sintering heat,** clinkering temperature	Sinterungstemperatur f	température f de frittage	температура спекания
S 343	**sintering plant**	Sinteranlage f	installation f d'agglomération par frittage	спекательная установка
S 344	**sinuous motion**	schlangenförmige (sinusförmige) Bewegung f	mouvement m sinusoïdal	змееобразное течение

S 345	sinusoidal loading	sinusförmige Belastung f	charge f sinusoïdale	синусообразная нагрузка
S 346	siphon	Saugheber m	siphon m	сифон, пульсометр
	site	s. C 607		
S 347	site connections, site traffic network	Baustellenverkehrsnetz n	installations fpl de transport au chantier	транспортная сеть строительной площадки
S 348	site facilities (installation, organization)	Baustelleneinrichtung f	aménagement m de chantier	оборудование строительной площадки
S 349	site regulations	Baustellenordnung f	instructions fpl de chantier	инструкция строительства (строительной площадки)
S 350	site structure	Baustellenstruktur f	structure f de chantier	структура строительства (строительной площадки)
	site traffic network	s. S 347		
S 351	six-tenth-rule	Zwei-Drittel-Regel f (Degressionsexponent)	règle f des deux tiers	правило «две третьих»
S 352	size classification	Klassieren n, Klassierung f, Trennen n nach Korn[grö-ßen]klassen	triage m, classement m	классификация [по крупности], сортировка
S 353	size distribution	Größenverteilung f	répartition f granulométrique	распределение по величине
S 354	size-distribution curve	Körnungskennlinie f, Kornverteilungskurve f	courbe f de répartition granulométrique	кривая рассева (гранулометрического состава)
S 355	size limit	Größenbereich m, Grenzgröße f	domaine m de grandeurs, grandeur f limite	диапазон размеров
S 356	size loss	Maßverlust m, Größenverlust m	perte f de mesure (taille)	потеря в размере
S 357	size of apparatus	Apparategröße f	dimensions fpl de l'appareil	размер (габариты) аппарата
S 358	size of separation, cut size	Trennkorngröße f	maille f de partage (coupure équivalente)	крупность разделения
S 359	size of sieve	Siebgröße f	numéro m de tamis	номер (размер) сита
S 360	size range	Korngrößenbereich m, Kornklasse f, Körnungsband n	fraction f (intervalle m) granulométrique	зернистость, класс крупности зерен
S 361	size reduction	Zerteilen n, Zerteilung f; Zerkleinern n, Zerkleinerung f	concassage m, broyage m, désintégration f	измельчение, дробление
	sizing	s. S 436		
	sizing drum	s. S 90		
S 362	sizing in grain groups	Sortierung f in Korngruppen	triage m fractionné, triage en intervalles granulométriques	сортировка по фракциям гранулометрического состава
	sizing machine	s. S 435		
S 363	skeleton rigidity	Skelettsteifigkeit f, Systemsteifigkeit f	rigidité f de l'ossature	жесткость скелета (системы)
S 364	slag, cinder, scoria, dross	Schlacke f	scorie f, laitier m	шлак
S 365	slag concrete	Schlackenbeton m	béton m de scorie (laitier)	шлаковый бетон, шлакбетон
S 366/7	slag skimmer	Schlackenabscheider m	séparateur m de laitier (scorie)	шлакоотделитель
S 368	slide face	Gleitfläche f	plan m (surface f) de glissement, glissière f	плоскость скольжения
	slide valve	s. G 102		
S 369	sliding friction	Gleitreibung f, gleitende Reibung f, Reibung f	frottage m (friction f) de glissement	трение скольжения, трение
S 370	sliding means	Gleitmittel n	lubrifiant m, agent m antifriction	средство скольжения
	slime	s. S 379		
S 371	slime pit, settling pond, sludge sump	Schlammbecken n, Absetzbecken n	bassin m de dépôt, bassin à schlamms	осадительный (осадочный) бассейн, грязевой амбар
S 372	slime separator	Schlammabscheider m, Schlammabsetzbecken n	bassin m de séparation de boue, bassin à schlamms	грязеотделитель, шламоотделитель
S 373	slip	Gleiten n, Rutschen n, Schlupf n, Gleitfähigkeit f	glissement m	скольжение
S 374	slippage factor	Rutschfaktor m	facteur m de glissement	фактор скольжения
S 375	slippage resistance	Gleitwiderstand m	résistance f au glissement	сопротивление скольжению
S 376	slip-stick motion	ruckhafte Gleitbewegung f	mouvement m de glissement et d'adhérence	скачкообразное движение скольжения
S 377	slit capillary	Schlitzkapillare f, Spalt m	tube m capillaire en forme de fente	щелевой канал (капилляр), щель
S 378	slop oil	Abfallöl n, Altöl n	résidu m d'huile, huile f usée	некондиционный (отбросный) нефтепродукт
S 379	sludge, slime, mud, slurry	Schlamm m	limon m, boue f, vase f, bourbe f	шлам, ил, грязь
	sludge activation	s. B 108		
	sludge activation plant	s. B 109		
S 380	sludge pump, sludger	Schlammpumpe f	pompe f à boue	шламовый насос
	sludge sump	s. S 372		
	sludge water	s. M 357		
	slugging	s. B 255		
	slurry	s. S 379		
	slurry pump	s. M 356		
S 381	slurry reactor	Suspensionsreaktor m, Sumpfphasenreaktor m	réacteur m à schlamm	шламовый (барботажный) реактор

S 382	**small lifting gears**	Kleinhebezeuge *npl*	organes *mpl* de levage petits	маломощные (малогабаритные) подъемные приспособления
S 383	**small-molecule fluid**	mikromolekulares fluides Medium *n*	fluide *m* micromoléculaire	микромолекулярная жидкая среда
S 384	**small-scale production**	kleintonnagige Produktion *f*	production *f* en petites quantités	малотоннажное производство
S 385	**small-scale test**	Modellversuch *m*	essai *m* sur modèle	модельный опыт
S 386	**small steam turbine**	Kleindampfturbine *f*	petite turbine *f* à vapeur	паровая турбина малой мощности
	smoke gas	*s.* F 273		
S 387	**smoke out/to**	ausdampfen	faire évaporer	отпаривать
S 388	**smooth flow**	Laminarströmung *f*	courant *m* laminaire, écoulement *m* laminaire	ламинарное течение
S 389	**smooth tube**	glattes Rohr *n*, Glattrohr *n*	tube *m* lisse	гладкая труба
S 390	**soaker**	Reaktionskammer *f*	chambre *f* de réaction	реакционная камера, реакционный змеевик
S 391	**softened water**	Süßwasser *n*, Weichwasser *n*	eau *f* douce	умягченная вода
	softener	*s.* P 207		
S 392	**softening**	Enthärtung *f*	adoucissement *m*	мягчение, умягчение, смягчение
S 393	**softening**, plasticization	Erweichung *f*, Weichwerden *n*	ramollissement *m*	размягчение, понижение
S 394	**softening point**, sagging point	Erweichungspunkt *m*	point *m* de ramollissement	точка размягчения, точка плавления
	soil/to	*s.* C 624		
	soil	*s.* C 627		
S 395	**soil temperature**	Bodentemperatur *f*	température *f* de sol	температура грунта
S 396	**sojourn probability**	Aufenthaltswahrscheinlichkeit *f*	probabilité *f* de séjour	вероятность местонахождения (пребывания)
S 397	**sojourn time**	Verweilzeit *f*	temps *m* de séjour	время пребывания
S 398	**solar energy**	Sonnenenergie *f*	énergie *f* solaire	солнечная энергия
S 399	**solid**	fester Körper *m*, Festkörper *m*, Feststoff *m*	solide *m*, corps *m* [solide], matière *f* solide	твердое тело (вещество)
S 400	**solid content**	Feststoffgehalt *m*	teneur *m* en matière solide	содержание твердого вещества
S 401	**solid feed rate**	Feststoffdurchsatz *m*	vitesse *f* (débit *m*) de matière solide	скорость [подачи] твердого вещества, скорость твердой фазы
S 402	**solidification**	Festwerden *n*, Erstarren *n*, Erstarrung *f*, Verfestigung *f*	solidification *f*, prise *f*, congélation *f*	затвердение, застывание
S 403	**solidification point**, congealing (setting, gel) point	Erstarrungspunkt *m*, Verfestigungspunkt *m*	point *m* de solidification	точка затвердевания, точка застывания
S 404	**solidification process**	Erstarrungsprozeß *m*	processus *m* de solidification	процесс затвердевания
S 405	**solidification range**	Erstarrungsbereich *m*	domaine *m* de solidification	область затвердевания
S 406	**solidification temperature**	Erstarrungstemperatur *f*	température *f* de solidification	температура затвердевания
S 407	**solid matter separation**	Feststofftrennung *f*	séparation *f* de la matière solide	сепарация твердого вещества
S 408	**solid mechanics**	Feststoffmechanik *f*	mécanique *f* des solides	механика твердых тел
S 409	**solid storage**	Feststofflagerung *f*	stockage *m* de matière solide	хранение твердого вещества
S 410	**solid surface**	Feststoffoberfläche *f*	surface *f* des solides	поверхность твердого вещества
S 411	**solid transport**	Feststofftransport *m*	transport *m* de matières solides	расход (транспорт) твердых веществ
S 412	**solubility**; solvent power	Auflösefähigkeit *f*, Löslichkeit *f*, Lösbarkeit *f*, Lösungsvermögen *n*	solubilité *f*, pouvoir *m* dissolvant	растворимость, способность к растворению
S 413	**solubility difference**, difference in solubilities	Löslichkeitsunterschied *m*	différence *f* de solubilité	разность растворимости
S 414	**solubility product**	Löslichkeitsprodukt *n*	produit *m* de solubilité	произведение растворимости
S 415	**solubilization process**	Lösungsprozeß *m*	processus *m* de dissolution	процесс растворения
	solubilizing power	*s.* D 403		
S 416	**solute phase**	Phase *f* des gelösten Stoffes	phase *f* du soluté	фаза растворенного вещества
S 417	**solution**	Lösung *f*, Auflösung *f*	solution *f*, décomposition *f*, dissolution *f*	раствор, растворение, решение
S 418	**solution addition**	Lösungszufuhr *f*, Lösungszuführung *f*	amenée (adduction) *f* de solution	добавление раствора
S 419	**solution charge**	Lösungsfüllmenge *f*	charge *f* de solution	количество раствора для зарядки системы
S 420	**solution composition**	Lösungszusammensetzung *f*	composition *f* de solution	состав раствора
S 421	**solution concentration**	Lösungskonzentration *f*	concentration *f* de solution	концентрация раствора
S 422	**solution consumption**	Lösungsverbrauch *m*	consommation *f* de solution	расход раствора
S 423	**solution contamination**	Lösungsverunreinigung *f*	contamination *f* de la solution	загрязнение раствора
S 424	**solution cooler**	Lösungskühler *m*	refroidisseur *m* de solution	охладитель раствора
S 425	**solution curve**	Lösungskurve *f*	courbe *f* de dissolution	кривая растворимости
S 426	**solution equilibrium**	Lösungsgleichgewicht *n*	équilibre *m* de dissolution	равновесие в растворе, равновесный раствор

S 427	solution escape	Lösungsverlust *m*	perte *f* de solution	вытекание раствора
S 428	solution feed-back	Lösungsrückführung *f*	recirculation *f* de solution	обратная подача раствора
S 429	solution pump	Lösungspumpe *f*	pompe *f* de (à) solution	насос для раствора
S 430	solution recirculation	Lösungsumlauf *m*	circulation *f* de solution	рециркуляция (циркуляция) раствора
S 431	solution variant	Lösungsvariante *f*	variante *f* de solution	вариант решения
S 432	solvent extraction	Lösungsmittelextraktion *f*	extraction *f* de (à) solvant	селективная экстракция, извлечение селективным растворителем
S 433	solvent phase	Lösungsmittelphase *f*	phase *f* de solvant	фаза растворителя
	solvent power	*s.* S 412		
	sorbate	*s.* A 203		
S 434	sorption	Sorption *f*	sorption *f*	сорбция, поглощение
S 435	sorter, sizing machine	Sortierer *m*	classeur *m*, trieur *m*	сортировочная машина, сортировщик
S 436	sorting, assorting, assortment, sizing, classification	Sortieren *n*, Sortierung *f*, Klassieren *n*, Klassierung *f*	calibrage *m*, triage *m*, assortiment *m*, classification *f*, classage *m*, classement *m*	сортировка, разборка, классификация, обогащение, разбирание, сортирование
S 437	sorting drum	Sichttrommel *f*	tambour *m* de triage	сортировальный цилиндр
S 438	sound absorption, sound attenuation, noise absorption	Schalldämpfung *f*	isolation *f* phonique, amortissement *m* de son	звукоизоляция
	sound control	*s.* S 443		
S 439	sound insulation, sound proofing	Schalldämmung *f*	isolation *f* acoustique, insonorisation *f*	звукоизоляция
S 440	sound level	Schallpegel *m*	niveau *m* acoustique	уровень шума
S 441	sound pressure	Schalldruck *m*	pression *f* acoustique	звуковое давление
	sound proofing	*s.* S 439		
S 442	sound propagation	Schallausbreitung *f*	propagation *f* du son	распространение звука
S 443	sound protection, sound control	Schallschutz *m*	amortissement *m* des bruits	звуковая защита, звукозащита
S 444	sound source, source of sound	Schallquelle *f*	source *f* sonore	источник звука
S 445	sound velocity, velocity of sound	Schallgeschwindigkeit *f*	vitesse *f* du son	скорость звука
S 446	sound wave	Schallwelle *f*	onde *f* sonore, ondulation *f* du son	звуковая волна
S 447	source of electricity	Elektrizitätsquelle *f*	source *f* d'électricité	источник электричества
S 448/9	source of energy, source of power, power source	Energiequelle *f*	source *f* énergétique (d'énergie)	источник энергии (питания), силовой источник
	source of heat	*s.* H 105		
S 450	source of losses	Verlustquelle *f*	source *f* des pertes	источник потерь
	source of power	*s.* S 448/9		
	source of sound	*s.* S 444		
	space filter	*s.* G 2		
S 451	space model	Raummodell *n*, räumliches Modell *n*	modèle *m* spatial	объемная (пространственная) модель
S 452	space required, spatial requirement	Raumbedarf *m*, Platzbedarf *m*	encombrement *m*, emplacement *m* nécessaire	потребность в объеме
S 453	space-time yield	Raum-Zeit-Ausbeute *f*	rendement *m* espace-temps	пространственно-временной выход
S 454	spacing	Bodenabstand *m*	distance *f* entre plateaux	расстояние между тарелками [колонки], расстояние, шаг
S 455	span	Stützweite *f*	portée *f*	размах, расстояние между опорами
	spare	*s.* S 893		
S 456	sparingly soluble, hardly soluble	schwer löslich	soluble difficilement	трудно растворимый
S 457	spark discharge	Funkenentladung *f*	décharge *f* à étincelles	искровой разряд
S 458	sparking	Funkenbildung *f*	formation *f* des étincelles	искрообразование, искрение
	spatial requirement	*s.* S 452		
S 459	special design	Sonderkonstruktion *f*	construction *f* spéciale	специальная конструкция
S 460	special equipment	Sonderausrüstung *f*	équipement *m* spécial	специальное оборудование
S 461	special transport	Spezialtransport *m*	transport *m* spécial	специальный транспорт
	specification	*s.* I 252; P 67		
S 462	specifications	Abnahmevorschriften *fpl*	prescriptions *fpl* pour la réception	инструкции по приемке
S 463	specific conductance (conductivity)	spezifische Leitfähigkeit *f*	conductivité (conductibilité) *f* spécifique	удельная электропроводность (проводимость)
S 464	specific enthalpy	spezifische Enthalpie *f*	enthalpie *f* spécifique	удельная энтальпия
S 465	specific exergy costs	spezifische Exergiekosten *pl*	frais *mpl* exergétique spécifique	удельные расходы эксергии
S 466	specific gravity	Wichte *f*, spezifisches Gewicht *n*	poids *m* spécifique	удельный вес
S 467	specific heat	Eigenwärme *f*, spezifische Wärme *f*	chaleur *f* propre (sensible, spécifique)	теплоемкость, удельная теплота
S 468	specific humidity	spezifische Feuchtigkeit *f*	humidité *f* spécifique	удельная влажность
S 469	specific price	spezifischer Preis *m*	prix *m* spécifique	удельная стоимость
	specific reaction rate	*s.* V 137		

S 470	**specific refrigerating effect**	spezifische Kälteleistung f	puissance f frigorifique spécifique	удельная холодопроизводительность
S 471	**specific resistance,** resistivity	spezifischer Widerstand m	résistance f spécifique	удельное сопротивление
S 472	**specific rotation**	spezifische Drehung f	rotation f spécifique	удельное вращение
S 473	**specific viscosity**	spezifische Zähigkeit f	viscosité f spécifique	удельная вязкость
S 474	**specific volume**	Eigenvolumen n	volume m propre	собственный (удельный) объем
S 475	**speed,** number of revolutions	Drehzahl f	nombre m de tours	число оборотов
S 476	**speed change gear**	Wechselgetriebe n	boîte f de vitesse	коробка передач
S 477	**speed variation**	Geschwindigkeitsänderung f	variation f de vitesse	изменение скорости
	spent lye	s. W 48		
S 478	**spherical tank**	Kugeltank m	réservoir m sphérique	шаровой (шарообразный) резервуар
S 479	**spider threads**	Spinnstrang m	échevau m de fil	прядь волокон
	spinneret	s. S 482		
S 480	**spinning bath**	Spinnbad n	bain m de filage	прядильная баня (ванна)
S 481	**spinning machine**	Spinnapparat m, Spinnmaschine f	métier m à filer	прядильный орган, прядильная фильера
S 482	**spinning nozzle,** spinneret	Spinndüse f	filière f	фильера
S 483	**spinning pump**	Spinnpumpe f	pompe f à filage	прядильный насос
S 484	**spinning solution**	Spinnlösung f	solution f à filer	прядильный раствор
	spiral blower	s. H 126		
S 485	**spiral cleaner**	Spiralscheider m, Wendelscheider m	séparateur m spiral	вихревой (спиральный) очиститель
	spiral condenser	s. C 355		
	spiral conveyor	s. C 684		
	spiral mixer	s. H 128		
S 486	**spiral tube**	Schlangenrohr n	serpentin m	змеевик
	splash water	s. D 506		
S 487	**split condenser**	einzeln angeordneter Kondensator m	condenseur m isolé (individuel)	отдельно монтируемый конденсатор
S 488	**splitting off**	Abspaltung f, Spaltung f	fission f	отщепление, расщепление, раскалывание
S 489	**splitting reaction**	Abspaltreaktion f	réaction f de fission	реакция отщепления
S 490	**splitting resistance**	Abtrennungswiderstand m	résistance f au détachement	прочность на расслоение
S 491	**splitting up**	Aufspaltung f (Benzin), Spaltung f, Kracken n	cracking m	расщепление
S 492	**spontaneous combustion (ignition)**	Selbstentzündung f	ignition (inflammation) f spontanée	самовозгорание, самовоспламение
S 493	**spontaneously inflammable,** self-igniting	selbstentzündlich	inflammable spontanément	самовозгорающийся
S 494	**spot analysis,** drop analysis	Tüpfelanalyse f	analyse f à la touche, analyse par goutte	капельный анализ
S 495	**spot cooling**	lokale Kühlung f	refroidissement m localisé	местное охлаждение
S 496	**spray/to**	verdüsen, versprühen	pulvériser	распылять
S 497	**spray chamber**	Diffusionskammer f	chambre f de diffusion	орошаемая камера распыленной жидкостью
S 498	**spray column**	Dispersionskolonne f (Extraktion), Sprühkolonne f	colonne f à dispersion (pulvérisation), colonne de distillation à pulvérisation	распылительная колонна
S 499	**spray condenser**	Berieselungsverflüssiger m, Rieselkondensator m	condenseur m à ruissellement	орошаемый конденсатор
	spray condenser	s. a. S 501		
S 500	**spray conduit**	Berieselungsleitung f	conduite f de ruissellement	оросительный трубопровод
S 501	**spray cooler,** spray condenser	Rieselkühler m, Berieselungskühler m	refroidisseur (condenseur) m à ruissellement	пленочная градирня
S 502	**spray cooler**	Sprühkühler m	refroidisseur m à pulvérisation	форсуночный охладитель
S 503	**spray cooling**	Sprühkühlung f	refroidissement m à pulvérisation	охлаждение орошением (разбрызгиванием)
S 504	**spray dryer**	Sprühtrockner m, Zerstäubungstrockner m	sécheur-pulvérisateur m, séchoir m à pulvérisation	распылительная сушилка
S 505	**sprayer,** spraying plant	Sprühanlage f	installation f de pulvérisation	разбрызгивающее устройство
S 506	**sprayer;** scrubber	Sprühapparat m	appareil m de pulvérisation, scrubber m	скруббер, распылительный аппарат
	sprayer	s. a. 1. A 599; 2. D 572		
S 507	**spraying**	Zerstäuben n (Flüssigkeiten), Verdüsung f, Verdüsen n	atomisation f, pulvérisation f	распыление, распыливание
	spraying nozzle	s. I 184		
	spraying plant	s. S 505		
S 508	**spraying process**	Sprühverfahren n	procédé m de pulvérisation	обрызгивание, распыление
S 509	**spray pipe**	Sprührohr n	tube m de pulvérisation	орошающая труба
S 510	**spray pump**	Sprühpumpe f	pompe f à pulvérisation	насос для орошения
S 511	**spray tower**	Sprühturm m, Rieselturm m	tour f à pulvérisation (ruissellement)	скруббер [с брызгалом], башня, орошаемая распыленной жидкостью
S 512	**spray-type evaporator**	Sprühverdampfer m, Berieselungsverdampfer m	évaporateur m à pulvérisation	оросительный испаритель
S 513	**spray-type heat exchanger**	Rieselwärmeübertrager m	échangeur m thermique à ruissellement	теплообменник оросительного типа

S 514	spray water	Sprühwasser *n*	eau *f* pulvérisée	орошающая вода
S 515	spray zone	Sprühzone *f*	zone *f* de pulvérisation	зона орошения
	spreader	*s.* D 443		
S 516	sprinkling	Benetzung f, Besprühung f, Befeuchten *n*, Berieselung *f*	arrosage *m*, ruissellement *m*	смачивание, обрызгивание, орошение
	sprung	*s.* C 811		
	squeezing out	*s.* P 392		
S 517	stability analysis	Stabilitätsanalyse *f*	analyse *f* de stabilité	анализ устойчивости
S 518	stability behaviour	Stabilitätsverhalten *n*	comportement *m* de stabilité	поведение [процесса] в устойчивом состоянии
S 519	stability condition	Stabilitätsbedingung *f*	condition *f* de stabilité	условие устойчивости
S 520	stability criterion	Stabilitätskriterium *n*	critère *m* de stabilité	критерий устойчивости
S 521	stack gas	Abgas n, Rauchgas *n*	gaz *m* d'échappement, gaz perdu (de fumée)	отходный (дымовой) газ
S 522	stage contactor	Etagenextraktor *m*	extracteur *m* à plusieurs étages	многоступенчатый экстрактор
S 523	stage of completion, completion degree	Komplettierungsgrad *m*	degré *m* de complétage	степень комплексности
	stage of construction	*s.* S 151		
S 524	stage of decomposition	Abbaustufe *f*	produit *m* de décomposition, degré *m* de dissolution	степень распада (разложения)
S 525	stage of mixing	Mischstufe *f*	étage *m* de mélange	смесительный каскад
S 526	stage of process	Verfahrensstufe *f*	stade *m* de procédé	ступень (подсистема) процесса
	stamp	*s.* I 60		
S 527	stanchion, tower, column	Säule f, Kolonne f, Turm *m*	colonne f, poteau *m* (*Bauwesen*), tour *f*	столб, стойка, колонна
S 528	standard air	Standardluft *f*	air *m* normal	стандартный воздух
S 529	standard capacity	Normalleistung *f*	puissance *f* normale	стандартная (нормальная) нагрузка
S 530	standard component	Baueinheit *f*	élément *m* de construction, baie *f*	элемент стандартного нормирования, [стандартизированный] элемент
S 531	standard component part	Normbauteil *n*	élément *m* normalisé	стандартная часть конструкции
S 532	standard condenser	Standardkondensator *m*	condenseur *m* standard	стандартный конденсатор
S 533	standard design	Normalausführung f, Standardausführung *f*	type *m* standard (normal)	стандартное исполнение, нормальный вариант
S 534	standard design series	Baureihe *f*	série *f* de fabrication	конструктивный ряд
S 535	standard deviation	Standardabweichung *f*	écart-type *m*	стандартное отклонение
S 536	standard element, normal element	Normalelement *n*	pile *f* standard	нормальный элемент
S 537	standard equilibrium potential	Standardgleichgewichtspotential *n*	potentiel *m* d'équilibre standard	стандартный равновесный потенциал
S 538	standard equipment	Standardausrüstung *f*	équipement *m* standard	стандартное оборудование
S 539	standard evaporator	Standardverdampfer *m*	évaporateur *m* standard	стандартный испаритель
S 540	standardization, uniformalization	Vereinheitlichung f, Unifizierung *f*	standardisation f, unification	стандартизация, унификация
S 541	standard load	Normalbelastung *f*	charge *f* normale	нормальная нагрузка
	standard measure	*s.* G 105		
S 542	standard method	Standardmethode f, Einheitsmethode *f*	méthode *f* normale	стандартный метод
S 543	standard rating	Normleistung *f*	puissance *f* standardisée	нормальная мощность
S 544	standard repair	Standardreparatur *f*	réparation *f* standard	стандартный ремонт
S 545	standards; rules	Normen *fpl*	normes *fpl*, règles *fpl*	стандарты, нормы
S 546	standard size	Standardgröße f, Einheitsgröße *f*	grandeur *f* standard	стандартный размер
S 547	standard solution, normal solution	Normallösung *f*	solution *f* standard (type, de réference)	нормальный раствор
S 548	standard solution	Titrierflüssigkeit *f*	solution *f* normale (de titrage)	титрованный раствор
S 549	standard specification	Normvorschrift f, Norm *f*	exigence *f* des normes (cahiers de charge), norme *f*	технические условия, нормы
S 550	standard substance	Bezugssubstanz *f*	substance *f* de référence	сравнительное вещество
S 551	standard system	Einheitssystem *n*	système *m* standardisé	единая (унифицированная, стандартная) система
S 552	standard type	Standardmodell *n*	modèle *m* standard	стандартная (унифицированная) модель
S 553	standard value; constant	Festwert *m*	constante f, valeur *f* fixe, paramètre *m*	параметр, константа, постоянная величина
S 554	standby cooler	Hilfskühler m, Reservekühler *m*	refroidisseur *m* auxiliaire	резервный охладитель
S 555	standby filter	Hilfsfilter n; Ersatzfilter n, Reservefilter *n*	filtre *m* auxiliaire	запасный (резервный) фильтр
S 556	standby reactor	Hilfsreaktor m; Reservereaktor *m*	réacteur *m* auxiliaire	запасный (резервный) реактор
S 557	standstill	Stillstand *m*	arrêt *m*	простой, остановка
S 558	Stanton number	Stanton-Zahl f, Stantonsche Kennzahl *f*	nombre *m* de Stanton	число (критерий) Стантона
S 559	starting; start-up	Inbetriebsetzung *f*	mise *f* en mouvement (marche), démarrage *m*	пуск, запуск

S 560	starting activity	Anfangsaktivität f	activité f initiale	начальная активность
S 561	starting behaviour	Anfahrverhalten n	comportement m au démarrage	поведение при пуске, приемистость
S 562	starting curve, warming-up curve	Anlaufkurve f	courbe f de démarrage	кривая набегания (притока)
S 563	starting loading, initial loading	Anfahrbelastung f	charge f de (au) démarrage	пусковая нагрузка
	starting material	s. R 63		
	starting operation	s. S 573		
	starting point	s. B 42		
S 564	starting section (of suction)	Ansaugquerschnitt m	section f d'aspiration	поперечное сечение органов всасывания
	starting stress	s. I 158		
S 565	starting value	Startwert m, Anfangswert m	valeur f initiale (de départ)	начальное значение
S 566	start reaction, initiating reaction	Startreaktion f	réaction f initiale	инициирующая реакция
	start-up	s. S 559		
S 567	start-up behaviour	Anfahrverhalten n	comportement m à la mise en marche, comportement au démarrage	поведение в период пуска
S 568	start-up control	Anlaufkontrolle f	contrôle m de démarrage	пусковая контроль, проверка перед пуском
S 569	start-up curve	Einlaufkurve f	courbe f de mise en œuvre	кривая разгона
S 570	start-up flow	Startströmung f	écoulement m initial (de début)	начальное течение
S 571	start-up period	Anfahrperiode f	période f de démarrage	пусковой период
S 572	start-up preparation	Anfahrvorbereitung f	préparation f de la mise en marche	подготовка к пуску
S 573	start-up process, starting operation	Anfahrregime n	transitoire m de démarrage	режим пуска
	state equation	s. E 220		
S 574	state of aggregation	Aggregatzustand m	état m physique (d'agrégation)	агрегатное состояние
S 575	state of equilibrium	Gleichgewichtslage f, Gleichgewichtszustand m	état m (position f) d'équilibre	положение равновесия
S 576	state of motion	Bewegungszustand m	état m de mouvement	состояние движения
S 577	state of steam	Dampfzustand m	état m de vapeur	парообразное состояние
S 578	state-time diagram	Zustands-Zeit-Diagramm n	diagramme m temps-état	диаграмма время-состояние
S 579	state value (variable)	Zustandsvariable f, Zustandsgröße f, Zustandsveränderliche f	variable f d'état	переменная состояния
S 580	static efficiency	statischer Wirkungsgrad m	rendement m statique	коэффициент полезного действия по статическим параметрам
S 581	static mixer	Statikmischer m	mélangeur m statique	статический смеситель
S 582	static operation recycle	Kreislauffahrweise f	fonctionnement m en circuit fermé	рециркуляционный режим, режим циркуляции, циркуляция, круговорот
S 583	static stability	Standsicherheit f, Stabilität f	stabilité f	стабильность, устойчивость
	station	s. L 219		
S 584	stationary behaviour	stationäres Verhalten n	comportement m stationnaire, régime m permanent	стационарное поведение
S 585	stationary flow, steady[-state] flow	stationäres Fließen n, stationäre Strömung f	courant (écoulement) m stationnaire	стационарное (установившееся) течение
S 586	stationary lifting gears	[stationäre] Hebezeuge fpl	appareils mpl de levage immobiles	стационарные подъемные приспособления
S 587	stationary operating point	Betriebspunkt m	état m de service stationnaire	рабочая точка, точка эксплуатации
S 588	stationary process	stationärer Prozeß m	procédé m stationnaire	стационарный (установившийся) процесс
	statistical model	s. E 114		
	steady flow	s. S 585		
S 589	steady product temperature	beständige (konstante) Produkttemperatur f	température f constante du produit	устойчивая температура продукта
S 590	steady state	stationärer (stabiler) Zustand m, Beharrungszustand m, Dauerzustand m	état m stable (stationnaire, stable de régime), régime m permanent	стационарное (длительное, постоянное, установившееся) состояние
	steady-state flow	s. S 585		
S 591	steam accumulator, heat accumulator, [steam] storage vessel	Dampfspeicher m	accumulateur m de vapeur	паровой аккумулятор
S 592	steam apparatus	Dampfapparat m	appareil m à vapeur	пароиспользующий аппарат
S 593	steam bath	Dampfbad n	bain m de vapeur	паровая баня
S 594	steam blower	Dampfgebläse f	soufflerie f à vapeur	паровая воздуходувка
S 595	steam blow-off pipe	Dampfabblasrohr n	tuyau m d'échappement à vapeur	патрубок для выпуска пара
S 596	steam boiler	Dampfkessel m	chaudière f à vapeur	паровой котел
S 597	steam bubble	Dampfblase f	bulle f de vapeur	пузырь пара
S 598	steam coil	Dampf[heiz]schlange f	serpentin m à vapeur	паровой змеевик
	steam condensation	s. V 81		

S 599	**steam consumption**	Dampfverbrauch *m*	consommation *f* de vapeur	расход пара
S 600	**steam content,** vapour content	Dampfgehalt *m*	teneur *m* en vapeur	паросодержание
S 601	**steam coolant**	Kühldampf *m*	vapeur *f* de refroidissement	охлаждающий пар
S 602	**steam-cooled**	dampfgekühlt	refroidi par la vapeur	охлаждаемый паром
S 603	**steam cooler**	Dampfkühler *m*	désurchauffeur *m*	паровой холодильник
S 604	**steam cracking**	Dampfkracken *n*	cracking *m* à la vapeur	крекинг с водяным паром
S 605	**steam cracking process**	Dampfkrackverfahren *n*	procédé *m* steam-cracking	крекинг-процесс с водяным паром
S 606	**steam cycle**	Dampfkreislauf *m*	circulation *f* de vapeur	паровой контур (цикл)
S 607	**steam cylinder**	Dampfzylinder *m*	cylindre *m* à vapeur	паровой цилиндр
S 608	**steam demand,** vapour demand	Dampfbedarf *m*	demande *f* (besoins *mpl*) en vapeur	потребность в паре
	steam density	*s.* V 85		
S 609	**steam desiccator**	Dampfentwässerer *m*	sécheur *m* de la vapeur, séparateur *m* d'eau	осушитель пара
S 610	**steam developer**	Dampfentwickler *m*	appareil-producteur *m* de vapeur	парогенератор, паровой котёл
S 611	**steam diagram**	Dampfdiagramm *n*	diagramme *m* de vapeur	диаграмма состояния пара
S 612	**steam distillation**	Dampfdestillation *f*, Wasserdampfdestillation *f*	distillation *f* par entraînement à la vapeur, distillation (entraînement *m*) à la vapeur d'eau	перегонка с водяным паром
S 613	**steam distillation plant**	Dampfdestillationsanlage *f*, Wasserdampfdestillationsanlage *f*	unité *f* de distillation à la vapeur d'eau	установка перегонки с водяным паром
S 614	**steam distribution**	Dampfverteilung *f*	distribution *f* de vapeur	распределение пара
	steam dryer	*s.* W 108		
S 615	**steam drying**	Dampftrocknung *f*	séchage *m* de (à) vapeur	сушка в паровых сушилках
S 616	**steam emulsion**	Dampfemulsion *f*	émulsion *f* à la vapeur	эмульсия с паром, паровая эмульсия
S 617	**steam exhaust**	Abdampfaustritt *m*	sortie *f* de la vapeur épuisée	выпуск отработавшего пара
S 618	**steam exhaust,** steam outlet	Dampfausströmung *f*	échappement *m* de la vapeur	истечение (выпуск) пара
S 619	**steam filter**	Dampffilter *n*	filtre *m* de vapeur	паровой фильтр
S 620	**steam flow**	Dampfstrom *m*	courant *m* de vapeur	поток пара, паровой поток
S 621	**steam flow**	Dampfdurchsatz *m*	débit *m* de vapeur	расход пара
S 622	**steam gauge,** steam pressure gauge	Dampfdruckmesser *m*, Manometer *n*	manomètre *m*	манометр
S 623	**steam-generating heat**	Dampfbildungswärme *f*, Dampfbildungsenthalpie *f*	chaleur *f* de vaporisation	теплота парообразования
S 624	**steam-generating plant**	Dampferzeugungsanlage *f*	appareil *m* vaporifère, générateur *m* de vapeur	котельная установка
S 625	**steam generation**	Dampferzeugung *f*	génération *f* de la vapeur, production *f* de vapeur	парообразование
S 626	**steam generator**	Dampferzeuger *m*, Dampfgenerator *m*	appareil *m* vaporifère, générateur *m* de vapeur	паропроизводитель, паровой генератор, генератор пара
S 627	**steam-heating apparatus (equipment)**	Dampfheizanlage *f*	installation *f* de chauffage à la vapeur	паровая отопительная установка
S 628	**steam humidification**	Dampfbefeuchtung *f*	humidification *f* à vapeur	увлажнение водяным паром
S 629	**steam humidifier**	Dampfbefeuchter *m*	humidificateur *m* à vapeur	паровой увлажнитель
S 630	**steam inlet**	Dampfeintritt *m*, Dampfeinlaß *m*	conduit *m* d'admission de la vapeur	вход (впуск) пара
S 631	**steam inlet connection**	Dampfeintrittsstutzen *m*	tubulure *f* à l'admission	объем поступающего пара
S 633	**steam jacket**	Dampfmantel *m*	manteau *m* (enveloppe *f*, chemise *f*) de vapeur	паровая рубашка
S 634	**steam jacket heating**	Dampfmantelheizung *f*	chauffage *m* par manteau de vapeur	паровой обогрев рубашкой, обогрев паровой рубашкой
S 635	**steam jet**	Dampfdüse *f*	jet *m* de vapeur, tuyère *f* à vapeur	паровой эжектор
S 636	**steam-jet compressor**	Dampfstrahlverdichter *m*	compresseur *m* à jet de vapeur	пароструйный компрессор
S 637	**steam-jet ejector**	Dampfstrahlejektor *m*, Dampfstrahlsauger *m*, Dampfstrahlsaugpumpe *f*, Dampfstrahler *m*	appareil *m* (pompe *f* aspirante) à jet de vapeur, pompe à diffusion, éjecteur *m* soufflant	пароструйный эжектор
S 638	**steam-jet pump,** injector	Dampfstrahlpumpe *f*	injecteur *m*, pompe *f* à jet de vapeur	пароструйный насос
S 639	**steam-jet refrigeration**	Dampfstrahlkühlung *f*	réfrigération *f* à jet de vapeur	охлаждение пароэжекторной холодильной машиной
S 640	**steam-jet refrigeration cycle**	Dampfstrahlkälteprozeß *m*	cycle *m* frigorifique à jet de vapeur	цикл пароводяной эжекторной холодильной машины
	steam liquid equilibrium	*s.* V 99		
S 641	**steam meter,** elaterometer	Dampfmengenmesser *m*	élatéromètre *m*, débitmètre *m*, mesureur *m* de débit	паромер
S 642/3	**steam-operated refrigeration system**	dampfbetriebene Kälteanlage *f*	installation *f* frigorifique à vapeur	холодильная установка с паровым приводом

	steam outlet	s. S 618		
S 644	steam outlet connection	Dampfaustrittsstutzen *m*	tubulure *f* de sortie de vapeur	паровыпускной патрубок
S 645	steam pipe	Dampfrohr *n*	tuyau *m* à vapeur, conduit *m* de vapeur	паровая труба
S 646	steam piping	Dampfleitung *f*	conduite *f* de vapeur	паропровод
S 647	steam plant	Dampf[kraft]anlage *f*	installation *f* (centrale *f* d'énergie) à vapeur	паровая машина, паросиловая установка
S 648	steam pot	Dampftopf *m*	chaudière *f*	пароприемник
S 649	steam power	Dampfkraft *f*	force *f* motrice à vapeur	сила пара
S 650	steam power	Dampfbetrieb *m*	exploitation *f* à vapeur, force *f* de vapeur	движение на паровой тяге
	steam pressure gauge	s. S 622		
S 651	steam quality	Dampfqualität *f*	qualité *f* de vapeur	качество пара
S 652	steam recovery	Dampf[rück]gewinnung *f*	régénération *f* de la vapeur	возврат пара
S 653	steam regulator	Dampfregler *m*	régulateur *m* de vapeur	регулятор притока пара
S 654	steam-saturated air	dampfgesättigte Luft	air *m* saturé de vapeur	воздух, насыщенный водяным паром
	steam separator	s. S 657		
S 655	steam space	Dampfraum *m*	espace *f* à vapeur	паровое пространство
	steam storage vessel	s. S 591		
S 656	steam superheater	Dampfüberhitzer *m*	surchauffeur *m*, surchauffeur de vapeur	пароперегреватель
	steam supply	s. V 121		
S 657	steam trap, steam separator	Kondenstopf *m*	séparateur *m* de vapeur, pot *m* de condensation	конденсационный горшок, водоотводчик
	steam trap	s. a. W 108		
S 658	steam turbine	Dampfturbine *f*	turbine *f* à vapeur	паровая турбина
S 659	steam valve	Dampfventil *n*	soupape *f* à vapeur	паровой вентиль
S 660	steel casting	Stahlguß *m*	fonte *f* (moulage *m*) d'acier, acier *m* coulé	стальное литье
S 661	**Stefan-Boltzmann law**	Stefan-Boltzmannsches Gesetz *n*	loi *f* de Stefan et de Boltzmann	закон Стефана-Больцмана
S 662	**Stefan-Maxwell equations**	Stefan-Maxwellsche Gleichungen *fpl*	équations *fpl* de Stefan et de Maxwell	уравнения Стефана-Максвелла
S 663	**step function**	Übergangsfunktion *f*, Sprungantwort *f*	réponse *f* indicielle	переходная функция
S 664	**stepping,** graduation	Abstufung *f*	échelonnement *m*, gradation *f*	расположение уступами, градация, оттенок
S 665	**step process**	Stufenprozeß *m*	procédé *m* étagé	многошаговый (ступенчатый) процесс
S 666	**step reaction,** successive reaction	Stufenreaktion *f*	réaction *f* étagée	ступенчатая реакция
S 667	**step reaction polymerization**	Stufenreaktionspolymerisation *f*	réaction *f* étagée de polymérisation	ступенчатая реакция полимеризации
S 668	**stepwise integration**	schrittweise Integration *f*	intégration *f* graduelle (pas à pas)	ступенчатая интеграция
S 669	**stick-slip process**	Haft-Gleit-Vorgang *m*	procédé *m* de «stick slip»	процесс прилипания-скольжения
S 670	**stiff-leg derrick,** erecting crane	Montagekran *m*	grue *f* de montage	монтажный кран
S 671	**still**	Blasenapparat *m*, Blase *f*	alambic *m*, cornue *f*	перегонный куб
	still	s. a. D 413		
S 672	**still head**	Destillationsaufsatz *m*	tête *f* de colonne de distillation	головная часть колонны
S 673	**still pot**	Destillationsvorlage *f*	récipient *m* de distillation	приемник перегонного аппарата
S 674	**stirred autoclave**	Rührautoklav *m*	autoclave *m* agité	автоклав с мешалкой
	stirred batch reactor	s. S 675		
	stirred-bead mill	s. B 30		
S 675	**stirred reactor,** stirred batch reactor	diskontinuierlicher Rührreaktor *m*	réacteur *m* agité [discontinu]	реактор перемешивания периодического действия
S 676	**stirred vessel,** continuous stirred vessel	kontinuierlicher Rührkessel *m*	réacteur *m* agité [continu]	непрерывный реактор перемешивания
S 677	**stirred vessel cascade**	Rührkesselkaskade *f*	cascade *f* de réacteurs agités	каскад реакторов перемешивания
S 678	**stirrer crystallization**	Rührerkristallisation *f*	cristallisation *f* à l'agitation	кристаллизация с перемешиванием
S 679	**stirrer speed**	Rührerdrehzahl *f*	vitesse *f* de l'agitateur	скорость вращения мешалки, число оборотов мешалки
S 680	**stirring apparatus**	Rührwerk *n*	mélangeur *m*, agitateur mécanique	устройство для перемешивания
S 681	**stirring device**	Rührorgan *n*	organe *m* d'agitation, dispositif *m* agitateur	смесительное устройство, мешалка
S 682	**stirring effect**	Rührwirkung *f*	effet *m* d'agitation	эффект перемешивания
S 683	**stirring intensity**	Rührintensität *f*	intensité *f* de l'agitation	интенсивность перемешивания
S 684	**stochastic elements**	stochastische Elemente *npl*	éléments *mpl* stochastiques	стохастические элементы
S 685	**stochastic optimizing method**	stochastisches Optimierungsverfahren *n*	méthode *f* stochastique d'optimisation	стохастический метод оптимизации

S 686	**stock,** provision, supply	Vorrat *m*	provision *f*, stock *m*, réserve *f*	запас
S 687	**stockpiling**	Reservehaltung *f*	stockage *m* de réserve	резервирование, резерв
S 688	**stock solution**	Stammlösung *f*	solution *f* fondamentale (de base)	основной (исходный) раствор
S 689	**stoichiometric coefficient**	stöchiometrischer Koeffizient *m*	coefficient *m* stoechiométrique	стехиометрический коэффициент
S 690	**stoichiometric equation**	stöchiometrische Gleichung *f*	équation *f* stœchiométrique	стехиометрическое уравнение
S 691	**Stokes' law**	Stokessches Gesetz *n*	loi *f* de Stokes	закон Стокса
S 692	**stopcock**	Absperrhahn *m*	robinet *m* d'arrêt	стопорный кран
S 693	**stoppage**	Betriebsstörung *f*, Betriebsstockung *f*, Betriebsunterbrechung *f*	perturbation *f* dans l'exploitation, dérangement *m* de marche, arrêt *m* de service	остановка, простой, неполадка
	stoppage	*s. a.* O 7		
	stopper	*s.* C 739		
S 694	**stopping,** putting out of action	Außerbetriebsetzung *f*	arrêt *m*, mise *f* hors service	остановка хода, прекращение работы
S 695	**stopping device**	Absperrarmatur *f*	robinet *m* (vanne *f*) d'arrêt	запорная арматура
S 696	**stopping system**	Absperrsystem *n*	système *m* d'arrêt	система блокировки
	stop up/to	*s.* B 136		
S 697	**storage,** accumulation	Speicherung *f*, Lagerung *f*	accumulation *f*, stockage *m*, emmagasinage *m*	накопление, аккумулирование, хранение
	storage	*s. a.* 1. I 175; 2. S 710		
S 698	**storage battery**	Elektrizitätsspeicher *m*	accumulateur électrique	аккумулятор электричества, аккумуляторная батарея
S 699	**storage capacity**	Speicherkapazität *f*, Speicherfähigkeit *f*	capacité *f* de stockage	аккумуляционная способность
S 700	**storage height**	Ladehöhe *f*	hauteur *m* d'entreposage, hauteur de stockage	высота складирования
S 701	**storage modulus**	Speichermodul *m*	module *m* de mémoire, module d'accumulation	модуль накопления
S 702	**storage pump**	Speicherpumpe *f*	pompe *f* pour l'usine d'accumulation	аккумулирующий насос
	storage room	*s.* S 711		
S 703	**storage silo**	Vorratsbunker *m*	trémie *f* (silo *m*) d'approvisionnement	бункер для хранения
S 704	**storage system,** warehousing	Bevorratungssystem *n*, Lagersystem *n*	système *m* d'emmagasinage, système de stockage, système d'entreposage	система-хранилище, система хранения
S 705	**storage tank**	Vorratstank *m*, Vorratsbehälter *m*, Sammeltank *m*	réservoir *m* principal (de stockage)	запасный бак, цистерна
S 706	**storage technology**	Lagertechnologie *f*	technologie *f* de stockage	технология хранения, складная технология
	storage vessel	*s.* S 591		
S 707/8	**store**	Reserve *f*, Bestand *m*, Vorrat *m*	réserve *f*	резерв, запас
S 709	**store**	Lagerraum *m*	magasin *m*, dépôt *m*, stock *m*	склад[ское помещение]
S 710	**store,** storage, reservoir	Speicher *m*	silo *m*, magasin *m*, entrepôt *m*, accumulateur *m*	емкость, хранилище
S 711	**store [room],** storage room	Vorratslager *n*	réservoir *m* de stockage	запас, склад, хранилище
S 712	**Stormer viscometer**	Stormer-Viskosimeter *n*	viscosimètre *m* de Stormer	визкозиметр Стормера
S 713	**straight-run naphta**	Destillatbenzin *n*	essence *f* de distillation directe	бензин прямой гонки
S 714	**straight-through cooling**	Durchflußkühlung *f*	refroidissement *m* à écoulement libre	проточное охлаждение
S 715	**straight-through reactor**	Durchlaufreaktor *m*	réacteur *m* continu	проточный реактор
S 716	**strain**	Dehnung *f*, Spannung *f*, Verzerrung *f*, Deformation *f*, Formänderung *f*	expansion *f*, allongement *m*, déformation *f*	растяжение, напряжение, деформация, изменение формы
	strain	*s. a.* 1. L 186; 2. L 197		
	strained	*s.* F 97		
S 717	**strain energy due to the change of volume**	Volumenänderungsarbeit *f*, Volumenänderungsenergie *f*	énergie *f* de changement de volume	полная работа, изменение объема
S 718	**strainer head**	Extrudersiebkopf *m*	tête *f* munie de tamis	экструзионно-фильтрующая головка
S 719	**strain rate tensor**	Dehnungsgeschwindigkeitstensor *m*	tenseur *m* de la vitesse d'expansion	тензор скорости растяжения
S 720	**strain tensor**	Dehnungstensor *m*	tenseur *m* d'expansion	тензор растяжения
	stray	*s.* L 59		
S 721	**stream factor,** utilization factor	Auslastungsfaktor *m*	facteur *m* d'utilisation	коэффициент использования установки
S 722	**stream function**	Strömungsfunktion *f*, Stromfunktion *f*	fonction *f* de courant	функция потока
S 723	**streaming flow**	laminare Strömung *f*	écoulement *m* laminaire	ламинарный поток, ламинарное течение
S 724	**streaming potential**	Strömungspotential *n*	potentiel *m* de l'écoulement	потенциал протекания
S 725	**streaming properties**	Strömungseigenschaften *fpl*	propriétés *fpl* de l'écoulement	свойства течения
S 726	**streamline**	Stromlinie *f*, Strömungslinie *f*	ligne *f* aérodynamique (de courant)	линия тока (течения)

	strengthened	s. C 511		
	strengthening	s. A 479		
S 727	strength of material	Werkstoffestigkeit f	résistance f de matériel	прочность материала
S 728	stress component	Spannungskomponente f	composante f de tension	составляющая напряжения
S 729	stress condition	Beanspruchungsbedingung f	condition f de sollicitation	условие эксплуатации (нагрузки)
S 730	stress deviator	Spannungsdeviator m	déviateur m des contraintes	девиатор напряжения
	stressless	s. F 342		
S 731	stress optical coefficient	optischer Spannungskoeffizient	coefficient m photo-élastique	оптический коэффициент напряжения
S 732	stress tensor	Spannungstensor m	tenseur m des tensions	тензор напряжения
S 733	stretching	Dehnung f	allongement m, expansion f	растяжение
S 734	stretching	Verstrecken n	étirage m, tirage m	растяжение
S 735	stringy liquid	zähe Flüssigkeit f	liquide m visqueux	вязкая жидкость
S 736	stripper	Stripper m, Abstreifer m	stripper m, colonne f de rectification	стриппинг-колонна
	stripping	s. E 405		
S 737	stripping column	Abstreiferkolonne f	colonne f de rectification (stripping)	перегонный аппарат для легких фракций
S 738	stripping factor	Abtriebsfaktor m	facteur m de strippage	коэффициент отгонки легких фракций
S 739	stripping off	Abstreifen n	strippage m	снимание, отгонка легких фракций, десорбирование
S 740	stripping section	Abtriebsäule f	colonne f (partie f inférieure de la colonne) de fractionnement	перегонная колонна
S 741	stripping steam	Strippdampf m	vapeur f de strippage	стриппинг-пар
S 742	strong solution	konzentrierte Lösung f	solution f concentrée	концентрированный раствор
S 743	structural analysis, structure analysis	Strukturanalyse f	analyse f de structure	анализ структуры
S 744	structural array (for the heat exchanger)	Strukturanordnung f, Strukturvariation f	arrangement m structural	структурное расположение (размещение)
S 745	structural change	Strukturwandlung f	changement m de structure	структурное превращение, изменение структуры
S 746	structural formula	Strukturformel f	formule f structurale	формула строения, структурная формула
S 747	structural-kinetic unit	strukturkinetische Einheit f	unité f structurale-cinétique	структурно-кинетическая единица
S 748	structural matrices	Strukturmatrizen fpl	matrices fpl structurales	структурные матрицы
S 749	structural model	Strukturmodell n	modèle m structural	модель структуры
S 750	structural property of chemical process system	Struktureigenschaft f verfahrenstechnischer Systeme	propriété f structurale des systèmes technologiques	свойство структуры химико-технологических систем
S 751	structural unit	Strukturelement n	élément m structural	структурный элемент
S 752	structure, construction	Konstruktion f, Bauen n, Errichtung f, Bauwerk n, Gebäude n	construction f, bâtiment m	конструкция, постройка, построение
	structure analysis	s. S 743		
	structure design	s. B 297		
S 753	structure diagram	Strukturdarstellung f	représentation f structurale	представление структуры
S 754	structure modelling, structure simulation	Strukturmodellierung f	modélisation f structurale	моделирование структуры
S 755	structure optimization	Strukturoptimierung f	optimisation f structurale	оптимизация структуры
S 756	structure parameter method	Strukturparametermethode f	méthode f des paramètres structurals	метод структурных параметров, интегрально-гипотетический метод
	structure simulation	s. S 754		
S 757	structure synthesis	Struktursynthese f	synthèse f structurale	синтез структуры
S 758	structure variant	Strukturvariante f	variante f structurale	вариант структуры
S 759	stuffing, packing	Dichtung f	joint m, garniture f, étoupage m	уплотнение, набивка
S 760	stuffing	Füllmasse f	pâte f de remplissage	наполнитель
	stuffing	s. a. P 8		
S 761	sub-assembly	Baugruppe f	unité f de montage	узел
	subbase	s. B 50		
S 762	subcooler	Nachkühler m	sous-refroidisseur m	переохладитель
S 763	subgroup	Nebengruppe f	sous-groupe m	подгруппа, побочная группа
S 764	sublimat	Sublimat n	sublimé m	возгон, сублимат, сулема
S 765	sublimation drying	Sublimationstrocknung f	séchage m par sublimation	сушка при температуре ниже 0°C, сублимационная сушка
S 766	sublimation pressure	Sublimationsdruck m	pression f de sublimation	давление пара возгонки
S 767	submerged condenser	Eintauchkühler m	condenseur m à immersion	погружной охладитель
S 768	submerging, dipping, plunging, diving	Tauchen n	plongement m	погружение
S 769	submersible pump	Tauchpumpe f	pompe f plongeante (immersible)	погружной насос
	subplant	s. S 779		
S 770	subprogram	Unterprogramm n	sous-programme m	подпрограмма
	subsequent charging	s. A 249		

S 771	subsequent cleaning	Nachreinigung f	nettoyage m postérieur	дополнительная очистка
	subsequent treatment	s. A 141; F 155		
S 772	subsidiary plant, additional plant	Nebenanlage f	installation f annexe	вспомогательная установка
S 773	substituted group	Substituent m	substituant m	заместитель
S 774	substitute product	Austauschprodukt n, Surrogat n	substitut m, produit m de substitution	продукт замены, [продукт-] заменитель
S 775	substitution	Substitution f, Vertauschung f, Ersetzen n	substitution f	замещение
S 776	substrate, substratum	Substrat n, Grundlage f	substrat m, base f	субстрат
	substrate	s. a. S 844		
	substratum	s. S 776		
S 777	subsystem	Teilsystem n, Untersystem n, Subsystem n	sous-système m	часть системы, подсистема
S 778	subtilization	Verflüchtigung f	volatilisation f	улетучиваемость
S 779	subunit, subplant, sectional plant	Teilanlage f	installation f partielle	часть установки, цех, отделение
S 780	succession	Aufeinanderfolge f, Reihenfolge f	succession f, suite f	последовательность, чередование
	successive reaction	s. S 666		
S 781	sucking action, suction	Sog m, Saugwirkung f	succion f, aspiration f, effet m d'aspiration	подсос, подсасывание, всасывающее действие
S 782	suction air	Ansaugluft f, Saugluft f	air m d'aspiration, air aspiré	засасываемый воздух
S 783	suction air plant	Saugluftanlage f, Absauganlage f	installation f pneumatique (d'aspiration)	вытяжная вентиляционная установка
S 784	suction bag filter	Saugschlauchfilter n, Saugfilter n	filtre m aspirateur à gaines	всасывающий рукавный фильтр
S 785	suction capacity	Ansaugleistung f	capacité f d'aspiration, volume m aspiré	мощность всасывания
S 786	suction chamber	Ansaugkammer f, Ansaugraum m	chambre f d'aspiration	всасывающая камера
S 787	suction conditions	Ansaugbedingungen fpl	conditions fpl d'aspiration	условия всасывания
S 788	suction connection	Ansaugstutzen m	tubulure f d'aspiration	патрубок всасывания, всасывающий патрубок
S 789	suction conveyor	Saugförderer m	convoyeur m pneumatique	пневматический всасывающий транспортер
S 790	suction cycle	Ansaugprozeß m	cycle m d'aspiration	цикл всасывания
S 791	suction duct, exhaust duct	Abzugsschacht m	puit m de soutirage	вытяжной канал
S 792	suction equipment	Ansaugstutzen n	système m d'admission	система всасывания
S 793	suction flask, filter flask	Saugflasche f	essoreuse f, bouteille f à filtrer	отсосная склянка
S 794	suction flue, suction tube	Ansaugrohr n, Saugstutzen m	tubulure f d'aspiration	всасывающая труба
S 795	suction gas	Sauggas n	gaz m aspiré	газ, полученный в генераторе под разрежением
	suction head	s. S 800		
S 796	suction line	Saugleitung f, Ansaugrohr n	conduite f d'aspiration, tuyau m d'admission	всасывающий трубопровод
S 797	suction loss	Ansaugverluste mpl	pertes fpl d'aspiration, pertes d'admission	потери всасывания (при всасывании)
S 798	suction nozzle	Saugdüse f	tuyère f d'aspiration	всасывающее сопло
	suction period	s. S 803		
S 799	suction pipe	Ansaugleitung f	conduite f (tuyau m) d'aspiration	всасывающий канал
S 800	suction pressure, suction head	Saugdruck m	pression f d'aspiration	давление при всасывании
S 801	suction pressure control	Saugdruckregelung f	régulation f de pression d'aspiration	регулирование давления всасывания
S 802	suction pump, lift pump	Saugpumpe f	pompe f aspirante (élévatoire)	всасывающий насос
S 803	suction stroke, suction period	Ansaugperiode f	période f d'aspiration	период всасывания
S 804	suction temperature	Ansaugtemperatur f	température f d'aspiration	температура всасывания
	suction tube	s. S 794		
S 805	suction valve	Ansaugventil n, Einsaugventil n	soupape m d'aspiration	всасывающий клапан
S 806	suction vapour	Ansaugdampf m, Saugdampf m	vapeur f d'aspiration	всасываемый пар
S 807	suction vapour throttling	Saugdampfdrosselung f	étranglement m de vapeur d'aspiration	дросселирование всасываемого пара
S 808	suction velocity	Absauggeschwindigkeit f (Vakuumpumpe)	vitesse f d'extraction	скорость отсасывания
S 809	suitability for spinning	Spinnbarkeit f	aptitude f à être filé	прядомость
S 810	sulfuric acid	Schwefelsäure f	acide f sulfurique	серная кислота
S 811	sulfuric acid production process	Schwefelsäuregewinnungsverfahren n	procédé m de production de l'acide sulfurique	способ получения серной кислоты
S 812	supercharged motor	Kompressormotor m	moteur m suralimenté	компрессорный двигатель
S 813	supercooling	Unterkühlen n, Unterkühlung f	surfusion f	переохлаждение
S 814	supercritical pressure	überkritischer Druck m	pression f hypercritique	сверхкритическое давление
S 815	supercritical temperature	überkritische Temperatur f	température f hypercritique	температура выше критической

S 816	superficial mass velocity	Oberflächenmassenstrom-dichte f	densité f de flux massique superficielle	плотность поверхностного массового потока
S 817	superfractionation	Feinfraktionierung f	fractionnement m fin	сверхчеткое фракциониро-вание
S 818	superheat area	Überhitzungsfläche f	surface f de surchauffe	поверхность зоны пере-грева
S 819	superheat conditions	Überhitzungsbedingungen fpl	conditions fpl de surchauf-fage	условия перегрева
S 820	superheat curve	Überhitzungskurve f	courbe f de surchauffe	кривая перегрева
S 821	superheated cooler	Heißdampfkühler m	désurchauffeur m	пароохладитель
S 822	superheated gas	überhitztes Gas n	gaz m surchauffé	перегретый газ
S 823	superheated steam	Heißdampf m, überhitzter Dampf m	vapeur f surchauffée	перегретый пар
S 824	superheater	Überhitzer m	surchauffeur m	перегреватель
	superheating	s. O 161		
S 825	superheating cycle	Überhitzungsprozeß m	procédé m de surchauffage	цикл перегрева
S 826	superheating plant	Überhitzeranlage f	installation f de surchauffe	перегревательная установка
S 827	superheat region	Überhitzungsbereich m	domaine m de surchauffe	зона перегрева
S 828	superheat stability	Überhitzungsbeständigkeit f	stabilité f à la surchauffe	стабильность перегрева
S 829	superhigh pressure steam	Höchstdruckdampf m	vapeur f à très haute pres-sion	пар сверхвысокого давле-ния
S 830	superposition principle	Superpositionsprinzip n	principe m de superposition	принцип суперпозиции
S 831	supersaturated air	übersättigte Luft f	air m sursaturé	пересыщенный воздух
S 832	supersaturated solution	übersättigte Lösung f	solution f sursaturée	пересыщенный раствор
S 833	supersaturated steam	übersättigter Dampf m	vapeur f sursaturée	пересыщенный пар
S 834	supersaturation, oversatura-tion	Übersättigung f	sursaturation f	пересыщенность
S 835	supplementary air	Luftzuschuß m, Luftzufuhr f	adduction f d'air	дополнительный воздух
S 836	supply/to	versorgen, einspeisen	alimenter	снабжать
S 837	supply	Versorgung f, Zuführung f, Beschickung f	alimentation f, distribution f	снабжение, обеспечение
	supply	s. a. S 686		
S 838	supply conduit	Beschickungsleitung f	conduit m d'alimentation	трубопровод подачи
S 839	supply connections	Versorgungsanschlüsse mpl	connexions fpl d'alimentation	питающие соединения, стыки питания
	supplying firm	s. C 646		
	supplying with coal	s. C 313		
S 840	supply pressure	Versorgungsdruck m	pression f de distribution, pression d'alimentation	свободный напор
S 841	supply state	Lieferungszustand m	état m de livraison	состояние поставки
S 842	supply stream	Extrastrom m	extracourant m	экстраток
S 843	supply system	Versorgungsnetz n	réseau m d'alimentation	питающая сеть
S 844	support, base, substrate	Unterlage f, Basis f	base f, appui m	подложка, основание
	support	s. a. C 328		
	supposition	s. A 588		
S 845	surcharged steam	ungesättigter Dampf m	vapeur f désaturée	ненасыщенный пар
	surface-active	s. S 886		
S 846	surface-active material	oberflächenaktiver Stoff m	produit m tensio-actif	поверхностно-активное ве-щество
S 847	surface activity	Oberflächenaktivität f	tensio-activité f	поверхностная активность
S 848	surface boiling	Oberflächenverdampfung f	ébullition f superficielle	кипение на поверхности
S 849	surface charge	Oberflächenladung f	charge f superficielle	поверхностный заряд
S 850	surface complex	Oberflächenkomplex m	complexe m superficiel	поверхностный комплекс
S 851	surface concentration	Oberflächenkonzentration f	concentration f superficielle	поверхностная концентра-ция
S 852	surface condensation	Oberflächenkondensation f	condensation f superficielle (de surface)	поверхностная конденсация
S 853	surface condenser	Oberflächenkondensator m, Oberflächenverflüssiger m	condenseur m à surface	поверхностный конденса-тор
S 854	surface cooler	Oberflächenkühler m	refroidisseur m à surface	поверхностный холодиль-ник
S 855	surface cooling	Oberflächenberieselung f	réfrigération à surface à ruis-sellement	поверхностное орошение
S 856	surface element	Flächenelement n	élément m de surface	элемент поверхности
S 857	surface energy	Oberflächenenergie f	énergie f superficielle	поверхностная энергия
S 858	surface expansion	Flächenausdehnung f	superficie f, étendue f	расширение площади
S 859	surface film	Oberflächenfilm m	film m superficiel (molécu-laire)	поверхностная пленка
S 860	surface finish	Oberflächenbeschaffenheit f, Oberflächenzustand m	constitution (nature) f de la superficie, état m de sur-face	состояние поверхности
S 861	surface force	Oberflächenkraft f	force f superficielle	поверхностная сила
S 862	surface friction	Oberflächenreibung f	frottement m superficiel	поверхностное трение
S 863	surface density	Oberflächendichte f	densité f surfacique	поверхностная плотность
S 864	surface density	Flächendichte f	densité f superficielle	поверхностная плотность
S 865	surface desiccation	Oberflächentrocknung f	dessiccation f superficielle	высыхание поверхностных слоев
S 866	surface diffusion	Oberflächendiffusion f	diffusion f surfacique	поверхностная диффузия
S 867	surface hardening	Oberflächenhärtung f	trempe f superficielle	поверхностная закалка (це-ментация)
S 868	surface hardness	Oberflächenhärte f	dureté f de la surface	поверхностная твердость

S 869	**surface heat exchanger**	Oberflächenwärmeübertrager *m*	échangeur *m* thermique de surface, échangeur de chaleur par surface	поверхностный теплообменник
S 870	**surface moisture**	Oberflächenfeuchte *f*	humidité *f* de surface	поверхностная влага
S 871	**surface of impingement**	Prallfläche *f*	surface *f* de rebondissement	стражающая поверхность
S 872	**surface of the screen**	Siebfläche *f*	surface *f* criblante	площадь решетки (грохота)
S 873	**surface pressure**	Oberflächendruck *m*	pression *f* superficielle	поверхностное давление
S 874	**surface protection**	Oberflächenschutz *m*	protection *f* superficielle	защита поверхности
S 875	**surface quality**	Oberflächengüte *f*	qualité *f* de surface	качество поверхности
	surface-reactive	*s.* S 886		
S 876	**surface refrigeration**	Oberflächenkühlung *f*	réfrigération *f* de surface, refroidissement *m* superficiel	искусственное охлаждение на поверхности
S 877	**surface resistance**	Oberflächenwiderstand *m*	résistance *f* superficielle (de surface)	сопротивление поверхности
S 878	**surface roughness**	Oberflächenrauheit *f*	rugosité *f* de la surface	поверхностная шероховатость, шероховатость поверхности
S 879	**surface structure**	Oberflächenstruktur *f*	structure *f* de la surface	поверхностная структура
S 880	**surface temperature**	Oberflächentemperatur *f*	température *f* de surface	поверхностная температура
S 881	**surface tension**	Oberflächenspannung *f*	tension superficielle	поверхностная натяжение (напряжение)
S 882	**surface texturing**	Oberflächengestalt[ung] *f*	configuration *f* superficielle	текстура (структура) поверхности
S 883	**surface treatment**	Oberflächenbehandlung *f*	traitement *m* de la surface	обработка (отделка) поверхности
S 884	**surface water**	Oberflächenwasser *n*	eau *f* de surface	поверхностная вода
S 885	**surfactant**	grenzflächenaktives Mittel *n*, Tensid *n*	agent *m* tensio-actif	поверхностно-активное средство
S 886	**surfactant,** surface-reactive, surface-active	oberflächenaktiv, grenzflächenaktiv	tensio-actif	поверхностно-активный
S 887	**surge bin**	Ausgleichsbunker *m*	trémie *f* tampon, soute *f* de compensation	выравнивающий (запасный) бункер
S 888	**surge drum**	Zwischengefäß *n*, Pufferbehälter *m*, Ausgleichsbehälter *m*	récipient *m* intermédiaire, chambre *f* d'équilibre	промежуточный (буферный) сосуд, ресивер
S 889	**surge drum,** surge vessel	Windkessel *m*, Pufferkessel *m*, Puffergefäß *n*, Zwischenbehälter *m*, Flüssigkeitsabscheider *m*	bouteille *f*, tampon *m*, réservoir *m* [à air]	сосуд для погашения пульсаций, отделитель жидкости
S 890	**surge separator**	Ausgleichsbehälter *m*, Pulsationsglätter *m*	réservoir *m* d'égalisation	гаситель пульсаций
S 891	**surge tank**	Beruhigungsbehälter *m*, Windkessel *m*, Zwischenbehälter *m*	réservoir *m* intermédiaire (à air)	камера гашения, уравнительный резервуар
	surge vessel	*s.* S 889		
S 892	**surplus load**	Mehrbelastung *f*, Überlast *f*	surcharge *f*	избыточная нагрузка
S 893	**surrogate,** spare	Ersatz *m*, Austauschstoff *m*	ersatz *m*, remplacement *m*	замена, заменитель, эрзац
S 894	**surrosion**	Gewichtszunahme *(durch Korrosion)*	augmentation *f* de poids	увеличение (прирост) веса
S 895	**surrounding area**	Umgebungsbereich *m*	zone *f* d'ambiance, domaine *m* ambiant	окружающая зона (область)
S 896	**susceptibility to aging**	Alterungsempfindlichkeit *f*	sensibilité *f* au vieillissement	чувствительность к старению
S 897	**susceptibility to corrosion**	Korrosionsanfälligkeit *f*	aptitude *f* à la corrosion	подверженность, предрасположенность к коррозии
	susceptible to corrosion	*s.* G 747		
S 898	**suspended**	suspendiert	en suspension	взвешенный, суспендированный
S 899	**suspending**	Suspendieren *n*	mise *f* en suspension	взвешивание, суспендирование
S 900	**suspension**	Aufschlämmung *f*, Suspension *f*	suspension *f*	суспензия, взвесь, пульпа
S 901	**swelling**	Quellen *n*, Quellung *f*, Aufquellen *n*	gonflement *m*	набухание
S 902	**swelling ratio,** elastic recovery ratio	Quellverhältnis *n*, Schwellung *f*	taux *m* de restitution élastique	фактор набухания, набухание, вспучивание
S 903	**swept volume**	Hubvolumen *n*	cylindrée *f*	рабочий объем
	swinging hammer	*s.* H 5		
S 904	**swinging screen**	Schwingsieb *n*	crible *m* oscillant, tamis *m* vibrant	качающееся сито
	swing sieve	*s.* S 260		
S 905	**switch board**	Schaltanlage *f*	installation *f* de distribution	распределительное устройство
	switch board	*s. a.* P 16		
	switch cabinet	*s.* C 648		
S 906	**switch element,** cubicle	Schaltzelle *f*	cellule *f* de coupure	подключаемый элемент, ячейка распределительного устройства
S 907	**switch gear**	Schaltgerät *n*	mécanisme *m* de couplage, appareillage *m* électrique	коммутационный аппарат

	English	German	French	Russian
S 908	switching, control, connection	Schaltung f	montage m, couplage m	включение, коммутация
	switching on	s. I 239		
S 909	switching signal, closing signal	Einschaltsignal n	signal m d'enclenchement	сигнал включения
	switch off/to	s. B 251		
S 910	synthesis gas	Synthesegas n	gaz m synthétique (de synthèse)	синтез-газ
S 911	synthetic fibres	Synthesefasern fpl	fibres fpl synthétiques	синтетические волокна
S 912	synthetic resin, plastics	Kunstharz n	résine f synthétique	искусственная смола
S 913	synthetic rubber	Synthesekautschuk m	caoutchouc m synthétique	синтетический каучук
S 914	system analysis	Systemanalyse f	analyse f de système	анализ систем, системный анализ, системотехника
S 915	system boundary	Systemgrenze f	limite f de système	граница системы
S 916	system capacity	Anlagenleistung f	capacité f de système	производительность системы
S 917	system design	Systementwurf m, Systemgestaltung f, Systemauslegung f	projection f de système	расчет (проектирование, синтез) системы
S 918	system-element-relations	System-Element-Relationen fpl	relations fpl système-élément	система-элемент-отношения, система-элемент-взаимосвязи
S 919	system engineering	Systemtechnik f	génie m de systèmes	системотехника
S 920	system for conversion of material	System n zur Stoffwandlung	système m de transformation de matière	система для превращения веществ
S 921	system function	Systemfunktion f	fonction f de système	функция системы
S 922	system hierarchy	Systemhierarchie f	hiérarchie f des systèmes	иерархия системы
S 923	system of drainage	Entwässerungssystem n	système m de drainage (canalisation)	дренажная (осушительная) система
S 924	system of units	Einheitensystem n	système m d'unités	система единиц
S 925	system parameter	Systemparameter m	paramètre m de système	параметр системы
S 926	system reliability	Systemzuverlässigkeit f	fiabilité f de système	надежность системы
S 927	system structure	Systemstruktur f	structure f de système	структура системы
S 928	system surroundings	Systemumgebung f	environnement m de système	среда, окружающая систему
S 929	system variable	Systemvariable f	variable f de système	переменная системы

T

	English	German	French	Russian
T 1	table filter	Planfilter n	filtre m plan	плоский фильтр
T 2	table of results	Ergebnistabelle f	table f des résultats	таблица результатов
	tabletting	s. P 85		
T 3	tail fraction	Endfraktion f	fraction f finale (de queue)	хвостовая фракция
T 4	tailings	Abfall m, Nachlauf m, Rückstände mpl; Destillationsrückstand m	déchets mpl, rebut m, résidu m	остаток дистилляции, хвосты
T 5	tailings disposal	Abfallbeseitigung f	enlèvement m des déchets	утилизация отбросов
T 6	tailings pump	Nachlaufpumpe f, Rückstandspumpe f (Destillation)	pompe f aux résidus	хвостовой насос, насос для отходов
T 7	tail run, tails, afterrun	Nachlauf m	eau f de retour, repasse f	хвостовой погон, хвосты
	tamp/to	s. B 136		
T 8	tangential annular flow	Tangentialringströmung f, Tangentialströmung f im Ringspalt	écoulement m annulaire tangentiel	тангенциальное течение в кольцевом зазоре
T 9	tangential force	Tangentialkraft f	force f tangentielle	касательная сила, окружное усилие
T 10	tangential pressure	Tangentialdruck m	pression f tangentielle	касательное давление
T 11	tangential stress	Tangentialspannung f	tension f tangentielle	касательное напряжение
T 12	tangential turbine	Tangentialturbine f	turbine f tangentielle	ковшовая турбина
T 13	tangential velocity	Tangentialgeschwindigkeit f	vitesse f tangentielle	тангенциальная скорость
	tank	s. 1. B 56; 2. R 300		
T 14	tank-air interface	Tankbelüftungsvorrichtung f	dispositif m d'aération du réservoir	приспособление для отходящих газов из резервуаров
T 15	tank cooler	Tankkühler m	refroidisseur m de tank (réservoir)	бак-охладитель
T 16	tank screen	Tanksieb n	tamis m de réservoir	сетчатый фильтр
T 17	tank unit	Tankanlage f	installation f de réservoirs	хранилище, танк
T 18	taper	Konizität f, Verjüngung f (z. B. einer Extruderschnecke)	conicité f, diminution f, contracture f	конусность
T 19	tapping point	Abgreifpunkt m	point m de prise	точка отвода
T 20	tap pressure	Anzapfdruck m	pression f de soutirage	давление отбираемого пара
T 21	tar separator	Teerabscheider m	séparateur m de goudron	дегтеотделитель
T 22	Taylor's vorticity transport theory	Taylorsche Wirbeltransporttheorie f	théorie f de transfert de tourbillon de Taylor	теория переноса вихрей Тейлора
T 23	tear-off	Abreißen n	déchirure f, rupture f	отрыв
	tear resistance	s. T 88		
T 24	technical evaluation	techn[olog]ische Bewertung f	évaluation f technique, estimation f technologique	техническая оценка, оценка по технологическим критериям

	technique		s. P 494	
T 25	technological diagram, flow sheet	technologisches Schema n	schéma m technologique	технологическая схема
T 26	technological property	technologische Eigenschaft f	propriété f technologique	технологическое свойство
T 27	technological structure	technologische Struktur f	structure f technologique	технологическая структура
T 28	technology of absorption	Absorptionstechnik f	technique f de l'absorption	абсорбционная техника, техника абсорбции
T 29	temperature amplitude	Temperaturamplitude f	amplitude f de température	амплитуда температуры
T 30	temperature anomaly	Temperaturanomalie f	anomalie f de température	температурная аномалия
T 31	temperature change	Temperaturänderung f	changement m de température	изменение температуры
T 32	temperature chart	Temperaturdiagramm n	diagramme m de température	температурная диаграмма
T 33	temperature coefficient	Temperaturkoeffizient m	coefficient m de température	температурный коэффициент
T 34	temperature compensation balance	Temperaturausgleich m	compensation f en température, équilibre m des températures	выравнивание температуры
T 35	temperature concentration curve	Temperatur-Konzentrationsverlauf m	évolution (courbe) f concentration-température	кривая температура-концентрация
T 36	temperature conditions	Temperaturzustand m, Temperaturbedingungen fpl	conditions fpl de température	температурный режим
T 37	temperature conductivity, thermal conductivity	Temperaturleitvermögen n	diffusivité f thermique	температурная проводимость
T 38	temperature contrast	Temperaturkontrast m, Temperaturunterschied m	contraste m (différence f) de température	температурный контраст
T 39	temperature control	Temperaturregelung f	régulation f de température	регулирование температуры
T 40	temperature control equipment	Temperaturregeleinrichtung f	équipement m de régulation de température	оборудование для регулирования температуры
T 41	temperature controller	Temperaturregler m	régulateur m de température, thermostat m	регулятор температуры
T 42	temperature correction factor	Temperaturfehler m	erreur m de température	температурная поправка, поправка на температуру
T 43	temperature curve, adiabatic	[adiabatische] Temperaturkurve f	courbe f de température adiabatique, adiabate f de température	адиабатическая кривая температур
T 44	temperature cycle	Temperaturzyklus m	cycle m de température	цикл изменения температуры
T 45	temperature decrease	Temperaturabnahme f	chute f de température, décroissance f de la température	понижение температуры
T 46	temperature degree	Temperaturgrad m	degré m de température	градус температуры
T 47	temperature dependence	Temperaturabhängigkeit f	dépendance f de la température	температурная зависимость
T 48	temperature detecting device	Temperaturfühleinrichtung f	appareil m controlleur de température	термочувствительное устройство
T 49	temperature deviation	Temperaturabweichung f, Temperaturunterschied m	différence f de température	отклонение (разность) температур
T 50	temperature distribution	Temperaturverteilung f	distribution f de température	распределение температуры
T 51	temperature disturbance	Temperaturstörung f	dérangement m de température	колебание температуры
T 52	temperature drop, temperature gradient	Temperaturgradient m, Temperaturgefälle n	gradient m thermique, chute f de température	перепад температуры, температурный градиент
T 53	temperature entropy diagram	Wärmediagramm n	diagramme m thermique	диаграмма теплосодержания
T 54	temperature equilibrium	Temperaturgleichgewicht n	équilibre m des températures	температурное равновесие
T 55	temperature extremes	Temperaturextreme fpl	extrêmes mpl de température	крайние значения температуры
T 56	temperature field	Temperaturfeld n	champ m de température	температурное поле
T 57	temperature fluctuation	Temperaturschwankung f	fluctuation (variation) f de température	отклонение (колебание) температуры
	temperature gradient		s. T 52	
T 58	temperature increase	Temperaturzunahme f	élévation (croissance) f de température	повышение температуры
T 59	temperature indicator	Temperaturanzeiger m	indicateur m de température	указатель температуры, термометр
T 60	temperature inertia	Temperaturträgheit f, Wärmeträgheit f	inertie f de température	тепловая инерция
T 61	temperature influence	Temperatureinfluß m	influence f de température	влияние температуры
T 62	temperature intervall	Temperaturintervall n	domaine (intervalle) m de température	интервал температур
T 63	temperature jump	Temperatursprung m	saut m de température	скачок температуры
T 64	temperature level, temperature step	Temperaturstufe f	niveau m de température	температурная ступень, уровень температуры
T 65	temperature limit	Temperaturgrenze f, Temperaturbegrenzung f	limite f de température	ограничение температуры
T 66	temperature line	Temperaturverlauf f	évolution f de température	ход температуры, температурный ход
T 67	temperature line	Temperaturkurve f	ligne (courbe) f de température	график изменения температуры

T 68/9	temperature measurement	Temperaturmessung f	mesure f de température, thermométrie f, pyrométrie f	измерение температуры
	temperature of combustion	s. C 442/3		
T 70	temperature of reaction	Reaktionstemperatur f	température f de réaction	температура реакции
T 71	temperature profile	Temperaturprofil n	profil m de température	температурный профиль, профиль температуры
T 72	temperature pulldown	Temperatur[ab]senkung f, Temperaturerniedrigung f	abaissement m de température	понижение температуры
T 73	temperature radiation	Temperatur[aus]strahlung f	radiation f thermique, rayonnement m calorifique	температурное излучение
T 74	temperature range	Temperaturbereich m	gamme f de température	область температур, температурная зона
T 75	temperature recorder	Temperaturschreiber m	enregistreur m de température	самопишущий термометр, термограф
T 76	temperature reduction	Temperaturreduzierung f	réduction f de température	понижение температуры
T 77	temperature regime	Temperaturregime n, Wärmeregime n	régime m thermique (de température)	режим температуры
T 78	temperature regulator	Temperaturregler m	thermostat m, dispositif m de réglage de température	автоматический регулятор температуры, терморегулятор, термостат
T 79	temperature requirements	Temperaturanforderungen fpl	conditions fpl nécessaires de température	требования к температуре
T 80	temperature responsive bulb	Thermofühler m, Thermofühlelement n	sonde f pyrométrique, palpeur m de température	термочувствительный. баллон
T 81	temperature rise	Temperaturanstieg m	élévation (croissance, hausse) f de température	повышение температуры
T 82	temperature scale	Temperaturskale f	échelle f thermométrique	температурная шкала
T 83	temperature stability	Temperaturkonstanz f	constance (stabilité) f de température	постоянство температуры
	temperature step	s. T 64		
T 84	temperature stratification	Temperaturschichtung f	stratification f thermique	температурная стратификация
T 85	temperature tolerance	Temperaturtoleranz f	tolérance f de température	допуск по температуре
	temperature variation	s. F 399		
T 86	temperature zone	Temperaturzone f	zone f de température	температурный пояс, температурная зона
T 87	tensile force, pulling (fractive) force	Zugkraft f	effort m de traction	сила тяги, тяговое усилие
T 88	tensile strength, tear resistance	Zerreißfestigkeit f, Zugfestigkeit f	résistance f à la rupture (traction)	предел прочности на растяжение, сопротивление разрыву
T 89	tensile stress, tension	Zugspannung f	effort m (tension f) de traction	напряжение растяжения, натяжение
	tension of steam	s. V 113		
	tension theory	s. T 110		
T 90	tensor analysis	Tensoranalysis f	analyse f tensorielle	тензорный анализ
T 91	terminal compression temperature	Kompressionstemperatur f, Verdichtungsendtemperatur f	température f terminale de compression	конечная температура сжатия
	terminal group	s. E 122		
T 92	terminal reactor	Endreaktor m	réacteur m terminal	концевой реактор
T 93	terminal value	Endwert m	valeur f finale	конечное значение
T 94	terminal velocity	Endgeschwindigkeit f	vitesse f terminale (finale)	конечная скорость
T 95	terminating reaction	Abbruchreaktion f	réaction f terminale	реакция обрыва цепи
	test/to	s. T 319		
T 96	test chamber	Testkammer f, Versuchsraum m	chambre f d'essai	испытательная камера
T 97	test column	Versuchskolonne f	colonne f d'essai, colonne d'expérimentation	опытная колонна
T 98/9	testing apparatus	Prüfgerät n	appareil m d'essai	контрольно-измерительный прибор
	testing wire	s. I 254		
	test line	s. I 254		
T 100	test period	Testzeitraum m, Prüfperiode f, Prüfzeit f	période f d'essai	контрольный срок, отрезок времени проверки
T 101	test plan	Versuchsplan m	plan m de recherche, plan d'expérience	план экспериментов, экспериментальный план
T 102	test pressure	Probedruck m	pression f d'épreuve	испытательное давление
T 103	test result, experimental datum	Versuchsergebnis n	résultat m expérimental (de test)	результат опыта
T 104	test run, trial run	Probebetrieb m, Versuchsbetrieb m	fonctionnement m d'essai	контрольный запуск, пробное испытание
T 105	test stand	Prüfstand m, Teststand m	banc m d'épreuve, plateforme f d'essai	испытательный стенд
	test time	s. O 75		
T 106	test value, data, instrument reading	Meßwert m, Versuchswert m	valeur f mesurée, valeur expérimentale	измеримая величина, замеряемое значение
T 107	T-fittings	T-Stücke f	pièces fpl (raccordements mpl) en T	тройники

T 108	theoretical Carnot value	theoretischer Carnot-Koeffizient *m*	valeur *f* (coefficient *m*) de Carnot théorique	теоретический коэффициент цикла Карно
T 109	theoretical tray	theoretischer Boden *m*	plateau *m* théorique	теоретическая тарелка
	theory of similitude	s. S 314		
T 110	theory of strain, tension theory	Spannungstheorie *f*	théorie *f* des tensions	теория напряжения
T 111	thermal arrest	Temperatur[halte]punkt *m*	arrêt *m* thermique	температурная площадка
	thermal balance	s. H 20		
	thermal capacity	s. C 28		
T 112	thermal chlorination	Heißchlorierung *f*	chloration *f* thermique	термическое хлорирование
T 113	thermal chlorination process	Verfahren *n* zur thermischen Chlorierung	procédé *m* de chloration thermique	процесс термического хлорирования
T 114	thermal compensation	Temperaturkompensation *f*	compensation *f* en température	температурная компенсация
T 115	thermal conductivity	Wärmeleitfähigkeit *f*, Wärmeleitzahl *f*	conductibilité *f* thermique	теплопроводность, коэффициент теплопроводности
	thermal conductivity	s. a. T 37		
T 116	thermal convection	Wärmekonvektion *f*	convection *f* thermique	тепловая конвекция
T 117	thermal cracking	thermisches Kracken *n*	cracking *m*, craquage *m* thermique	термический крекинг
T 118	thermal decomposition	thermische Zersetzung *f*	décomposition *f* thermique	термическое разложение
	thermal diffusion	s. T 141		
T 119	thermal diffusivity, coefficient of thermometric conductivity	Temperaturleitzahl *f*, Temperaturleitfähigkeit *f*, Temperaturleitvermögen *n*	diffusivité *f* thermique, conductivité *f* thermométrique	коэффициент температуропроводности (температурной проводимости), температуропроводность
	thermal dilatation	s. T 122		
T 120	thermal energy	thermische Energie *f*	énergie *f* thermique	термическая энергия
T 121	thermal equilibrium	Wärmegleichgewicht *n*	équilibre *m* thermique	тепловое равновесие
T 122	thermal expansion, thermal dilatation	Wärmeausdehnung *f*, thermische Ausdehnung *f*	dilatation (extension) *f* thermique	тепловое расширение
T 123	thermal expansion coefficient	thermische Ausdehnungszahl *f*	coefficient *m* de dilatation thermique	коэффициент теплового расширения
T 124	thermal grade	thermischer Wirkungsgrad *m*, [thermischer] Gütegrad *m*	rendement *m* thermique	термический коэффициент полезного действия
T 125	thermal motion	thermische Bewegung *f*	mouvement *m* thermique	тепловое движение
T 126	thermal power station	Wärmekraftanlage *f*	station (centrale) *f* thermique	теплосиловая установка
T 127	thermal process	thermischer Prozeß *m*, Wärmevorgang *m*, Wärmeprozeß *m*	procédé *m* thermique	тепловой процесс
T 128	thermal property	Temperaturverhalten *n*	comportement *m* thermique	термическое свойство
T 129	thermal radiation	Wärmestrahlung *f*	rayonnement *m* thermique, radiation *f* de la chaleur	тепловое излучение, тепловая радиация
T 130	thermal resistance	Wärmedurchgangswiderstand *m*	résistance *f* à la conductibilité de chaleur	термическое сопротивление
T 131	thermal shaping	thermische Bearbeitung *f*, Wärmebehandlung *f*	traitement *m* thermique	термическая обработка
T 132	thermal shock	Temperatursturz *m*	chute *f* brusque de température	резкое падение температуры
T 133	thermal stress	Wärmespannung *f*	effort *m* de tension thermique	тепловое (температурное) напряжение
T 134	thermal structure	Temperaturstruktur *f*	structure *f* thermique	термическая структура, стратификация по температуре
T 135	thermal treatment	Wärmebehandlung *f*, thermische Behandlung (Aufbereitung) *f*	traitement *m* thermique	термическая обработка
T 136	thermal unit, calorie	Wärmeeinheit *f*	unité *f* de chaleur, calorie *f*	единица тепла, тепловая единица
T 137	thermal value, heat value	Wärmewert *m*	valeur *f* thermique	тепловой эквивалент, теплота сгорания
T 138	thermoanalysis	Thermoanalyse *f*	analyse *f* thermique	термоанализ
T 139	thermocouple	Thermoelement *n*	couple *m* thermo-électrique, thermocouple *m*	термопара, термоэлемент
T 140	thermocycle	Wärmeprozeß *m*, Wärmekreisprozeß *m*, thermischer Prozeß *m*	thermocycle *m*, procédé *m* thermique	тепловой цикл
T 141	thermodiffusion, thermal diffusion	Thermodiffusion *f*	thermodiffusion *f*	термодиффузия
T 142	thermodynamic analysis	thermodynamische Analyse *f*	analyse *f* thermodynamique	термодинамический анализ
T 143	thermodynamic efficiency	thermodynamischer Wirkungsgrad *m*	rendement *m* thermodynamique	термодинамический коэффициент полезного действия
T 144	thermodynamic equilibrium	thermodynamisches Gleichgewicht *n*	équilibre *m* thermodynamique	термодинамическое равновесие
T 145	thermodynamic functions	thermodynamische Funktionen *fpl*	fonctions *fpl* thermodynamiques	термодинамические функции
T 146	thermodynamic potential	thermodynamisches Potential *n*	potentiel *m* thermodynamique	термодинамический потенциал

T 147	thermodynamic process	thermodynamischer Prozeß *m*	processus *m* thermodynamique	термодинамический процесс
T 148	thermodynamic properties	thermodynamische Eigenschaften *fpl*	propriétés *fpl* thermodynamiques	термодинамические свойства
T 149	thermodynamic relationship	thermodynamische Beziehung *f*	relation *f* thermodynamique	термодинамическое соотношение
T 150	thermodynamics	Thermodynamik *f*, Wärmelehre *f*	thermodynamique *f*	термодинамика
T 151	thermoeconomic evaluation	thermoökonomische Bewertung *f*	évaluation *f* thermoéconomique	термоэкономическая оценка
T 152	thermoelectric heat exchange device, Peltier heat exchanger	Peltier-Wärmeübertrager *m*	échangeur *m* de chaleur thermoélectrique (Peltier)	термоэлектрический теплообменник
T 153	thermoelectric liquid cooler	Peltier-Flüssigkeitskühler *m*, thermoelektrischer Flüssigkeitskühler *m*	refroidisseur *m* de liquide thermoélectrique	термоэлектрический охладитель жидкости
T 154	thermoelectric power	Thermokraft *f*	force *f* thermo-électrique	термоэлектродвижущая сила
T 155	thermoelectric refrigerating element	Peltier-Kühlelement *n*	élément *m* frigorifique thermoélectrique	термоэлектрический элемент охлаждения
T 156	thermoelectric refrigeration	Peltier-Kühlung *f*	rèfrigération *f* thermoélectrique d'effet Peltier	термоэлектрическое охлаждение
T 157	thermoelectric stress	Thermospannung *f*	tension *f* thermo-électrique	термонапряжение, температурное напряжение
T 158	thermoelectric water chiller	Peltier-Wasserkühler *m*	refroidisseur *m* d'eau thermoélectrique	термоэлектрический водоохладитель
T 159	thermoforming	Warmformen *n*, Thermoformen *n*	façonnage *m* à chaud	термическое формование
T 160	thermograph	Temperaturschreiber *m*	thermographe *m*	самописец температуры, термограф
T 161	thermomechanical shaping	thermomechanische Bearbeitung *f*	transformation *f* (façonnage *m*) thermomécanique	термомеханическая обработка
T 162	thermonuclear reaction	thermonukleare Reaktion *f*	réaction *f* thermonucléaire	термоядерная реакция
T 163	thermopile	Thermosäule *f*	pile *f* thermo-électrique	термоэлектрический столбик, термоэлектрическая батарея
T 164	thermoplastic	Thermoplast *m*, thermoplastischer Kunststoff *m*	thermoplastique *f*	термопласт, термопластичная пластмасса, термопластичный материал
T 165	thermoregulator, thermostat	Thermoregler *m*, Thermostat *m*	thermostat *m*, thermorégulateur *m*, automate *m* thermostatique	терморегулятор, термостат
T 166	thermoset material	Duroplast *m*	matière *f* thermodurcissable	дуропласт, термореактивная пластмасса
T 167	thermosiphon cooling	Wärmeumlaufkühlung *f*	refroidissement *m* par thermosiphon	термосифонное охлаждение
	thermostat	*s.* T 165		
T 168	thermostatic liquid level control	thermostatischer Flüssigkeitsregler *m*	régulateur *m* thermostatique de niveau de liquide	регулятор уровня жидкости с использованием терморегулирующего вентиля
T 169	thick/to, to concentrate	verdicken	épaissir, congeler	сгущать
	thickener	*s.* C 523		
	thickening	*s.* C 515		
T 170	thickening agent	Verdickungsmittel *n*	épaississant *m*, agent *m* épaississant	сгуститель, агент сгущения, сгущающий агент
T 171	thickness of layer	Schichtdicke *f*	épaisseur *f* de la couche	толщина слоя
T 172	thick-walled	dickwandig, starkwandig	à paroi épaisse	толстостенный
T 173	thin-film dryer	Dünnschichttrockner *m*	séchoir *m* à couche mince	сушилка для высушивания в тонком слое
T 174	thin-film evaparator	Filmverdampfer *m*, Dünnschichtverdampfer *m*	évaporateur *m* à film	пленочный испаритель
T 175	thin film heat exchanger	Filmwärmeaustauscher *m*, Filmwärmeübertrager *m*	échangeur *m* de chaleur à couches minces	пленочный теплообменник
T 176	thin-layer evaporator	Dünnschichtverdampfer *m*	évaporateur *m* à couche mince	пленочный выпарной аппарат
T 177	thin-walled	dünnwandig	à paroi mince	тонкостенный
T 178	third law of thermodynamics	dritter Hauptsatz *m* der Thermodynamik	troisième principe *m* de la thermodynamique	третье начало термодинамики
T 179	Thomas-Gilchrist process, basic Bessemer process	Thomas-Verfahren *n*	procédé *m* Thomas	томасовский процесс
T 180/1	Thomson effect	Thomson-Effekt *n*	effet *m* de Thomson	эффект Томсона
T 182	thread formation	Fadenbildung *f*	formation *f* de fil	волокнообразование
T 183	thread shaping	Fadenformung *f*	formage *m* de fil	формование волокон
T 184	three-dimensional design	dreidimensionaler Entwurf *m*	projet *m* tridimensionnel	трехмерный чертеж, проектирование в трехмерном пространстве
T 185	three-dimensional heat flow	dreidimensionale Wärmeströmung *f*	transport *m* de chaleur tridimensionnel	трехмерный процесс теплопередачи
T 186	three-dimensional storage	dreidimensionale Speicherung *f*	emmagasinage *m* tridimensionnel	запоминание в трехмерном пространстве
T 187	three-dimensional stress state	räumlicher Spannungszustand *m*	état *m* de tension tridimensionnel	пространственное напряженное состояние

T 188	three-dimension matrix	Dreidimensionsmatrix f	matrice f à trois dimensions	трехмерная матрица
T 189	three-pass heat exchanger	dreigängiger Wärmeübertrager m	échangeur m thermique à trois passages	трехходовой теплообменник
T 190	three-phase system	Dreiphasensystem n	système m de trois phases	система, состоящая из трех фаз, трехфазная система
T 191	three-stage compression	dreistufige Verdichtung f	compression f à trois étages	трехступенчатое сжатие
T 192	three-step cooling system	dreistufiges Kühlsystem n	système m de réfrigération à trois étages	трехступенчатая холодильная установка
T 193	three-way tap	Dreiwegehahn m	robinet m à trois voies (fins, orifices)	трехходовой кран
T 194	threshold energy	Einsatzenergie f	énergie f en action	пороговая энергия
	throttle	s. C 250		
T 195	throttled air	gedrosselte Luft f	air m étranglé	дросселированный воздух
T 196	throttled refrigerant	gedrosseltes Kältemittel n	fluide m frigorigène détendu	дросселированный холодильный агент
T 197	throttled steam	Drosseldampf m, gedrosselter Dampf m	vapeur f étranglée	мятый пар
T 198	throttle valve, butterfly valve	Drosselventil n	soupape f d'étranglement, papillon m	дроссельный вентиль (клапан)
	throttling	s. C 251		
T 198a	throttling process	Drosselprozeß m	processus m d'étranglement	процесс дросселирования
T 199	throttling valve	Entspannungsventil n	soupape f de détente	дроссельный вентиль
T 200	throughput	Durchfluß m, Durchsatz m, Durchsatzleistung f	débit m, capacité f	расход, производительность
T 201	throughput ratio	Durchsatzverhältnis n	rapport m de débit	коэффициент рециркуляции
T 202	thrust force, pushing force	Schubkraft f	puissance f de poussée, force f de poussée	поперечная (касательная, сдвигающая, перерезывающая) сила
T 203	thumb rule	Faustformel f	formule f approximative	приближенная (упрощенная) формула
T 204	time-cost optimization	Zeit-Kosten-Optimierung f	optimisation f temps-coûts	оптимизация зависимости затрат от времени
T 205	time-delayed	zeitverzögert	retardé	с запаздыванием, с временной задержкой
T 206	time-dependent	zeitabhängig	en fonction du temps	зависящий от времени
T 207	time-dependent behaviour	zeitabhängiges Verhalten n	comportement m dépendant du temps	нестационарное поведение
T 208	time factor	Zeitfaktor m	facteur m de temps	коэффициент времени
	time of delivery	s. D 15		
T 209	time of discharge	Ausflußzeit f	temps m de décharge	время истечения
	time of evaporation	s. E 293		
T 210/1	time of experimentation, duration of test	Versuchsdauer f	durée f de l'essai	продолжительность опыта
T 212	time of filtering	Filterdauer f, Filtrationsdauer f	temps m de filtration	время фильтрации, продолжительность фильтрации
T 213	time required for extraction	Extraktionsdauer f	durée f de l'extraction, temps m d'extraction	время экстракции, продолжительность экстракции
T 214	time response	Zeitverhalten n	comportement m dynamique (en fonction du temps)	динамическое (нестационарное) поведение, динамика, временная характеристика
T 215	time-temperature curve	Zeit-Temperatur-Kurve f	diagramme m temps-température	кривая в координатах время-температура
	tip	s. D 557		
T 216	titration analysis	Titrationsanalyse f	titrage m, analyse f volumétrique	титровальный анализ
T 217	tolerance factor	Toleranzfaktor m	facteur m de tolérance	фактор стабильности
T 218	tolerance index	Toleranzindex m	indice m de tolérance	индекс разнотолщинности
T 219	tolerance limit	Toleranzmaß n	marge f de tolérance	предельный размер
T 220	tolerance system	Toleranzsystem n	système m des tolérances	система допусков
T 221	tolerance unit	Toleranzeinheit f	unité f de tolérance	единица допусков
T 222	tonnage	Tonnage f	tonnage m, port m	тоннаж, грузовместимость судна в тоннаж
	top	s. C 808		
	top gas	s. B 121		
T 223/4	topological procedure	topologische Vorgehensweise f	procédure f topologique	топологический подход [к расчету]
	topping-up	s. F 71		
T 225	top pump around reflux	Kopfrücklauf m	reflux m en tête	верхнее орошение
T 226	torch, flare	Fackel f	torche f	факел
T 227	torque	Drehmoment n	moment m de torsion, couple m de rotation	вращательный момент
T 228	Torricelli's law	Torricellisches Gesetz n	loi f de Torricelli	закон Торричелли
T 229	torsion	Torsion f, Verdrehung f, Drall m	torsion f	кручение
T 230	torsional stress	Torsionsspannung f	tension f de torsion	напряжение при кручении
T 231	total absorption	Totalabsorption f	absorption f totale	полное поглощение
T 232	total cooling	Gesamtkühlung f	refroidissement m total	общее охлаждение
T 233	total cross section	Gesamtquerschnitt m	section f totale	общее сечение

T 234	total energy	Gesamtenergie f	énergie f totale	полная энергия
T 235	total enthalpy	gesamte Enthalpie f	enthalpie f totale	общая энтальпия
T 236	total heat	Gesamtwärme f	chaleur f totale	полная теплота *(парообраз-ования)*
T 237	total heat balance	Gesamtwärmebilanz f	bilan m thermique total	общий тепловой баланс
T 238	total heat leakage	Gesamtwärmeabfuhr f	dissipation f de chaleur totale	общий теплоотвод
T 239	total heat load	Gesamtwärmelast f	charge f thermique totale	общая тепловая нагрузка
T 240	total heat rejection	Gesamtwärmeentzug m	soutirage m de chaleur totale	общий отвод тепла
T 241	total heat rejection	Gesamtwärmeabführung f	transmission f de chaleur to-tale	общая отдача тепла
T 242	total heat transfer	Gesamtwärmeübertragung f	transfert m global de chaleur	общая теплопередача
T 243	total load	Gesamtbelastung f	charge f totale	общая (полная) нагрузка
T 244	total loss	Gesamtverlust m, Totalver-lust m	perte f totale	общая потеря
T 245	total output	Gesamtleistung f	effet (débit) m total, puis-sance f totale	полная (суммарная) мощ-ность
T 246	total quantity	Gesamtmenge f	quantité f totale	общее количество
T 247	total reflection	Totalreflexion f	réflexion f totale	полное внутреннее отраже-ние
T 248	total reflux	Totalrücklauf m	reflux m total	полная флегма, полное орошение
T 249	total refrigerating capacity	Gesamtkälteleistung f	capacité f frigorifique totale	общая холодопроизводи-тельность
T 250	total refrigeration load	Gesamtkältelast f	charge f totale de réfrigéra-tion	общая тепловая нагрузка холодительного оборудо-вания
T 251	total refrigeration require-ments	Gesamtkältebedarf m	demande f totale de réfrigé-ration	общая потребность в хо-лоде
T 252	total synthesis	Totalsynthese f, Gesamtsyn-these f	synthèse f totale	синтез из элементов
T 253	total temperature differential	Gesamttemperaturdifferenz f	différence f de température totale	общая разность темпера-тур
T 254	total water content	Gesamtfeuchte f, Gesamtwas-sergehalt m	teneur m total en eau	общее влагосодержание
	tower	s. S 527		
T 255	tower loop reactor	Kreislaufturmreaktor m	tour f de réaction à recircula-tion	башенный реактор с цирку-ляцией
T 256	tow in/to	einschleppen, einbringen	introduire	заносить, вводить
T 257	toxicity	Toxizität f, Giftigkeit f	toxicité f	токсичность, ядовитость
T 258	trace of the pipeline	Rohrleitungsführung f	disposition f de tuyautage	трассировка трубопроводов
T 259	tracer element	Spurenelement n	élément m de trace	микроэлемент
T 260	tracer particle	Tracerteilchen n	particule f de traceur	индикаторная частица
T 261	traction	Zug m, Ziehen n	tension f, traction f, tirage m	натяжение
T 262	traffic route, communication	Verkehrsweg m	chemin m de communication	коммуникация, путь сооб-щения, трасса
T 263	tramp screen, vibrating grate	Schüttelrost m	grille f à secousses	виброгрохот, вибрационный грохот
T 264	transfer, transport, convey-ance	Förderung f	transport m, manutention f	транспортировка, транс-порт, откатка, доставка
	transfer	s. a. T 286		
T 265	transfer area	Übertragungsfläche f	surface f de transfert	площадь (поверхность) пе-реноса
T 266	transference number	Überführungszahl f, Übertra-gungszahl f	nombre m de transfer	число переноса
T 267	transfer function	Übertragungsfunktion f	fonction f de transfert	переходная функция
T 268	transfer moulding	Spritzpressen n	moulage m par transfert	литье под давлением
	transform/to	s. C 157		
T 269	transformability	Umformbarkeit f	transformabilité f	преобразуемость
	transformation	s. C 158		
	transformation point	s. T 277		
T 270	transformation product	Umwandlungsprodukt n	produit m de transformation	продукт превращения
T 271	transformer, converter	Umformer m, Transformator m, Konverter m, Wandler m, Umsetzer m, Umwand-ler m	convertisseur m, transforma-teur m	умформер, преобразова-тель, трансформатор, конвертер
T 272	transient, unsteady	instationär	non stationnaire	неустановившийся
T 273	transient heat-flow	instationäre Wärmeströmung f	flux m de chaleur non sta-tionnaire	нестационарный тепловой поток
T 274	transient refrigeration load	veränderliche Kältelast f	charge f de réfrigération va-riable	переменная тепловая на-грузка холодильного обо-рудования
T 275	transient response	Übergangsverhalten n	stade m transitoire	поведение в переходном состоянии
T 276	transition element	Übergangselement n	élément m de transition	переходный элемент
T 277	transition point, transforma-tion point	Umwandlungspunkt m	point m de transformation	точка превращения
T 278	transition pores	Übergangsporen fpl	pores mpl de transition	переходные поры, мезо-поры
T 279	transition reaction	Übergangsreaktion f	réaction f de transition	переходная реакция
T 280	transition regime	Übergangszustand m	stade m transitoire, régime m de transition	переходное состояние, пе-реходный режим

T 281	transition state theory	Theorie f des Übergangszustandes	théorie f de l'état de transition	теория переходного состояния
T 282	transition temperature	Übergangstemperatur f, Umwandlungstemperatur f	température f de transition (transformation)	переходная температура, температура превращения
T 283	translation	Translation f	translation f, mouvement m de translation	поступательное движение, параллельный перенос, трансляция
T 284	translatory velocity	Translationsgeschwindigkeit f	vitesse f de translation	скорость поступательного движения
	transmission	s. P 68		
T 285	transpiration cooling	Transpirationskühlung f	refroidissement m par transpiration	охлаждение транспирацией
T 286	transport; transfer	Transport m	transport m	транспортировка, перевозка, перенос, передача, транспорт
	transport	s. a. T 264		
T 287	transportability	Transportfähigkeit f	transportabilité f	транспортабельность, провозная способность
T 288	transportation	Beförderung f	transport m, traction f	доставка, транспортирование
	transportation line	s. C 152		
T 289	transportation system	Transportanlage f	installation f de transport	подъемно-транспортное устройство
T 290	transport-induced	transport-induziert	induit par le transport, résultant du transport	индуцированный транспортом
T 291	transport optimization	Transportoptimierung f	optimisation f de transport	транспортная оптимизация
T 292	transport system	Transportsystem n	système m de transport	транспортная система, система перевозок
T 293	transport technology	Transporttechnologie f	technologie f de transport	транспортная технология
T 294	transport vessel	Transportbehälter m	container m, caisse f mobile, conteneur m	сосуд для транспортировки
T 295	transverse finning	Querberippung f	nervure f transversale	поперечные ребра
T 296	transverse flow	Querströmung f, Querstrom m	courant m transversal, écoulement m en travers	поперечное течение, поперечный поток
	trap	s. 1. O 33; 2. S 216		
	travelling-belt screen	s. B 81		
T 297	travelling grate	Wanderrost m	grille f mobile (mécanique)	подвижная (механическая) решетка
	traversal	s. P 68		
T 298	tray	Kolonnenboden m, Siebblech n, Siebwasserbehälter m, Siebwassersammelbecken n	plateau m	тарелка, лоток, желоб, подсеточная яма
T 299	tray efficiency	Bodenwirkungsgrad m (einer Destillationskolonne)	efficacité f d'un plateau, efficacité de plateau	коэффициент полезного действия тарелки
T 300	tray elements	Bodenelemente npl	éléments mpl de plateau	элементы тарелки
T 301	treated air	aufbereitete Luft f	air m traité	обработанный воздух
T 302	treated water	aufbereitetes Wasser n	eau f traitée	обработанная вода
T 303	treating with vapour	Bedampfen n	vaporisation f	паровакуумный метод
	trial run	s. T 104		
T 304	triaxial stress state	dreiachsiger Spannungszustand m	état m général de contrainte	трехмерное напряженное состояние
T 305	trickle down/to	abrieseln	épandre	стекать по капле
T 306	trickling filter	Tropfkörper m (Abwasser)	lit m bactérien	капельный фильтр
T 307	trickling tower	Rieselturm m	tour f à ruissellement	орошаемая башня
T 308	triple point	Tripelpunkt m	point m triple	тройная точка
T 309	triturate/to	verreiben	triturer, broyer	растирать в порошок
	trommel	s. C 905		
	trouble recording	s. D 446		
	trouble shooting	s. D 445		
T 310/11	trough point	Tiefpunkt m	point m bas	низшая точка
	Trouton law (rule)	s. T 313		
T 312	Trouton's formula	Troutonsche Formel f	formule f de Trouton	формула Трутона
T 313	Trouton's rule, Trouton rule (law)	Troutonsche Regel f	règle f de Trouton	правило Трутона
T 314	Trouton viscosity	Troutonsche Viskosität f	viscosité f de Trouton	вязкость Трутона
T 315	true boiling point	wahrer Siedepunkt m	point m d'ébullition réel	истинная температура кипения
T 316	true boiling point curve	wahre Siedekurve f	courbe f de distillation réelle	кривая истинных температур кипения
T 317	true solution	echte Lösung f	solution f vraie	истинный раствор
T 318	true temperature	tatsächliche (echte) Temperatur f	température f réelle (vraie)	действительная температура
T 319	try/to, to experiment, to test	versuchen	essayer, tenter, expérimenter	испытать, опробовать, экспериментировать
T 320	tube bunch (bundle)	Rohrbündel n	faisceau m tubulaire	пучок труб
T 321	tube condenser	Röhrenkondensator m	condenseur m tubulaire (à tubes)	трубный конденсатор
T 322	tube evaporator	Röhrenverdampfer m	évaporateur m tubulaire	трубчатый испаритель
T 323	tube insulation	Rohrisolation f	isolation f de tube, isolation du tuyau	изоляция трубы

T 324	tube mill	Rohrmühle *f*	tube *m* broyeur	трубная мельница
	tube reactor	s. T 330		
T 325	tube viscometer	Röhrenviskosimeter *n*, Kapillarviskosimeter *n*	viscosimètre *m* à tube	капиллярный вискозиметр
	tubing	s. C 88		
T 326	tubular centrifuge	Röhrenzentrifuge *f*	centrifugeuse *f* tubulaire	трубчатая центрифуга
T 327	tubular coil	Röhrenkondensator *m*	condenseur *m* tubulaire	трубчатый конденсатор
	tubular cooler	s. C 356		
T 328	tubular heat exchanger	Röhrenwärmeübertrager *m*	échangeur *m* thermique tubulaire, échangeur de chaleur à tubes	трубчатый теплообменник
T 329	tubular jacket	Rohrmantel *m*	manteau *m* (paroi *f*) de tube	кожух трубы
T 330	tubular reactor, tube reactor	Rohrreaktor *m*, Röhrenreaktor *m*	réacteur *m* tubulaire	трубчатый реактор
	tumbling mixer	s. R 387		
T 331	tunnel dryer	Kanaltrockner *m*, Tunneltrockner *m*	canal *m* de séchage, séchoir *m* tunnel	канальная (коридорная, туннельная) сушилка
T 332	tunnel furnace (kiln)	Tunnelofen *m*, Kanalofen *m*	four *m* tunnel (à canal)	туннельная печь
T 333	turbidimeter, nephelometer	Trübungsmesser *m*	néphélomètre *m*	нефелометр
T 334	turbidity factor	Trübungsfaktor *m*	facteur *m* néphélométrique	фактор мутности
T 335	turbine agitator, turbine mixer	Turbinenrührer *m*	agitateur *m* turbinaire	турбинная мешалка, смеситель с турбинной мешалкой
T 336	turbine engine	Turbinenmotor *m*	turbopropulseur *m*	турбодвигатель, быстроходный электродвигатель
	turbine mixer	s. T 335		
T 337	turbine output	Turbinenleistung *f*	rendement *m* de la turbine	мощность турбины
T 338	turbine plant	Turbinenanlage *f*	installation *f* à turbines	турбинная установка
	turboblower	s. C 140		
T 339	turbocompressor	Turboverdichter *m*, Turbokompressor *m*	turbosoufflante *f*, turbocompresseur *m*	турбокомпрессор
T 340	turbodryer	Turbinentrockner *m*	turbosécheur *m*	турбосушилка
T 341	turbogenerator	Turbogenerator *m*	turbo-alternateur *m*, turbogénératrice *f*	турбогенератор
T 342	turbulence, vorticity	Turbulenz *f*, Verwirbelung *f*	turbulence *f*	турбулентность, вихревое движение
T 343	turbulence characteristic	Turbulenzcharakteristik *f*, Turbulenzeigenschaft *f*	caractéristique (propriété) *f* de turbulence	свойство (характеристика) турбулентности
T 344	turbulence effect	Turbulenzeinfluß *m*	effet *m* turbulent (de turbulence)	эффект турбулентности
T 345	turbulent diffusion	turbulente Diffusion *f*	diffusion *f* turbulente	турбулентная диффузия
T 346	turbulent energy transport	turbulenter Energietransport *m*	transport *m* turbulent d'énergie	турбулентный перенос энергии
T 347	turbulent flow	turbulente Strömung *f*	écoulement *m* turbulent	турбулентное течение
T 348	turbulent momentum transport	turbulenter Impulstransport *m*	transfert *m* turbulent de la quantité de mouvement	турбулентный перенос импульса
T 349	turnover, reaction, conversion	Umsatz *m*, Umsetzung *f*, Umwandlung *f*, Konvertierung *f*	conversion *f*, taux *m* de conversion, transformation *f*, décomposition *f*	обмен, обменная реакция, реакция обменного разложения, превращение, степень превращения, конверсия
T 350/1	turnover-time curve	Umsatz-Zeit-Verlauf *m*	évolution *f* du taux de conversion en fonction de temps	кривая степень превращения-время
	twin-screw extruder	s. D 553		
T 352	twist nozzle	Dralldüse *f*	buse *f* de tors	сопло для придания вращения
T 353	two-phase area	Zweiphasengebiet *n*	domaine *m* de deux phases	двухфазная область
T 354	two-phase cooling	Zweiphasenkühlung *f*	refroidissement *m* de deux phases	двухфазное охлаждение
T 355	two-phase flow	Zweiphasenströmung *f*	écoulement *m* en deux phases	двухфазный поток
T 356	two-phase mixture	Zweiphasengemisch *n*	mélange *m* à deux phases	двухфазная смесь
T 357	two-phase system	Zweiphasensystem *n*	système *m* de deux phases	система, состоящая из двух фаз, двухфазная система
T 358	two-stage centrifugal pump	zweistufige Kreiselpumpe *f*	pompe *f* centrifuge à deux étages	двухступенчатый центробежный насос
T 359	two-stage circuit	zweistufiger Kreislauf *m*	circuit *m* à deux étages	двухступенчатая схема
T 360	two-stage crushing	zweistufige Zerkleinerung *f*	broyage *m* biétagé, désintégration *f* en deux étages	двухстадийное измельчение
T 361	two-stage cycle	zweistufiger Prozeß *m*	cycle *m* à deux étages, procédé *m* à deux étappes	двухступенчатый цикл
T 362	two-stage expansion	zweistufige Entspannung *f*	détente *f* à deux étages	двухступенчатое расширение
T 363	two-stage refrigeration	zweistufige Kühlung *f*	réfrigération *f* à deux étages	двухступенчатое охлаждение
T 364	type design, construction	Bauart *f*	structure *f*, type *m*, conception *f*	конструкция, строение
T 365	type model	Typenmodell *n*	modèle *m* type	типовая модель

U

	ultimate analysis	s. E 69		
U 1	ultimate partial pressure	Endpartialdruck m	pression f partielle finale	конечное парциальное давление
U 2	ultracentrifuge	Ultrazentrifuge f	ultracentrifugeuse f	ультрацентрифуга
U 3	ultrahigh vacuum	Höchstvakuum n	ultravide m	сверхглубокий вакуум
U 4	uncertainty, indeterminacy	Unbestimmtheit f, Unsicherheit f	indétermination f, incertitude f	неопределенность
U 5	uncondensed vapour	nichtkondensierter Dampf m	vapeur f non condensée	неконденсированный пар
U 6	under-drive centrifuge, pendulum-type hydroextractor	Pendelzentrifuge f	essoreuse f oscillante	маятниковая центрифуга
U 7	underground deposit, deposit	Untergrundspeicher m, Kaverne f	dépôt m souterrain (au sous-sol)	подземное хранилище
	underground pump	s. G 157		
U 8	undersize	Unterkorn n	grain m de grosseur inférieure	нижний (подрешетный) продукт
U 9	unfiltered air	ungefilterte Luft f	air m non filtré	нефильтрованный воздух
U 10	uniaxial extension	einachsige Dehnung f	allongement m uniaxe	одноосевое растяжение
U 11	uniaxial loading	einachsige Belastung f	charge f uniaxe	одноосевая нагрузка
U 12	uniflow cooling	Gleichstromkühlung f	refroidissement m à courant parallèle	прямоточное охлаждение
	uniflow current	s. P 21		
	uniformalization	s. S 540		
U 13	uniform cooling	gleichmäßige Kühlung f	refroidissement m uniforme	равномерное охлаждение
U 14	uniform dehydration	gleichmäßige Trocknung f	déshydratation f (séchage m) uniforme	равномерное осушение
U 15	uniform temperature distribution	gleichmäßige Temperaturverteilung f	distribution f de température uniforme	равномерное распределение температур
U 16	uninflammable	unentzündlich, nicht brennbar	ininflammable, non inflammable	невозгораемый, невоспламеняемый
	union	s. A 587		
U 17	unit, assembly	Aufbaueinheit f, Baueinheit f	élément m, unité f de construction	стандартный узел
U 18	unit assembling	Blockmontage f	montage m monobloc	блочный монтаж
U 19	unit capacity charging	Leistungsbelastung f	charge f au cheval-vapeur	удельная нагрузка
U 20	unit cell	Einheitszelle f, Elementarzelle f, Elementarkörper m	cellule f unitaire (élémentaire)	единичная клетка (ячейка), элементарная ячейка, элементарное тело
U 21	unit cell vector	Basisvektor m	vecteur m de base	базисный вектор
U 22	unit charge	Ladungseinheit f	unité f de charge	единица заряда
U 23	unit design	Blockprojektierung f	projection f d'unité	блочное проектирование, проектирование по основным узлам
U 24	unit diagram	Einheitsdiagramm n	diagramme m unitaire	стандартная (унифицированная) диаграмма
U 25	unit impulse	Einheitsimpuls m	impulsion f unitaire	единичный импульс
U 26	unit impulse function	Einheitsimpulsfunktion f	fonction f d'impulsion unitaire	единичная импульсная функция
U 27	unit matrix	Einheitsmatrix f	matrice f unité	единичная матрица
U 28	unit of area	Flächeneinheit f	unité f de superficie	единица площади
U 29	unit of measure	Maßeinheit f	unité f de mesure	единица измерения
U 30	unit of notation	Einheitsmaß n	mesure f standardisée	стандартный (типовой) размер
U 31	unit of quantity	Mengeneinheit f	unité f de quantité	единица массы (количества)
U 32	unit of surface	Oberflächeneinheit f	unité f de surface	единица поверхности
U 33	unit of work	Arbeitseinheit f	unité f de travail	единица работы
U 34	unit operation	Grundoperation f	opération f unitaire (de base)	основная операция, [основной] процесс
U 35	unit pressure	Einheitsdruck m	pression f normale	давление на единицу поверхности
U 36	unit process	Grundprozeß m, Grundverfahren n	procédé m de base	основной процесс
U 37	unit step function	Einheitssprungfunktion f	fonction f indicielle	единичная ступенчатая функция
U 38	unit vector	Einheitsvektor m	vecteur m unité	единичный вектор
U 39	universal gas constant	universelle Gaskonstante f	constante f universelle des gaz	универсальная газовая постоянная
U 40	unloading, discharge	Ausladung f, Entladung f	déchargement m	разгрузка, выгрузка
U 41	unloading cycle	Entleerungsprozeß m	procédé m de décharge	цикл разгрузки
U 42	unpremixed	nicht vorgemischt, ohne Vorvermischung	sans prémélangeage	непредварительно смешанный, без предварительного смешения
U 43	unsaturated air	ungesättigte Luft f	air m non saturé	ненасыщенный воздух
U 44	unsaturated steam	ungesättigter Dampf m	vapeur f non saturée	ненасыщенный пар
	unstable	s. I 247		
U 45	unstationary behaviour	instationäres Verhalten n	comportement m transitoire, état m de transition	нестационарное поведение
U 46	unsteadiness	Unstetigkeit f	discontinuité f	разрывность, нерегулярность

	unsteady	s. T 272		
U 47	unsteady flow	nichtstationäre Strömung f, nichtstationäres Fließen n	écoulement m non stationnaire, mouvement m non permanent	нестационарное течение
U 48	unsteady heat conduction	nichtstationäre (instationäre) Wärmeleitung f	conduction f thermique non stationnaire	нестационарная теплопроводность
U 49	unsteady heat transfer	instationäre Wärmeübertragung f	transfert m de chaleur non stationnaire	нестационарная теплопередача
U 50	unsteady viscous flow	nichtstationäre (instationäre) viskose Strömung f	écoulement m visqueux non stationnaire	нестационарное вязкое течение
U 51	unwinding	Abwickeln n	déroulement m	смотка, скатывание, разматывание
U 52	upflow	Aufwärtsstrom m, Aufwärtsströmung f	écoulement m montant, courant m ascendant	восходящий поток, восходящее течение
	upgrade/to	s. I 61		
U 53	upgrading	Anreicherung f, Verstärkung f	enrichissement m, concentration f	обогащение, укрепление
U 54	upstream component	aufschwimmender Bestandteil m, Stromauf-Komponente f, vordere (obere) Komponente f	composante f nageante (en amont)	верхний (выплывающий) компонент
U 55	upward flow	Aufwärtsströmung f	écoulement m vers le haut, flux m ascensionnel	поток, направленный вверх
U 56	urea synthesis	Harnstoff-Synthese-Verfahren n	synthèse f d'urée	метод синтеза мочевины
	usability	s. U 63		
U 57	use, utilization	Verwendung f, Benutzung f, Gebrauch m, Nutzung f	emploi m, utilisation f, usage m	применение, употребление, использование
U 58	used oil; waste oil	Altöl n	huile f usagée	отработанное масло
U 59	used oil preparation	Altölaufbereitung f	traitement m de l'huile usée	переработка отработанного масла
U 60	used oil regeneration	Altölregenerierung f	régénération f de l'huile usée	регенерация отработанного масла
U 61	useful capacity	Nutzkapazität f	capacité f utile	полезная емкость (производительность)
U 62	useful current	Nutzstrom m	courant m actif (utile)	полезный ток
U 63	usefulness, adaptability, usability, utility	Verwendbarkeit f, Einsetzbarkeit f, Eignung f, Brauchbarkeit f, Benutzbarkeit f, Benutzungseignung f	utilisabilité f, aptitude f, applicabilité f, utilité f, aptitude à l'usage	используемость, пригодность, годность, полезность
	useful performance	s. A 129		
U 64	useful work	Nutzarbeit f	travail m utile	полезная работа
	use of energy	s. E 173		
	user	s. C 608		
U 65	use temperature	Gebrauchstemperatur f, Anwendungstemperatur f	température d'utilisation, température à l'emploi	температура употребления (применения)
U 66	use value	Gebrauchswert m	valeur f de service	потребительная стоимость
	utility	s. U 63		
U 67	utility value-cost optimization	Gebrauchswert-Kosten-Optimierung f	optimisation f valeur de service-frais	оптимизация соотношения затрат и стоимости
U 68	utility value-cost relation	Gebrauchswert-Kosten-Relation f	relation f valeur de service-frais	соотношение между стоимостью и затратами
U 69	utilization, wear	Abnutzung f, Verschleiß m	usure f, détérioration f	износ
	utilization	s. a. U 57		
U 70	utilization coefficient (factor)	Ausnutzungskoeffizient m, Ausnutzungsgrad m	coefficient (facteur) m d'utilisation	коэффициент использования
U 71	utilization of capacity	Leistungsausnutzung f	utilisation f de puissance, exploitation f de puissance	использование мощности
U 72	utilization of waste heat	Abhitzeverwertung f, Abhitzeverwendung f	valorisation f de la chaleur perdue, utilisation f de la chaleur d'échappement	использование отходящего тепла, утилизация тепла
U 73	utilization plant	Verwertungsanlage f	installation f d'utilisation	установка для утилизации
U 74	utilize/to	verwerten, verwenden	utiliser, valoriser	использовать, применять
U 75	U-tube	U-Rohr n	tube m en U	U-образная труба

V

	vacuum	s. L 256		
V 1	vacuum apparatus	Vakuumapparat m	appareil m à vide	вакуум-аппарат
V 2	vacuum chamber	Vakuumkammer f	chambre f à vide	вакуумная камера
V 3	vacuum cleaner	Vakuumreiniger m	purificateur m à vide	пылесос, вакуумный очиститель
V 4	vacuum column	Vakuumkolonne f	colonne f à vide	вакуумная колонна
V 5	vacuum cooler	Vakuumkühler m	refroidisseur m sous vide	холодильник мгновенного испарения
V 6	vacuum cooling	Vakuumkühlung f	réfrigération f sous vide	вакуумное охлаждение
V 7	vacuum crystallization	Vakuumkristallisation f	cristallisation f sous vide	кристаллизация в вакууме
V 8	vacuum crystallizer	Vakuumkristallisationsapparat m	cristallisoir m à vide	вакуум-кристаллизатор
V 9	vacuum degasifier	Vakuumentgaser m	installation f de dégazage sous vide	вакуумный деаэратор

V 10	vacuum dehydration	Vakuumtrocknung f	déshydratation f à vide, séchage m sous vide	сушка в вакууме
V 11	vacuum distillate	Vakuumdestillat n	distillat m sous vide	вакуумный дистиллят
V 12	vacuum distillation	Vakuumdestillation f, Unterdruckdestillation f, Vakuumabtreibung f	distillation f sous (dans le) vide, distillation à pression réduite	перегонка в вакууме
V 13	vacuum distillation apparatus	Vakuumdestillationsapparatur f	appareil m de distillation sous vide	вакуум-перегонный аппарат
V 14	vacuum distillation column	Vakuumdestillationskolonne f	colonne f de distillation sous vide	вакуум-дистилляционная колонна
V 15	vacuum distillation plant	Vakuumdestillationsanlage f	unité f de distillation sous vide	установка дистилляции под вакуумом
V 16	vacuum distillation residue	Vakuumdestillationsrückstand m	résidu m de distillation sous vide	остаток от вакуумной перегонки
V 17	vacuum-distilled	vakuumdestilliert	distillé sous vide	дистиллировано под вакуумом
V 18	vacuum drum filter, rotary vacuum filter	Vakuumtrommelfilter n	filtre m à tambour sous vide	[вращающийся] вакуумный барабанный фильтр
V 19	vacuum dryer	Vakuumtrockenapparat m, Vakuumtrockner m	séchoir m à vide	вакуум-сушильный аппарат, вакуумная сушилка, вакуум-сушилка
V 20	vacuum drying	Vakuumtrocknung f	séchage m à vide	сушка в вакууме
V 21	vacuum-drying plant	Vakuumtrockenanlage f	unité f de séchage sous vide	вакуумная сушилка
V 22	vacuum ejector, air ejector	Vakuumejektor m, Strahlsauger m, Strahlsaugpumpe f	éjecteur m, pompe f à jet, injecteur m	эжектор, водоструйный насос, эжектор для откачки воздуха
V 23	vacuum evaporating	Vakuumverdampfung f	évaporation f au vide	возгонка под вакуумом
V 24	vacuum evaporation plant	Vakuumverdampfungsanlage f	appareil m d'évaporation à vide, installation f d'évaporation sous vide	вакуум-выпарная установка
V 25	vacuum evaporator	Vakuumverdampfer m	évaporateur m au vide	вакуумный выпарной аппарат
V 26	vacuum extraction still	Vakuumextraktionsanlage f	installation f d'extraction dans le vide	вакуум-экстракционная установка
V 27	vacuum filter	Vakuumfilter n	tambour (filtre) m à vide	вакуум-фильтр
V 28	vacuum filtration	Vakuumfiltration f	filtration f sous vide	вакуумная фильтрация
V 29	vacuum flash column	Vakuumflashkolonne f	colonne f de vaporisation instantanée sous vide	колонна однократного испарения в вакууме
V 30	vacuum forming	Vakuumverformung f, Vakuumformung f	déformation f sous vide, formage m par le vide, moulage m par vide	вакуумное форм[ир]ование, формирование под вакуумом
V 31	vacuum freeze dryer	Vakuumgefriertrockner m	sécheur m par congélation sous vide	вакуумная сушилка вымораживания
V 32	vacuum gas oil	Vakuumgasöl n	gas-oil m [obtenu par distillation] sous vide	вакуумный газойль
V 33	vacuum gas oil desulphurization	Vakuumgasölentschwefelung f	désulfuration f du gas-oil sous vide	обессеривание вакуумного газойля
V 34	vacuum jacket	Vakuummantel m	manteau m à vide	вакуумная рубашка
V 35	vacuum line	Vakuumleitung f	conduit m à vide	вакуумная линия
V 36	vacuum oil	Vakuumöl n	huile f vacuum	вакуумное масло
V 37	vacuum oil refining plant	Vakuumölreinigungsanlage f	unité f de raffinage du gas-oil sous vide	вакуум-установка для очистки нефти
V 38	vacuum plant	Vakuumanlage f	installation f sous vide	вакуумная установка
V 39	vacuum plate evaporator	Vakuumplattenverdampfer m	évaporateur m à plaques sous vide	вакуумный плиточный испаритель
V 40	vacuum process	Vakuumverfahren n	procédé m au vide	вакуумирование, процесс под вакуумом, вакуумная технология
V 41	vacuum pump	Vakuumpumpe f	pompe f à vide	вакуум-насос
V 42	vacuum receiver	Vakuumvorlage f	récipient m à vide	приемник вакуум-аппарата
V 43	vacuum rectification	Vakuumrektifikation f	rectification f sous vide	перегонка под вакуумом
V 44	vacuum redistillate	Vakuumredestillat n	redistillat m sous vide	вакуумный редистиллят
V 45	vacuum refrigerating system	Vakuumkälteanlage f	installation f frigorifique à vide	вакуумная холодильная установка
V 46	vacuum regulator	Vakuumregler m	régulateur m de vide	регулятор вакуума
V 47	vacuum residue	Vakuumrückstand m	résidu m sous vide	кубовой остаток (вакуум-колонны)
V 48	vaccuum residue desulphurization	Vakuumrückstandsentschwefelung f	désulfuration f du résidu sous vide	обессеривание остатка от вакуумной перегонки
V 49	vacuum residue desulphurization catalyst	Vakuumrückstandsentschwefelungskatalystor m	catalyseur m de désulfuration du résidu sous vide	катализатор обессеривания остатка от вакуумной перегонки
V 50	vacuum residue desulphurization process	Vakuumrückstandsentschwefelungsprozeß m	procédé m de désulfuration du résidu sous vide	процесс обессеривания остатка от вакуумной перегонки
V 51	vacuum rotary dryer	Vakuumtrommeltrockner m	sécheur m à vide rotatif	вакуум-барабанная сушилка
V 52	vacuum siphon	Vakuumflüssigkeitsheber m	siphon m à vide	вакуумный гидроподъемник
V 53	vacuum still	Vakuumdestillationsblase f	alambic m (vase m distillatoire) sous vide	вакуум-перегонный аппарат

V 54	vacuum sublimation apparatus	Vakuumsublimierapparat *m*	appareil *m* de sublimation sous vide	вакуум-аппарат для возгонки
V 55	vacuum tank	Vakuumkessel *m*	bassin *m* à vide, réservoir *m* sous vide	вакуум-сборник
V 56	vacuum technique	Vakuumtechnik *f*	technique *f* de vide	вакуумная техника
V 57	vacuum thickening	Vakuumeindickung *f*, Vakuumaufkonzentrierung *f*	concentration *f* sous vide	концентрирование в вакууме
	vacuum-tight	*s.* H 133		
V 58	vacuum trap	Vakuumvorlage *f*	réservoir *m* à vide	приемник вакуум-аппарата, вакуумная ловушка
V 59	valve tray	Ventilboden *m*	plateau *m* à clapets	вентильная тарелка
V 60	valve tray column	Ventilbodenkolonne *f*	colonne *f* à plateaux à clapets	колонна с вентильными тарелками
V 61	vaporizability, volatility	Verdampfbarkeit *f*, Flüchtigkeit *f*	évaporabilité *f*, volatilité *f*	летучесть, испаряемость
V 62	vaporizable, volatilizable, volatile	verdampfbar, verdunstbar	évaporable, volatilisable	испаряющийся
	vaporizable	*s. a.* G 45		
	vaporization	*s.* E 278/9		
V 63	vaporization plant	Bedampfungsanlage *f*	appareil *m* pour métallisation au vide	установка для нанесения [слоя] испарением
V 64	vaporization point	Verdampfungspunkt *m*	point *m* de vaporisation	точка испарения, температура образования пара
V 65	vaporization pressure	Verdampfungsdruck *m*	pression *f* d'évaporation	давление кипения
V 66	vaporize/to, to evaporate, to flash, to volatilize	verdampfen, verdunsten, verflüchtigen	vaporiser, évaporer, volatiliser	испарять, упаривать, отпаривать, испаряться, улетучиваться
	vaporize/to	*s. a.* G 54		
V 67	vaporized condition	dampfförmige Komponente *f*	composante *f* à l'état de vapeur	парообразное состояние
	vaporizer	*s.* 1. E 297; 2. R 133		
V 68	vaporizing, evaporating	Sieden *n*, Verdampfen *n*	évaporation *f*, vaporisation *f*	упаривание, уваривание, испарение
	vaporizing	*s. a.* G 46		
V 69	vaporizing unit	Verdampfungsanlage *f*	unité *f* d'évaporation	испарительная установка
V 70	vapour	Brüden *m*, Wrasen *m*	vapeur *f* chaude	влажный пар
V 71	vapour absorption refrigerating system	Dampfabsorptionskälteanlage *f*	installation *f* frigorifique à l'absorption de vapeur	пароабсорбционная холодильная установка
V 72	vapour adsorption	Dampfadsorption *f*	adsorption *f* de vapeur	паровая адсорбция
V 73	vapour-blast process	Bedampfung *f*	vaporisation *f*, métallisation *f* au vide	нанесение испарением
V 74	vapour bleeding tap	Dampfentnahmestutzen *m*	tubulure *f* de soutirage de vapeur	патрубок для отвода пара
V 75	vapour capacity factor	Dampfbelastungsfaktor *m*	facteur *m* de charge de vapeur	нагрузка по пару, паровая нагрузка
V 76	vapour chart	Dampftafel *f*, Wasserdampftafel *f*	table *f* de la vapeur d'eau	таблица свойств водяного пара
V 77	vapour composition	Dampfzusammensetzung *f*	composition *f* de la vapeur	состав пара
V 78	vapour compression	Dampfverdichtung *f*	compression *f* de vapeur	сжатие пара
V 79	vapour compression system	Dampfkompressionsanlage *f*	système *m* de compression de vapeur	паровая компрессионная установка
V 80	vapour compressor	Dampfkompressor *m*	compresseur *m* à vapeur	паровой компрессор
V 81	vapour condensation, steam condensation	Dampfkondensation *f*	condensation *f* de vapeur	конденсация пара
V 82	vapour consumer	Dampfabnehmer *m*, Dampfverbraucher *m*	consommateur *m* de vapeur	потребитель пара
	vapour content	*s.* S 600		
V 83	vapour cycle	Dampfkreisprozeß *m*	cycle *m* à vapeur	паровой цикл (контур)
V 84	vapour degreasing	Dampfentfettung *f*	dégraissage *m* à vapeur de solvant	обезжиривание парами растворителя
	vapour demand	*s.* S 608		
V 85	vapour density, steam density	Dampfdichte *f*	densité *f* de vapeur	плотность пара
V 86	vapour deposition, coating	Aufdampfen *n*	vaporisation *f*, projection *f* de vapeurs	покрытие напылением
V 87	vapour diffusion	Dampfdiffusion *f*	diffusion *f* de vapeur	диффузия пара
V 88	vapour enthalpy	Dampfenthalpie *f*	enthalpie *f* de vapeur	энтальпия пара
V 89	vapour entropy	Dampfentropie *f*	entropie *f* de vapeur	энтропия пара
V 90	vapour fraction	dampfförmige Fraktion *f*	fraction *f* de vapeur	паровая фракция
V 91	vapour header	Dampfsammler *m*, Dampfbehälter *m*	collecteur *m* de vapeur	коллектор пара, паросборник
V 92	vapour heat carrier	dampfförmiger Wärmeträger *m*	caloporteur *m* à l'état de vapeur	парообразный теплоноситель
V 93	vapour-heated evaporator	dampfbeheizter Verdampfer *m*	bouilleur *m* à vapeur	испаритель, обогреваемый паром
V 94	vapour inlet temperature	Dampfeintrittstemperatur *f*	température *f* de vapeur à l'admission	температура пара на входе
V 95	vapour-jet unit	Dampfstrahlanlage *f*	appareil *m* à jet de vapeur	пароструйный прибор
V 96	vapour leak	Dampfaustrittsstelle *f*, Dampfleckstelle *f*	fuite *f* de vapeur	утечка паров
V 97	vapour lift pump	Gasblasenpumpe *f*	siphon *m* thermique	термосифон

V 98	vapour-liquid contacting zone	Dampf-Flüssigkeits-Kontakt-gebiet n	zone f de contact liquide-vapeur	зона контакта пара и жидкости, парожидкостная контактная зона
V 99	vapour liquid equilibrium, steam liquid equilibrium	Dampf-Flüssigkeits-Gleichgewicht n	équilibre m vapeur-liquide	равновесие пар-жидкость
V 100	vapour lock	Dampfblasenbildung f	formation f de tampon de vapeur, vapour-lock m	образование паровой пробки
V 101	vapour motion	Dampfbewegung f	mouvement m de vapeur	движение пара
V 102	vapourous	dampfförmig	à l'état de vapeur	парообразный
V 103	vapourous fluid	dampfförmiges Medium n	fluide m à l'état de vapeur	парообразная среда
V 104	vapour outlet temperature	Dampfaustrittstemperatur f	température f de vapeur à l'échappement	температура пара на выходе
V 105	vapour output	Dampfaustritt m	sortie f (échappement m) de la vapeur	выход (выпуск) пара
V 106	vapour percolation	Dampffiltration f	percolation f de vapeur	фильтрация пара
V 107	vapour permeability	Dampfdurchlässigkeit f	perméabilité f à la vapeur	проницаемость пара
V 108	vapour phase	Dampfphase f	phase f gazeuse (à l'état de vapeur)	паровая фаза
V 109	vapour-phase adsorption	Dampfphasenadsorption f	adsorption f en phase vapeur	адсорбция в паровой фазе
V 110	vapour-phase alkylation process	Verfahren n zur Dampfphasenalkylierung	procédé n d'alkylation en phase vapeur	процесс алкилирования в паровой фазе
V 111	vapour-phase chlorination	Dampfphasenchlorierung f	chloration f en phase vapeur	хлорирование в паровой фазе
V 112	vapour-phase desorption	Dampfphasendesorption f	désorption f en phase vapeur	десорбция в паровой фазе
V 113	vapour pressure, vapour tension, tension of steam	Dampfdruck m	pression (tension) f de la vapeur	давление (упругость) пара
V 114	vapour-pressure diagram	Dampfdruckkurve f, Dampfdruckdiagramm n	diagramme m de vapeur	диаграмма давления паров
V 115	vapour-pressure thermometry	Dampfdruckthermometrie f	thermométrie f de pression de vapeur	измерение температуры по упругости насыщенного пара
V 116	vapour refrigerant	dampfförmiges Kältemittel n	frigorigène m à l'état de vapeur	парообразный холодильный агент
V 117	vapour resistance	Dampfdichtheit f	étanchéité f à la vapeur	паростойкость
V 118	vapour return	Dampfrückführung f	retour m de vapeur	возврат паров
V 119	vapour separator	Dampfabscheider m	séparateur m de vapeur	пароотделитель
V 120	vapour suction	Dampfansaugung f	aspiration f de la vapeur	всасывание пара
V 121	vapour supply, steam supply	Dampfversorgung f	approvisionnement m de vapeur	подача пара, снабжение паром
V 122	vapour temperature	Dampftemperatur f	température f de vapeur	температура пара
	vapour tension	s. V 113		
V 123	vapour throughput	Dampfdurchsatz m	débit m de vapeur	пропускание (расход) пара
V 124	vapour-tight seal	dampfdichter Abschluß m	fermeture f étanche à la vapeur	паронепроницаемый затвор
V 125	vapour velocity	Dampfgeschwindigkeit f	vitesse f de vapeur	скорость пара
V 126	variable	Einflußgröße f	coefficient m de l'influence	переменная
V 127	variable costs	variable Kosten pl	frais mpl variables	эксплуатационные затраты
V 128	variable volume	veränderliches Volumen n	volume m variable	переменный объем
V 129	variant comparison	Variantenvergleich m	comparaison f de variantes	сравнение (сопоставление) вариантов
	variation	s. 1. C 158; 2. L 59		
V 130	variational principle	Variationsprinzip n	principe m de variation	вариационный принцип
V 131	variation range	Variationsbereich m	domaine m de variation, zone f de changement	область изменения
V 132	varying temperature	veränderliche Temperatur f	température f variable	переменная температура
V 133	vat	Badbottich m	cuve f de bain	чан, бак
V 134	vector optimization	Vektoroptimierung f	optimisation f vectorielle	векторная оптимизация
V 135	vector quantity	Vektorgröße f	grandeur f vectorielle	векторная величина
V 136	vector representation	Vektordarstellung f	représentation f vectorielle	векторное представление
V 137	velocity constant, specific reaction rate	Geschwindigkeitskonstante f	vitesse f spécifique de la réaction, constante f de vitesse	постоянная (константа) скорости
V 138	velocity distribution	Geschwindigkeitsverteilung f	distribution f des vitesses	распределение скорости
V 139	velocity gradient	Geschwindigkeitsgradient m	gradient m de vitesse	градиент скорости
V 140	velocity-hold-up	Geschwindigkeits-hold-up n	retenue f de vitesse, hold-up m de vitesse	задержка скорости (в колонне)
	velocity of detonation	s. Q 27		
V 141	velocity of dissolution	Auflösungsgeschwindigkeit f	vitesse f de dissolution	скорость растворения
V 142	velocity of emission	Emissionsgeschwindigkeit f	vitesse f d'émission	скорость излучения (испускания)
	velocity of migration	s. M 214		
	velocity of sound	s. S 445		
V 143	velocity profile	Geschwindigkeitsprofil n	profil m de vitesse	профиль скоростей
V 144	velocity through ports	Durchgangsgeschwindigkeit f	vitesse f de passage	скорость прохождения
V 145	vena contracta	vena f contracta, verengter Flüssigkeitsstrahl m	jet m de liquide rétréci	суженная струя жидкости, суженная жидкая струя
V 146	vented extruder	Entgasungsextruder m, Entgasungsschneckenpresse f	boudineuse f à vis dégazée	дегазационный экструдер
V 147	vent filter	Entlüftungsfilter n	filtre m d'aération	фильтр сапуна
	vent gas	s. W 33		
	ventilating system	s. V 152		

V 148	ventilation	Entlüftung f, Ventilation f	ventilation f, aération f, purge f de l'air	откачка (отсос, спуск) воздуха, вентиляция
V 149	ventilation duct	Lüftungsschacht m	chèminée f d'aération	вентиляционный шахтный ствол
	ventilation equipment	s. V 152		
V 150/1	ventilation factor	Lüftungskoeffizient m	facteur m de ventilation	коэффициент вентиляции
V 152	ventilation installation, ventilating system, ventilation equipment	Entlüftungseinrichtung f	désaérateur m, appareil m de ventilation	вентиляционное устройство
	ventilation stock	s. E 327		
	ventilator	s. B 147		
V 153	ventilator cooler	Ventilatorkühler m	condenseur m à ventilateur	охладитель с вентилятором
V 154	venting pressure	Entlüftungsdruck m	pression f de désaérage	давление деаэрации (вентилирования)
V 155	Venturi fluided bed	Venturi-Wirbelschicht f	couche f fluidisée Venturi	кипящий слой Вентури
V 156	Venturi scrubber	Venturi-Wäscher m	scrubber m Venturi	скруббер Вентури
V 157	Venturi tube	Venturi-Düse f, Venturi-Rohr n	Venturi m, tube m de Venturi	труба Вентури
V 158	vent valve	Entlüftungsklappe f, Luftventil n, Luftklappe f, Lüftungsventil n	clapet m (ventouse f) d'aération	выпускной (вентиляционный) клапан
	vertical cooler	s. S 293		
V 159	vertical tube, lift tube, rising pipe	Steigrohr n	tuyau m élévateur (ascendant)	стояк, сифонный литник
	vertical tube	s. a. I 75		
V 160	vertical tube furnace	Vertikalrohrofen m	four m à tubes verticaux	вертикальная трубчатая печь
V 161	vibrating feeder	Vibrationseinspeisevorrichtung f, Schüttelspeiser m	alimentateur m oscillant (vibrant), gouttière f à secousses	вибрационный (качающийся) питатель
	vibrating grate	s. T 263		
V 162	vibrating grate unit	Schüttelrostanlage f	unité f de grilles à secousses	решетка с качающимися колосниками
V 163	vibrating mill	Vibrationsmühle f	vibreur m	вибромельница
	vibrating screen	s. S 260		
	vibrational energy	s. V 165		
V 164	vibration dryer	Vibrationstrockner m	séchoir m à vibrations	вибрационная сушилка
V 165	vibration energy, vibrational (wave) energy	Schwingungsenergie f	énergie f oscillatoire	энергия колебания
V 166	vibration frequency	Schwingungszahl f, Frequenz f	fréquence f, nombre m des oscillations	число колебаний
V 167	vibration mixer	Vibrationsrührer m	agitateur m à vibrations	вибрационная мешалка
V 168	vibrator	Vibrator m, Rüttelvorrichtung f	vibr[at]eur m, dispositif m à secouer	вибратор, вибрационное устройство (приспособление)
V 169	vibrator conveyor	Vibrationsspeiser m	transporteur m par vibration, alimentateur m vibrant	вибрационный питатель, вибропитатель
V 170	visco-elastic	viskoelastisch	visco-élastique	вязкоэластичный, вязкоупругий
V 171	visco-elasticity	Viskoelastizität f	visco-élasticité f	вязкоэластичность, вязкоупругость
V 172	viscometer	Viskosimeter n	viscomètre m, viscosimètre m	вискозиметр
V 173	viscoplastic	viskoplastisch	visco-plastique	вязкопластический
V 174	viscoplastic body	viskoplastischer Körper m	corps m viscoplastique	вязкопластичное тело
V 175	viscous dissipation	viskose Dissipation f	dissipation f visqueuse	диссипация вязкого течения
V 176	viscous flux	viskoser Fluß m, viskose Strömung f, viskoses Fließen n	flux (écoulement) m visqueux	вязкое течение
V 177	viscous liquid	viskose Flüssigkeit f	liquide m visqueux	вязкая жидкость
V 178	vitrification	Verglasung f, Übergang m in den glasartigen Zustand, Glasfluß m	vitrification f	застеклование, переход в стеклообразное состояние
V 179	void fraction	Lückenvolumen n	volume m du vide	доля объема пустот сыпучего материала
	volatibility	s. V 61		
	volatile	s. V 62		
V 180	volatile in steam	dampfflüchtig	entrainable à la vapeur	летучий с паром
	volatilizable	s. V 62		
	volatilization	s. E 278/9		
	volatilize/to	s. V 66		
	voltage drop	s. P 276		
V 181	volume contraction, contraction in volume	Volumenkontraktion f	contraction f volumique, diminution f du volume	сжатие объема
V 182	volume element	Volumenelement n	élément m de volume	объемный элемент, элемент объема
V 183	volume expansion, overage	Volumenerweiterung f, Volumenvergrößerung f	augmentation f de volume	объемное расширение
V 184	volume flow	Volumenabfluß m	débit m volumétrique	объемный расход
V 185	volume governor	Mengenregler m	régulateur m de volume	регулятор объема (массы)

V 186	volume of water	Wasserinhalt *m*	cubage *m* d'eau	содержание воды
V 187	volume pulsation	Volumenpulsation *f*	pulsation *f* de volume	объемная пульсация
V 188	volume rate of flow	Volumenstrom *m*	débit *m* volumique	объемный поток
V 189	volume ratio	Volumenverhältnis *n*	rapport *m* volumétrique	объемное [со]отношение
V 190	volume stability	Volumenbeständigkeit *f*	constance *f* de volume	постоянство объема
V 191	volume-stable reaction system	volumenbeständiges Reaktionssystem *n*	système *m* de réactions à volume constant	система реакций без изменения объема
V 192	volume storage tank	Volumenspeicher *m*	réservoir *m* volumique	емкостный баллон
V 193	volumetric analysis, volumetry	Maßanalyse *f*	analyse *f* volumétrique, titrage *m*	объемный анализ
V 194	volumetric efficiency	Liefergrad *m*	rendement *m* volumétrique	коэффициент подачи
V 195	volumetric flow rate	Volumenstrom *m*	débit *m* volumique	объемный поток
V 196	volumetric proportions in mixture	Mischungsvolumenanteile *mpl*	parties *fpl* en volume de mélanges	объемные пропорции в смеси
V 197	volumetric strain	volumetrische Dehnung *f*	expansion *f* volumétrique	объемное растяжение
V 198/9	volumetric suction capacity	Saugleistung *f*	volume *m* aspiré	объемная производительность по всасыванию
	volumetry	s. V 193		
V 200	volume weight	Volumengewicht *n*	poids *m* spécifique	объемный вес
V 201	vortex cooler	Wirbelkühler *m*	refroidisseur *m* vortex	охладитель с вихревой трубой
V 202	vortex cooling	Wirbelkühlung *f*	refroidissement *m* vortex	вихревое охлаждение
	vortex flow	s. E 13		
V 203	vortex point, fluidizing point	Wirbelpunkt *m*	point *m* de fluidisation	критическая точка перехода слоя в виброкипение, точка перехода в ожижение
V 204	vortex point velocity	Wirbelpunktgeschwindigkeit *f*	vitesse *f* de fluidisation	скорость точки ожижения
V 205	vortex sifter, vortical (centrifugal dust) collector	Wirbelsichter *m*, Zyklon *m*, Wirbelabscheider *m*	cyclone *m*, séparateur *m* à force centrifuge	вихревой сепаратор, циклон
V 206	vortex vacuum pump	Wirbelpumpe *f*	pompe *f* tourbillonnaire	вихревой насос
	vortical collector	s. V 205		
	vorticity	s. T 342		
V 207	vulcanizability	Vulkanisierbarkeit *f*	vulcanisabilité *f*	вулканизуемость
V 208	vulcanization	Vulkanisierung *f*	vulcanisation *f*, sulfuration *f* du caoutchouc	вулканизация

W

W 1	wake, eddy current, whirling stream	Wirbelstrom *m*	courant *m* parasite, stillage *m*	вихревой [по]ток
W 2	wall effect	Wandeffekt *m*, Wandphänomen *n*, Wandeinfluß *m*	effet *m* de paroi	стеночный (пристенный, пристеночный) эффект
W 3	wall heat flux	Wandwärmestrom *m*	flux *m* calorifique mural	стенный тепловой поток
W 4	wall pressure	Wanddruck *m*	pression *f* aux parois	пристенное давление
W 5	wall temperature	Wandtemperatur *f*	température *f* de paroi	температура стенок
	warehousing	s. S 704		
W 6	warm-air dryer	Warmlufttrockner *m*	séchoir *m* à air chaud	сушилка, работающая на теплом воздухе
W 7	warm-air furnace	Warmluftofen *m*	four *m* à air chaud	печь с горячим дутьем
W 8	warm-air heating	Luftheizung *f*	calorifère *m*, chauffage *m* à air	воздушное отопление, обогрев горячим воздухом
W 9	warming	Erwärmen *n*	échauffement *m*	нагревание
	warming-up curve	s. S 562		
W 10	warm-up rate	Warmlaufgeschwindigkeit *f*	vitesse *f* d'échauffement	скорость прогрева
W 11	warping	Verziehen *n*, Werfen *n*	gondolage *m*, gauchissement *m*	скорчивание
W 12	washability curve	Waschkurve *f*	courbe *f* de lavage	кривая обогатимости
W 13	wash bottle	Spritzflasche *f*	fiole *f* à jet	промывалка
W 14	wash column	Waschkolonne *f*	tour *f* de lavage	промывная колонна
W 15	washer	Wäscher *m*, Waschapparat *m*, Reinigungsapparat *m*, Berieselungsturm *m*, Rieselturm *m*	laveur *m*, appareil *m* laveur, scrubber *m*	скруббер, очистное сооружение
W 16	washing, purification, cleaning	Läuterung *f*	épurement *m*	мойка, осветление, очистка
W 17	washing, rinsing, flushing, elution	Spülung *f*, Waschen *n*	curage *m*, écurage *m*, rinçage *m*, lavage *m*, épuration *f*	промывка, продувка, промывание, споласкивание, мокрообогащение
W 18	washing, wet purification	Naßreinigung *f*	épuration *f* à l'eau, épuration par voie humide	мокрая очистка, влажная чистка
	washing	s. a. E 98		
W 19	washing liquid	Waschflüssigkeit *f*	liquide *m* laveur	промывочная жидкость
W 20	washing tower, scrubber	Waschturm *m*, Skrubber *m*	tour *f* de lavage, scrubber *m*	скруббер
W 21	washing trommel	Läutertrommel *f*	tambour *m* de purification	моечный барабан
	wash water	s. R 355		
W 22	waste, waste products	Abprodukte *pl*, Abfall *m*	produits *mpl* secondaires, déchet *m*	отходящие (отработанные) продукты, продукты на выброс
	waste	s. a. 1. L 235; 2. R 237		

W 23	**waste alkali**	Abfallauge *f*	alcali *m* épuisé	отработанная щелочь
W 24	**waste chamber**	Abfallverschlag *m*, Müllbunker *m*	chambre *f* de déchets	угарная камера, угарный бункер
	waste channel	*s.* D 487		
W 25	**waste cleaner**	Abfallreinigungsmaschine *f*	nettoyeuse *f* pour déchets	машина для очистки угаров (отходов)
W 26	**waste disposal**	Abproduktbeseitigung *f*, Abfallbeseitigung *f*	élimination *f* des déchets	удаление отходов
	waste disposal system	*s.* W 58		
W 27	**waste energy**	Abfallenergie *f*	énergie *f* non utilisée	отходящая энергия
W 28	**waste fat**	Abfallfett *n*	déchets *mpl* de graisse	утильный жир
W 29	**waste fibre**	Abfallfaser *f*	déchet *m* de fibre	отходное волокно
W 30	**waste-free process**	abproduktfreier Prozeß *m*	procédé *m* sans déchets	безотходный процесс
W 31	**waste-free technology**	abproduktfreie Technologie *f*	technologie *f* sans déchets	безотходная технология
W 32	**waste fuel**	Abfallbrennstoff *m*	combustible *m* de déchets	топливный отброс
W 33	**waste gas,** vent (exhaust) gas	Abgas *n*	gaz *m* brûlé, (perdu, d'échappement, de fumée)	отходящий (отходный, отработанный) газ, абгаз
	waste gas	*s. a.* B 121		
W 34	**waste gas analysis**	Abgasanalyse *f*	contrôle *m* du gaz brûlé, analyse *f* des gaz d'échappement	анализ отходящего газа
	waste gas chimney	*s.* E 327		
	waste gas collector	*s.* E 328		
	waste gas feeder	*s.* E 11		
W 35	**waste gas flue**	Abgaskanal *m*	canal *m* d'échappement	дымоход, газоотводящий канал
W 36	**waste gas superheater**	Abgasüberhitzer *m*	surchauffeur *m* à gaz d'échappement	перегреватель, работающий на уходящих газах, газовый перегреватель
W 37	**waste gas utilization**	Abgasverwertung *f*	récupération *f* des gaz perdus	использование отработавшего газа
W 38	**waste gate**	Abfallschleuse *f*	égout *m* de déchets	сбросной шлюз
	waste heat	*s.* L 246		
W 39	**waste-heat boiler,** economizer	Abhitzeofen *m*	four *m* à récupérer la chaleur perdue	печь с котлом-утилизатором, экономайзер
W 40	**waste-heat boiler**	Abhitzekessel *m*	chaudière *f* de récupération	котел-утилизатор, экономайзер
W 41	**waste-heat engine**	Abwärmekraftmaschine *f*	moteur *m* utilisant la chaleur perdue	тепловой двигатель, работающий с использованием отходящего тепла
W 42	**waste-heat flue**	Abhitzekanal *m*	carneau *m* de chaleur d'échappement	котел-утилизатор
W 43	**waste-heat kiln**	Abwärmeofen *m*	four *m* à récupérer la chaleur perdue	печь для утилизации отходящего тепла
W 44	**waste-heat loss**	Abwärmeverlust *m*	perte *f* de la chaleur perdue	потеря [отходящего] тепла
W 45	**waste-heat plant**	Abwärmeverwertungsanlage *f*	installation *f* de récupération de chaleur	установка для использования отходящего тепла
W 46	**waste-heat recovery**	Abhitzerückgewinnung *f*	récupération *f* de la chaleur d'échappement	использование отходящего тепла
W 47	**waste-heat utilization**	Abwärmeverwertung *f*	utilisation *f* de la chaleur perdue	использование отходящего тепла, утилизация неиспользованного тепла
W 48	**waste liquor (lye),** spent lye	Ablauge *f*	lessive *f* résiduaire	отбросный (отработанный) щелок
W 49	**waste material**	Altstoff *m*, Abfall *m*	vieux matériaux *mpl*, matériel *m* usagé, déchet *m*	отработанный материал, отходы
	waste oil	*s.* U 58		
W 50	**waste pipe**	Abfallrinne *f*, Abfallrohr *n*	tuyau *m* à déchets	сбросное русло, сбросной лоток
W 51	**waste pipe,** escape	Ablaufrohr *n*	tuyau *m* de décharge (purge)	спускная (отточная) труба
W 52	**waste product**	Abfallprodukt *n* *(unverwertbar)*	déchet *m*	отход, отброс[ный продукт]
	waste products	*s.* W 22		
W 53	**waste shaker**	Abfallreinigungstrommel *f*	tambour *m* de nettoyage pour les déchets	угарная пылевыколачивающая машина
	waste steam	*s.* E 338		
W 54	**waste steam utilization**	Abdampfverwertung *f*	utilisation *f* de la vapeur épuisée	утилизация отходящего пара
W 55	**waste stream**	Abfallstrom *m*	débit *m* de déchets	ток отпускания, сбросной поток
W 56	**waste treating facility**	Abfallaufbereitungsanlage *f*	unité *f* de traitement des déchets	установка обработки отходов
W 57	**waste treatment**	Abfallbehandlung *f*	traitement *m* des déchets	переработка (обработка) отходов
W 58	**waste treatment facilities,** waste disposal system	Entsorgungssystem *n*	système *m* d'évacuation et de traitement des déchets	система для обезвреживания отходов
	waste tube	*s.* O 151		
W 59	**waste utilization**	Abproduktverwertung *f*, Abproduktnutzung *f*	utilisation *f* des sous-produits	утилизация отработанных продуктов
W 60	**waste utilization plant**	Abfallverwertungsanlage *f*	installation *f* de l'utilisation des déchets	установка для утилизации отбросов

W 61	wastewater	Abfallwasser n, Abwasser n	eau f résiduaire (usée, d'égout)	сточная вода
W 62	wastewater column	Abwasserkolonne f	tour f de traitement des eaux usées	колонна для обработки сточных вод
W 63	wastewater disposal, sewage disposal	Abwasserbeseitigung f	traitement m des eaux d'égouts	удаление сточных вод
W 64	wastewater stripping column	Abwasserstrippkolonne f	colonne f de strippage pour les eaux usées	стриппинг-колонна для сточных вод
W 65	water absorption	Wasseradsorption f	adsorption f d'eau	абсорбция воды
W 66	water backflow, back water	Wasserrücklauf m	reflux m d'eau	обратный ток воды
W 67	water bath	Wasserbad n	bain m d'eau, bain-marie m	водяная баня
W 68	water calender	Wasserkalander m, Naßkalander m	calandre f à eau (essoreur)	водяной каландр
W 69	water chiller	Wasserkühler m	refroidisseur m de l'eau, réfrigérant m à eau	холодильник с водяным охлаждением
W 70	water circulating plant	Wasserumwälzanlage f	installation f de circulation d'eau	циркуляционная установка для воды
W 71	water circulation	Wasserumlauf m	circulation f d'eau	обвоз (циркуляция) воды
W 72	water collection system	Wassersammelanlage f	installation f de collecte pour les eaux	водосборное сооружение
W 73	water column	Wassersäule f	colonne f d'eau	водяной столб
W 74	water column pressure	Wassersäulendruck m	pression f de la colonne d'eau	давление водяного столба
W 75	water conditioning	Brauchwasseraufbereitung f	conditionnement m d'eau	кондиционирование воды
	water conditioning	s. a. W 102		
W 76	water content	Wassergehalt m	teneur m en eau	содержание воды, влажность
W 77	water cooler	Wasserwärmetauscher m, Wasserwärmeübertrager m	réfrigérant m à eau	водяной теплообменник
W 78	water cooling, water jacket	Wasserkühlung f	refroidissement m par eau	водяное охлаждение
W 79	water drain	Wasserabfluß m	déversoir m, vidange f d'eau	отток (сток) воды
	water drift	s. D 503		
	water feed	s. I 189		
W 80	water filter	Wasserfilter n	filtre m d'eau	водяной фильтр
W 81	water flow control	Wassermengenregelung f	régulation f du débit d'eau	регулирование расхода воды
W 82	water for industrial use	Betriebswasser n	eau f industrielle (de service)	вода для технических нужд, промышленная вода
W 83	water gas	Wassergas n	gaz m à l'eau	водяной газ
W 84	water glass	Wasserglas n	verre m soluble, silicate m de sodium	растворимое (жидкое) стекло
W 85	water grades	Wassergüteklassen fpl	classes fpl de qualité de l'eau	классы качества воды
W 86	water ingredient	Wasserinhaltsstoffe mpl	ingrédients mpl de l'eau	примеси, содержащиеся в воде, состав примесей воды
W 87	water-in-oil emulsion	Wasser-in-Öl-Emulsion f	émulsion f du type «eau dans l'huile»	эмульсия типа «вода в масле»
W 88	water jacket	Kühlwassermantel m	chemise f [réfrigérante] d'eau	водяная рубашка, водоохлаждаемый кожух
	water jacket	s. a. W 78		
W 89	water jet	Wasserstrahl m	chasse (veine) f d'eau, jet m fluide	струя воды
W 90	water jet apparatus	Wasserstrahlapparat m	appareil m à jet d'eau	водоструйный прибор (аппарат)
	water jet aspirator (injector)	s. G 121		
W 91	water level	Wasserniveau n Wasserstand m	niveau m d'eau, hauteur m des eaux	уровень воды
W 92	water miscibility	Wassermischbarkeit f	miscibilité f à l'eau	смешиваемость с водой
W 93	water network	Wassernetz n	canalisation f d'eau	гидрографическая сеть
W 94	water pipe	Wasserrohr n	tube m (tuyau m, conduite f) d'eau	водяная труба
W 95	water pollution	Wasserverunreinigung f	pollution f de l'eau	загрязнение воды
W 96	water power system	Wasserkraftsystem n	système m hydroélectrique	гидроэнергетическая система
W 97	water processing	Wasseraufbereitung f	traitement m de l'eau	обработка воды
W 98	water processing plant	Wasseraufbereitungsanlage f	installation f de l'épuration de l'eau	установка подготовки воды
W 99	waterproof, watertight	wasserdicht	étanche à l'eau	водонепроницаемый
W 100	waterproof layer, watertight layer	Wassertrennschicht f	couche f séparatrice d'eau	водонепроницаемый слой
W 101	water pump	Wasserpumpe f	pompe f à eau	водяной насос
W 102	water purification, water conditioning	Wasseraufbereitung f, Wasserreinigung f	conditionnement m d'eau, épuration f de l'eau	водообогащение, очистка воды
W 103	water quench column	Wasserquenchkolonne f	tour (colonne) f de trempe à l'eau	колонна для резкого охлаждения водой
W 104	water recooling	Wasserrückkühlung f	réfrigération f d'eau de retour	обратное охлаждение воды
	water removal	s. D 489		
W 105	water sample, water testing	Wasserprobe f, Wasserdruckprobe f	essai m (épreuve f) hydraulique	испытание водой
W 106	water separation ability	Wasserabscheidevermögen n	pouvoir m désémulsionnant	деэмульгирующая способность

W 107	water separator	Wasserabscheider m	séparateur m de l'eau	водоотделитель
W 108	water separator, steam trap, steam dryer	Dampftrockner m, Dampftrockenapparat m	sécheur m de vapeur, étuve f à vapeur	сушилка с паровым обогревом, паросушитель, паровая сушилка
W 109	water softening	Wasserenthärtung f	adoucissement m de l'eau	мягчение воды
W 110	water-soluble	wasserlöslich	soluble dans l'eau	растворимый в воде, водорастворимый
W 111	water spray	Naßfilter n	filtre m humide	влажный фильтр
W 112	water storage tank	Wasserspeicher m	réservoir m d'eau	водохранилище
W 113	water stream	Wasserstrom m	courant (écoulement) m d'eau	водоток, [по]ток воды
W 114	water stream energy	Wasserstromenergie f	énergie f de courant d'eau	энергия текучей воды
W 115	water supply	Wassereinspeisung f, Wasserversorgung f, Wasserwirtschaft f	alimentation f d'eau, aménagement m (distribution f) des eaux	питание (снабжение) водой, водное хозяйство
W 116	water suspension reactor	Wassersuspensionsreaktor m	réacteur m de suspension à l'eau	водосуспензионный реактор
	water testing	s. W 105		
	watertight	s. W 99		
	watertight layer	s. W 100		
W 117	water tower	Wasserturm m	château m d'eau	водонапорная башня
W 118	water-tube boiler	Wasserrohrdampferzeuger m, Wasserrohrkessel m	chaudière f à tubes d'eau	водотрубный котел
W 119	water-tube-type cooler	Wasserröhrenkühler m	refroidisser m à tubes d'eau	трубчатый радиатор с водяными трубками
W 120	water utilization	Wassernutzung f	utilisation f de l'eau	водопользование
W 121	water utilization plant	Wassernutzungsanlage f	installation f de l'utilisation d'eau	установка для водопользования
W 122	water-vapour cycle	Wasser-Dampf-Kreislauf m	cycle m eau-vapeur	пароводяной контур
W 123	water wash	Wasserwäsche f	lavage m à l'eau	промывка водой
W 124	water wash column	Wasserwaschkolonne f	tour f de lavage à l'eau	колонна для промывки водой
W 125	water wheel generator	Wasserkraftgenerator m	générateur m hydroélectrique	гидросиловой генератор, гидрогенератор
W 126	waterworks	Wasserwerk n	usine f de distribution d'eau, entreprise f hydraulique	водопроводная станция
W 127	water yield	Wasserergiebigkeit f	rendement (débit) m d'eau	дебит воды
W 128	watt density	Heizleistung f	rendement m calorique	теплопроизводительность
W 129	watt-second	Wattsekunde f, Joule n	watt-seconde f, joule m	ватт-секунда
	wave energy	s. V 165		
W 130	wave filter	Wellensieb n	filtre m d'onde	волновой фильтр
W 131	wave mechanics	Wellenmechanik f	mécanique f ondulatoire	волновая механика
	weak gas	s. P 267		
W 132	weak liquid	arme Lösung f	solution f faible	слабый раствор
W 133	weak point	Schwachstelle f	point m faible	узкое (слабое) место
W 134/5	weak vapour	armer Dampf m	vapeur f pauvre	слабый газ, слабый пар
	wear	s. U 69		
W 136	wear allowance, allowance for wear	Abnutzungsreserve f	réserve f à l'usure	запас на износ
W 137	wear behaviour	Abnutzungsverhalten n, Verschleißverhalten n	comportement m d'usure	поведение во время износа
W 138	wear degree, degree of wear	Verschleißgrad m, Abnutzungsgrad m	degré (indice) m d'usure	степень износа (изнашивания)
W 139	wear failure	Verschleißausfall m	défaillance f par usure	отказ в стадии износа
W 140	wearing process	Abnutzungsvorgang f	usure f, dégradation f	процесс износа
W 141	wearing surface	Abnutzungsfläche f, Verschleißfläche f	surface f d'usure	поверхность износа
W 142	wearing test	Abnutzungsprüfung f, Verschleißtest m	essai m d'usure	испытание на износ
W 143	wear out/to	verschleißen	user	изнашивать
W 144	Wegstein method	Wegstein-Methode f	méthode f Wegstein	метод [итерации] Вегштейна
	weight	s. L 186		
W 145	weight bridge	Abfüllwaage f	peseuse f	расфасовочные весы
W 146	weighted average molecular weight	gewichtete mittlere Molmasse f	masse f moléculaire moyenne pondérée	взвешенная средняя мольная масса, взвешенный средний молекулярный вес
W 147	Weissenberg effect	Weißenbergscher Effekt m	effet m Weissenberg	эффект Вейсенберга
W 148	wet-and-dry-bulb psychrometer	Psychrometer n, Verdunstungsfeuchtigkeitsmesser m	psychromètre m	психрометр
W 149	wet crushing	Naßzerkleinerung f	broyage m à l'eau	мокрое дробление
W 150	wet dust separation	Naßentstaubung f	dépoussiérage m humide	обеспыливание мокрым способом
W 151	wet grinding	Naßmühle f, Naßmahlung f	concasseur m à l'eau, broyage m humide	мокрый помол, мокрое измельчение
W 152	wet process	Naßverfahren n	procédé m humide	мокрый способ
	wet purification	s. W 18		
W 153	wet screening	Naßklassieren n	triage m humide, classement m à eau	гидравлическая классификация

W 154	wet steam	Naßdampf m	vapeur f humide (directe, mouillée)	влажный (насыщенный) пар
W 155	wettability	Benetzbarkeit f	mouillabilité f, aptitude f au mouillage	смачиваемость
W 156	wettability properties	Benetzungseigenschaften fpl	propriétés fpl de mouillabilité	свойства смачиваемости
W 157	wetted-wall column	Dünnschichtkolonne f, Rieselkolonne f	colonne f à ruissellement	колонна с орошаемыми стенками
W 158/9	wetted-wall column, film reactor	Dünnschichtreaktor m	réacteur m à couche mince	пленочный реактор
	wetted-wall tower	s. F 7		
W 160	wetting agent	Benetzungsmittel n	agent m mouillant	смачиватель
	wetting agent	s. a. M 300		
W 161	wetting drum	Befeuchtungstrommel f	tambour m à humecter, cylindre m de mouillage	барабан для смачивания (увлажнения)
W 162	wetting surface	Benetzungsfläche f (Füllkörperschüttung)	surface f de mouillage	поверхность смачивания
W 163	wet transport	Naßförderung f	transport m hydraulique	гидравлический транспорт
W 164	wet-type air cooler	Naßluftkühler m	refroidisseur m d'air [type] humide	мокрый воздухоохладитель
W 165	wet-type dust collector	Naßentstauber m, Gaswäscher m	scrubber m, laveur m de gaz	скруббер, мокрый пылеуловитель
W 166	wet vapour region	Naßdampfgebiet n	région f (domaine m) de vapeur humide	область влажного пара
	whirling stream	s. W 1		
W 167	Whitman two-film theory	Whitmansche Zweifilmtheorie f	théorie f des deux films de Whitman	двухпленочная теория Вайтмана
W 168	Wilson cloud chamber, cloud chamber	Wilson-Nebelkammer f, Nebelkammer f	chambre f [de détente] de Wilson	камера Вильсона
W 169	winch, capstan, jack, hoist	Winde f	treuil m, cric m	лебедка, ворот, домкрат, мотовило
W 170	winding machine	Fördermaschine f	machine f d'extraction	подъемная машина
W 171	windup equipment	Aufwickelvorrichtung f	rouloir m, enrouloir m	устройство для намотки
W 172	wire coating	Drahtummantelung f, Drahtbeschichtung f	enrobage m de fil, revêtement m de fil	обкладка проволоки, нанесение слоя на проволоку
W 173	wire gauge	Drahtdurchmesser m, Drahtdicke f	calibre m du fil	диаметр (толщина, калибр) проволоки
W 174	wire-mesh tray	Drahtgewebeboden m	tamis m, toile f métallique	тарелка из проволочной сети
	wiring diagram	s. C 255		
	wiring system	s. C 254		
W 175	withdrawal	Entnahme f	prise f, soutirage m	отбор, забор, взятие
W 176	withdrawal of steam	Dampfentnahme f	soutirage m de vapeur	отбор пара
W 177	withdrawing tank, off-take tank	Entnahmebehälter m	réservoir m de prise	водозаборный резервуар
W 178	Wolfe-Ostrowski method	Wolfe-Ostrowski-Methode f	méthode f de Wolfe-Ostrowski	метод [итерации] Вольфа-Островского
W 179	workability	Bearbeitbarkeit f	usinabilité f, aptitude f d'être usiné	способность поддаваться обработке
	workable	s. O 55		
W 180	work capacity	Arbeitskapazität f, Arbeitsfähigkeit f	capacité f de travail	объем работ, рабочая мощность
W 181	work cyclic/to, to work intermittent	absatzweise arbeiten	travailler en batch	прерывно работать
W 182	working capacity	Arbeitsvermögen n	pouvoir m de travail, énergie f	рабочая мощность, производительность [труда]
W 183	working conditions	Betriebszustand m, Betriebsbedingungen fpl	conditions fpl d'exploitation	рабочие условия, условия эксплуатации
	working cycle	s. R 417		
W 184	working diagram	Arbeitsdiagramm n	diagramme m de travail	рабочая (индикаторная) диаграмма, диаграмма работ
W 185	working drawing	Ausführungsunterlagen fpl	documentation f d'exécution	проектная документация, исполнительные чертежи
W 186	working expenses	Betriebsunkosten f	frais mpl d'exploitation	издержки производства, производственные затраты
W 187	working intensity	Arbeitsintensität f	puissance f, intensité f de travail	интенсивность работы, производительность труда
W 188	working line, operating line	Arbeitsgerade f, Arbeitslinie f	droite f opératoire, courbe f caractéristique	рабочая линия
W 189	working load	Arbeitsbelastung f	charge f de travail	рабочая нагрузка
W 190	working load, live load	Nutzlast f	charge f (poids m) utile	полезная нагрузка
W 191	working medium	Arbeitsmedium n	fluide m de travail	рабочая среда
W 192	working method	Arbeitsmethode f, Betriebsweise f	mode m opératoire, méthode f de travail	метод работы, рабочий метод
W 193	working plan, detailed drawings	Ausführungszeichnung f	dessin m d'exécution	рабочий чертеж
W 194	working pressure	Arbeitsdruck m	pression f de travail (service)	рабочее давление
W 195	working reactor	Produktionsreaktor m	réacteur m technique (de production)	производственный реактор
W 196	working steam	Arbeitsdampf m	vapeur f de travail	рабочий пар

W 197	working temperature	Arbeitstemperatur f	température f de travail	рабочая температура
	working temperature	s. a. O 71		
	working volume	s. P 170		
	work intermittent/to	s. W 181		
	works	s. P 188		
	worm conveyor	s. S 103		
W 198	worm drive (gearing)	Schneckenantrieb m, Schneckengetriebe n	engrenage m à vis sans fin	червячная передача, червячный привод
W 199	worm thread	Schneckengewinde n	filet m hélicoïde	червячная резьба

Y

Y 1	yield, output	Ertrag m, Ausbeute f	rendement m, produit m	выход, производительность, доход, выручка
	yield	s. a. P 541		
	yield of crude product	s. C 868		
Y 2	yield of purified product	Reinausbeute f	rendement m de produit pur	выход очищенного продукта
Y 3	yield point (strength)	Streckgrenze f, Fließgrenze f	limite f d'allongement, limite d'étirage, limite d'écoulement	предел текучести (прочности на растяжение)
	Young's modulus	s. M 292		

Z

Z 1	zero adjustment	Nulleinstellung f	mise f à zéro, mise (ajustage m) au point zéro	установка на нуль
Z 2	zero charge, zero power	Nullast f	charge f nulle	нулевая нагрузка
Z 3	zero method	Nullmethode f	méthode f de réduction à zéro	нулевой метод [измерения]
Z 4	zero point	Nullpunkt m	zéro m, point m neutre	нулевая (нейтральная) точка
Z 5	zero position	Nullage f	position f zéro	нулевое положение
	zero power	s. Z 2		
Z 6	zero series	Nullfolge f	suite f de zéros	нулевая последовательность
Z 7	zero value	Nullwert m	valeur f nulle	нулевое значение
Z 8	zero velocity	Nullgeschwindigkeit f	vitesse f nulle (zéro)	нулевая скорость
Z 9	zone cooler	Zonenkühler m	refroidisseur m à zones	ярусный холодильник
Z 10	zone melting	Zonenschmelzen n	fusion f zonale	зонная плавка
Z 11	zone of breathing (deaeration, degasing)	Entgasungszone f	zone f de dégazage	зона выделения летучих веществ
Z 12	zone of dispersion	Streuungsbereich m	domaine m de dispersion	диапазон дисперсии

German Index

A

Abätzen E 258
Abbau D 73, D 373, R 177
Abbaugrad R 178
Abbauhammer P 233
Abbauprodukt D 47
Abbauprozeß D 46
Abbaustufe S 524
Abbautemperatur C 825
Abbauverluste E 380
Abbeizmittel P 141
Abbeizung C 286
Abbiegen B 85
Abbildungsgerät I 32
Abbildungsgleichung I 31
Abbildungsvariable P 143
Abbindefähigkeit S 237
Abbindewärme H 90
Abblasedruck B 156
Abblasehahn B 154
Abblaseklappe B 130
Abblasekolonne B 146, D 204
Abblaseleitung B 143
Abblaserohr B 155
Abblasesystem B 144
Abblasetank B 145
Abblaseventil B 130, B 157
abbrennbar C 424
abbrennen B 125
Abbrennen B 126
Abbruchreaktion T 95
Abdampf E 338
Abdampfabsorptionskältema-
  schine E 340
Abdampfaustritt S 617
Abdampfdruck E 344
Abdampfeintritt E 341
Abdampfkessel E 268
Abdampfprobe G 162
Abdampfrohr E 342
Abdampfrohrnetz E 343
Abdampfrückstand E 291
Abdampfschale E 270
Abdampfspannung E 344
Abdampfverwertung W 54
Abdeckfläche C 809
Abdeckplatte C 810
Abdeckung C 808
abdestillieren D 405
abdichten S 114
Abdichtfläche S 117
Abdichtung S 116
Abdichtungsmaterial S 115
Abdichtwalze S 119
Abdrosselung C 251
Abdrücken P 433
abfackeln B 317
Abfall D 292, R 237, T 4, W 22,
  W 49
Abfallaufbereitungsanlage W 56
Abfallauge W 23
Abfallbehandlung D 387, W 57
Abfallbeseitigung W 26
Abfallbeseitigungsanlage D 388
Abfallbrennstoff W 32
Abfallenergie W 27
Abfallerzeugnis B 332
Abfallfaser W 29
Abfallfett W 28
Abfallgrenze C 835
Abfallöl S 378
Abfallprodukt B 332, W 52
Abfallreinigungsmaschine W 25
Abfallreinigungstrommel W 53
Abfallrinne 50
Abfallrohr D 342, D 476, W 50
Abfallsammelanlage R 238
Abfallschleuse W 38
Abfallstrom W 55
Abfallvernichtung R 239

Abfallvernichtungsanlage R 240
Abfallverschlag W 24
Abfallverwertung R 173
Abfallverwertungsanlage W 60
Abfallwasser W 61
abfetten D 75
Abfiltrieren F 109
abfließen F 248
Abfließen D 488
abfließendes Produkt E 30
Abfluß D 342, O 111
Abflußbecken B 84
Abflußdruck O 52
Abflußprozeß C 717
Abflußhahn D 328
Abflußkanal D 487
Abflußkoeffizient R 422
Abflußkrümmer D 338
Abflußkurve R 60/1
Abflußleitung D 339, D 494
Abflußmenge D 345, R 55
Abflußmengenlinie O 109
Abflußneigung D 337
Abflußöffnung D 119
Abflußregelung O 116
Abflußregler O 117
Abflußrinne G 163
Abflußrohr D 334/5, D 342,
  S 250
Abflußschieber O 121
Abflußventil D 348, D 497
Abflußverhältnis O 120
Abfragetaste R 291
Abfühlkontakt S 186
Abführung E 19
Abführung fühlbarer Wärme
  S 185
Abführungssektion C 861
Abfüllanlage F 74, R 4
Abfülleitung P 70
Abfüllen F 495, R 3
Abfüllen von Flüssiggasen
  B 217
Abfüllmaschine B 216
Abfüllwaage W 145
Abgabe D 148
Abgabedruck D 155
Abgabetemperatur D 158
Abgang E 30
Abgas F 273, S 521, W 33
Abgasanalyse W 34
Abgasanlage S 330
Abgasbehandlung O 12
Abgasbrüden E 338
Abgaskanal W 35
Abgaskessel F 274
Abgasleitung E 324
Abgasreinigung E 331
Abgassammler E 328
Abgasschornstein E 327
Abgasstutzen F 275
Abgastrockner E 329
Abgasüberhitzer W 36
Abgasverwertung W 37
Abgasvorwärmer E 11
Abgaswaschkolonne O 10/1
abgefiltert F 97
abgehender Strom E 32
abgeschlossenes System C 294
abgesetzt P 322
Abgießen D 29
abgleichen B 15
Abgleichen B 26
Abgreifpunkt T 19
Abgrenzungsmethode D 213
Abhitze L 246
Abhitzekanal W 42
Abhitzekessel E 268, F 274,
  W 40
Abhitzeofen W 39
Abhitzerückgewinnung W 46

Abhitzeverwendung U 72
Abhitzeverwerter R 247
Abhitzeverwertung U 72
Abklärgefäß D 30
Abklingzeit D 32
Abkochung D 39
Abkühldauer E 692
Abkühler C 698
Abkühlgeschwindigkeit C 241
Abkühlgrad C 709
Abkühlkurve C 708
Abkühlmittel Q 24
Abkühlperiode P 599
Abkühlprozeß C 717
Abkühltemperatur R 225
Abkühlung C 700
Abkühlversuch C 721
Abkühlwärme H 74
Abkühlzeit C 692
Ablagerung D 176, D 557, E 85
Ablassen D 350
Ablaßgeschwindigkeit D 499
Ablaßhahn B 154
Ablaßkanal S 250
Ablaßleitung D 494
Ablaßstutzen D 342
Ablaßventil D 328
Ablauf O 111, P 580
Ablaufpumpe D 344
Ablaufrohr D 476, W 51
Ablaufschräge D 337
Ablaufstutzen D 475
Ablauftrichter D 493
Ablaufwasser D 333
Ablauge W 48
Ableitung B 242
Ableitungsrinne D 496
Ableitungsverlust L 238
Ablenkbarkeit D 60
Ablenkkammer D 61
Ablenkplatte B 10, D 63
Ablenkung D 222
Ablesefehler R 122
Ablesegenauigkeit A 82
Ablesegerät S 317
Ablieferungstermin D 16
Abluft E 321
Abluftgebläse E 325/6
Abluftkanal A 392
Abluftrohr A 391
Abluftstrom E 322
Abluftwärmerückgewinnung
  E 323
Abnahme D 292
Abnahme der flüchtigen Be-
  standteile D 223
Abnahmedruck F 136
Abnahmeleistung P 305
Abnahmeprüfung A 68
Abnahmeversuch A 68
Abnahmevorschriften S 462
Abnutzung A 5, D 212, U 69
Abnutzungserscheinungen
  S 313
Abnutzungsfläche W 141
Abnutzungsgrad W 138
Abnutzungsprüfung W 142
Abnutzungsreserve W 136
Abnutzungsverhalten W 137
Abnutzungsvorgang W 140
Abprall B 223
Abproduktbeseitigung W 26
Abprodukte W 22
abproduktfreier Prozeß W 30
abproduktfreie Technologie
  W 31
Abproduktnutzung W 59
Abproduktverwertung W 59
abpumpen P 608
Abreibung A 5
Abreißen T 23

Abrieb A 5
Abriebfestigkeit A 3
Abriebversuch A 4
Abriebwirkung A 2
abrieseln T 305
absacken S 1
Absacktrichter S 2
Absackvorrichtung B 12
Absackwaage B 13
absättigen S 38
absatzweise arbeiten W 181
Absauganlage S 783
Absauggeschwindigkeit S 808
Absaugluft E 320
Absaugmaschine E 334
Absaugpumpe E 337
Absaugverfahren E 335
Absaugvorrichtung E 333
Abschaltelement D 356
abschalten B 251
Abschaltleistung B 259
Abschaltung D 355
Abschaltzeit D 480
Abschätzung A 528
Abscheider D 30, S 216,
  S 239
Abscheideventil D 497
Abscheidung D 177, E 85
Abscheidungspotential D 178
Abscheidungsprodukt D 175
Abscheidungsverfahren M 200
abschlämmen E 96
Abschlämmung E 97
Abschleudern C 141
abschmieren G 147
Abschmieren L 299
abschrecken Q 19
Abschreckkolonne Q 20
Abschrecköl Q 25
Abschrecktemperatur Q 26
Abschreckturm Q 20
Abschreckung Q 22
absedimentieren S 158
Absetzbecken S 239, S 241,
  S 371
Absetzbehälter S 239, S 242
Absetzen D 29, S 240
Absetzer S 239
Absetzgeschwindigkeit R 49,
  S 156
Absetztank S 239
Absetzzentrifuge S 154
Absetzzone S 157
Absolutdruck A 13
absolute Abweichung A 7
absolute Empfindlichkeit A 14
absolute Feuchtigkeit A 11
absoluter Druck A 13
absoluter Feuchtegehalt A 12
absoluter Wärmeeffekt A 10
absolutes Filter A 8
absolutes System A 15
absolute Temperatur A 16
absolute Verzögerung A 6
absolute Viskosität A 17
absolute Wärmetönung A 10
Absorbens A 21
Absorber A 25, A 30
Absorberabgas A 29
Absorberkopf A 28
Absorbersumpf A 26
absorbierbar A 19
Absorbierbarkeit A 57
absorbieren A 18
Absorbieren A 31
Absorption A 34
Absorption der leichten Anteile
  L 87
Absorptionsanlage A 49
Absorptionsapparat A 25
Absorptionseinheit A 54

G

N

French Index

A

abaissement cryoscopique
L 251
abaissement de la pression de
vapeur L 252
abaissement de température
T 72
abaissement du point de congé-
lation F 357
abaque N 51
abord A 73
abrasion A 5
absorbable A 19
absorbant A 21
absorber A 18
absorbeur A 25, A 30, A 32,
R 135
absorbeur à garnissage P 6
absorbeur à lit fixe F 184
absorbeur à serpentins C 359
absorbeur à un étage S 327
absorbeur d'ammoniaque A 464
absorbeur de réfrigérant R 205
absorbeur principal M 17
absorptif A 56
absorption A 31, A 34
absorption adiabatique A 157
absorption à simple effet S 328
absorption d'ammoniaque
A 465
absorption de chaleur H 17
absorption de gaz G 4
absorption d'énergie E 127
absorption de réfrigérant R 206
absorption des fractions lé-
gères L 87
absorption d'humidité M 297
absorption d'ions I 331
absorption principale M 18
absorption propre I 88
absorption thermale (thermi-
que) H 17
absorption totale T 231
absorptivité A 57
accélérateur A 64
accélérateur d'inflammation
C 145
accélération A 60
accélération centrifuge C 124
accélération de convergence
C 672
accélération de la pesanteur
G 146
accélération de la sédimenta-
tion A 62
acceptation A 67
accepteur A 69
accès A 73
accessibilité A 71
accessoire de levage L 211
accouplement C 300
accrochage D 372
accroissement de pression
P 427
accumulateur S 710
accumulateur de chaleur H 103
accumulateur de froid C 381
accumulateur d'énergie E 168
accumulateur de vapeur S 591
accumulateur d'information
D 13
accumulateur électrique S 698
accumulation A 76, I 175, P 2,
S 697
accumulation de boue A 77
accumulation de chaleur H 18,
H 107
accumulation d'énergie E 128
accumulation des erreurs C 299
accumulation d'humidité M 298

acétométrie A 89
achèvement C 462
achever C 457
acide carbonique solide D 532
acide sulfurique S 810
acidification A 88
acidimétrie A 89
acidité A 90
acidorésistance A 100
acier coulé S 660
action A 102
action de chargement I 131
action de prémélange P 317
action secondaire S 136
activant A 117
activation A 113
activation de boue B 108
activer A 105
activeur A 117
activité A 113
activité catalytique C 91
activité initiale S 560
additif A 140
addition A 140, F 322
additivité A 146
additivité logarithmique L 222
adduction d'air A 378, S 835
adduction de chaleur isothermi-
que I 354
adduction des eaux d'égouts
S 247
adduction de solution S 418
adhérence A 147, S 120
adhésif A 151, F 191
adhésion A 147
adhésivité S 120
adiabate A 164
adiabate de condensation C 890
adiabate de température A 180
adiabate d'expansion A 165
adiabatique A 156
adjacence A 184
admission F 241, I 188
admission de chaleur H 109
admission de chaleur isothermi-
que I 354
admission de l'air A 349
admission d'humidité M 310
admission du gaz G 90
adoucissement S 392
adoucissement de l'eau W 109
adsorbabilité A 201
adsorbable A 202
adsorbant A 208, A 209
adsorbate A 203
adsorbeur A 208, A 232
adsorbeur à charbon activé
C 179
adsorption A 211
adsorption à lit fixe F 185
adsorption à lit fluidisé F 286
adsorption chimique C 209
adsorption cyclique P 106
adsorption d'eau W 65
adsorption de vapeur V 72
adsorption d'humidité M 299
adsorption discontinue P 106
adsorption en phase vapeur
V 109
aérage A 236
aérateur A 238
aération A 236, V 148
aéré A 240
aérer A 241
aérobie A 242
aérochauffeur A 371
aérodynamique A 243, A 245
aéroréfrigérant d'air A 432
affinage R 193
affinage au vent B 124

affinage par soufflage B 124
affinerie S 89
affluant I 138
affluence I 139, I 187
afflux I 139
agent A 140
agent absorbant A 21
agent anticorrosif A 511
agent antifouling A 514
agent antifriction S 370
agent antigel A 515
agent antimousse A 513
agent auxiliaire de filtration
A 628
agent corrosif C 752
agent d'agglomération A 263
agent d'attaque M 134
agent de basification B 47
agent de chauffe H 56
agent de dispersion D 378
agent de dissolution M 134
agent de flottation F 215
agent de l'humectation M 300
agent de réaction R 125/6
agent de refroidissement primaire P 463
agent de séparation S 195
agent dessiccateur S 299
agent de transfert de chaleur
H 38
agent de transport M 135
agent épaississant T 170
agent frigorifique C 701
agent frigorifique gazeux G 27
agent frigorifique inorganique
I 220
agent frigorifique primaire
A 463
agent humidificateur M 300
agent moteur P 588
agent mouillant W 160
agent moussant F 305
agent moussant d'extinction du
feu F 307
agent réfrigérant Q 24
agent tensio-actif S 885
agent troublant D 556
agglomérat A 262
agglomération A 264
aggloméré A 262, A 266
agglomérer A 261
agglutinant A 151
agglutination A 267
agitateur A 284, A 285, A 288
agitateur à ancres A 495
agitateur à bande hélicoïdale
H 128
agitateur à racloires S 75
agitateur à vibrations V 167
agitateur à vis R 347
agitateur incorporé B 302
agitateur mécanique S 680
agitateur turbinaire P 15, T 335
agitation A 282, M 245, S 256
agiter A 278, R 1
agrégat A 269
agrégation A 76
aile[tte] agitatrice A 285
ailette de refroidissement C 712
ailette hydrométrique F 245
air additionnel A 615
air aspiré S 782
air auxiliaire A 615
air chaud H 192
air circulé R 142
air comprimé C 478, H 148
air d'aspiration S 782
air de basse pression L 257
air de by-pass B 324
air d'échappement E 321
air de circulation R 142

air de combustion C 427
air de l'humidification D 6
air d'entrée I 190
air de réfrigération (refroidissement) C 703/4
air de séchage D 534
air de sortie E 320, L 63
air détendu E 355
air d'évacuation E 321
air d'impact R 16
aire C 858
aire interfaciale I 276
air entraîné E 196
air entrant I 190
air épuré C 283
air étranglé T 195
air évacué E 320
air filtré F 98
air forcé A 291, D 150
air froid C 368
air humide H 203
air liquide L 146
air non condensé N 57
air non filtré U 9
air non saturé U 43
air normal S 528
air primaire D 150, P 456
air purifié C 282
air réfrigéré C 693
air résiduel R 303
air saturé de vapeur S 654
air sec D 534
air secondaire S 129
air sortant E 321
air soufflé B 118, D 150
air sursaturé S 831
air traité T 301
ajoutage A 140
ajustabilité R 123
ajustage A 196, C 583, R 124
ajustage au point zéro Z 1
ajuster A 191
ajusteur R 253
ajutoir C 583
alambic D 432, F 208, S 671
alambic sous vide V 53
alcali épuisé W 23
alcalinité A 454
aléatoire R 21
algorithme de calcul C 12
algorithme de Sargent-Wester-
berg S 37
algorithme d'identification I 15
alignement A 184
alimentateur F 31, F 43
alimentateur à disques tour-
nants R 337
alimentateur à vis sans fin
S 107
alimentateur d'acide A 86
alimentateur oscillant (vibrant)
V 161
alimentation C 181, I 131, I 189,
S 837
alimentation binaire B 95
alimentation d'air A 378
alimentation d'eau W 115
alimentation de chaudière
B 173
alimentation d'énergie A 142
alimentation de réacteur R 113
alimentation directe D 309
alimentation en charbon C 313
alimentation en énergie A 142,
P 307
alimentation maximum P 77
alimentation multiple M 372
alimentation par pulsation
P 601
alimentation partielle P 41
alimentation pulsée P 601

Russian Index

A

абгаз W 33
абразивное действие A 2
абсолютная влажность A 11
абсолютная вязкость A 17
абсолютная задержка A 6
абсолютная система A 15
абсолютная температура A 16
абсолютная теплотворная
способность A 10
абсолютная чувствительность
A 14
абсолютное влагосодержание
A 12
абсолютное давление A 13
абсолютное отклонение A 7
абсолютный фильтр A 8
абсорбент A 21
абсорбер A 25, A 30, A 32,
R 135
абсорбер с неподвижным
слоем F 184
абсорбер холодильного
агента R 205
абсорбирование A 31
абсорбировать A 18
абсорбирующая жидкость
A 47
абсорбирующая способность
A 57
абсорбирующая среда A 21
абсорбирующий A 56
абсорбционная башня (ко-
лонна) A 32
абсорбционная кювета A 36
абсорбционная способность
A 57
абсорбционная техника T 28
абсорбционная трубка A 53
абсорбционная установка
A 49, A 54
абсорбционная холодильная
машина A 51, E 340
абсорбционное масло A 24
абсорбционное охлаждение
A 52
абсорбционное равновесие
A 41
абсорбционные потери A 48
абсорбционный влагопоглоти-
тель A 40
абсорбционный охладитель
A 37
абсорбционный процесс A 50
абсорбционный фильтр A 23
абсорбция A 34
абсорбция аммиака A 465
абсорбция влаги M 297
абсорбция воды W 65
абсорбция газа G 4
абсорбция ионов I 331
абсорбция легких фракций
L 87
абсорбция холодильного
агента P 206
аванкамера F 26
аварийная защита A 74, D 5
аварийное отключение E 104
аварийный анализ H 11
аварийный выключатель E 102
аварийный сигнал A 450
авария B 253, C 87, D 444
авария производства B 254
автокатализ A 607
автокаталитическая реакция
A 608
автоклав A 609, B 66, P 437
автоклав для полимеризации
P 259
автоклав с мешалкой A 286,
P 438

автоматизация процесса
P 496
автоматизированное проекти-
рование C 507
автоматический мешконапол-
нитель B 12
автоматический питатель F 30
автоматический приемный
бункер A 611
автоматический регулиру-
ющий клапан с пневматиче-
ским управлением A 396
автоматический регулятор
температуры A 78
автоматическое регулирова-
ние A 610
автоокисление A 612
агглютинация A 267
агглютинирующий A 268
агент R 66
агент желатинизации A 263
агент сгущения T 170
агент, образующий основа-
ние B 47
агитатор A 283
агломерат A 262
агломератор A 266
агломерационная печь S 341
агломерация A 264, S 339
агломерирование A 264
агломерировать A 261
агломерирующая печь A 265
агрегат A 269
агрегатная форма P 140
агрегатное состояние P 140
адгезив A 151
адгезионная способность
A 152, A 154
адгезия A 147
аддитив A 140
аддитивное свойство A 145
аддитивность A 146
адиабата A 164
адиабата конденсации C 890
адиабата расширения A 165
адиабатическая абсорбция
A 157
адиабатическая десорбция
A 166
адиабатическая колонна
A 159
адиабатическая кривая темпе-
ратур A 180
адиабатическая мощность
A 174
адиабатическая отдача тепла
A 172
адиабатическая работа сжа-
тия A 162, I 347
адиабатический A 156
адиабатический коэффициент
полезного действия A 169
адиабатический реактор
A 177
адиабатическое дросселиро-
вание A 182
адиабатическое изменение
I 346
адиабатическое изменение
состояния A 158
адиабатическое насыщение
A 178
адиабатическое охлаждение
A 163
адиабатическое повышение
температуры A 181
адиабатическое расширение
A 170
адиабатическое сжатие A 160
адиабатная диаграмма A 167
адиабатная кривая A 164

адиабатная оболочка A 175,
A 183
адиабатная сушилка A 168
адиабатный A 156
адиабатный показатель A 171
адиабатный процесс A 176
адсорбат A 203
адсорбент A 209
адсорбер A 208, A 232
адсорбер с активированным
углем C 179
адсорбированная вода A 233
адсорбированное вещество
A 203
адсорбируемый A 202
адсорбирующая способность
A 201
адсорбирующее вещество
A 209
адсорбирующий агент A 209
адсорбтивная коагуляция
A 235
адсорбционная вода A 233
адсорбционная емкость A 234
адсорбционная осушительная
установка A 215
адсорбционная пленка A 204,
A 221
адсорбционная способность
A 210, A 234
адсорбционная установка
A 231
адсорбционная холодильная
установка A 229
адсорбционное равновесие
A 219
адсорбционное явление A 212
адсорбционный анализ C 253
адсорбционный индикатор
A 224
адсорбционный катализ A 213
адсорбционный осушитель
A 216
адсорбционный процесс
A 228
адсорбционный слой A 205
адсорбционный способ A 228
адсорбционный эффект
A 217
адсорбция A 211
адсорбция влаги M 299
адсорбция в паровой фазе
V 109
адсорбция в псевдоожижен-
ном слое F 286
адсорбция на неподвижном
слое F 185
азеотропная дистилляция
A 650
азеотропная жидкость A 651
азеотропная колонна A 649
азеотропная смесь A 652
азеотропная среда A 651
азеотропная точка A 653
азотнотуковое удобрение
N 32
аккумулирование S 697
аккумулирование тепла H 107
аккумулированный холод
A 75
аккумулирующий насос S 702
аккумуляторная батарея
S 698
аккумуляторный бак A 80
аккумулятор холода C 381
аккумулятор электричества
S 698
аккумулятор энергии E 168
аккумулирующая способ-
ность S 699
аккумуляция A 76

аккумуляция влаги M 298
аккумуляция тепла H 18
активатор A 117
активация A 113
активированный уголь A 106
активированный угольный
фильтр A 107
активировать A 105
активная поверхность A 119
активная турбина A 104
активность A 113
активность катализатора C 91
активный ил A 108
активный уголь A 106
акцептор A 69
алгоритм вычисления C 12
алгоритм идентификации I 15
алгоритм решения проблем
P 492
алгоритм Саржента и Вестер-
берга S 37
алифатический A 453
альтернатива C 249
аммиачный абсорбер A 464
аммиачный испаритель A 469
аммиачный компрессор A 467
аммиачный раствор G 101
аммиачный ресивер A 470
амортизация A 472
амплитуда температуры T 29
анализатор A 494
анализ возможных аварий
H 11
анализ возмущений D 450
анализ затрат и прибыли B 89
анализ крупности зерен P 62
анализ надежности R 265
анализ отходящего газа W 34
анализ причин аварии D 445
анализ процесса O 80, P 495
анализ работы установки O 80
анализ разгонки A 487
анализ размерностей D 285
анализ размеров частиц P 62
анализ сжиженного газа L 157
анализ систем S 914
анализ структуры S 743
анализ технологии O 80
анализ устойчивости S 517
анализ чувствительности
S 188
аналитическая структура
A 489
аналитический метод A 491
аналитическое выражение
A 490
аналитическое моделирова-
ние A 493
аналитическое соотношение
A 492
аналоговая вычислительная
машина A 481
аналоговая вычислительная
техника A 482
аналоговый компьютер A 481
аналого-цифровая вычисли-
тельная машина A 483
аналого-цифровая техника
A 485
аналого-цифровое моделиро-
вание A 484
анаэробная ферментация
A 480
анод P 212
аномальная вязкость A 509
антиадгезионный A 150
антикоррозионная окраска
R 428
антикоррозионное средство
A 511
антикоррозионный защитный
слой A 512

Ю

Я

EMOTIONAL FORCES
IN THE FAMILY

CONTRIBUTORS

Joost A. M. Meerloo, M.D.
High Commissioner for Welfare to Netherlands Government, World War II; Faculty, Columbia University and The New School of Social Research, New York City.

Lucie Jessner, M.D.
Professor of Psychiatry and Director of the Child Psychiatric Section, University of North Carolina School of Medicine, Department of Psychiatry.

Claire M. Ness, M.D.
Director, Cleveland Guidance Center; Member, Board of Directors, American Orthopsychiatric Association; Member, Board of Directors, Cleveland Welfare Federation; Consultant for Children's Services, Cleveland.

Sidney Berman, M.D.
Faculty, Washington Psychoanalytic Institute; Assistant Clinical Professor of Psychiatry, George Washington School of Medicine; Attending Staff, Pediatric Psychiatry, Children's Hospital, Washington, D.C.; Consultant, Laboratory of Child Research, National Institute of Mental Health.

Nathan W. Ackerman, M.D.
Associate Clinical Professor of Psychiatry, Columbia University; President, Association for Psychoanalytic Medicine; Fellow, New York Academy of Medicine.

D. Griffith McKerracher, M.D.
Professor of Psychiatry, University of Saskatchewan; Chief, Department of Psychiatry, University Hospital.

Judd Marmor, M.D.
Clinical Professor of Psychiatry, School of Medicine and Visiting Professor of Social Welfare, University of California at Los Angeles; Training Analyst and Past President, Institute for Psychoanalytic Medicine of Southern California.

Bertram Schaffner, M.D.
Psychoanalyst, New York City; President, William Alanson White Society, New York City; Consultant in Mental Health, Health Services, United Nations; Associate, Graduate Seminar on Communications, Columbia University; Editor, Conferences on Group Processes, Josiah Macy, Jr. Foundation.

Lawrence S. Kubie, M.D.
Director of Training, The Sheppard and Enoch Pratt Hospital, Towson, Md.; Clinical Professor of Psychiatry, Yale University School of Medicine (on leave of absence); Faculty, New York Psychoanalytic Institute (on leave of absence); Lecturer in Psychiatry, College of Physicians and Surgeons of Columbia University, New York City (on leave of absence).

EMOTIONAL FORCES
IN THE FAMILY

Edited by

SAMUEL LIEBMAN, M.D.

Medical Director, North Shore Hospital, Winnetka, Ill.;
Clinical Assistant Professor of Psychiatry,
University of Illinois College of Medicine

J. B. LIPPINCOTT COMPANY
Philadelphia · Montreal

Preface

THIS VOLUME, the fifth in a series based upon lectures given at
the North Shore Hospital in Winnetka, Ill., broadens the focus
of inquiry to include the individuals involved in significant interpersonal relationships with the patient. In previous volumes we
have directed our attention primarily to how the individual might
react in various situations that were stressful, the management of
various psychiatric problems, growth and development, and
finally, last year, the *Emotional Problems of Childhood*.

It is obvious to everyone that the patient does not operate in
a vacuum. He has daily contact as a child with his parents, his
playmates, his siblings, school teachers, his Church, etc. As he
develops, his contacts with the members of the family continue
and play an important role in his personality development. As
he goes to work, gets married and assumes the responsibilities of
life, he is confronted with new roles.

In this series we have especially sought for forces in the family
as they impinge upon the person and how the individual responds
to these forces. We have even tried to look into the future.

In this volume, as in previous ones, the Board of Directors of
the North Shore Hospital and the individual authors have
assigned all royalties to the American Psychiatric Association.

SAMUEL LIEBMAN, M.D.

v

Contents

1

The Development of the Family in the Technical Age

Joost A. M. Meerloo, M.D., Ph.D.

INTRODUCTION

FORMATION AND DEVELOPMENT of the family are subjects for anthropologic investigation. The effect of outside influences on family and family life is of psychiatric and psychological concern. The subject of this chapter is the development of the family under the influence of 20th-century technologic developments.

Since the family unit of parents and children is the prototype and the nucleus of man's greater social experiment in living together, it is worthwhile to investigate the effect that the technical era is exerting on the subtle relations between marriage partners and between parents and their children.

Consideration of the changes in the family due to our technical and scientific age evokes a host of experiences with parents in distress, which, in their multiplicity, may confuse the mind. That is why, from the beginning, I desire to limit my subject by asking specifically: What strange intrusions and coercions did I find that were caused or influenced exclusively by new technologic developments? What happens to marriage and to the family unit when the machine age takes over and, according to Lenin's formula, "the

1

government of men is to be replaced by the administration of things"?

True, such limitation of subject usually implies the stressing of the more negative and disintegrative aspects of a problem, but it also helps to bring these aspects more sharply into focus. The technical invasion of the family through collective coercion definitely exists, and we shall try to observe this phenomenon clinically. Mere mention of those aspects is no therapy, but it can help to prevent the disturbing effect of inadvertent social pressures on people.

I shall divide my subject into 3 areas:

1. The technical invasion: the process of technicizing

2. The invasion by psychological concepts: the process of psychologizing

3. The invasion by the confusing semantics of love: the process of glamourizing.

THE TECHNICAL INVASION
(THE PROCESS OF TECHNICIZING)

Our technical age, with its many material advantages and greater luxuries, is affecting modern man and his basic human relations. He had better become well aware of this so that he may be able to correct technology's inadvertent persuasions. Machines and impersonal institutions push him into a new kind of conformism and technical collectivization, the consequences of which we cannot as yet foresee. In general, we may say that our 20th-century situation causes a process of deindividualization and a depersonification through emphasis on the merging of man with fellow man. Yet, despite all this emotional publicity for "togetherness," it is still the individual who creates culture and moral responsibility. Various technical devices, especially our highly developed means of communication such as the press, radio, movies and television, put new mass imprints on the mind from which it is difficult to escape, especially since many people are not even aware of their being influenced by such unobtrusive coercive action. From being unique and individual, people are fast becoming votes in public-opinion polls in which every opinion is statistically equal in value.

Western civilization used to exhibit a clear trend toward individualization; this began in the Renaissance with the discovery of the unique value of the individual, which resulted in his maturation as a step *beyond* collective consciousness. We may say that in the 20th century a reversal has taken place. Both technical and political developments attach less value to the individual and his unique qualities. He has become a cog in a wheel, which is itself a small part of a bigger machine. In the process of automation, human relations are threatened with disruption because the machine takes away the idea of individual human service. For instance, people no longer understand that selling a product, repairing a piece of furniture or treating a patient requires a personal act of dedication. People have learned to look at these as being automatic actions. The dime is thrust in the slot, and something comes out of the machine. Nobody cares. People have become automatons, smiling automatically, but the heart is missing. There is no true service, no true love, no true hatred, no pleasure in giving service to a fellow being. While creating tools, man himself became one.[3]

The communistic ideology especially, and technology too, show trends toward seeing man as a mere object whose value is determined only by his place in the master plan. If people object to being molded into this schedule, brutality or political oppression are called on to level off their egos and push them back into conformity. The Nazis called this form of psychological pressure *Gleichschaltung*—leveling and equalization. The Russians call it collectivization, and we speak of it as a trend toward greater conformity. The White Collar Workers, the Organization Man and the threat of the Gray Flannel Mind have become clichés.[2]

This tendency to deindividualization and depersonalization, exerted by the collective and conformistic trends under the effects of mechanical rules, has made tremendous inroads in the family structure. It has led to the devaluation of individual action and commitment, these true tests of man's honesty and wisdom. I do not have to cite the extreme examples of family disruption in totalitarian societies, in which children were forced by coercive suggestion to spy on their parents and even to betray them, in which the state took these children out of their homes and imprinted on their minds disloyal behavior and crude prejudices. The risk of such a form of collectivization is that man's conscience gradually will fail to recognize personal responsibility and commitments. Consequently, his conscience will atrophy and no longer provide the roots from which morals and ethics may spring. A creeping unobtrusive mental coercion exists, inherent in the tremendous forces that cause the social revolution of our epoch. Of this inadvertent persuasion I shall give some examples.

Are people aware of the way that current technology is, in a willy-nilly fashion, molding their minds? Technic influences man's philosophic attitude toward life. Daily it propounds that the shortest and easiest way is the best. It calls for efficiency and for more labor-saving, magical gadgets and is therefore in conflict with the psychological rule that toil, resistance, challenge and difficulty modify and mold the personality. Radio and television seduce people to become passive receivers of external stimuli at the cost of internal stimulation and creativity. Healthy, strong egos are not formed by passivity and facilitation, by luxuries and leisure time. The personality has to grow by accepting challenge.

How technology has quietly intruded into the equilibrium of the home and into the parent-child relationship can be illustrated by a form of neurosis that I shall call television apathy—the unwillingness of the child to have personal relations other than with the spellbinding, fascinating TV screen. I have seen children between 4 and 6 years of age who could communicate with the TV screen but not with their parents. True, the parents started the problem by being fixed to the TV screen themselves, hardly speaking to one another because of the hypnotizing effect of this new toy. The mother works in a factory or an office during the day; at noon the children go alone to the automat to exchange their dimes for food. Automatic tools have taken over the function of giving affection. Between children and their parents has crept a technical, mechanical world that keeps them far apart psychologically. No wonder the children refuse to learn to read in school. They crave a warmer, direct *verbal* communication, which is lacking so badly at home.

The technical age also has changed the dinner pattern of many a family. The modern household is dominated by schedules of every member of the family. Buses, schools and commuter trains interfere with the quiet breakfast gathering of the family, especially when the mother also has an outside job. A survey of the breakfast habits of school children (according to E. Karpoff in the *New York Times Magazine,* March 2, 1958) showed that more than 30 per cent skipped at least one breakfast. He says: "In our time-possessed society, everybody has to squeeze minutes from the family time." For many families the daily communion of being together at the dinner table no longer takes place. The only time that this still may happen is at Thanksgiving, Christmas or Passover. This is the picture that Dicks calls the Atomized Family.[1]

Such a broken family circle is bound to lead to disrupted communication and a dependence on technical sources of verbal exchange. We may also put it this way: The new technical age interferes with the family as a communicative community in the service of spontaneous relief of mutual tensions. In several cases the lack of parental communication leads to a simple reading block in the child. Reading not only creates a greater distance from the spoken word but also demands extra activity in learning. What we are apt to call "reading block" consists, in the majority of cases, of the child's unconscious refusal to progress from spoken communication and to enter into a lonely dialectic relation with the printed word. It is his passive resistance to the parents' addiction to gadgets and automats. In 1958, 4,000 of 16,000 seventh graders from the New York City schools could not be promoted because of this reading block. In 25 per cent of the children the reading ability was at the fourth-grade level or lower.* This points up the size of the problem.

In some instances I could urge the father to play Scrabble (a word game) with the children instead of watching TV. In several cases the child overcame his reading block as a result of a changed, warmer attitude on the part of the father. Mutual relation and communication are needed not only in the service of love but also in the service of growing self-esteem and ego formation.

Besides the intrusion into family communication, another disturbing interference by radio and TV exists. The daily dose of excitement, fear, murder stories, criminal sadism and crooning

---

* New York Times, June 24, 1958.

self-pity prevents the building up of inner values that belong *exclusively* to the realm of the family.

A school survey in a middle-class suburb of Buffalo found—according to *Time* magazine, March 24, 1958—that kindergarten tots are at their TV sets 50 per cent of their classroom time. However, as the pupils grow up to the 6th grade, they devote nearly equal time to school and TV.

A school official concluded: "Television is changing American children from irresistible forces into immovable objects." This stealer of time hypnotizes many people, especially children, into indiscriminate watching. Older people are often seized with panic and depression, provided that they are sensitive enough to understand the cynical attitudes to which our television exposes its public. Not long ago one of our medical journals† related suicides to such anxiety-provoking behavior. There is much talk of mental hygiene, but as yet the subject of daily toxic mental contagion has not been put under close scrutiny.

Another "instrument" that has brought distortion into the family is the automobile. Not only has it put unneeded emphasis on speed, causing man's healthy habit of walking and hiking to deteriorate so that he is no longer in direct touch with nature, but the gadget has also become a special tool to provoke latent neurotic tendencies in the family. The car becomes the symbol of family glamour. In many analytic explorations I also found a special Sunday neurosis related to the car. This syndrome came to the fore through painful memories of dreaded week ends, when the kids were thrown into the back of the car, with the parents occupying the front seat.

Bickerings during most of the outing made the distracted, conflict-prone mate also accident-prone. Sitting behind the wheel of an omnipotent gadget doubtlessly brings out all the driver's latent hostility. Fathers curse at other drivers, mothers berate their husbands' performance behind the wheel, while the kids are perforce dragged along with the whims of their parents' pleasures. I could describe a variety of Sunday-afternoon dramas with car sickness and worse things happening because of the driver's obsessive unwillingness to stop.

---

† J.A.M.A. **167**:497, 1958.

This reversal of what ought to have been a happy family reunion into a hated, repetitious occurrence can be laid at the door of technology with its effect on the family.

Other factors detrimental to a stable family life are cramped living quarters, the telephone, often used by the hour and continually intruding into intimate conversation while substituting for personal visits, and, last but not least, the institution of the baby sitter as substitute parent.

Technology provokes in nearly everybody an infantile magic-thinking and an increased dependency need. It suggests passivity and a withdrawal from communal activity. It takes away man's genuine interest in accepting responsibility, unless a quest for power or domination of others is the motivation. Technology can change people into social insects with geared patterns. It teaches them a new, cold ritual of gadgets, knobs and handles, replacing former healthy family habits of making and doing things oneself and together.

The mutual communication with its stealthily growing relationships between men and gadgets, between persons and machines, has now become so intertwined and related that none of us can escape their influence any more. The mass media of communication—the headlines, the slogans and the advertising jingles —daily and deeply modify our personal norms and evaluations without most of us becoming aware of their subtle suggestions and coercions.

The technical era tends to transform relations between man and man into relations between men and things, men and machines, men and institutions. The political party, the business, the factory, the hospital, the state, the service corps have all become impersonal but omnipotent justifiers for everything that takes place.

All this does not imply that we have to do away with the fruits of technology, but, in realizing better its negative psychological effects, we can restore normal circumstances and prevent technical gadgets from becoming relation-spoiling intruders in the family. We have to assure parents again and again that *they* are the best educators the child can possibly have. They are because of their biologic ties. No lecture, book or TV can replace the parental influence on the child. No official educator is willing to adopt the endurance and the tolerance of most parents. Those psychologists who have caused parents to be fearful of their task have not served those parents well but have introduced a strained, overintellectualized atmosphere into the family circle.

THE INVASION BY PSYCHOLOGICAL CONCEPTS (THE PROCESS OF PSYCHOLOGIZING)

As technology took away awareness of and awe for the immense universe, popularization of psychology took away the awareness of the tremendous dimensions of man's inner space. In this age of technical growth and intellectual expansion, half-understood concepts—often with a frightening effect on both marriage partners—subtly invade the family. Let me give an example of such dependency on bookish concepts.

An engineer and his wife came to talk to me about their marital troubles. They loved each other very much, yet, at the same time, they could not stand each other. There was a continual alternation between being "lovey-dovey" together and indulging in mutual violent outbursts. Both came from deprived families and had found in each other a mutual solace for the affection they had not received at home. In this case of mutual dependency some emotional outburst had to come. Yet, it was only after the wife started to study psychology from various books and dragged her husband into an intellectual explanation of mutual shortcomings that their defenses against pent-up hostility broke down. At first it seemed so easy to shoot explanatory labels as psychological bullets at each other instead of indulging in the usual vituperation, but it was impossible to displace the inner fury entirely with sophisticated words, and finally all hell broke loose. After this, mutual remorse made itself felt, and the marriage partners talked and talked for hours in psychoanalytic terminology to demonstrate each other's good will—until the next storm came up.

At first I saw only the husband, and I insisted that he stop psychologizing with his wife, that he leave to the expert the interpretations when needed. From that time on, the mutual psychological interception stopped, and some release was found in more enjoyable and pleasurable common endeavor. After having counseled the husband, I saw the wife for a few visits. The insight that each used the other as a scapegoat for the darts they would like to have shot at their parents improved their relations and eventually led to better mutual appreciation.

I use this case to show how an imposed intellectualization of emotions, induced by radio lectures or popular books, can influence a family—not by stimulating mutual understanding and tolerance but by sabotaging attempts at mutual adjustment, thereby promoting a breakdown in marital relationships. Many people try to read beyond their psychological understanding. Consequently, an attempt is made to replace the affectionate deed with the not understood psychological cliché. Such "psychologizing" as a substitute for mature action is a typical phenomenon of our time. The delusion of explanation replaces the appropriate act. Words, words and mere words are produced rather than good will and good action. Sex itself is expressed in words instead of affection.

A typical mental danger is hidden in this form of verbal sexuality. Unobtrusively, in these psychologizing circles people regress to an infantile magic atmosphere, preoccupied with the unwholesome need to tear one another apart verbally. They participate in a verbal promiscuity, in which the frontiers of the ego get lost and depersonification takes place. The cult of reserve is no longer practiced; mental privacy has become unknown. I have known patients who lost themselves and their ego strength in those verbal promiscuous circles and couples who lost their mutual affection and faithfulness under the effects of fortuitously imposed primitive participation.

Technical terms and clichés are often abused. Think, for instance, of the word "neurotic," which has practically lost its original meaning and has come to express almost exclusively a feeling of dislike. I have seen people pushed into homosexuality by the wrong use of the words "neurotic," "sissy," and "fairy." Several of the persons whom I treated had been forced to accept, albeit in an overtly masochistic way, the verdict that oversophisticated peers imposed on them. They were, so to speak, pushed into homosexual circles and into homosexuality. It took a long therapeutic struggle to lead them away from this social verdict.

Perhaps the strangest example of distortion through popularized psychological concepts is the woman's compulsion to attain climax and orgasm with every sexual act. Such women read about it, dream about it and then feel deprived and inferior because they do not regularly experience the golden ecstasy promised by the book. Thus, in many families, the sexual encounter has become an act of prestige, a technical investment, rather than an act of love and affection.

Once a woman came to me for consultation because her "lack of orgasm" had led her husband into starting an extramarital affair. The couple had been married for 3 years and had always had enjoyable sexual relations. There was no reluctance on the wife's side; she was a warm, dedicated partner. On the contrary, it was the husband who was often cold and aloof. The wife wanted to come for treatment to fulfill her husband's desires and emotional investment in her having a complete vaginal orgasm. Strangely enough, the first interview brought out the information that before her marriage she had felt orgasm several times in relations with another young man. Only after I had met the husband, with his sophisticated emphasis on the wife's orgasm, did it become apparent that he was a latent homosexual, continually identifying with his wife's feelings. He did not want children, thereby causing his wife to be careful and anxious in the sexual act. Accusing her of not being sexually adequate was a defensive denial on his part. His extramarital affair meant to him the need to show his masculinity. In this tragedy of errors he never discovered that his extramarital mate was in treatment for frigidity; she was able to hide this from him. The marital problem could not be solved because he insisted that his wife, rather than he himself, be treated. Finally, the couple obtained a divorce, and the wife remarried. She now has satisfactory relationships, sexual and otherwise. I have met women who believed that they had no orgasm because their husbands told them so. I have seen men who were impotent because their mates unconsciously wanted them to be.

A purely intellectual emphasis on orgasm has caused much sorrow in many marriages in which there was at first a belief in great mutual love and satisfaction. As we know from deeper analysis, orgasm is the result of a mutual relationship in which unconscious images from the past may have an inhibiting action.

In removing the sophistication from the relationship, and displacing the emphasis from sexual prestige to mutual affection and tolerance, much orgastic and libidinal pleasure returns where misunderstood psychology had distorted the feelings.

Man's inner need to play with the concepts of *jealousy* and *envy* and of pain and injustice done to him paints one of the saddest pictures of our epoch. Usually the feeling of marital jealousy is either overromanticized or underplayed, although often a couple suddenly feel compelled to act as the glamorous people on movie screen or stage. By and large, people are more tolerant than they pretend to be, and an instance or mere fantasy of unfaithfulness of one of the marriage partners often becomes a provocation for the prevalence of infantile feelings of unfairness, long forgotten in the victim's mind.

A wife who came to ask for help for burning feelings of jealousy that left her sleepless and completely unhappy not only had tolerated the relationship between her husband and his girl friend but, as came out later, had also even unwittingly promoted the situation. The romantic anticipation—learned from books—of being devoured by real sensations of passion and envy had disturbed this rather infantile marriage. The husband, poor man, did not even quite know what to do with the girl who had been cleverly "played" into his hands by the wife. All he wanted was to get rid of the girl as soon as he could without feeling too guilty about her. Of course, he had confessed to his wife and had beaten his breast. 'The whole hoax had been half-consciously schemed by the wife in order to cover up her own feelings of shortcoming and to gain extra power over her husband. Every time he confessed she grew in self-admiring greatness and at the same time re-experienced through him the humiliations she had suffered from her mother. She had acted as the Biblical Sarah did to Abraham, only she misused the situation to humiliate her husband.

Many of our authors, with their dramatic stories of jealousy and revenge, lost love and murder, have influenced our minds in the wrong way. *Othello* is a mere historical incident compared with their violent tragedies. In daily life people tolerate and forget and are usually too unaware of the intensity of their feelings, depending on their wish either to dominate or to suffer and wallow in self-pity. They play with fantasies of passion and unfaithfulness and act them out from time to time, but real love and passion are as dead as a doornail.

In the case I described above, years of therapeutic fighting with the infantile feelings of revenge and injustice were required before the wife could become a genuinely passionate lover for whom jealousy no longer existed because she felt sure of herself.

My point is that as a result of overdramatized suggestions from outside on how to behave passionately in cases of sexual betrayal and unfaithfulness, people are persuaded to re-enact these emotion-laden scenes. For them, however, this acting out of movie scenes represents a defense against their lack of passion and empathy, a defense against boredom and substitutes for non-existent self-esteem.

No movie, play or television story, not even a novel, describes the continuity of a happy marriage: the abundance of attention, the constant exchange of little signs of affection, the rhythmic, simultaneous growing away from and growing toward each other. The movie usually finishes with glamour and ecstasy—at long last the hero gets his heroine. As in a fairy tale, the young couple is supposed to live happily ever after. However, very few adults teach youngsters that marriage is not a glamorous state but something much deeper and more enjoyable than that. Where literature finishes, life begins. In marriage, togetherness or conflict can start around a toothbrush or a misused kitchen towel; from this point on, every little spark can enrich life or increase conflict.

THE INVASION BY CONFUSING SEMANTICS OF LOVE (THE PROCESS OF GLAMOURIZING)

Whereas in the first part of this discussion I showed how the technical age suggests cold relations and in the second part how the loss of awe for subtle inner processes intruded into the family, in this last part I would like to scrutinize some subtle suggestions given by our era of social and technologic predominance. The glamourization of human love and wisdom and the need for spectacular events takes away confidence in the redundancy and the continuity of life.

I desire to call attention to an exalted and overromanticized concept of love with which people usually become acquainted through the movies and tortured literature. This "love" wrongly presents and emphasizes the hasty attainment of goals that in reality take long years of steady companionship and endurance to ripen and are therefore—naturally—beyond reach for a young, just-beginning marriage relationship.

Couples unable to recreate within a short time this suggested infantile magic picture of eternal romantic love and unwilling to exert the energy and the tolerance of initial mutual adjustment begin to feel disgruntled; they separate and seek a divorce. They continue their eternal search for what exists only in fairy tales. It is no coincidence that so many of our love songs are songs of *Weltschmerz,* sung with a crooning, petulant voice, a chant of deprivation, as if no constant, ecstatic human relationship were possible.

As I mentioned before, our modern means of communication, especially the movie screen and TV, have a peculiar effect on our patterns of wooing and courting. They have to be condensed in short moments with technical perfection. The language of silence no longer exists.

"Making love" has become for many a technical act to be executed without sentiment—often in the back of the car. Youngsters in junior high school are already forced into aggressive, precocious dating habits without either wanting them or being ripe for them. Yet, it is at this age that feelings of mental distance, privacy and reserve are built up. In these cases it is the precocious sexual habits that prevent the building up of the personality.

At the risk of rushing into dangerous areas of semantic confusion, I now want to describe what our technical age can do and has done to the subtle, age-old concept of marital love and marital bliss that generally evokes embarrassment in people when spoken about.

Nothing is more difficult to define than mature and ripened love. Even to psychologists, "love" can be an embarrassing word and is omitted in many studies on marriage relations. Sometimes it means purely physical attraction; at other times it means pure self-love and the adoration of all those who strengthen that self-love. The phrase "I love you" has so many meanings. Most of all it reveals the mere quest for affection, the need to be liked, so deeply engrained in our teen-agers.[4] But the essential language of love is very simple and mostly unspoken. Lovers have no need of words to communicate their feelings.

Yet, the simple phrase "I love you," crooned so pitifully in many love songs, has brought much confusion. It may be an empty expression, repeated daily without meaning; or it may be a barely audible murmur, full of surrender. It never is the word "love" itself that conveys the real meaning. Sometimes it means "I desire your body sexually." It may mean "I hope that you love me" or "I hope that I shall be able to love you." Often it indicates "Maybe a love relationship will develop between us," or the phrase can cover up "I hate you!"

Many times it is a wish for some emotional exchange, a mutual quest for self-esteem: "I want your admiration in exchange for mine." Both partners use their marriage to dispel their feelings of inferiority.

To declare love is to ask for love. A declaration of love is mostly a request or even a demand: "I desire you" or "I want you to gratify *me*" or "I want your protection." What in harsh English idiom is called "making love" is mostly a continual quest for affection or a routine commitment.

Very often people love others for the emotional investment that they have in them. Sometimes it expresses the need for security and tenderness, for parental affection invested in the new partner. "I love you" may mean "My self-love goes out to you because you are like me." But it may also express mere submissiveness: "Please take me as I am" or "I feel guilty about you; I want, through you, to correct the mistakes I have made with other loves." "I love you" may also mean a self-sacrifice or a masochistic wish for dependency: "Because of vaguely hidden feelings of guilt, I want to play the martyr and punish myself with your permanent company."

However, love also can be full of affirmation of the beloved, a taking up of the responsibility for a mutual exchange of feelings.

The problem of evaluation—of giving value—is related to the

way that people are conditioned and trained to love. The capacity to love and to give value to the other is rooted in the early exchange of warmth, affection and love between mother and baby and the gradual maturation of this love relationship to the point where love is no longer only a yearning for *receiving* gifts and affection. Yet, the first condition is that the infantile need for love and affection has to be fulfilled. If this does not take place, the child may grow into a person who uses his partner in order to lose himself in old conflicts, and the word "love" becomes for him a faulty semantic expression for physical and emotional narcosis, with all the infantile ambivalence and mutual destructivity involved. The sexual act for him is for the most part a defiant gesture and an attempt to solve hidden aggression, often becoming a weapon in the search for power.

Love-hunger is a neurotic sickness that may be acted out in various ways and may finally develop into a defensive compulsion to "love," which really means the need to *be* loved only.

Monogamy is the limitation of polygamous urges rampant in everybody. Yet as soon as mates feel responsible for the emotions which they provoke in each other, the monogamus limitation is the only solution. Monogamy starts when people do not want to hurt each other's feelings any more. In their mutual relations they do not feel compelled to love each other at all times, but they are committed not to hurt each other.

Consequently, most extramarital affairs are based on the acting out of some hostility rather than on love. Often it is more the hidden need to hurt the former partner than love toward the new mate that prompts the new attachment. This process can go so far that the inner contrasting feelings demand a resistant and hating new mate. The choice of extramarital relations is often prompted by an inner need to discharge pent-up aggressive feelings more easily.

Yet, "being in love" is a state of creation. It goes far beyond the sexual reflex action. It is creating dreams—man in love evokes creative, inspiring images. Love creates the partner and makes him lovely. The youth catches the loved one in a dream and eternalizes her in an image. True, the dream may be based on earlier love and happiness, but it is still a re-creation.

I saw a demonstration of this creative "capacity for loving" early in my career. I had been practicing only a short time when a husband and a wife came to me asking for assistance in a problem concerning their 17-year-old daughter. The girl was pregnant. The shocked parents both belonged to the higher civil servant class, and they believed that their dignified status in society would be destroyed if the facts about their unmarried daughter were made public. What they required of me was an attest to prevent the continuation of the pregnancy through a psychiatric suggestion to the obstetrician. The mother wept and thought at that moment, it seemed to me, more of her own impending shame than of her daughter's well-being. I promised to talk to the girl.

The daughter arrived at my office the next day accompanied by her boy friend. At first she was heatedly opposed to any psychiatric meddling, but when I reassured her she composed herself and told me about her long, close friendship with the boy. A carpenter who studied evenings, the boy belonged to the so-called "lower class," but she had confidence in him and in his eventual success. A year before this they had planned to marry, but her parents would not permit it. The pregnancy was the girl's method not only of obtaining her ends but also of fulfilling the love relationship with the boy. It was the warmest relationship either had ever experienced, and it had been growing from the time they had first met at the age of 12. A lonely girl, this child of two working parents had found her mate and a nearly ideal companionship very early in life. The boy and the girl both spoke of the pregnancy as a happy, desirable occurrence, and it seemed to me that never before had I met such mature children. They were full of reverence for each other.

My consultations with the girl's parents were difficult, especially with the mother who, not wanting to surrender her injured pride and supposed shame, did not change her mind about her daughter until I was able to explain her own guilt and her responsibility for the young couple's seemingly unsocial actions.

The girl was permitted to marry her young lover, and for years afterward I was able to witness their wholesome, simple happiness, without the intruding suggestion that they had to love each other forever.

I realize that writing and speaking about "love" has its dangerous implications. While I am writing my notes, I hear a harsh, petulant voice on the radio singing a song and using the same semantic term to express something very sexual but at the same time something mechanical without subtlety of emotion. In a culture in which radio voices croon, the word "love" refers more to self-pity than to giving affection. If, however, there is enough loving capacity in people, they can clarify and simplify their knowledge and insight into others. First it can be shared with their mate and beloved, who shares with them their insights in an almost unspoken manner.

Then they can exchange knowledge with people outside their family circle, provided that those people are ready to read and hear what is written and spoken between the lines. However, without their love and sympathy, that, too, will be impossible. This is the impasse in speaking and writing on marriage problems.

SUMMARY

In the development of the family in the technical age, three areas of significant effect can be distinguished:

1. The technical invasion describes mainly the coercive and corroborative action of the new means of communication leading to an unobtrusive change in family relations. Television apathy and reading block are dealt with as clinical examples of such technologic coercion.

2. The invasion by psychological concepts paints the lack of reserve and privacy resulting from misunderstood popularization and oversophistication of psychological concepts and the way people use psychologizing in an aggressive way. This is clinically illustrated by the overemphasis on sexual orgasm, the misjudgement of marital jealousy and a lack of mutual commitment.

3. The invasion by the confusing semantics of love calls attention to the precocious imprints on sensitive minds of overglamourized concepts of eternal love, leading to a confusion of marital attitudes, lack of loving tolerance and difficulties in mutual adjustment.

As in many other chapters in human behavior, awareness of these new encroachments on our happiness and peace of mind gives us at the same time the weapons with which to face these difficulties and to maintain peace and understanding within the family, this basic unit and nucleus of all social harmony.

REFERENCES

1. Dicks, H. V.: Strains within the family *in* Proceedings of National Association of Mental Health, London, 1954.
2. Eisenstein, V., *et al.*: Neurotic Interaction in Marriage, New York, Basic Books, 1956.
3. Meerloo, J. A. M.: Cultural Stress in the Atomic Age *in* What's New, North Chicago, Ill., Abbott Labs., 1958.
4. Remmers, H. H., and Radler, D. H.: Teenage attitudes, Scientific American vol. 198, 1958.

2

The Role of the Mother
in the Family

Lucie Jessner, M.D.

NATURE AND SOCIETY both assign the role of mother to the woman. In this role she biologically assures the continuation of the human species, overstepping the finiteness of the individual's life, forming "a link in the long chain of historical evolution."[8] The growing body of the girl is predestined for menstruation and thus prepared for pregnancy. Sociologically, the mother is expected to rear her child not only for survival and fitness in a physical sense but also as a member of a family, a community, the world. In this way the mother takes part in the historical process of civilization by her influence on the personality of the next generation. Most cultures have given the mother a leading role in the human drama. In art, religious and secular, the mother-child theme is ever recurring. Folklore, legends and fairy tales portray the mother as the protectress from outer and inner dangers, who is endowed with the quality of mercy and the capacity to forgive, as the self-sacrificing defender and rescuer of the child. Or else she is represented as a witch, a spider or Medea who kills her children to revenge herself on their father. In some myths one sees the wish to get along without her (Athena is born from the head of Zeus). Stories of animal mothers who defend their young with incredible strength and self-sacrifice— or devour them—have a special emotional appeal. We can recognize

19

in these portraits the projections of universal infantile images of the mother, which we have "forgotten," that is, repressed, but they can return to memory destructively during psychosis or therapeutically during psychoanalytic treatment.

Psychologically, the role of the mother carries the potential fulfillment of the wishes, the fantasies and the anxieties that have evolved in the little girl from childhood. The expectations precede the experience of being a mother but are stirred up and accentuated by the events of pregnancy, delivery and the presence of the infant. In contrast with animals and with women in primitive societies, women in our civilization have a certain degree of choice in accepting the role of the mother through birth control or by delegating the care of the child to substitutes: the grandmother, the nurse or the maid.

The biologic, sociologic and psychological aspects of the maternal role are interdependent. Nor can the mother be seen fully without the other actors on the scene: the child with his special endowment, the father, the grandparents, the neighbors, the community and its representatives (teachers, ministers, social workers, physicians, etc.). All are forces that can facilitate or obstruct mothering.

This consideration leads to another determinant: a mother in our society is supposed—and as an individual inclined—to play other roles with or without change of costume. She has to function also as a wife, being sexually, emotionally and intellectually attractive to at least her husband. She has to make a home and keep it clean. She has to have a personality of her own and be a responsible citizen as well. She has to perform not only in the nursery but also in the living room, the bedroom, the kitchen and on the outside.

Aside from these simultaneous functions of the mother, we have to consider the dimension of time. The role of the mother changes, of course, with the phase of development of her children, calling for different aspects of mothering. In contrast with most animals, which care for their offspring only as long as they are completely helpless, the human mother maintains a bond with her child, although the importance of her role as a mother gradually diminishes. There is increased freedom, but there are also tragic overtones for most mothers. "One of the masochistic experiences of motherhood arises from the fact that the child's emotions develop centrifugally, away from the mother, while the mother remains tied to him and must renounce him."[8] The Madonna is most frequently pictured smiling at the infant on her lap or bemoaning the death of her son.

This general description of the mother role sounds somewhat overwhelming for any one person to fulfill. It is no surprise that conflicts, anxieties and a sense of failure often darken the picture. Physicians, ministers, social workers and teachers are faced with mothers struggling with their role. Outspokenly or by innuendo, they are requested to help. On the other hand, women raised children for centuries without much fuss and continue to do so in other cultures.

Why should this be so? The reason will be found in the tremendous social and psychological superstructure of our complex civilization and its insatiable desire for improvement. Ours is a highly dynamic culture in contrast with the more static nature of "primitive" ones. This means less reliance on traditional attitudes and models for the mother's role and therefore insecurity. Our standards have prolonged the economic and professional dependency of the child. Psychiatric and anthropologic investigations have shown that the quality of child-rearing—in particular the attitude of the mother—is a decisive factor in the development of personality. Increasing knowledge in this field encroaches on the role of the mother from two sides:

1. The belief in the 19th century that constitution predestines the human being has been shaken by findings which demonstrate that learning and environmental influences modify the original endowment in its adaptation. Even behavior considered to be characteristic of a certain species of animal has been shown to alter with changes in early experiences; e.g., Kuo[31] reported that a majority of cats reared in isolation from other cats did not kill rats when exposed to them, and that cats reared with rats could not be induced to kill the rats. Most of us have observed dog and cat behave "unproverbially" when brought up together. In a study of children who were blind, due to retrolental fibroplasia, and suffering from severe behavior problems,[1] it was shown that

the adaptation of these children could be improved decisively by the mother's or therapist's response to the special needs of the blind child. The most dramatic example of the potential modification of a child—blind, deaf-mute, withdrawn and hostile—probably is Helen Keller.[13] On the other hand, inadequate mothering, a disturbance of the mother-child relationship, is more often than not found in the life history of psychotic and neurotic patients.[19,21,22] In such cases mothers usually are described as rejecting, overprotective, domineering or emotionally ill during the patient's early years. The detrimental effects of the lack of mothering during infancy and of prolonged early separation from the mother on the physical, intellectual and emotional development of the child have been documented convincingly by René Spitz [26,27,28] and John Bowlby.[5]

2. Another challenge of the mother role stems from the realization that the assumption of "maternal instinct," either in the sense of an innately determined drive or of a "pattern of complex behavior released at childbirth,"[2] is not altogether correct.* Differences from one culture to another in the basic characteristics of the mother's behavior toward her offspring demonstrate that social variables determine maternal behavior in large measure.[2] Observations of mothers in child-psychiatric clinics impress one with the wide variety of maternal attitudes even in a relatively homogeneous culture. This means that the maternal instinct is subject to many factors that modify it and determine the degree of "motherliness."

In contrast with "motherhood"[8] and "mothering,"[3] concepts referring to the actual care of the child, "motherliness" designates the special emotional quality of a woman that responds to a child's helplessness and his needs. This quality is related to—although not identical with—what Helene Deutsch has described as the characteristics of the feminine woman, namely, the harmonious interplay between the narcissistic wish to be loved and the masochistic readiness to accept pain in the process of loving

---

* In regard to the maternal behavior of the chimpanzee, Yerkes and Tomilin[30] state: "Structurally determined patterns of activity are neither adequate nor dependable, but instead require facilitation, modification and supplementation through experience."

and giving. There are mothers who have many children and are not "motherly," whereas some women who never have had children show genuine motherliness.[3]

If we ask what makes one woman more motherly than another, we shall find components from different areas which contribute to that quality.

A *physiologic* factor is involved, although I think that it has not yet been sufficiently identified. David Levy,[18] studying three groups of mothers, "high maternal," "low maternal" and a middle group, found a significant positive correlation between maternal behavior and duration of menstrual flow. The majority of subjects with 4-day periods or less belonged to the group of "low maternal." The majority with periods 6 days or longer were classified in the "high maternal" group. Therese Benedek[4] and others found a relation between motherliness and blood levels of prolactin and estrogen. However, that hormonal processes are not exclusively the source of maternal behavior is evident from our clinical experience and is supported by animal experiments; e.g., Scott[24] reports that a female lamb was experimentally separated from its mother for the first 9 days of life and fed artificially. Placed with the flock after that, it was driven off by other lambs, never suckled and soon turned to grazing alone. After she bore her own lamb she showed little concern for it. In consequence of the mother's wandering away, the lamb was often neglected and would call for the ewe loudly but in vain. This lamb also became independent but unusually fearful and nervous as well. Apparently, the lack of experience in being mothered interfered grossly with the sheep's ability to act as a mother and with the normal development of her own lamb.

In human beings the particular ways of mothering are, to some extent, also determined by cultural patterns. In speaking here only of the American mother, we can see shifting attitudes in different historical periods; in frontier days the mother, among tough men, had to be "the cultural censor, the religious conscience, the aesthetic arbiter and the teacher.[9] In puritanic times the molding of the child toward early suppression of sensual wishes for the sake of morality seemed to be the mother's main function. Then came a period of emphasis on scientific methods of child-rearing under the influence of medicine and pediatrics, rightly proud of achieving a tremendous decrease in infant mortality. General hygiene and preference for bottle versus breast feeding, with measured quantities at regular intervals, became important functions of the mother. However, around the turn of the century the leading pediatricians—e.g., Pfaundler in Vienna—recognized that nutrition and protection from germs were not sufficient for the survival of the infant, but that emotional interplay with the mother was another necessary ingredient. We now call it T.L.C. (tender loving care). The great variation from one child to another in appetite, need and self-regulatory tendencies (as pointed out by Benjamin Spock[29] among others) led to a more individualized and less rigid approach to the child as a way to avoid feeding problems. The hazard of anoxia for the long-screaming hungry infant has been demonstrated by Margaret Ribble.[23] However, the most intense request for a change in the mother's disciplinary role comes from psychiatry. Freud's recognition that neurosis in the adult stems from unresolved conflicts in his childhood had an appeal for mothers who wanted to spare their children the suffering from restrictions that they themselves had experienced and were afraid to arouse resentment against themselves as the prohibiting parents similar to their own. Misunderstanding and misapplication of psychoanalytic viewpoints led to a standard of complete permissiveness and avoidance of frustrations by all means in order to prevent inhibitions and hostility. Fortunately, the pendulum that swung from rigidity to indulgence has left these extremes. The mother's role now involves the application of the right amount of restraint in accordance with the child's stage of development in an atmosphere of love and understanding. This, we believe, allows the greatest mutual enjoyment between mother and child, but it is also more difficult for the conscientious mother to know whether she is doing "the right thing" than it was in a former generation with religious, moral and medical principles telling exactly which path to follow. Another confusing element comes from the fact that these modifications of the mother role not only followed each other in rapid succession but that they also are coexistent, each maintained by different groups of people. Attempts to re-establish confidence in the mother role have been made by Spock[29] and Hilde Bruch,[7] for example, and can be done by the general practitioner and the pediatrician to guide and to reassure perplexed mothers. The role of the mother also varies from one socioeconomic stratum to the other, differs in rural from urban communities and is modified by religious beliefs. Physicians, social workers and others have to be aware that our middle-class concepts cannot be applied universally. One special variation occurs when the mother shares her role with another woman, that is, if she works or belongs to a social environment that assigns much of the child-rearing to servants. A recent study in our department[16] highlights the compensating or conflicting influence of the grandmother as participant in the mother role. Another pattern evolves in the care of the white child by a Negro woman in the South, when the biologic mother keeps a certain distance and a special prestige, while the child grows up in bodily and emotional closeness to his Negro nurse, whom one of my Boston analysands, reared in Georgia, called "the dark roots of my existence." Phyllis Greenacre[11] has discussed the importance of this early experience for the maternal role of the female child.

Physiologic and cultural elements can hardly be separated from psychological forces, which we regard as the most relevant determinant of motherliness and its potential disturbance. They merge particularly when we regard the mother's role within her specific element, her family. The relationship between husband and wife has a decisive bearing on the form of maternal behavior, even before parturition, strengthening or inhibiting the wish for a child. The child may be the natural outcome of mutual love, the unquestioning acceptance of the result of marriage, the desire for family life or the enjoyment of children. In contrast with these attitudes that support the mother's role are interfering ones: the child may appear as an intruder in the parent's life. The mother of a young schizophrenic girl told me that while driving to the maternity hospital she had clung to her husband, crying, "From now on we will never be alone." This feeling remained, and she was aware that she had to force herself to love the child, in which she did not succeed. Two years after the delivery she became pregnant again, "so that my daughter would have a companion," hoping that this would decrease the child's demands for parental affection. Other women, either out of narcissistic concern or because they sense their husbands' feelings, are afraid to lose their beauty and sexual attraction and thus the love of their partner. Whereas exclusive concentration on marital love or *la grande passion* might interfere with wholehearted acceptance of the mother role, we see situations in which the husband becomes the "fifth wheel," while the mother turns all her affection to the child.

Weston LaBarre[15] points out the problem in human beings that, in contrast with other mammals, sexual life and the upbringing of the young go on simultaneously.

> In wild animals, breeding and maternal care operate in alternation and do not occur within the same span of time. That is, the sexuality of wild animals is ordinarily seasonal. The sexes breed and separate; the offspring are born when the female is alone; and the dependency of the young is over in a season. The female's roles as protective mother and as breeding mate do not occur during the same time period; for when the next breeding season comes around, the young usually have departed.

In an amusing story, "My Oedipus Complex,"[20] Frank O'Connor describes a little boy's bitterness when his father returns from war and displaces him in his mother's affections and in her bedroom. He announces that when he grows up he is going to marry her and have lots of babies. His mother assures him that they will have one soon, which pleases him no end. However, when the baby comes, the father sleeps with him, "very bony but better than nothing. . . . After turning me out of the big bed, he had been turned out himself. Mother had no consideration now for anyone but that poisonous pup. . . . I couldn't help feeling sorry for Father. I had been through it all myself." Not infrequently the very maternal mother withdraws her feelings from her husband or treats him as another child in the household, thus depriving her child of a forceful father figure, so essential as a model of identification for the boy and as an image of masculinity for the girl. If the mother represents the only source of authority, she may become so dominant that the son is unable to achieve masculine identity and lives in passive submission or persistent rebellion against authority. Another disturbance of motherliness may occur in women who feel that their own dependency needs are not fulfilled by their husbands. Pregnancy and delivery increase such needs. Many women are not able to give to the child unless they receive care and affection for themselves. An unhappy marriage in which the child is either desired because he might improve it or dreaded because he might interfere with separation are among other factors interfering with motherliness.

The child himself—because of his own characteristics or because of a special meaning that he carries for the mother—also conditions her role. The sex of the child, particularly of the first-born, may determine whether or not the mother will develop all the empathy that she is capable of; e.g., for a mother who resents being a female, a daughter might be something second-rate. The same type of woman may admire but also envy her son because he is better equipped than she. The mother of a 9-year-old, extremely anxious boy pushed him to hold his own with his peers, saying, "If I had had a body like his, I would have stood up for myself," despising in him the weakness that she disliked in herself. However, when he got angry at home and kicked the furniture she was furious at such aggressive behavior, which never would have been tolerated in a girl. Whereas she felt incompetent with her son, she was a happy, understanding mother for her daughter. Special characteristics of the child—the degree to which he responds to tenderness or demands attention, or his activity type—arouse different feelings in the mother. That defective infants are a special challenge for any mother goes without saying. The adaptation of a physically or mentally crippled child depends to a high degree on the capacity of the mother to accept him and to respond to his special needs. Also, slight variations within the "normal" range affect the mother's role. In a study of asthmatic children[12] it became evident that the majority of them had shown from infancy a special sensitivity to the mothering that they received and a particular need for continuing the earliest symbiotic union with the mother; this resulted in a painful conflict between such infantile longings and the ego ideal to become independent. Thus, what we see as the "asthmatic mother" is an outcome not only of her own personality but also of the child's provocative requirements. Inquiry into the early history of patients with "psychosomatic illness" seems to me to be a fruitful way to get to the psychobiologic roots of these disorders. Not infrequently a mother of an emotionally disturbed child—described, for example, as rejecting or dominating—is an entirely different person in relation to her other children.

In other cases the mother's function is hampered not by the actual attributes of the child but by a displacement of feelings onto him, which she harbors for or against some key figure in her life, past or present. As an example of this, Nic, a 6-year-old boy, was referred from the pediatric service to child psychiatry. He had been hospitalized several times each year since the age of 1 year because of chronic diarrhea and dehydration. Each time his symptoms disappeared quickly when he was on the ward or in a convalescent home, but they promptly recurred on discharge. The patient's mother, Mrs. I., spontaneously said that Nic and she annoyed each other, so that both had to run to the bathroom when they were together for any length of time. Mrs. I. was in despair, sensing that she had something to do with Nic's illness. She herself, at the age of 12, had lost her mother, who died in childbirth delivering a boy. Mrs. I. and her older sister brought up their brother, sacrificing their free time and their savings. Mrs. I. gave up her scholarship in college for the education of her brother. He was a brilliant student, but he alienated himself from his less-educated sisters, married clandestinely and left them for good. Mrs. I., at the age of 30, married on the rebound from this disappointment and took a job, which she enjoyed. When she became pregnant she hoped that she would have a girl, and she cried when told that the baby was a boy. When the child was 10 months old Mrs. I. felt obliged to give up her job in order to be a real mother to him. Shortly afterward his symptoms began. In psychotherapeutic interviews it became clear that bringing up Nic meant to her a repetition of raising her younger brother, and that she anticipated that Nic would be just as ungrateful as he and would leave her. Mrs. I. was able to see that she had identified her child with her brother and had displaced her feelings. Accepting Nic as a person in his own right enabled her really to become his mother.

The mother of an inhibited girl with learning difficulties was extremely hostile to the child until she realized that she hated in her daughter all the characteristics that she found objectionable in herself. The child represented to her a bad alter ego.

Of course, the mother's role is modified by the child's developmental stage. Some women experience the greatest satisfaction during pregnancy and then after parturition feel empty and depressed beyond the "baby blues." Others find the greatest fulfillment in the care of the infant but find mothering difficult when the child begins to become a separate individual. The negativistic phase of the 1-year-old, for instance, the task of toilet-training, the intense love and hatred of the 4- and the 5-year-old or the rebellion of the adolescent require flexibility and modifications of the mother role.

Such flexibility, combined with a consistent loyalty toward the child, depends largely on the mother's past experiences and in particular on her relationship to her own mother from infancy. Affection for and identification with her are most favorable for a maternal attitude in girls, usually manifest already in childhood —through intensity and tenderness in playing with dolls, through the interest in taking care of younger children and in the choice of a profession (nurse, teacher, etc.).

Too early and too exclusive a mothering attitude in a young girl brings, on the other hand, the danger that she may be prevented from developing her own personality and with it the maturity necessary for the mother role. To give an example: A diabetic boy of 13 was referred to child psychiatry because he was unable to give himself insulin and to test his urine, was enuretic and depressed. The patient felt weak and threatened by the apprehension of his early death. At the same time he was aware of being a rather husky boy with the wish to go hunting and fishing. In his fantasies he rescued people sicker than himself. In reality he shied away from any outside activity and felt safe only near his mother. Being governed by his fear made him feel ashamed and unmanly. His mother was even more afraid than he that he might die. When she was 12 years old her younger sister was killed in an accident, and her mother became psychotic. She took care of her younger siblings and her father, whom she nursed until he died. In her 30's she married a man with diabetes who depended on her care. The patient, her only child, was born when she was 41. She was always anxious about his health, extremely devoted, fostering his dependency on her. He never was bladder-trained because she thought that this would be too difficult for him, even before the diabetes was diagnosed at the age of 6½.

This mother displays what David Levy has called "maternal overprotection,"[17] an attitude that he considers in its pure form an exaggeration of normal maternal behavior. In its neurotic form it represents a reaction-formation against hostile impulses toward the child. They are unacceptable and arouse guilt feelings. As a defense against them the mother must reassure herself incessantly that she preserves the child. "Overprotection" can be extremely indulgent or possessive, can be displayed in infantilization, extreme bodily contact or too intense affection.

Guilt feelings, tension and anxiety may have their influence on the child already during gestation as studies, particularly those of Sontag, at the Fels Research Institute for the Study of Human Development,[25] suggest: "Oxygen requirements, carbon dioxide level in the blood, and many hormonal and metabolite levels in the blood are dramatically influenced by emotion. Placental permeability permits these humoral changes in the mother to become environmental changes for the fetus." Sontag has "observed a relationship between the emotional state of the mother and the activity level of the fetus, as well as the emotional state of the mother and the physical state and gastrointestinal function of the new-born infant. . . . The same observations . . . have been noted by many obstetricians." That children's fears can be due not—or not so much—to a threatening situation but to the mother's anxieties has been shown particularly in Anna Freud's[10] reports of children's behavior during the bombings in England. "A child in the infant stage of one, two, three, four years of age will shake and tremble with the anxiety of its mother." The induced anxiety of the child in turn may increase the mother's insecurity.* Also,

---

* The use of so many "mays" and "mights" in this presentation is unpleasant but unavoidable. There is not a one-to-one cause-and-effect relation: The outcome of an unfavorable condition depends on many variables, e.g., factors which neutralize or counteract—stemming either from qualities of the mother or other key figures in the environment, or from the child's personality. Behavior, like symptoms, is overdetermined.

hypochondriac concerns in children may stem to a large extent from the mother's apprehension, which the child senses even if she does not speak about it.

It has been mentioned that a positive identification with her own mother is likely to give a woman confidence in taking over the mother role. If, however, the relationship with her mother was a hostile or a strongly ambivalent one, we see most often that a woman tries to treat her child in a different, even opposite, way than she was treated, but to her surprise she finds herself at times repeating the pattern which she has so eagerly tried to avoid. This shows that at some stage she did incorporate her mother—that is to say her "bad" mother has become a part of herself. Consequently, she is not free to act on her own convictions. Disharmony in her relationship with her mother also makes it more difficult for her to tolerate hatred and ambivalent feelings in the child. We often see this type of mother getting depressed if the child, although given all the love and the material things that she may have resentfully missed in childhood, still bursts out with "I hate you" or "I want another mommy." Women with a romanticized image of the "ideal mother" cannot tolerate the unavoidable ambivalence of the child.

Even the most exquisite mother cannot spare the infant a certain amount of deprivation and pain, thus arousing disappointment and rage. His own hostility in turn leads the child to expect retribution. Much depends on the mother's capacity to help the child in this conflict, so that love becomes stronger than hatred and aggression. The intensity and the effect of these contradictory experiences in the early child-mother relationship are the basis of Melanie Klein's theory of emotional illness in children and adults.[14]

Some women still retain a longing for the intimacy and the closeness to their mothers that they either missed or never could give up. As mothers, they frequently create such a symbiotic union with the child and try to perpetuate it, thus curbing the child's potential to become a separate being. This prolonged and enforced dependency in turn evokes rage and aggressive behavior in the child. A part of the rebellion so typical of adolescents is the young person's drive to resolve even the relatively "normal" and necessary dependency on his parents. This struggle is aggravated when the mother is unable to relinquish her need to possess the child.

At the other end of the spectrum we see mothers for whom the child represents a threat to their existence as an independent being. As was mentioned earlier, a woman's role is not exclusively in her family. Human beings are bisexual to a varying degree. So-called masculine strivings for creativity, for intellectual achievement, for participation in community or world affairs may—but do not have to—interfere with the role of mother.

A young outstanding woman scientist brought her 5-year-old daughter for psychotherapy. She was tense and defensive as well as self-accusing because she felt that her profession prevented her from giving sufficient attention to this very sensitive child. She had cut down her hours of work on several occasions to spend more time with her little girl. In doing so she became irritable, and the child's symptoms became worse. In working with this mother it became clear that the conflict between guilt feelings for neglecting the child and resentment against the girl for interfering with her role as a scientist disturbed the basic relationship between them. Getting "permission" to follow her career, she became less tense and was able to understand the child. The decisive factor was not the amount of time spent with the child but the empathy in the relationship.

This point leads us back to the complications created by our present pattern of the role of the mother at her best. We emphasize the importance of the love and the intimacy which the child needs from his mother. Therefore, conditions most favorable for such a relationship have been suggested: Natural childbirth, for instance, not only spares the infant the potential damage through drug effects but it also gives the mother the actual experience of birth. Helene Deutsch[8] states:

Woman's active part in the delivery process, her lasting pride in her accomplishment, the possibility of rapid reunion with her child, and some degree of gratification of that primary feminine quality that assigns pain a place among pleasure experiences in the psychic economy are precious components of motherhood, and an effort should be made to preserve them.

Indeed, many mothers, who after previous deliveries under anesthesia underwent parturition without pharmacologic support, experience great happiness and a feeling of achievement.

There are others whose fears and fantasies about delivery cannot be alleviated by relaxing exercises; they require anesthesia. It seems important not to arouse guilt feelings in them for having failed at the first step. It is similar with the suggestion of rooming-in, which allows the mother to have her baby with her from the start, undisturbed by outside visitors. It also allows the possibility of caring for him, assisted by an experienced nurse, rather than starting on her own with an infant she hardly knows. This is a good suggestion for women who can take it and enjoy it. But for those who need to be taken care of themselves after the stress of delivery, this arrangement will defeat its purpose. Breast feeding is, from a psychiatric viewpoint, preferable to bottle feeding because it permits the most intimate bodily contact between mother and child; the symbiosis that existed during pregnancy finds here a continuation instead of a complete break of this very special union. However, here also the personality of the mother ought to be taken into account. Apprehensions may make the mother tense and feeding a painful procedure for her and the child. In that case, a bottle given with ease and confidence is a preferable condition for establishing a good relationship and for the enjoyment of feeding and being fed. Similarly, self-demand feeding is for most babies an excellent arrangement for dealing with their individual appetites. However, some mothers become anxious if there is no order to reassure them that the infant is getting enough nourishment. Avoidance of tensions, insecurity and anxieties in the young mother being more important than the best of conditions, we ought either to try to allay the apprehension or to reassure the mother that she can still fulfill her role.

We have come now to preventive aspects. It was pointed out that in giving mothers the best advice we have at this point, we need to be aware that not every one can accept it and that it is important not to increase anxieties and guilt feelings. On the other hand, the recognition mentioned earlier that mothering is not only an instinctual process best left alone also suggests that we assist expectant mothers to prepare inwardly for their role. In her intensive study, *Patterns of Mothering*,[6] Sylvia Brody demonstrates the importance of experience for adequate mothering.

The primiparas were earnest to give their infants the best kinds of care, but their relative incompetence and their inner uncertainty was inescapable. Most of the multiparas were more competent . . . more self-assured and more self-reliant.

She came to the conclusion that maternal skill can be learned.

In the case of a mother who feels normal anxiety because of inexperience or because of unanticipated conditions within her infant, it may seem easier to extend help. In the case of a mother whose anxiety is neurotic, it may be more difficult for her to make use of it. . . . In matters of diet and antisepsis, no physician would any longer say that a mother knows best. . . .

In matters of mental health and the development of personality we know at present less than in the field of hygiene and preventive medicine. However, over the last 30 years investigators from several disciplines (psychiatry, pediatrics, sociology, psychology, anthropology and social work) have been motivated by intense curiosity to search for the roots of the widespread major and minor emotional disturbances we are plagued with and by the desire to forestall psychological deformities. Thus a nucleus of knowledge has emerged and is growing. One of the most relevant findings is the importance of the mother's role in the psychological condition and the outcome of children. It seems imperative to put our knowledge at the disposal of mothers. This may mean more than information; it may include guidance, support or therapy for the mother. It also calls for early detection of vulnerabilities in the child for which modifications of mothering are indicated.

That sounds like an ambitious project. In support of its worthwhileness, let me quote from Plato's *Symposium* something that Diotima tells Socrates:

See you not how all animals, birds, as well as beasts, in their desire of procreation, are in agony when they take the infection of love, which begins with the desire of union; whereto is added the care of offspring, on whose behalf the weakest are ready to battle against the strongest even to the uttermost, and to die for them, and will let themselves be tormented with hunger or suffer anything in order to maintain their young.

She answers the question "Can you tell me why?" with the words:

The mortal nature is seeking as far as is possible to be everlasting and immortal: and this is only to be attained by generation, because generation always leaves behind a new existence in the place of the old.

REFERENCES

1. Allan, J.: Unpublished data.
2. Ausubel, D. P.: Theory and Problems of Child Development, New York, Grune, 1958.
3. Benedek, Therese: Psychobiological aspects of mothering, Am. J. Orthopsychiat. 26:266, 1956.
4. ———: Psychosexual Functions in Women, New York, Ronald, 1952.
5. Bowlby, J.: Some pathologic processes set in train by early mother-child separation, J. Ment. Sc. 99:265, 1953.
6. Brody, Sylvia: Patterns of Mothering, New York, Internat. Univ. Press, 1956.
7. Bruch, Hilde: Don't Be Afraid of Your Child, New York, Farrar, Straus & Young, 1950.
8. Deutsch, H.: The Psychology of Women, vol. 2, New York, Grune, 1945.
9. Erikson, E. H.: Childhood and Society, New York, Norton, 1950.
10. Freud, A., and Burlingham, D.: War and Children, New York, Medical War Bks., 1943.
11. Greenacre, P.: Child wife as ideal, Am. J. Orthopsychiat. 17:167, 1947.
12. Jessner, L., Lamont, J., Long, R., Rollins, N., Whipple, B., and Prentice, N.: Emotional impact of nearness and separation for the asthmatic child and his mother, Psychoanal. Study Child 10:353, 1955.
13. Keller, Helen: Teacher Ann Sullivan Macy, New York, Doubleday, 1955.
14. Klein, M.: The Psychoanalysis of Children, New York, Norton, 1932.
15. La Barre, W.: The Human Animal, Chicago, Univ. Chicago Press, 1954.
16. La Barre, M., Ussery, L., and Jessner, L.: Significance of grand-mothers in the psychopathology of children, Am. J. Orthopsychiat. In press.
17. Levy, D. M.: Maternal Overprotection, New York, Columbia Univ. Press, 1943.

18. ———: Psychosomatic studies of some aspects of maternal behavior, Psychosom. Med. 4:223, 1942.
19. Mahler, M. S.: On child psychosis and schizophrenia, Psychoanal. Study Child 7:286, 1945.
20. O'Connor, F.: My oedipus complex *in* The Stories of Frank O'Connor, published by Alfred A. Knopf, Inc., New York; appeared originally in Today's Woman.
21. Rank, B., and McNaughton, D.: A clinical contribution to early ego development, Psychoanal. Study Child 5:53, 1950.
22. Rank, B., Putnam, M. C., and Rochlin, G.: The significance of emotional climate in early feeding difficulties, Psychosom. Med. 10:279, 1948.
23. Ribble, M. A.: The Rights of Infants, New York, Columbia Univ. Press, 1943.
24. Scott, J. P.: Social behavior, organization and leadership in a small flock of domestic sheep, Comp. Psychol. Monograph 18:1, 1945.
25. Sontag, L. W.: Fetal Adaptation *in* Symposium on Organic Evolution, Bull. No. 7, New Delhi, India, National Institutes of Science, 1955.
26. Spitz, R. A.: Anaclitic depression, Psychoanal. Study Child 2:313, 1946.
27. ———: Hospitalism. An inquiry into the genesis of psychiatric conditions in early childhood. I, Psychoanal. Study Child 1:53, 1945.
28. ———: Hospitalism. A follow-up report, Psychoanal. Study Child 2:113, 1946.
29. Spock, B.: The Common Sense Book of Baby and Child Care, New York, Duell, Sloan & Pearce, 1957.
30. Yerkes, R. M., and Tomilin, M. I.: Mother-infant relations in chimpanzee, J. Comp. & Physiol. Psychol. 20:321, 1935. (Quoted by Brody, S.[6])
31. Zing Yang Kuo: The genesis of the cat's responses to the rat, J. Comp. Psychol. 11:1, 1930. (Quoted by Brody, S.[6])

3

The Role of the Father
in the Family

Claire M. Ness, M.D.

THE HAUNTING BIBLICAL QUESTION "Am I my brother's keeper?"
serves to remind us that man has long puzzled over his role and
responsibilities in the family. Concerns over interpersonal relationships are not new; only the contemporary scene with its social
complexities is new, reshaping the old questions. In our changing
culture, the search for emotional adjustment continues with
greater urgency.

This discussion will concentrate on fathers in the drama of
everyday living.

In considering a subject of such emotional significance for all of
us, it is appropriate to acknowledge that no one can be entirely objective. The observer is always a part of that which is observed. To
this discussion it is inevitable that we bring our personal selves as
well as convictions and thinking from our special fields of interest.
I propose to speak from the experience orientation of child guidance
and child psychiatry.

Uniting the specialties of child psychiatry, clinical psychology
and social work, the child guidance clinic concentrates attention on
the child who is having personality and adjustment problems. The
objective is to understand the complex and tangled forces in a child's
life in order to affect and modify these for his ultimate benefit.

To such clinics come parents with a wide range of problems and differing capacities to use help. Many are confused, anxious and guilty, because they feel that somehow they have failed as parents or they would not be seeking outside help. They often bring their child with the hope and the expectation that the clinic will advise and direct them how to solve their problems. Essentially concentrated on the child, and with a nonprejudicial approach, the clinic has these aims with parents: by accepting their conscious attitudes and thinking, to help them to release and relieve feelings about themselves and in relation to their child and to give them better awareness of causes and effects of the child's problems so that they may redirect their emotional energies toward more satisfying relationships in the family. This work with parents on the external forces affecting the child is an essential corollary to the therapeutic and direct help to the child to deal with his inner stresses and fantasy life. These aims are implemented by the prevalent conviction in child guidance clinics that most parents earnestly wish their children to be happy and socially well adjusted.

Children in conflict have taught us that they are essentially no different from happier, better-adjusted children in their need for the basic elements required for emotional growth. They have mirrored for us that the child needs parents who can give him acceptance, protection, necessary restraints and freedom to mature. Studies of disturbed children have told us that these essentials cannot be provided successfully by one parent alone. The interlocking nature of parental attitudes and influences is constantly demonstrated. Mother and father together provide the framework within which each has unique and always complementary functions.

This interrelationship of influences always has been recognized in child guidance. However, the belief in the father's importance has not always been translated into practice. The stated reasons have been based on fact, fantasy and expediency; workers said that fathers were too busy, that they were much less involved, and the mother-child relationship was more important. Now, though it may require special considerations in planning interview appointments, more and more clinics try to involve fathers actively from the beginning of the exploratory diagnostic phase through the treatment period. It has been observed that many fathers are surprised by the clinic's interest in their role in the family. Many initially say that they have no particular information to give; the mother has a closer tie to the child; what can they possibly contribute? This common attitude reflects the traditional thinking that the father's place in the home is bound up almost entirely in his role as breadwinner.

Giving fathers the assurance that they are more involved than has been recognized frequently evokes their positive response that they want to feel more influential. Many welcome the opportunity to express their opinions and philosophy and to discuss their feelings about themselves and their children. It is the experience in our guidance center that 91 per cent of our parent-referred cases have the active participation of fathers. This demonstration that fathers can be involved and can aid greatly in the promotion of better family harmony supports the conviction that in the broader population fathers are not generally conscious of their prominent role in influencing the family milieu. Because fathers may have tended to agree with the popular concept that they are isolated from the family, they need more encouragement to believe that their presence and influence is a constant factor in the family's adjustment and in the child's emotional development.

Another conviction which has emerged from studies and treatment of emotionally disturbed children is that children, however absorbing, are only one facet of their parents' lives. We have learned that if we are to be of assistance to parents we need to respect them first as individuals with interests, activities, associations and aspirations. Upon the substructure of their unique and individual selves, parenthood is superimposed. It is equally true of both parents that our treatment efforts are unsuccessful if we unrealistically expect parents to submerge themselves completely into parental roles.

A father's role begins before the birth of his children when he faces the prospect of parenthood and becomes aware of feelings about this. The common expectations that it would be fulfilling and enriching to have children may be supplanted by doubts and anxieties. The expectant father may have realistic reasons to worry about his increased financial burden. He may not be ready for this experience in terms of his occupational ability to support a family. More importantly, he may not be ready emotionally to share his wife with another person even though that person is his own child. Pregnancy, confirmed, forces some kind of adjustment to the idea of fatherhood. It stimulates fantasies of what the child will be like, what rearrangements of his family life will be necessary. Almost always it brings to the fore some preference or denial of preference for the sex of the unborn child. Our society gives an elevated social status to the expectant mother. In contrast, the expectant father may well have uncomfortable feelings of isolation and of being left out of the creativeness of the process. Balancing such feelings can be pride in his virility, tenderness toward his wife and pleasant anticipation. All these are important feelings and attitudes which somehow become woven later into the fabric of his relationship with his child.

I have wondered if expectant fathers could not benefit from more concerted efforts to give them both information and support. To some extent they have been included in parental classes provided in urban centers. The Children's Bureau reports that in 1956 only 55,000 expectant parents had the benefit of such instructions, and classes were held in only 39 states. Of course, many expectant fathers receive advice, encouragement and information from the attending physician, but perhaps more could be done if it were recognized more generally that expectant fathers, too, have normal anxieties.

For father, the birth of his child removes the vestiges of fantasy and makes fatherhood a reality. Widely known are the remarkable advances in medicine in the management of childbearing and postnatal care, and of certain comfort to fathers is the reassurance that childbirth is no longer dangerous. It could be postulated that fathers no longer fit the caricature of the past in which they traditionally needed greater attention from doctors and nurses than did the mother. However, with all the rational reasons for confidence, the birth event stimulates a wide variety of feelings which need our recognition and understanding. Again, the father may feel isolated from the circumstances which so vitally affect him. Although some hospitals now permit his presence during parturition, the practice is not supported widely. I cannot hazard the proportion of men who might wish to have this experience, and there may be many and good reasons for excluding fathers from the birth room. However, we know that practices traditionally held in many hospitals do not contribute to father's prestige and status. He is usually treated like any other visitor, limited by rules for the aseptic protection of the newborn. As admirable and important as such limitations may be, they can frustrate both his personal and paternal feelings at a time when these might better be encouraged. My belief is that father's concept of himself as a person and as a father is influenced by many things. Wherever it is possible to foster his status, it seems advisable to do so.

I wish to turn now to a consideration of father's role in the psychosocial development of the child, as we have come to see this in the practice of child psychiatry.

In the first few months of his child's life, father's psychological role in the family is more prominent in relation to the mother's needs than to the child's. If he adapts well to the changed circumstances in the home and is able to be of emotional support to his wife, this significantly strengthens the emotional climate in which the child is nurtured. He is not isolated from the child, but his contribution is overshadowed by the mother's dramatic and close symbiotic relationship with the child. The submerging of the father's role at this time may be influenced partially by his practical role as breadwinner, which takes him away from physical closeness to the child. However, more specifically it is due to the infant's emotional need for the mother. Child psychiatry holds that the very roots of the child's personality are laid down at this time when he is passively dependent and helpless, and it is believed that the gratifications the baby receives from his mother form the basis for his later important capacity to give love as well as to receive it.

In these months when father is a sustaining and supporting force, he must come to terms with his concepts of what are feminine and masculine tasks in the management of the home and baby care. It is widely acceptable now that fathers participate actively in many of the tasks traditionally assigned to women.

In *Baby and Child Care* which has become a highly respected guide for millions of parents, Dr. Spock states, "a man can be a warm father and a real man at the same time." He suggests that fathers should share some of the routines of baby care, pointing out that these early experiences pave the way for later friendliness and closeness to the child. In this advice, we can see the obvious advantages and foundations for togetherness in the family. At the same time we may be concerned somewhat that roles and functions should not become too diffuse and desexualized. Since we believe that both fathers and mothers provide the living demonstrations for their children of what constitutes masculinity and femininity, it is important that in many small and everyday matters, father is not seduced into being mother's helper. It may appear contradictory to stress the importance of father's interaction, then seemingly suggest that father hold himself aloof from household responsibilities. Actually there is no real contradiction, father's role and interaction being enhanced by preserving his strong masculinity. How individuals incorporate these concepts into daily life is always dependent upon their own unique life situations.

Generally, through the first 3 years of the child's life, it is believed that the father continues his dynamic role in the family, with slowly increasing direct importance to the child. Father is becoming identified and familiar to the baby, who responds more to him as an individual. As interactions increase, father is realistically many different selves to the child. He gives affection and attention, but he is also at times necessarily depriving and thwarting to help the child give up his egocentricity. With increasing confidence in his security and growing independence, the baby begins to exhibit the capacity to love father, too, but also to resist and resent him. In the toddler stage, the child incorporates into his own being both positive and negative feelings which he acts out in all directions both to gain acceptance and to express his new-found aggression. Dynamically, father's role in this developmental phase is to supplement and implement mother's role, to present to the child reasonable restraints, to allow the testing of limits and to understand that aggressiveness now is the child's new and valuable tool.

Since this phase of the child's emotional development usually coincides in our culture with toilet-training efforts, we see how much the psychological mechanisms of the child require unity and consistency on the part of parents. With awareness that he has control over his functions, the baby realizes that he has a powerful weapon with which to please or displease his parents. It takes the child's time and the parents' patience to work out this psychological problem. Physicians have an important function in helping parents to understand that too early or urgent pressure for cleanliness may defeat the desired result. Though I do not see in this phase any particular uniqueness in father's psychological influence, I believe that the baby's eventual mastery of his bodily functions comes about because he wants his father's approval as well as his mother's.

Father's role becomes more specific in the next phase of the child's psychosocial development. In time, the onset and the resolution of this phase are roughly between 2 and 5 years. Herein occurs the child's faltering steps to know and accept his own sexual identity, a process in which father plays a very significant part. The primary attachment to the mother must be weakened, its bonds loosened, if children are to grow toward a later healthy heterosexual adjustment. Father, in the triangular situation, becomes the target of a small son's fear and hostility, the recipient of a little daughter's affection and wish for greater attention. Widely known as the Oedipal conflict, it has been repudiated by those who interpret that it ascribes adult sexual motivations and actions to children. Gradually the concept has been commonly accepted that the core of sexuality is present from birth on; it undergoes highly observable changes in adolescence, more subtle differentiation in the Oedipal phase.

In clinical practice one sees this psychological process dramatically expressed in the ways little children speak of their parents or act out feelings in play. The conflict is close to the surface and there for anyone to see. We hear the little boy talk admiringly of his father's strength and bigness as he mauls and destroys the image of father in doll play. We hear him say that he will marry his mother when he grows up. He wants to be more important to his mother than he thinks father will allow. He indulges in fantasies that father will go away, then he is fearful that his hostility is perceived and he will be punished. Failing to estimate reality, he believes that his hostile thoughts may destroy his father. Many examples can be drawn not only from clinical experience but from the stories of parents. One mother said that her 4-year-old John was very upset when his father was taken to the hospital. As she prepared the next day to leave home to visit the father, John inquired tearfully whether she, too, were going to the cemetery. He was certain that his father was dead and could not be reassured until father telephoned to say that he was better and would be coming home soon. One father reported that he was puzzled by his son's need to kiss him many times before he left for work. He perceived that this anxious sort of affection occurred only when father had to leave the child alone with its mother. Thinking about this he said, "I know he's jealous of me—why should he act as though I were jealous of him?"

Unquestionably, these manifestations of preference are handled intuitively and well by many parents, but it is not at all unusual for fathers to curb this hostility. Often seemingly amused, they add fuel to the fire by encouraging the boy to be a little man and take care of mother. Fantasies of being powerful, fears and nightmares occur commonly in this period, and father's maturity is tested as the child finds many ways to reject or depreciate him. Obviously, when parents have neurotic problems in the area of sex, the child's resolution of his identification may be jeopardized. If he has a domineering mother and a passive father, the child may be influenced toward weak masculinity; a complex variety of personality characteristics come from this period. A father needs to be able to take the hostility although he may be disappointed that his son regards him as a burden and an obstacle. If the emotional climate of the family does not permit freedom of expression, the child may turn too quickly to identify with the big man who is his father. Because castration fears in the child are a prominent part of this phase, father helps best if he avoids harsh physical punishment or threats of this for nonconformity. He can use other measures at this time of the child's great vulnerability to fears of castration. Father's friendliness and consistency are particularly important. This does not mean excessive attention or compliance with every wish of the child. It suggests only that the child needs to like the kind of person his father is, so that eventually he will wish to emulate him and gradually repress the earlier unrealistic wish to replace him.

In his relationship to a small daughter, father is less threatened. This time he is the favored one, and his role requires that he help the child to turn again to the mother. The conflict for the little girl is that she fears that the mother will abandon or reject her for preferring the father. This too shows up in countless behavioral ways. Becoming aware of sexual differences, the girl is often confused about her physical make-up, feels damaged because she has no penis. She, like the boy, is especially vulnerable to family tensions or disagreements, often feels responsible if parents quarrel. Eventually, aided by both parents, she represses her desires to have exclusive possession of the father and becomes willing to share him with the mother. It is a difficult task for father to discriminate between being affectionate and seductive with his small daughter. It is hard to know how to balance permissiveness with firmness with his son. If father respects and likes his own sexual role, it is easier for him to guide his son into identification with him and to permit the little girl to identify with her mother.

With identification problems more settled, there ensues another period, labeled latency, when father's role seems to churn around the dilemma of discipline. Dr. Jocelyn in *The Happy Child* believes that the term "latency" is euphemistic and suggests that, contrary to its implications, latency is a period of great psychological growth. My own experience corroborates this view. Parents together are both catalysts and reactors as the child begins to move beyond the close confines of home. What needs in the child determine father's psychological role? I believe that essentially they are the child's unspoken cry for continuing gratification reinforced by reasonable control so that he may know the limits of his independence. In contrast with the permissiveness, advocated by individuals who have misinterpreted children's needs for expressive freedom, sensible restrictions and limits are not only desirable but necessary. In many families disciplinary measures are relegated to father without his participating in setting up the house rules. Communication between parents around their expectations of children could smooth out some discord and make father's role as arbitrator and judge more plausible.

Father's role in this period has its emotional compensations. He has a natural reason for pride and gratification as the child exhibits his increasing self-confidence and developing capacity to achieve and to conform to new demands in his expanding world. In many families, this is a time of comparative serenity.

In these years before adolescence there can emerge the symptoms of emotional problems created by new external stresses or carried over from earlier years. Children in this age group make up the large majority of cases referred to child guidance centers. I mention this because most often the responsibility of knowing about and using community resources is assigned to mothers. I suggest that fathers could well incorporate in their role a broader knowledge of the community environment. They need to be curious about the social institutions and the social mores which impinge upon their children in their life outside the home. Does the school system meet modern high standards? Are recreational centers adequate and supervised? Are there special facilities for assistance to emotionally disturbed children? Do hospitals consider a child's emotional reaction to illness and separation? Is the Juvenile Court humanistic or legalistic? These are some of the questions which fathers can ask, and then join with other individuals in securing positive answers. Improvements in community welfare are sometimes the work of individuals. Their continuation can always be traced to enlightened citizenry who support them through their elected officials or through united influence.

The turbulence of the adolescence stage of development calls on father to maintain a steady balance and capacity to understand that tumultuous physical, physiologic and psychological forces are converging upon his child. It helps if he can remember his own sensitivity and struggle for independence, his own rebellion against "old-fashioned" parents, his devotion to the gang and to the ideas of his peers about such things as clothes, haircuts and behavior. It helps, too, if he recalls his own sexual misgivings and anxieties. Most helpful, in his relationship with the adolescent, is a large measure of confidence that his child can and will harmonize his inner and outer conflicting forces.

Our society has created special burdens for the adolescent. In an attempt to overcome past rigidities, it has come to advocate extreme permissiveness. At a time when the youth is under severe stress, wanting yet rejecting independence, our tendency has been to weaken controls, to give him too much freedom. In adolescents in guidance centers we see the frantic search for acceptance, the desperate need to govern themselves, to master their impulses and anxieties and to conform to the ideals of their group. In all of this their references to the father show a reawakened need to find their own sexual identity, to reconcile their conflictual aggressive feelings. They want father's love and control but must fight against it.

Recently, in the news, appeared a candid statement by J. Edgar Hoover writing on Juvenile Delinquency. He said that children now are the victims of a society which has substituted indulgence for discipline. "They are the victims of a breakdown of authority and moral standards in the home, in the neighborhood and—too frequently—in the community." Sometimes called an advocate of a "get tough" policy, Mr. Hoover challenges adults first to "get tough" with themselves.

The same paper quoted from an article by Dr. C. Fischer from *The Journal of Pediatrics* of November, 1957. This quote follows:

Our youth now love luxury. They have bad manners, contempt for authority, disrespect for older people. Children nowadays are tyrants. They no longer rise when their elders enter the room. They contradict their parents, chatter before company, gobble their food, and tyrannize their teachers.

Of interest is the fact that this is not a quotation from a recent writer, but these are the words of Socrates, written in the 5th Century B.C.

A third article referred to Bishop Hazen Werner addressing a national conference on family life in Chicago, sponsored by the Methodist Church in which Bishop Werner said:

The father needs to come back into the home. For a long time he ran the show in the family. Nobody seems to be running it now. You can't have five or six individuals running the home and still have order.

Surely Bishop Werner, Mr. Hoover and Socrates reflect the same concern that adults have always had about the capacities of the younger generation. No one would deny that adults, and importantly fathers, have an obligation to affect improved social attitudes and circumstances toward a better milieu. However, I wish also to emphasize my belief in the presence of an innate force stimulating emotional growth and counterbalancing destructive influences.

Vast sociologic changes in our culture make it virtually impossible to delineate or isolate father's role except in general ways. Accustomed to using statistics as a basis for formulating opinions, we can find in them some foundation for further consideration of father's role in our times. However, we need to remember that generalizations always obscure the individual who, massed with other individuals, makes up the statistics on which we base our opinions. The interpretation of statistics is always influenced by our individual frame of reference.

What about the effect of urbanization upon the father's role? This has been said to be one factor weakening family cohesiveness in that children are less dependent upon father's teaching guidance and control. Here statistics point also to some conclusions but raise many questions for consideration. We know that schools, recreation centers, nurseries, and child care agencies have assumed greater degrees of responsibility for children. How much is father's delegation of responsibilities healthy and constructive? How much do economic pressures upon him and occupational competition enter into father's willingness to share with others the guidance and the care of his child? My personal reaction is that there are far more benefits accruing to children from the availability of such socializing resources than there are dangers of upsetting family cohesion. In this connection I think of the protective services which social measures and community agencies often provide to aid families to stay together despite poverty, illness or temporary difficulties. Urbanization resulting from rapid industrialization may well have brought increased tension for fathers in their occupational competition. But I feel the tendency to look back upon the "good old days" with nostalgia for a different culture denies our many increased opportunities for personal growth and emotional satisfactions. Fathers of past days probably had less time and opportunity for sharing family life than is now true despite our mental image of past family cohesiveness.

In the matter of physical health and the changing American family, there is irrefutable evidence of a striking decline in mortality. In 1900 more than a quarter of the children born in this country faced the prospect of becoming orphans by the time they reached 18 years. Now this is true of only 7 per cent of all infants. What are the implications of this for fathers in their role as husbands and nurturing parents of children? Here a pleasant and optimistic assumption can be drawn from statistics. Fathers now can have greater assurance that their life spans will encompass the dependency years of their children. Relative to the studies of mortality it also may be said that family life itself seems to promote health. It should contain some comfort, compensatory to fathers with their heavy responsibilities, that married men can expect to live longer than single, divorced or widowed men. Especially is this true in the period of the married man's life prior to age 45 when most families have young children under their care. The conclusion then can be upheld that fathers not only live longer to discharge their paternal responsibilities but can reasonably expect to be healthier and less susceptible to intercurrent illness than are men without family responsibilities.

Due to medical advances, immunization procedures, safer surgical procedures, antibiotics and other pharmaceutical discoveries, fathers today generally have less cause for anxiety over their children's physical health. Of course, cases of individual tragedy do occur, and children's accidents are increasing. Father's concern for an ill child may still bring sleepless nights, and the necessity to lend strength to mothers in the care of sick children, but again our hypothetical father has benefited greatly from the social force of improved medical practice.

Using another example, statistics on the remarkable uptrend of the employment of married women can give an intimation of broad effects upon father's role in the family. We know that in 1900 only 5.5 per cent of married women were in our labor force; by 1957 almost 28 per cent were working. The social sanction for men to participate more in household tasks could well be one effect of this social change. But two additional sets of statistics can suggest that the employment of women has not altered the traditional pattern that fathers are the major breadwinners, mothers the homemakers. These are seen in government releases that (1) in 63 per cent of our families father has the chief occupational role and (2) less than one fourth of our 10.8 million working mothers have children under school age.

My final emphasis is that social forces have created new conditions to which fathers and mothers both bring the essentially unchanged influences related to their biologically determined natures; they bring also their innate and acquired capacities and feelings which affect the development of children and the harmony of the family. Neither one can or should "run the show." Fathers continue to provide the vital and dynamic half of the whole of parenthood.

4

The Role of Children
in the Family

Sidney Berman, M.D.

THE ROLE CONCEPT as it applies to a child in his family opens up complex vistas for the investigation of the child's relation to his family and the multiple concurrent factors which enter into and determine the way the child's behavior becomes structured. There has been a shift in the historical perspective of child development, with a special emphasis on the psychodynamics of family life, which has broadened the study of child behavior. The family is conceived of dynamically as a transactional field of adaptional processes in which the child is observed in relation to his total environment. This includes what occurs simultaneously in his internal environment and his external environment. The latter is made up of his family and the social structure in which the family is a basic unit.

The child is exposed to a maze of psychological forces, both from within and from the world without, which are brought to bear upon and determine the form that the child's personality will assume. In some large part, the child's growth and development is derived from given psychological factors inherent in the human organism. It is recognized that the child goes through specific phases of development and that this is an intrinsic process characteristic of all humans, but the way in which these phases of development unfold depends on the environmental factors which are brought to bear upon the child. In other words, the child is exposed to a continuous interplay among his inner psychic operations, the transactional relationships within the family and cultural factors which determine his behavior. Therefore, in order to understand the various roles which a child assumes in the long and arduous process of socialization, it is necessary to evaluate all of these factors as they impinge upon the child's growth and development.[1]

This discussion is divided into 4 parts in which will be considered the relationship of the child to the culture, the relationship of the child to the family, the nature of role structure and the function and the technics which influence role patterning of the child.

RELATIONSHIP OF THE CHILD TO THE CULTURE

In this society the family has the responsibility for the care and the rearing of children, and it stands at the crossroads between the individual and his culture. Therefore, what transpires among the individuals within the family will assume the highest level of significance in shaping the psychological organization of the child, the nature of his position in his family and the manner in which he adapts to community life. The family is a complex operational system which is intimately related structurally and functionally to all the other units of the social system. Therefore, the members of the family will be influenced by a number of cultural factors. These include the nature of the geographic location where the family resides, the class stratification of the family, the ethnic grouping of the family, the occupational system, the educational and the religious systems. All of these will have a direct bearing on child-rearing practices, since the cultural characteristics will be transmitted to the children through the family.

Moreover, cultural attitudes which exert a psychological effect of the highest significance in shaping our behavior are created. These cultural attitudes are described as value orientations, and they define the generalized or the organized socially determined concepts which influence behavior. In any given society certain values prevail, and these govern the way in which the majority of individuals operate in a given social system. These value orientations may be divided into 5 different categories. The first defines man's relationship to nature. In our society we express an intense desire to master the forces of nature, and our children are being trained to achieve this objective. In contrast, some societies live in harmony with nature and others feel subjugated to the forces of nature. The second category defines man's place in his culture in a time sense. We constantly stress a future time orientation as the main point of reference. Long-range future goals are structured for ourselves and our children. This strongly influences the way in which we plan our future for "bigger and better things" and place high value on change. Our children are being trained in such an educational atmosphere, more in terms of what we want them to be like rather than in terms of what they are like. They are the ones to whom we entrust our destiny and the fate of our future.

The third category defines our relational orientation, that is, the nature of our family ties. In this society the dominant orientation has become individualistic, and a weakness in our collateral or lineal relationships has occurred. Children grow up with very little intimate contact with their relatives in this society. We also are expected to take care of ourselves, and when we cannot, we create social institutions other than our family to do so. The fourth value orientation determines what is good and proper or bad and improper in our relation to our fellowmen. This refers to the innate human-natue orientation, that is, the nature of the impulse by which action is determined. We look upon our children and ourselves as evil and perfectible, and we stress the need for constant control and discipline in order to achieve goodness. This strongly influences the manner in which the child structures his self image in terms of a good or bad person and emphasizes to the child the need to ge "good." Last, our culture also communicates to us the modality of human activity which is most highly prized. We stress achievement, and success is measured by our accomplishments, especially in material things. Our children also emphasize this as a virtue. Thus we see that the functional aspects of our society exert a most powerful influence on all of us, although we are hardly aware of them; yet they play a dominant part in determining the way we socialize our children. These 5 points of reference need to be kept in mind in evaluating the value orientations which have an effect on role patterning of children.

THE RELATIONSHIP OF THE CHILD
TO THE FAMILY

The child also is intimately influenced by a complex of biologic and psychological reactions which arise from within the family among its individual members. The psychodynamics of family life occur as a transactional process. This refers to the interpenetration and mutually reverberating and reciprocal effects of processes operating in the family. These processes attempt to achieve some degree of complimentarity or balance among its members and maintain the equilibrium of the family as a social unit.

Among the many interrelated functions which the family provides to its members and to the social system, two appear to be fundamental and basic to the family. These two functions also define the relationship of the family to the child. One of these is the legitimatization of parents in order to regulate the reproductive processes and introduce new members into the society in an orderly manner. The legitimatization of parents endeavors to preserve, regulate, maintain and perpetuate our social organization. In this way the rearing of children within families as social units can be assured.

The other basic function is the socialization of children. Once a child has arrived, the family is designated to provide the emotional and the physical climate which will prepare the child for membership in the social order. The process of socialization is a family function of fundamental importance. As you know, the very nature of the child's biologic and psychological organization, that is, his prolonged immaturity, makes the child particularly susceptible to the tensions created by his external and internal environment. The infant's need for a mothering one is absolute and basic for his survival. A child brings very little predetermined behavior into the world with him. Because of this he is dependent on his parents to meet his primary growth and developmental requirements and to establish an atmosphere of security. It also takes a great deal of time for the child to integrate his psychological and biologic functions, so that he eventually learns how to adapt to the full range of social role expectations.

At birth and during the early years a child is extremely impressionable. However, he tends to learn more through the modality of anxiety than by means of an objective appraisal of his experiences. Experiences once learned in this context cannot easily be unlearned or relearned. As a result, they will become enduring patterns of adaptation which the child will use as a point of reference in his reaction to the world about him. Because of this, the child requires a stable continuity of parental relationships if the process of socialization is to proceed in an optimum manner. Wise and kindly care and guidance by parents facilitates the child's capacity to avoid the extremes of psychological pain or anxiety which interfere with development. This in turn permits the child to expand his adaptational capacities in terms of social roles so that he may arrive at a rewarding level of integration between himself, his family and the social system.

There are 2 more functions of the family in which the child is directly involved. In the one the family supplies the child's needs, and in the other the child serves as a psychological need of the parents. In either case the effect these have on the reciprocal relationship between the child and the parents is of significance in structuring the child's role patterns.

The family supplies the essential physical, emotional and social necessities which nurture healthy living and provide personal satisfaction among all of its members. These consist of the satisfaction of hunger, the management of the eliminative functions, the avoidance of physical pain, the satisfaction of sexual needs as defined in the broadest sense and the need to avoid physical loneliness. Also, there is the need for mastery, which provides a feeling of security and accomplishment. The gratification of these needs serves the purpose of self-preservation, propagation and unification of life. As for the child, the basic needs seem simple enough to provide. However, in obtaining them, they are subject to the most complex vicissitudes and transformations. How these needs and responses to them are joined by the family and the social system will vary strikingly with the specific stages of the child's development. As a result, the child's behavior will be structured by the triad of intrapsychic forces, family attitudes toward the child's needs and social expectations. Therefore, one observes in the child shifting role patterns in keeping with the specific stages of development until adulthood is reached. Because of this the child-family relationship requires a degree of flexibility in order that the child move through his specific phases of development under optimum circumstances and evolve role patterns appropriate to these stages of development.

The child also serves a psychological need of the parents. This function which is structured by family life is frequently overlooked. Children provide a means by which parents endeavor to relive their own lives. This implies that there is a need for the psychological reproduction by the parents of their own life experiences through the lives of their children. In other words, the child is the instrumentality by means of which parents relive their own life experiences. This seems to parallel the phenomenon of biologic reproduction. The way parents reproduce their life experiences through a child is of major importance in determining how the parents structure reality for the child. This may strongly affect the child's cognitive functions and his role patterns. The process is entirely outside of awareness. It may result in serious child-parent problems, since the parents' need for the child is relative only to their own life experiences. The outcome may be for better or worse. This phenomenon has a unique paradoxical quality. On the one hand the child sees his parents as the need-gratifying objects who determine his sense of well-being. However, the child also is subject to turbulent psychological forces in the parents which impel the parents to use the child to relive their own lives. When the experience is gratifying to both participants, it leads to a new phase of integration for the parents. As we were loved and cared for, so too, do we administer to our children, giving to them that which we once had, cherished and sought. However, where the reliving mobilizes anxieties in the parent which were experienced as the result of his own early childhood deprivations, the child becomes the object of displeasure and disappointment. As one mother put it, "I actually wanted no children because I felt there was nothing in it for me. All my life no one took care of me." With the best of intentions, we often unwittingly attempt to structure the lives of our own children in keeping with our own aspirations and disappointments. A father formulated it in terms of "I know it is proper to say it matters not if I had a boy or a girl, but I really wanted a boy so that I could name him after me," with the hope that the child would give him immortality and attain all those self-centered wishes the father strived for. It is also seen in a negative way, as when some couples avoid having children because they fear that a child would be subject to conflicts similar to their own. This unique phenomenon is significant in determining the child's role in the family, because it may strongly influence the way in which parents distort reality for the child and impair the child's cognitive functions.

ROLE STRUCTURE AND FUNCTION

It is the responsibility of the family to develop in children culturally appropriate roles and to structure an adequate fit between the child and society. The role concept requires clarification, since it is not always used in a clearly definable sense. In a transactional situation such as exists in the family, the role of any member refers to a goal-directed pattern of interaction between that member and others for the given position that member maintains in relation to others in the family group. The role concept is a convenient and unique way of forming the mental impression of a dynamic link which unites the intropsychic processes and social influences, so that they are comprehended simultaneously as we evaluate the given behavior assumed by anyone.

This is a new and difficult way of formulating human behavior in a multidimensional frame of reference. It presents a whole trend in the way of thinking which is not in accord with our accustomed modes of thought. Yet it can prove to be a useful way of joining or uniting the intrapersonal systems of the id, the ego, and the superego with the interpersonal system of the family in which the former derives many of the characteristics from the latter. It also provides a method whereby cultural value orientations can be described simultaneously from the point of view of the individual and from that of society. In addition, the role concept may furnish a knowledgeable conceptual scheme for the study and the development of child-rearing technics unique for our culture and in accord with our cultural goals. This might introduce a workable way of applying known mental health principles to child-rearing practices not only in a scientific frame of reference but also in a practical manner.

The analysis of a role may be used as an evaluative procedure in understanding the child's behavior in relation to his family. The child must reconcile his inner feelings in regard to a range of life situations in which he learns his behavior through reciprocal transactions with his parents. Consequently each role response is accompanied by a set of complimentary or counter-role responses of the others involved, and all role responses are influenced in varying degree by a number of forces which operate both within and outside of the awareness of the participants.

Now, I shall present a complex series of statements which will require further elaboration. Any given role may be analyzed on the basis of the following 6 points of reference. What is the goal in terms of its explicit and implicit objectives? The goal may be for the purpose of seeking gratification or may be chosen for the sake of defense. What is the allocative basis of the role? For example, is it an ascribed role, such as son, or an achieved role, such as student? What cultural value orientations enter into and influence the response? What is the biologic organization of the individual? What instrumental aspects are used in the expression of the role? What cognitive-normative methods are employed to help the child in the role adjustment? For the child needs information in order to know what is expected of the child in terms of his behavior.

Let us apply these 6 points of reference to two 13-year-old boys in the educational role. Both families show a concern about the education of their children. The first boy adjusts well. The explicit goal is the integration of knowledge toward a professional career. The allocative basis describes the role as an achievement role, specifically the role of student. The cultural value orientation places high priority on his being good, preparation for the future and producing well. Biologically he possesses a superior intellectual endowment. The instrumentality is the school setting. The cognitive-normative directives are communicated to the youngster in keeping with the environmental realities. Now, as to the other boy, he adjusts poorly to his role as a student. The goal of education appears secondary to his need to distract and annoy his teachers. The allocative basis of his role is a distorted one as far as the student role is concerned. Instead he adopts the character role of being a prankster. There is a discrepancy of his cultural value orientations because he is "bad," interested in his immediate wishes and is nonproductive. However, he has the same intellectual endowment of the first boy. The cognitive-normative data he uses has not been defined for him in terms of role expectations. His mother has infantalized him, seduced him and considered his behavior cute at home, and his father has abdicated his role as parent, so that the child feels himself to be special and expects to be treated this way. The analysis of a role in this manner defines at the same time the discrepancies in role adjustment as well as the proper goals in terms of role expectations. It also demonstrates that a person's right to a role will be challenged when it threatens the function of the culture and that pressures will be put to bear upon such a person to achieve a fit in his adjustment or be subject to the consequences.

Time permits only a brief description of the general role categories. These have been classified into 4 principle groups, the ascribed, the achieved, the adopted and the assumed roles. Each of these may be subdivided into more specific role patterns. The analysis of these roles furnishes a way to evaluate the position of the child in the family at any given point in his development and defines the manner in which he fits into the family and cultural patterns.

An ascribed role is one which the culture expects the person to take, and very little deviation is allowed to modify it. The roles in this category consist of attributes of a general nature which belong to all individuals. These assigned attributes apply to age, sex, and body management functions at a biologic level and to kinship, class and ethnic relationships at a semibiologic level. For example, a child will be confronted with serious problems if he attempts to change age or sex roles. Also, his position is fixed in terms of his kinship and ethnic relationships. Society holds that these roles must be responded to with proper behavior and expects these roles to be learned in the home. The role of the child is shaped or patterned by the parents, and this holds whether or not the parents are aware of their influence.

Achieved roles are those which are acquired through effort. A person must do something about attaining these roles. Of course, they need to be made available by society, and they must be within the sphere of interest and the ability of the individual. The specific roles in this category are educational, occupational, religious, political, recreational and intellectual. The children will respond to these role patterns in keeping with family aspirations and the transmitted cultural value orientations. At times we observe that the family aspirations and the cultural expectations are contradictory. Today the educator and sages in the affairs of state want all of our children to go to college. However, in many families, other values are conveyed to the children directing them to seek occupational gratification, such as that of a journeyman in a print shop or a foreman in a garage, so that the child feels this is just as valuable.

The adopted roles are even more specific in nature. Some are transitional, such as the role of traveler or sick person. However, we are more concerned with the more fixed adopted roles, that is, the character roles. These behavioral patterns are not structured as syndromes in a clinical sense. Rather they are analyzed in the context of the specific behavior observed. We are interested in the specific unique features which a person displays in a given situation. For example, a child may be the honest one, the exhibitionist, the deceiver, the bed wetter, the leader, the follower, the prankster and the like. The purpose of this approach is to evaluate the behavior in a transactional setting by means of the given 6 points of reference already mentioned. Such an analysis is illustrated by a charming and attractive girl, 11 years old, who suddenly refuses to go to school. In a psychiatric clinical sense, this problem would be classified as a school phobia. Of course, this classification is purely descriptive. Actually she is obstinate about doing anything which will take her from her home, and she assumes an unyielding dependency role. Her explicit goal is to stay close to her mother, and she wishes her mother to care for her external and internal needs. Her value orientations are totally present-time oriented, with a need to be cared for rather than to produce. The allocative role is a character role which is adopted. Biologically she possesses a brilliant mind but functions in an immature way with a need to avoid accepting the biologic changes of puberty. The instrumentality used is one in which she remains close to home and to mother. The cognitive data structures sex as violent, with the fear of losing mother. Mother had encouraged her daughter's dependency as a way of achieving gratification she could not obtain from her husband. Violent quarrels occured between husband and wife, which caused the daughter to equate sexuality and growing up with horror and caused her to fear that something horrible would happen to her mother. Therefore, sexual development was felt by this child as gruesome and something she had to repudiate. It will be observed that behavior is not studied in terms of an investigation of intropsychic processes alone. Instead, the evaluation is made in terms of the whole complex of role relations, with an analysis of the transactional processes to which the child is subjected.

The last major role category consists of assumed roles. This is related to the fantasy life of everyone. However, it is a dominant characteristic of the child's way of adjusting. The problem is one of degree, that is, whether the fantasy is considered to be imagined or real. If it is the latter, it then becomes an adopted role, which will lead to serious distortions of adaptation. All children resort to egocentric thought processes as a way of coping with the child's inner and outer world of reality. Nevertheless, the cognitive-normative analysis of the role by the parent strongly influences the course the fantasy will assume in the child and its ultimate fate. A child who persistently assumes, in fantasy, that she is a boy probably will end up crippled in her ability to achieve motherliness. Whereas, the little girl who plays at mother and cares for dolls indicates that she endeavors to acquire mothers' characteristics and in the process of socialization seeks to achieve the quality of motherliness.

TECHNICS FOR ROLE PATTERNING OF CHILDREN

This logically brings us to the technics parents and others use in order to structure the role responses of children. What I have to say may appear so banal and so well-known to everyone that you may take for granted or tend to minimize the significance of these little things which are called complimentary or counter-role responses of parents to the behavior of children. Yet these responses of parents will determine for better or worse the course the process of socialization of the child will follow.

Whenever disequilibrium occurs among the members of the family, a complex transactional process will be set up in an attempt to achieve stability or bring about a re-equilibrium. The role responses of a child always initiates complimentary role responses in others in the family, particularly the parents, in an attempt to resolve tension.

Spiegel and Kluckhohn have been pursuing the investigation of the concept of social role in analyzing the way the individual members of the family adjust to each other. In a paper on the resolution of role conflict within the family, Spiegel[2] offers a classification of the technics used by parents in the restoration of equilibrium within the family. The value of such a classification is in the fact that it endeavors to structure what we take for granted, and it provides a rationale for appraising what goes on between ourselves and our children in the process of socialization.

These technics may be divided into 3 different groups. First, parents may attempt to induce or force the children into certain role patterns. Direct or indirect pressures may be applied to the child in an attempt to get the child to conform. The second group are technics oriented about modifying the role response of the child on a mutually acceptable basis. The last group are responses which may lead to a serious dislocation of the role patterns, so that the behavior becomes more or less distorted. These various technics will be considered in a very condensed way, not in terms of evaluating their merits but rather to call attention to the specific things parents do in order to influence the child's behavior. You will also observe that these technics may overlap each other or occur in combination. Nor are these technics all inclusive or definitive.

The induction technics may consist of coercion, coaxing, postponing, evaluating, masking or reversing the role as methods used to force the child to change his given position.

Perhaps coercion is the most common technic which parents use. It may be either physical or verbal or both and represents a hostile aggressive counter-role response on the part of the parent. As an example, a child may assume the role of a provocateur around the house. Mother may respond with "I will tell your father when he comes home." This, by the way, may place father in the unwilling role of being a villain. The counter-role response of mother carries a verbal and possibly a physical threat of punishment. In another example, a boy adopted the role of voyeur and witnessed the exhibitionism of a little girl. His father came upon the scene and thrashed him furiously. He submitted to this punishment and never dared to explore this interest again except by means of distorted sexual fantasies. The prevailing attitude of both parents blocked this boy's attempt to master this body management problem, and the child looked upon these feelings as evil yet tempting and to be repudiated. Eventually this boy reached adolescence and showed a complete psychological decompensation when he had to face the resurgence of his powerful sexual drives. Whenever coercion is used, a child may respond with defiance or with futility expressed as submission. Coercion may end in a lack of complimentarity, which perpetuates the disequilibrium within the family, or the role conflict may be settled by the submission of the child in which he accepts the complimentary role enforced by the parent. This latter solution may or may not lead to future problems.

Coaxing is a maneuver whereby parents promise, ask, plead, beg, and tempt the child so that he will fulfill the parent's desires. The child is inveigled to do the parent's bidding. However, this may tend to create a false sense of power in the child and establish in him the conviction that his omnipotent wishes will prevail. One may hear parents resort to the manipulations of awards, to be granted in the present or the future, consisting of attitudes or material benefits ranging from "a pretty please" to "you will get an automobile if you get passing grades." The child's answer may be either in the form of a counter-induction response such as hostile defiance, thus perpetuating the disequilibrium, or in the form of compliance in order to be assured the gratification. I recall a mother who would buy her child any material thing in order to get him to be more responsive and compatible. This only intensified his greed and hostility and did not resolve role conflict. He felt unloved by her and believed her gestures to be hypocritical. In this he was right, for she actually was masking another problem—one with her husband.

Postponing is a common technic used as a seemingly passive way of dealing with role conflict between child and parent. There is the hope that by delaying action, the child will change his mind or that intervening factors will arise to alter the situation. "It's too soon to decide." "We will see." "Let me think it over." "We will have to ask daddy when he gets back." These are the types of statements parents use with the hope that the child will change his mind. Often this technic works, but if the child feels hostile because of the postponement or for other reasons, he may attempt to gain his objective by being provocative.

Evaluation probably enters into every relationship which occurs between children and parents. It consists of such responses on the part of parents as approval or disapproval, praising, blaming, or shaming and comparing unfavorably. The evaluation may be overt or implied. It is used as a value judgement which has the quality of reward or punishment. As far as the child is concerned, evaluation always creates discomfort, for if the child is good he is expected to stay good, which is no easy task, and if he is not, he will be expected to change. As a result, he may respond with embarrassment and discomfort whether he is praised or blamed. I saw a child who took on the role of infant when her baby sister was born. She resorted to persistent thumb sucking and bed wetting. The parents attempted to shame her. The humiliating nature of their evaluation intensified her feeling of being unloved and also intensified her defiance. A more common observation is one in which parents frequently judge the school performance of one child by what another child is doing and in a self-defeating way aggravate the problem.

Masking invariably colors the transactional relations in a family, due to the presence of deceit or dishonesty. This is a technic which refers to the withholding of correct information or the substitution of incorrect information in order to manipulate a situation. Parents disguise their motives and may use little white lies in order to achieve their objectives. This is seen frequently in relation to sex information and what transpires between the parents themselves. The child gets unclear and confused information which will cause him to fall back on his fantasy life as a way of attempting to clarify his relationship to those about him. The inappropriate withholding of data always creates tension and hostility, with a disorganization of reciprocal role relationships. A child may attempt to unmask the parent, thereby embarrassing him. The child also may annoy his parents constantly in order to get information, or he may conform, knowing that what he has been told is not correct. A girl of 11 was told by her father that he would take her out for dinner while her sister, aged 8, had her friends in for a birthday party. In that way the older girl would have fun, too. This child responded with "you just want to get rid of me, get me out of the house," and she was right. The mother had told the father to do just that. Masking invariably creates an atmosphere of distrust and insecurity which restricts the development of reciprocal role relationships, and in that sense, it prevents the achievement of a high degree of complimentarity in the family.

Still another maneuver used in order to control the behavior of the child is to resort to role reversal, which is different from role clarification. In role reversal a parent may say "I think I see what you mean, but—" or "put yourself in my place—" or "I know you want to, but—" or "I see why you want to do this, and I will let you, but—." Such a stand may be used with the hope of intimidating the child or getting the child to postpone action. It may even mask the parent's real feelings, so that the child will go ahead and be hurt and "learn the hard way." The child also may be asked to put himself in the place of the parent, as an attempt to make him be sorry for his behavior, and conform out of a sense of guilt.

Role modification technics are different from role induction technics in that they are based on a reciprocal role relationship, between the members of the family, which endeavors to establish harmonious communicative procedures and achieve a measure of insight or understanding. These technics include humor, arbitration, exploration, compromise and role clarification. They are used to maintain or achieve complimentarity or equilibrium within the family.

Humor implies an appreciation of the other person's point of view. Humor refers to the ability to adapt one's self to the exigencies of a situation with a minimum of anxiety or discomfort. In humor the parent is able to reveal the quality in a situation that appeals to a child's sense of the ludicrous or exposes the incongruous side of a situation. As an example, a boy, 6 years old, says to his mother that he wants to be married to her. The mother may respond in one of several ways. She may feel flattered and be seductive, thereby reinforcing a role discrepancy the child is attempting to assume. On the other hand, she may say he is quite a young man, but he is daddy's and mother's son, and when he grows up he will find a lovely girl of his own to marry. In another incident, a child tells his mother no one loves him and she replies with a "yes, yes, yes, you poor underprivileged character, and you know, you just know, that is not the truth. I never heard such nonsense." Humor may be applied to a wide range of conflict situations as a means of lessening the intensity of the disequilibrium and in combination with other role modification methods usually proves effective.

Arbitration is a situation in which another person is chosen to settle the differences between the parties concerned by exploring the problem to effect a solution. However there is the danger that it might crate a split between the husband and wife concerning their real position in the matter. Nevertheless, there are times when a third party will see more clearly the nature of the problem and may present a solution which is mutually acceptable. "Go speak to your father about it," is different from "Let's discuss this with daddy and come to some reasonable solution about the matter." There is a danger, in that parents may mask their real feelings and use the situation as an indirect way to express hostile feelings toward each other, but when this is done in good faith, the child's response to the final decision will be more in keeping with the realities of the situation. Obviously, the more information the third party has available to him, the better is the chance for him to avoid conflict and to provide a solution satisfactory to the other role partners.

Exploration is a technic used by the parents and child in an attempt to reach a solution as to the nature of the role a child will assume through the study of alternative methods for handling the situation. Possible solutions are proposed and rejected until the participants come to an agreement. An example of this is the case of a child who wishes to use the family car for the recreational role. The parents and the child search for ways to handle this situation, since they also need the car. It may end up with the child getting the car or the family driving the child where he wishes to go or another parent providing the transportation and the like. The same type of exploration may go on as to the time the child will return home. Exploration implies a mutual respect for the wishes and goals of the other and a willingness to search for a solution satisfactory to all concerned.

Compromise refers to a settlement being reached in role relationships wherein mutual concessions are made. It means the settlement of different complimentary roles than those from which each one had started and a change in the goals each one initially wanted. Children often make unreasonable demands from which parents may negotiate with them until a reasonable solution is reached by all the participants. This does not imply submission in order to achieve "a little peace and quiet." Rather it refers to the existence of the mutual capacity to appraise the position taken by the participants and a capacity to come to terms by mutual concession, so that complimentarity is reached.

Role clarification appears to be the most successful method in establishing the highest degree of complimentarity in the family. It refers to the parent's ability to structure an appraisal of the problem in terms of the reality factors involved. This implies an understanding of the child's role position and an ability to evaluate it appropriately, using as a guide the personal, family and cultural role expectations. Although a child may not respond initially with agreement, eventually mutual agreement is reached because of the unwavering firmness, consistency and appropriateness of the position taken by the parents. The parents are the ones who determine what is in keeping with the proper response to a given situation, and it is their cognitive-normative analysis of a situation which will influence the course of the child's behavior.

The degree to which the parents recognize accurately the roles assumed by the child, and the nature of their complimentary role responses will determine how the child will manage a specific role situation.

I recall a girl, 5 years old, who insisted on wearing boys clothes and refused to play with the doll house and the dolls mother bought her. The mother was bewildered by the response of her daughter and felt helpless and disappointed in the face of her daughter's behavior. When the mother achieved cognitive data which permitted her to understand why this child resented being a girl, she felt less anxious about the matter. She also could tell her daughter, in different ways, on appropriate occasions, that she understood why her daughter wanted to be physically like her older brother and that it was mere nonsense to think that boys had more. This simply was not true because only she could have babies, and furthermore, she was a very attractive and charming girl, exactly what father and mother wanted. Not only that, she also could grow up to be a fine mother and have her own family some day. For several days thereafter, this child wanted reassurance and asked her mother to repeat her statements in order to evaluate the mother's sincerity and to be assured of her own physical and psychological status. Then for the first time she began to play with dolls and the doll house which was given to her by her mother.

It is observed in all these role modification technics that the process of consolidation goes hand in hand with them if there is going to be real change. When there is a more adequate redistribution of goals, this is followed by an associated redistribution of rewards. Consolidation comes with gratification or mastery of experiences, with learning how to work through new roles and discovering how to synthesize them. Consolidation brings together a closer union of the members of the family, and with this, the family functions at a higher level of complementarity.

Only a brief statement is possible about role dislocation. Roles may be repudiated by parents, as when they want to escape from being parents. Parents also may resign their roles and turn them over to substitutes. In addition, some parents frankly refuse to accept their roles, so that others need to replace them. Obviously, children subjected to such counter-role responses in relation to their needs are seriously handicapped in their adaptation because they have no one to turn to in order to develop role patterns relatively free from conflict.

Parents also may transpose to their children their own early life experiences by reliving their serious problems through their children. Such responses may be transitory or permanent, and the latter may lead to serious distortions in the child's development. Lastly, a parent may attenuate the role of the other parent by excluding that parent from his rightful role in the family, or by segregating the parent through aligning the children against him. Children in situations such as these are forced into narrow and rigid roles which impair their cognitive-normative functions and severely restrict their normal psychosocial development. This may be pressed to the point where, by attrition, no roles are left to the child which permit him to adapt to the inner and outer pressures. Then the child has no recourse but to resort to deviant behavior as an attempt to survive the untenable external world in which there does not seem to be a place for the development of gratifying role patterns.

Children function as a barometer reflecting the psychodynamics of family life. Their roles are complex and structured to fulfill their own needs, the needs of the family and the needs of society. Therefore, children exhibit an infinitely complicated interplay of biologic, psychological, family and cultural factors in which the child's role patterning represents a fusion of all these functions. Thus the role of a child in the family cannot be understood except in this broader context of cultural value orientations, the psychodynamic aspects of family life and the motivational processes which occur in the individual. Out of such a study it may be possible to formulate technics by means of which mental health principles may be applied on a more scientific basis, to both the family and society.

REFERENCES

1. Kluckhohn, F., and Spiegel, J. P.: Integration and conflict in family behavior (Report No. 27), Topeka, Kansas, Group for the Advancement of Psychiatry, 1954.
2. Spiegel, J. P.: The resolution of role conflict within the family, Psychiatry 20:1-16, 1957.

5

Emotional Impact of In-Laws
and Relatives

Nathan W. Ackerman, M.D.

Recognizing the gaps in present-day understanding of human development, breakdown and illness, it would be hard to know whether an attempt to assay the emotional impact of in-laws and relatives is a useful piece of intellectual daring or downright foolishness. At the outset I candidly confess that for this problem I can offer no definitive solution. At the very most I can aspire to clarify its meaning, delineate its depth and breadth and suggest a conceptual frame within which it might be possible to pursue useful forms of study.

The psychic bond of an individual with a family relative obviously holds the power to exert either a harmful or a healing influence; the effect on emotional health can be plus or minus. In a family in which the father is an alcoholic and a deserter, the son may form a compensatory attachment to a maternal uncle who provides a more satisfactory paternal image, a masculine ideal. The relationship of son and uncle neutralizes the harmful influence of the alcoholic father. What is involved is an equilibrium between two competing emotional forces, the bond of son with father, and a bond of son with uncle, both influenced by the son's attachment to mother. At the opposite pole there is a broad range of relationships with a relative which may exert a negative or injurious influence. For example, a man in marital conflict may say, "I love my wife, but I can't stand my mother-in-law." Or, one is reminded of the aphorism, "A mother-in-law who is inside and takes sides soon finds herself outside."

In clinical practice the range and the complexity of such family problems is legion: a depressive illness in a newly married woman, excessively dependent on her mother; the outbreak of an acute phobic neurosis in a young wife whose husband romances with his attractive mother-in-law; or a young woman who develops a postpartem psychosis after disappointing her parents-in-law with a girl baby rather than a boy baby. It is equally easy to call to mind analogous forms of disturbance involving aunts, uncles, grandparents, etc. The role of grandparents is especially problematic. Their experience and wisdom, once respected, are now disparaged. Grandparents are measured only for their nuisance value. Their advice is unwelcome; it is often resented, if not actually condemned. Yet, grandparents are exploited for their economic help or as convenient and unpaid sitters for the baby. Relatives may be used as a source of compensatory love and protection to offset a fear of injury from some other part of the family, or, at the other end, a relative may be used as a scapegoat, a target for hostilities displaced from other family relationships. The negative emotional influence of a relative is often alluded to euphemistically as part of the precipitating situation. Also, there are the curious emotional involvements with married friends who are taken into the family nest almost as if they were part of the family. A variant of this is seen in some suburban communities where two sets of marital couples exchange partners in sexual relations or even trade spouses legally.

In this connection we ought to be aware that the attachment of a family member to a professional person, a psychiatrist, a case worker, or even a clinic or a hospital, often carries a transference meaning. The bond with the professional helper need not always personify a tie to a parent; it may symbolize an attachment to an uncle, a grandparent, etc. Such a relationship provides emotional support or a channel for the release of displaced hostilities. It may also set up an influence which competes with that of other family bonds, in this way stirring divided loyalties. Such competing bonds affect the fate of personal conflict. The salient question is the extent to which the core conflict expresses itself in varying combinations of dependency, shifting and competing identifications, jealousy, hostility and emotional alienation, or acts as a spur to healthy growth and maturation. The specific influence of a relative may be explored within the frame of the group structure of the family, the role adaptation within that group and the personal history of the individual member. The emotional impact of the relative varies with such factors as status, age, sex, etc., and the influence on the family member varies with the chronologic age and the maturational achievement of the member. Whatever the pattern may be, there is always the fundamental value of the family bond, the powerful emotional undercurrent that pervades and colors all such situations: "When in need, look to your family"—"If your family doesn't care about you, nobody else will." Thus, despite any and all distortions in family relations, there is always the basic tie of kinship and the persistent sentiment of loyalty to family.

It is, I suppose, easy enough to take one's point of departure from a clinical hunch, to describe 1 or 2 dramatic case histories and from these to offer some tentative, theoretical speculations. Individual case study is par excellence the method of the clinician. It holds a place of respect in all the healing professions. More than that, the clinical hunch deriving from case history is the inspirational source of exploratory, empirical research. However, the fact remains that clinicians encounter different cases which they study in different ways. Therefore, it cannot be surprising that clinicians, however discerning they may be, draw from these empirical experiences varying and discrepant meanings. Up to a point, each of these clinicians may be right. Each of them, through a process of selective perception, highlights particular phases of the problem. In essence, it is one man's experience and views matched against another's, but this is a piecemeal process. It is exploratory. It is on the way to conceptualizing the nature of the problem; it uncovers some parts but misses others. By itself it is a method which, scientifically viewed, is incomplete. It does not encompass those other procedures so essential to the documentation of scientific truth: replication of observations and control studies by which hunches and hypotheses are tested and verified. The clinical case study approach, however fruitful in the hypothesis finding stage, is by itself insufficient for the establishment of a scientific law.

If, in confronting the problem of the emotional impact of relatives, we aspire to the spirit and eclat of science, we must at this moment concede that we are ill-equipped for the task. Surely, we have not yet accumulated the necessary and relevant information. We hardly can claim to possess the appropriate conceptual formulations, the scientific tools and the specific methods of research which might lead to reliable generalizations. In other words, we are a far distance from being able to achieve the power of prediction.

The emotional impact of in-laws and relatives is simply a special case of a larger category, the significance of family in influencing successful or failing patterns of adaptation, and thus influencing states of illness and health. In-laws are related by marriage, relatives by blood. The emotional effect of such persons may or may not be significantly patterned by this difference. This is a question of the way in which these relatives are perceived by family members, contingent on the family type, and the vicissitudes of individual development within that family type.

For the special problem of the emotional impact of relatives, we must mark out the significant points of reference. Relatives are persons occupying specific positions in the family constellation. The emotional expectations attached to such persons depend on the status and the role of a given relative within a given family. The relative has a definitive image of himself related to his family position, which in turn, is related to the images which every other family member hold of him. There is continuous interplay between the relative's image of self as relative, and the reciprocal images which other family members hold of this same person. The corresponding relationship patterns are molded by this ongoing interplay between image of self and image of other. It is self-evident that these processes are conditioned by the idiosyncratic features of a particular family group, its internal emotional organization, and its external adaptation to the surrounding community. There is the further factor of change over time affecting the individual family member, the vicissitudes of family role adaptation and the life of the family as a whole.*

Of course, within this context, it is important to know to what extent a given relative is perceived as belonging inside or outside the family, that is, as being a member of the in-group or the out-group. Since by definition a relative occupies a fringe position, he may be felt to be inside the family at one time and outside at another time. A relative living under the same roof may be emotionally excluded or exiled from the family, whereas a relative living in another physical home may, nonetheless, be psychically included within the nuclear family group.

Now then, how may we approach this question of the emotional impact of relatives? Is this a scientific posing of the problem? What impact? On Whom? Surely, the question calls for sharper formulation. We must break it down to *what, where, when, how* and *why*. This is not a 1-person problem; it is a relationship of 2 or more people. The emotional influence is circular, not linear. Therefore, we must conceptualize this entity as a circular, interpenetrating process in which relative and family member alternate the roles of sender and receiver of influence. We cannot examine meaningfully only one side of the equation. We must scrutinize the relative both as sender and receiver of influence and the family member also as receiver and sender. The validity of this viewpoint is confirmed in the principle of examining a mother-child pair as a 2-way phenomenon. The child influences the mother, the mother influences the child. Moving a step further, since we now evaluate the dynamics of mother-child interaction within the broader frame of transactional processes of the family as a whole, we cannot do less with the interaction of a family member with a relative. It matters not whether we are concerned with the emotional impact of mother-in-law, brother-in-law, aunt, uncle, or grandparent; the issue is the same. The phenomenon must be viewed in a group context.

The pertinent questions are multiple. How does the emotional joining and interaction of an individual with a relative affect the role behavior of each partner in this pair? What is the influence

---

* See Ackerman, Nathan: Chapter on Role and Personality *in* The Psychodynamics of Family Life, New York, Basic Books, 1958.

of the family group on the emotional identity of this pair? What is the effect of the pair on the identity of the family as a whole? Does it set up opposing authority representations? Does it alter the alignment of family relationships? Or is the bond with the relative exploited, so that hostile emotion against the authority representations in the family is displaced on to this relative? Does this bring about an emotional rift in the family, or does it alter previous configurations of family rift? These are only a few of the issues which are germane to the problem. Ultimately we reach down for an answer to the narrower question: the specific emotional influence of a relative on the particular family member who happens to be our primary patient.

The problem is a complex one. In order to understand it clearly, a conceptual framework is required, a tentative theory within which to study the relevant questions.* Therefore, I offer in condensed form a tentative theory of family dynamics by which it may be possible to understand the emotional interconnections of disturbance in one person with the psychosocial configuration and mental health functioning of a particular family group.

For this purpose I have devised a group of core concepts which attempt an operational formulation for the dynamics of family process, the who, what and how of family life, and the corresponding functional patterns of family relationships. Within this conceptual framework one can strive for more reliable correlations of individual and family behavior. The essence of the theory is represented in the following schematic diagram, a 6-pointed star in which one set of triangular variables is superimposed upon another.

One phase of this conceptual scheme deals with identity and differentiation in the on-going relations of individual and family; the other phase deals with the stabilization of behavior, which is influenced both by the internal processes of personality and interpersonal adaptation between family members.

The concept, identity, subsumes the image of self and associated strivings, expectations and values. The concept, stability, involves

---

* See The Psychodynamics of Family Life, Basic Books, N. Y., 1958, Chapters on "Family Identity, Stability and Breakdown," "Clinical Aspects of Family Diagnosis" and "Behavioral Disturbances of the Contemporary Family."

the continuity of identity in time, the control of conflict, the capacity to change, learn and achieve further development, in effect, the quality of adaptability and complementarity in new role relationships. The concept, identity, refers to the direction and the content of striving, while stability refers to the organization and the expression of behavior in action. The questions, *who, what* and *how,* and the corresponding formulations of identity and stability, can be applied to the functioning of individual, family pair, and family group. A joined pair of persons or group may be conceived as possessing a unique identity just as does the individual. Psychic identity changes as it evolves through time. It answers the question: *Who am I* or *who are we* in the context of a given life situation. It orients personal strivings to relations with others. At any given point in time, the individual has an image of his personal identity and his family identity, both continuously influenced by the images which outside persons hold of these same identities. The identity of a family pair or group refers to elements of joined psychic identity. This is a segment of shared identity, represented in layers of joined experience and enacted in the reciprocal or complementary role behaviors of these joined persons. It is this process which determines the manner in which elements of sameness and difference among the personalities of family members are held in a certain balance.

Psychological identity and stability of behavior must be considered together. I have already alluded to the component processes which mold stability: the continuity of identity, the control of conflict, the capacity to change, learn and achieve further development; i.e. adaptability and complementarity in new role relations.

Stability in its first phase epitomizes the capacity to protect the continuity and the integrity of identity, under the pressure of changing life conditions. It insures the intactness and the wholeness of personal behavior in the face of new experience. This is the conservative phase of stability. The other aspect of stability must provide for accommodation to new experience, learning and further development. It represents the potential for change and growth. Effective adaptation or homeostasis requires a favorable balance between the protection of sameness and continuity and the need to accommodate to change. It requires preservation of the old coupled with receptivity to the new, a mixture of conservatism and readiness to "live dangerously." The capacity to change and grow is a feature of family life as well as a feature of individual development.

Stabilization of behavior is influenced by the capacity to cope with conflict. The control of conflict is a special dimension relevant to the relations of individual and family. The failure to find effective solutions leads to adaptive breakdown and illness. Within the individual, pathogenic conflict and the vicissitudes of control and restitution correspond most closely to the bases for clinical psychiatric diagnosis. If we trace the fate of salient conflicts within the individual and between family members, we can trace the relations between adaptive breakdown and illness in one individual and pathogenic disturbance in the family pattern.

In this connection it is of special importance to define the individual's capacity to achieve complementarity in family role relationships. The term, complementarity, refers to specific patterns of reciprocity in family relationships that provide satisfaction, avenues of solution of conflict, support for a favored self-image and buttressing of crucial forms of defense against anxiety. Complementarity may be differentiated further as positive or negative.

Positive complementarity is that form which promotes emotional growth of the relationship and the interacting individuals. Negative complementarity is more static; it signifies a buttressing of defense against pathogenic anxiety but does not, additionally, provide the potential for further emotional growth. In this sense negative complementarity mainly neutralizes the destructive effects of conflict and anxiety and barricades family relationships and vulnerable members against trends towards disorganization. Contemporary patterns of family conflict may potently affect the outcome of individual conflicts internalized at earlier periods of development. It should be remembered, too, that families as groups have characteristic forms of defense just as do individuals. Therefore, it is necessary to trace the interplay between individual and family group defense against the disintegrative effects of conflict.

Within this conceptual structure, it seems possible to outline the pathogenic areas of family conflict, those which push the stabilizing or homeostatic functions of the individual toward decompensation and thus aggravate the tendency toward disorganization, regression, breakdown of communication and emotional alienation; also, at the opposite pole, to define the potential in the reciprocity of family role relations for providing paths for solution of conflict, establishing effective compensation or complementarity, fostering support of new levels of identification, thus promoting health and growth. An approximation of these opposite trends may enable us to view more clearly the balance of forces within individual and family that predispose to illness or health and positive development.

Within this theoretical scheme let us be as clear as possible concerning the concept of family identity. Family identity is not, and cannot be, a fixed or pure thing. It is crystalized out of the fluid, on-going processes of multiple competing and co-operating partial identity representations. When 2 people marry and set themselves to create a new family group with offspring of their own, they bring with them as individuals a personal identity linked to the identities of their original families. At some levels, these identity representations conflict; at others, they collaborate, and they may even merge. Therefore, family identity represents a fluid, continuously evolving psychosocial process. It pertains to a dominant identity pattern composed of joined goals, values and strivings in a life context where there is perpetual competition of partial identities and values. In this sense family identity cannot be a clear, single, unadulterated entity. It is the result of a dynamic evolution of ever-new integrations of shared identity created out of competing components of identity. It is in the very nature of the family phenomenon, that each individual member must reconcile his personal identity and values with the shifting representations of family identity. This is deeply influenced by the on-going processes of integration of the individual's personality into changing family roles. In this context one can readily understand the significance of family in molding the individual's success or failure in coping with conflict, either new or old.

On this background it is clear that the union of a family member with a relative may affect family functioning either toward greater unity of family identity or toward deeper disunity. Thus the emotional pairing of a family member and a relative may exert a positive or negative influence on family equilibrium and the evolving patterns of family identity. Of special significance is the potential of a union of member and a relative for setting up a pair identity and corresponding strivings and values which compete with other partial representations of family identity. In the long view, this may have integrative or disintegrative effects. It may undermine the influence of other family representations or serve as a kind of referee and controller of family conflicts. This especially may be the case where a family group is threatened by a critical emotional split such as, for example, a rift down the middle, the mother joining the son, the father joining the daughter. In such a case the link of a family member with a relative may protect a particular family from total disintegration.

In this setting a relative may conceivably fulfill the unintended role of psychotherapist for family disturbance. In fact, a professional person, whether psychiatrist or case worker, is called upon precisely to serve the function of surrogate for a family relative of good will. He may, in fact, pinch-hit for a grandparent, serving as guide, mentor, etc. Thus the union of family member and relative may act as a defense, an antedote to disorganizing tendencies within the family group. He may also function as a catalyzer of forces moving toward higher levels of personality integration and higher levels of positive emotional health. On the other hand, the union of member and relative may play the part of aggravating the tendencies to breakdown of family unity. In effect then, the importance of this pairing is that it may influence the homeostatic equilibrium of the family identity in either a minus or plus direction. It may serve as a neutralizing agent in pathogenic conflict or may aggravate it.

The specific effects upon an individual family member may be direct or indirect: the immediate personal impact of the relative on the family member and the indirect influence of the pair which is exerted through the more complex interactional processes of the family as a whole. On the side of pathology, when a family is emotionally divided and split into opposed factions, the pair relationship of a family member and relative can play a most potent part in determining the ultimate fate of conflict, intra-psychic or interpersonal. When one part of the family is set against another, the emotional union with a relative has the power to make or break the stability and the health of an individual family member. It can precipitate illness or protect emotional stability.

To make the issue concrete let us take the following example: at some levels of family experience the emotional union of a child with a grandparent may crystalize a set of goals, values, an ego ideal, which clashes with the child-rearing standards of the parental couple; or the relationship of a married woman with her favorite sister may pit one kind of identity and life striving against a competing bond between the woman's husband and father. However, at other levels of family functioning there may be a shift, so that these partial representations of family identity merge. It is in this sense that the integration of a dominant family identity is a continuously evolving process.

In the mental health field today, the hot breath of pressure to grapple with the vexing, complicated processes of family life is burning our faces. The call to explore these questions is a commanding and irresistible one. The core of the special problem of the emotional impact of a relative rests in the processes of identification and homeostasis. Relatives are extensions of family. They are identity bridges with the bigger world outside. They are the agents which link homeostatic balance in the internal life of the family with homeostasis of family with the larger community. To do justice to this challenge, we must be clear about two things: the climate of contemporary family life and the conceptual trends which presently emerge in the realm of behavior theory.

The features of family life sensitively echo the radical change that characterizes modern society. As Halbert Dunn expresses it, our world is a shrinking world, a more crowded world, an older world, and a world of mounting social and political tensions. Industrialization, urbanization, increased social mobility, have brought with them a push toward the collective man. The balance of power between individual and group is shifting. Society imposes upon the individual a greater obligation, but allows less freedom, less privacy. As the world changes, so do families and people change. Society is experiencing a critical integrative disturbance. This becomes reflected, inevitably, in a parallel integrative disturbance in individual behavior. It is the mark of our time that people are cut off from the secure traditions of the past, they cannot foresee the future, they hang precariously in the present moment. People experience a great difficulty in pulling themselves together. They experience themselves in parts rather than as intact, whole beings. They have a distressing struggle in synthesizing a personal identity which is congruous with the requirements of modern life. Under these social conditions it is difficult to know what is real or not real, what is appropriate or inappropriate. People are chronically unsure; they feel lost, alone, disoriented. The tasks of problem solving, decision making, and action, become especially difficult. In the attempt to feel more whole and more complete, people seek to join a part of their identity with that of other persons. Problem solving, decision making, and action, to this extent, are less a function of the individual and more a function of shared and joined identity with other persons. Conflict is no longer snuggly contained within the one person but is rather externalized and lived out in the zone between the person and the environment. It is this externalization of conflict which brings into prominence such defenses as projection, magic thinking, substitute of aggression for anxiety and "acting out." All forms of action are some combination of real and unreal perception, appropriate and inappropriate feeling. In this social context there is potentially an element of "acting out" in all forms of action, and "acting out" is itself a function, not of the isolated individual but rather of 2 or more persons with a component of shared identity.

On the one hand there is the special problem of an integrative disturbance in the behavior of individuals, and on the other hand there is a tendency to emotional alienation and isolation in the functioning of both individuals and family units.

In this social setting the structure of the family shifts from the larger unit, the tribe or clan, to the smaller, procreative group, the nuclear family unit. The trend toward alienation and atomization deeply encroaches upon the inner life of the nuclear family group. There is mounting tension and conflict among the parts of the family group itself. Witness the effects of the social, cultural change on the family entity in Nazi Germany, Soviet Russia and China. The influence seems to be in the direction of piecing apart the relations of man and wife, parent and child. In some parts of the world, one part of the family is set strongly against another. This dismembering effect on the unity of the family is of the greatest consequence for those forces which have to do with effective or failing adaptation. In our time psychiatric disablement is significantly influenced by this atomizing aspect of contemporary family life, which isolates the individual, sets man against woman, child against parent, and challenges each individual to find security not within the family but somewhere outside.

In a subtle manner this special tension has spurred in family life a defense reaction, a trend toward compensatory family togetherness. It is as if people sense the insidious danger of fragmentation of family unity and respond with a strong effort to fortify family cohesion. Yet this spontaneous effort to safeguard the sanctity of the home is itself fickle. In some places it works; in other places it fails. It also brings new complications. In describing this family change some authorities, like Mowrer, refer to a trend toward disintegration of the family; others, like Talcott Parsons, speak of the "disintegration of transition," which expresses the belief that the family will really not be destroyed but will reorganize itself in the new form to fit the conditions of a changing society.

In any case the essential functions of family are profoundly altered. The family no longer epitomizes the center of experience for work, for education, for religious practices, for nursing of the sick. Instead, it is becoming the final fortress within which people hope to protect their emotional and ethical integrity. Ernest Burgess calls it "the companionship family." Erich Fromm characterizes it as the "psychic agency of society," a specialized agency for the protection of emotional health. It is on this background of radical transformation of family pattern that we must consider the interrelations of culture, family configuration, individual development, and the special question of the emotional impact of relatives. What is universal in the relations of the individual family member with a relative? What is prone to vary in these relations among different cultures and in the varying subcultures of our society.

A corollary trend which bears on this question is the changing climate in the behavioral sciences. The modes of thought and the fashions of behavior theory are in a high state of flux. Definitions of illness and health are undergoing rapid change. The dynamics of the individual are now placed within the broader framework of the dynamics of the group. Intrapsychic functions are correlated with interpersonal events. We no longer dichotomize individual and family. We find ourselves on the threshold of a new conceptual adventure, the building of an integrated theory of family which encompasses the dynamic processes of individual behavior. This development is in keeping with field theory and with the urge to make use of a holistic, dynamic approach to behavior as against an atomistic, mechanized one. We now make use of a comprehensive biopsychosocial model of personality, rather than an individualistic, biologistic, mechanistic model. This shift in theoretical orientation requires us to rethink some basic problems in human behavior, the continuity of human experience at the levels of individual, family and community; the essential unit, of past, present and future time in the molding of adaptation; the merging of old and new in human development; the special problem of homeostasis and the related questions of learning and plasticity of adaptation.

Of course, life stress and the tensions of family relations are surely nothing new. However, what is new is the unique quality of stress that prevails in family life, the extraordinary pace of change in family and community patterns. Of particular significance to clinicians is the fact that such tensions are filtered through the organizing patterns of the family. The response to such stress can be defined only through the integrated network of family relationship processes. Therefore, the approach to mental health through the dynamics of family life is of central importance.

6

The Impact of Aging
in the Family

D. Griffith McKerracher, M.D.

TODAY, THE TROUBLES OF THE AGED disturb the family. The elderly
individuals, their children and their grandchildren will all suffer
unless we can understand and resolve their problems.

We know that the difficulties of age spring from three chief
sources: (1) conflict among the generations, (2) ill health and (3)
the hurdles which modern living has created for the aged.

Parents and children always have spiced their mutual affection
with disagreement—from the day when the first young man stalked
in anger from his father's cave and moved into a cave of his own.

However, during the past century, as our way of living became
more complex, we have seen these troubles grow worse. Where
formerly the difficulty of getting on with one's offspring and of adjusting to failing health were all important, now the elderly individual must also struggle with where to live, how to pay and what
to do with their time.

No members of the family group can be in difficulty without repercussions on the remainder. Euripedes said, "The gods visit the
sins of the fathers on the children." But we know now that children also acquire their parent's anxiety. So, we shall study what
irks the aged and suggest what might help.

No one who can read remains totally ignorant of the problems of the aged. Usually, the older one is, the greater one's interest becomes. Even if you are more concerned about the young, you will learn much from studying the aged, for their difficulties are but exaggerations of the problems facing mankind and hence the family.

My own interest stems not only from feeling that "age with its stealing steps hath clawed me in its clutch" but also from my clinical experience in the field of psychiatry, for in the mental illness of the aged, those things about the elderly individual which so bewilder the young and the middle-aged stand forth in bold relief. Under family stress personality quirks may even blossom into psychotic symptoms.

In my discussion "the family" will include not only parents and children living together but also grandparents, whether living in the same dwelling or as an extended part of the group. Although the aging process starts in the cradle, here we shall restrict the use of the term to the latter part of life—the age of 65 and over. This we do for the sake of clarity, being mindful of the limitations of dating old age so definitely.

UNDERSTANDING THE PROBLEMS
OF THE ELDERLY INDIVIDUAL

We would understand better the effects of aging if we knew "what makes people tick." So we need a workable, conceptual framework of behavior. Many conflicting theories exist, but fortunately most agree on the nature of human drives. Man struggles for security, love, recognition, achievement and physical satisfaction. Let us examine how the elderly individual keeps the other family members from meeting these needs.

When frustrated, we become anxious. We all have ways of defending against anxiety, and families, like individuals, also develop their own patterns of keeping down tension. Three things determine how successful these defenses will be: (1) the emotional maturity of its members, (2) the degree of the stress and (3) outside help.

How does age create stress? Let us look at a family during the passing of a generation. Look first at the middle-aged head of a household. Life for him is then at its best, for the prime of life is the time of life when needs are most readily met. Usually, by now, his income is big enough. The love which his dependent children hold for him has yet to feel the blows of adolescent revolt. He still basks in the recognition of his peers as, with pride, he surveys his continuing accomplishments.

However, look again after 25 years. Our middle-aged man is now an old man. The love which his children bore for him was shaken rudely during their successful struggle for independence—a period which left a trail of battle scars destined never to heal completely. He is no longer the family pivot, for now their love has shifted to their own small children whom they cherish with a biologically determined intensity. They retain for the aging parent only a culturally dictated sense of responsibility—a frail tie indeed. Based on the commandment to "Honor thy father," the affection of the second generation for the first will stand only limited strain. So, within the family circle, the dependent aged can never compete on even terms with the dependent young—they wage a losing battle, for the young have a future, and they do not.

HOW DOES THE ILL HEALTH OF THE AGED AFFECT THE FAMILY?

Physical breakdown must occur, whether suddenly or gradually, since age is always the final lap in the race of life, with death standing on the finish line. Therefore, ill health ranks high among the problems of the aged; it upsets the equilibrium of the family. Of all persons over 65, half are incapacitated by sickness. Illness confines 10 per cent of the elderly to their homes and another 3 per cent to institutions. In America this last group alone comprises 500,000 individuals, half of whom are in mental hospitals, the remainder in nursing homes and homes for the aged. Not all of the institutionalized aged have children but many do. Here the sickness of the elderly individual has distant repercussions in the family setting. Families vary greatly in their reactions to older members being institutionalized. Some are satisfied as long as the older member is in a quiet church home for a physical illness, others feel shame and guilt, especially if the aged parent is in a state hospital.

Any illness handicaps the older person's effort to adjust—for one thing his pain makes him anxious. Often the illness directly or indirectly affects his nervous system disturbing his emotions and his intellect; cerebrovascular disease especially confuses the old, distressing the family.

Illness cuts earnings and makes many old people dependent. They often react to this dependence with irritability and stubbornness. Disfigurement of the aged by illness threatens self-esteem. The social contacts of the elderly person, limited by his own sickness and by the death of his friends, throw him back on the mercy of his family for companionship; this he may resent in a spectacular fashion.

The sickness of the aged parent generally annoys his middle-aged children. Its impact on the second generation differs from the anxiety felt over the illness of a child. It usually hurts the family financially, as the aged, short on comprehensive insurance, become less able to pay bills. A sick grandparent limits the family's mobility and, since most American families love to go and to do, responsibility for the sick creates a restraint seldom graciously received. As the level of annoyance and frustration rises in the family, a feeling of guilt usually follows, especially in the presence of subconscious wishes for the ill~~ess to end with "The Great Release."

While this picture of rising family anxiety, rejection and guilt in face of the older member's illness is common, it is not universal. Many, if not most, families effectively meet the challenge of this problem, providing adequate care for their sick aged. None the less, the impact on the group of the sickness of the aged never can be lightly dismissed.

HOW DOES MODERN LIVING THREATEN THE AGED?

From the problems common to all family life, let us turn now to those which result from modern living. The industrial revolution almost ruined the standing of the aged in the family; let's see how this happened. Formerly, the home was the center of living, and the aged one was the head of the house; the farm life of the last century illustrated this best; but increasing industrialization moved the center of things from the house to the factory. So to others outside the home, the senior member of the family transferred the job of providing the necessities of life, of protecting property, of nursing the sick and, above all, of education. Until the center of gravity shifted from the home, the oldest member of the family knew most about these essential activities. Only he could pass on useful information to the next generation; he held the key to the family's future, but, when the factory took over, others assumed this role. Thus the head of the family lost much of his value to the group; he lost the security which his learned position had given him; he lost esteem; he lost his sense of importance and of achievement; and, not too surprisingly, having lost worth, he lost love.

So, in the new scheme of things, the advantage passed from the aged to the young. The complexity of the new machinery was mastered more easily by the young mind, for it required scientific education which, in turn, made learning from books essential. Mathematics eclipsed manual skill; hence the practical experience of the old was of little value. Since today's living is geared to the new knowledge, youth now holds the high cards. So, despite the carping criticisms leveled at modern education, it does equip for today's living. Thus this was the first blow dealt to the aged by social change; the patriarchal life of the family was replaced by a new democracy in which the contribution of the former leaders had become much less significant.

The second blow was the relative increase in the number of aged, compared to children. Science also contributed to this, since disease control kept more old persons alive, while birth control reduced the size of families. So, quite understandably, youth became glamorized and the aged devalued. This greater worth of the young inspired legislation to protect the child, and family agencies, while claiming interest in all ages, concentrated on the welfare of children.

The shift from home to factory-based economy brought other important changes affecting the aged and family life. New ways replaced traditional methods, and tradition lost its former position of respect. New ideas increased efficiency, so old ideas fell like tenpins. With the drop in the influence of tradition, the influence of religion became less. Similarly, the moral code of the past lost ground, and what the aged had considered wrong now seemed right. It became more acceptable for the young to dance, to smoke, to drink liquor, to form sex habits previously taboo, even divorce became respectable. However, traditions, religious structure and moral codes have always protected the patriarchal family and given great support to the aged in family relations. As these vanished, the aged lost ground.

Thus the social changes from the industrial upheavals have reversed the importance of age and youth, especially in middle-class families. Where formerly the aged were an asset, now they can be a handicap, and the expectation of a useful old age has changed to a dread of becoming a burden. Hence, the social changes of recent years have created problems for the aged—disturbing to the second generation.

Where the old person lives, affects his impact on the family group. Most aged live with spouses in quarters which they control and, all things being equal, this is best for everyone. Sometimes it costs more than the old person can afford, and the family has to supplement, to the annoyance of all concerned. When the aged are feeble, housekeeping chores may be a burden. This distresses the offspring and may lead to overhasty action to bring the aged to live in the family home.

When one of the elderly married couple dies, and the survivor lives alone, the situation changes little as far as the family is concerned. Here still remain the advantages of the separate dwelling and independence for both generations. True, loneliness and isolation may enter the picture, but for most aged these are not as bad as hostility; for the elderly individual can maintain dignity even if lonely. However, when old persons live at great distance from their children, anxiety and loneliness do increase with time. Then the threat of illness, with expensive travel, enters the picture.

The next alternative is for the aged to live in the home of a married offspring. In the past, this was the common practice, although, when elderly individuals were less numerous, and more revered, this arrangement was better than now; at that time houses were larger and families lived closer together. Today it is fraught with potential difficulty, since the old parent and the middle-aged child retain their scarcely buried conflicts of former years. These are plowed up through daily contact. In such situations mother and daughter will often clash noisily, while even more threatening is the quieter hostility between mother and daughter-in-law. The old man living with his married offspring has problems all his own. Nor do the children in such a household escape these tensions. For the child is often either the pawn or the weapon in the battle between the first and second generations. Today's smaller homes with their lack of privacy increase the difficulties, for the TV set has crowded Grandma right out of the chimney corner. Thus the virtue of having the first and second generations live together comes chiefly from the money saved and in having at hand nursing care for the feeble aged.

Except for the cost, the boarding home has fewer drawbacks. Sometimes old persons even prefer to board in hovels rather than share better quarters with their children.

One is often deceived by the apparent calm of the old folks's home. Here the relief from no longer living with children and grandchildren sometimes obscures other problems. The financial burden may exceed the resources of the elderly individual and of the family; privacy is rare in such communal life. Suitable homes for the aged located so as to maintain family ties are scarce. Many of today's nursing homes for the sick aged are overcrowded, cheerless places with little medical care. The poor ones bring small comfort to the patient and to the other members of the family who often deprive themselves to pay the bill.

Today's state hospitals, which house so many old, do great violence to the aged and to the family group. Distant, overcrowded, understaffed, dilapidated concentration areas, unfortunately they provide a distinctive but unhappy symbol of our civilization. May the justifiable guilt they generate, prick our collective conscience towards constructive action!

Old people suffer from a lack of money, and this disturbs the family's equilibrium. Although this century's pension legislation has partly readjusted this imbalance, the aged still have much less money than has the second generation. Probably 25 per cent have less than a satisfactory minimum, while between 5 and 10 per cent are in serious want. This financial insecurity of the aged augments family tensions.

The source of the older person's income affects his role in the group. To own his own business or other property brings him the greatest prestige, but the 20th Century's expanding and fluctuating economy tends toward early transfer of the control of the family's business to the second generation. While this may be wise, it usually hurts the old man. Not only does he lose power, but also the recipients often feel much guilt which they express in ill-disguised hostility. Recently, I tried to persuade a prosperous butcher to accept responsibility for his aged father. He had attributed his own resistence to his memories of being abused as a child by this father. I thought a simpler, truer explanation was that he had paid his father nothing for the family business.

So the change from the patriarchal to the democratic family pattern removes control earlier from the first generation. This early shift of responsibility places the old man in a situation comparable to the lameduck status of a president in his last term of office who, having lost his bargaining power, is now more lightly regarded. While most parents today spend to educate their children, rather than leave a property legacy, education, too, may create a barrier between the aged and their adult offspring even greater than does a transfer of property. Whatever the source of income, the financial state of most older persons today troubles the family. When the funds come from wages the possibility of job loss threatens security. Since pensions die with the pensioner they add little to his negotiating power. Social aid has low prestige value at home and often increases the anxiety of the elderly person. Yet direct family help, accounting as it does for 15 per cent of all income of the aged, may increase resentment both of the giver and the recipient. While it may seem wrong to put a price tag on the older person's position in the family, experience shows that it is realistic. Much of the trouble of the old man today comes from the disappearance of his financial power.

How has downgrading the role of the elderly individual in industry affected the family? Since the shift from an agricultural to an industrial economy, the aged have had a harder time. When people cultivated crops, the old man made his best contribution, whereas he fared worse in cultures based on hunting. Today's mobile, competitive economy with high pressure selling resembles the hunter's life and so the aged do badly. In industry today, the accent features change. Old jobs are discontinued—new ones created. The cry is for employees to learn new habits quickly without much help from past experience—so elderly persons suffer.

The complicated modern machines create hazards for those who are slow of foot and eye. All this pushes the aged slowly, but inevitably, out of the employment stream, while even those with adequate competence come under the axe of forced retirement. So, less than 40 per cent of people over 65 are now employed— a third of these work for themselves.

Now, what happens to the aged when no longer employable? This disturbs some less than others, especially women, content with their domestic skills. For them, self-esteem is not usually tied to a steady job. Also, some older persons relish retirement, especially if they have absorbing interests and are financially independent of the family. However, a great many face retirement with apprehension; at the very least many people want part-time employment. Idleness is tedious and threatening to most old folk, for the loss of income not only disturbs security but also diminishes prestige in the family group. Not only the spouse but also the second and the third generation feel the effects when the old head of the family loses his job. Then the part which occupation normally plays in maintaining morale is lost, hurting the spirit of the whole family.

HOW SHALL WE PLAN TO MEET THE IMPACT OF AGE IN THE FAMILY?

We have tried to understand the reactions in the family to the problems of aging; now we must use our understanding. To make a scientific attack on these problems, we need a plan. This plan must advance the welfare of the family as a whole (including the aged). It must be democratically formed, not only by those it aims to help but also by those who have responsibility for family welfare. Here I include Governmental and private agencies, and I would also involve physicians, social workers, sociologists, psychologists and the other professional people who have studied the family.

We must help the aged (and their families) learn how to deal with their own difficulties; for the elderly individual, like all others, gains by successfully solving his own problems. We should assist only where needed, for most aged people now get on fine with their own children. Only about one fifth are not able to do so and need our help.

HOW CAN WE HELP THE ILL AGED?

Prompt return to the community, rather than permanent hospitalization should be the goal for all sick aged. Better hospital conditions at less cost would spare the family stress and crippling expense. Of all people, the aged need the most and the best medical care, yet they are the least able to pay for it. We always must treat remedial disabilities regardless of age; the family, too, will benefit if the old have proper visual, hearing and dental aids. Where money is short, state or private groups must step in, for elderly persons need complete health coverage.

The family doctor should control and co-ordinate the health program. Properly equipped, he could assist the family in complete management of the old person's illness. He needs a good hospital for the acute sick; the mentally ill should have the same standard of treatment as the physically ill. This means that the mentally sick, as well as the physically sick, should be treated by the family doctor in the community general hospital. This he can do with adequate training and with proper support from psychiatric specialists, for none is better placed than he to help the family through all stages of illness in hospital and in the community.

Today we decry the tendency to shove people into institutions —especially big institutions. We oppose permanent banishment of the sick from their homes. Cozin of Oxford, with his home-care program, leads the attack against a passive role for the sick aged. He is a successful exponent of the doctrine of rehabilitation for all convalescent elderly persons, including the mentally ill.

The stigma of the state hospital probably can be dealt with best by trying to eliminate that institution altogether. The full possibilities of treating all the mentally sick aged in general hospitals and in the community have never been explored adequately.

In Saskatchewan we are studying a plan to replace the old Provincial mental hospitals by 300-bed regional units attached to the local general hospitals and of the same standard. Here, along with the other mentally and physically ill persons, the aged psychotics would receive treatment for acute sickness—then sent back to the community as soon as possible.

Usually the aged, when convalescing from any type of illness, can return to their former setting with spouse, offspring or landlady. While the convalescent patient often worries the family, an adequate home-care program would lessen this anxiety. Inevitably, we must organize home care or else banish the sick to institutional life for long periods. Subsidized home care enables the family to look after its own sick aged people. Home-care programs provide visiting hospital physicians who work in close co-operation with the family doctor. They also make available visiting nurses, special duty nurses and housekeepers. Trained volunteers can help in special circumstances. With such aid the family can keep the sick aged without undue burden. When I have asked families to take home convalescing elderly persons, they usually will if assured of prompt backing when in trouble.

Nursing homes become necessary when direct return home is impossible. Here the state must correct a major deficiency; it must provide more and better nursing homes at costs the family of the aged can meet. These should be situated for close contact with the family and the elderly individual should leave the nursing home to return to the community as soon as able.

When the aged become ill the treatment plan always should include re-employment if at all feasible. Like other sick members of the family, they should have sick-time insurance, with this designed to encourage early return to work.

Planning for this re-employment should provide some part-time work in sheltered workshops. This will require the co-operation of employers and of unions. However, the good from gainful occupation alone makes the effort worthwhile, for the family always will profit when the sick older person can look forward to a future life of usefulness rather than idleness. The partially disabled should be able to work for pay in their own homes; trained volunteers could help organize this. No problem of aging presents a greater threat to the family equilibrium than illness. We know how to cope with this but do little about it; however, action should start now.

HOW CAN WE HELP THE OLD ADJUST?

The family suffers when anything restricts the social life of the aged. When the lonely, isolated older person concentrates too much on the gloomy side of his problems, this hurts the group about him. So we must give the aged person a reasonably attractive role to replace the one he has lost. Through industrial progress we have shown how the shift from the patriarchal to the democratic family has shattered the prestige which the aged used to have. So we must now make an honored place for elderly persons and grant them a chance to reach it, for age is really a part of normal human experience—just as vital as youth and in some ways richer. Like all others, the aged have definite social needs— needs for contacts with people and for absorbing interests not directly related to the business of earning a living.

Let us consider 4 important outlets for man's social drives—the church, cultural and community interests and recreation. Better than the others, the church has supported the aged one during his losing battle with the glamorized young; it will continue to support him. I would like to stress the potential benefits to the aged of cultural activities. That the aged can learn we know; in fact, they have to learn to adjust to their new situation. Adult education for the 65-year-old person, could sharpen his wits and increase his contribution to the family. The arts provide especially attractive possibilities here. He should cultivate his talent, however meagre, in music, painting or writing. Mutual cultural interests would strengthen his ties subtly with the rest of the family—ties which we must preserve. Such interests also would give him outside contacts.

Community activities could interest many aged. Some who now feel useless should volunteer to help in service clubs and family agencies. Politics offer an interesting challenge to others, since the stake of the aged in governments is greater now. More elderly voters means more political power for the older politician. This he could use, not only to the advantage of the pensioner but also to help work out other problems which today confront the whole family. The need for better recreation faces all who now have more leisure time. Senior citizens clubs, properly run, could supply companionship for the more gregarious. So while the aged have suffered socially by civilization's switch from family to factory, we could do much to fill up the gaps. Only this will keep the family from paying too much for its new found freedom.

WHERE SHOULD THE AGED LIVE?

As age increases, where the older person lives concerns the family. Although, the aged person himself usually can decide where he will be, yet sometimes the family must try to alter his decision. All must realize that what is decided today may have to change with tomorrow's circumstances, and even then this decision is never final. In deciding whether or not the aged will live with the family, we must consider many things. These include the health of all concerned, the financial state, the past family frictions and many other factors.

While each decision must fit special circumstances, yet one general principle is important; namely, experience has shown better results when the aged live near, but not with, their adult children. The reasons for this have been discussed before; in summary, they are that tensions from conflicts rise with proximity, whereas anxieties from guilt go up in proportion to the geographic distance between the members of the first and second generations.

If the aged must live with an offspring, it augurs better when the member of the second generation is single or widowed, and, where the older person himself controls the dwelling, the chance of harmony improves. Finally, it helps when the offspring's work keeps him away all day. Such arrangements minimize the risk of touching sore spots—unhealed remnants of the struggles of bygone adolescent days. Where the aged must live with a married offspring, then usually the old person has to assume the role of guest. If there is a choice, for obvious reasons it is better to live with a married daughter than a married son—that daughter-in-law again!

Structuring the situation well may prevent unnecessary tensions when 3 generations live together. The older person should have an area, however small, that he can call his own. He should have definite planned responsibilities. The mutual expectations should be realistic, understood and *met*. Obeying helpful rules provides useful discipline for all ages. Where the older person lives near the married child, his place in the family still must be kept open for him. This means mutual responsibilities and reciprocal assistance. Grandmothers can mind babies. To make possible more of this type of living for the aged, governments should construct imaginatively designed, low-cost housing units. Inability to pay should never, by itself, bar any aged person from suitable living quarters.

WHAT ARE THE COMMUNITY ALTERNATIVES TO LIVING WITH ONE'S FAMILY?

A boarding home may provide a better family atmosphere than living with one's own children. This is especially true when the family live near. I think we should use more boarding homes which house from 1 to 4 aged guests. Until now, lack of money has limited their use, but if elderly individuals had more funds, this would provide the incentive for more homes to open. Some complain of the crowded and unclean state of many boarding homes for the aged. True, such should have government inspec‧ tion; nonetheless, the inspector should realize that friendliness has often greater value that cleanliness.

The final alternative is the Home for the Aged. Although its true place will be established only after more experience, it now seems less desirable than self-contained quarters for most aged, but better than living with married children. Whether or not it has a greater potential than the boarding home, we have yet to find out. At any rate, let us keep its guest load at less than 50. Also, look closely at its provision for social and health needs. It should be situated so as to permit easy communication with the family. Until now, of all homes for the aged, those run by religious bodies seem to provide the best environment.

DO THE AGED NEED MORE MONEY?

A large enough income would help the older person regain his influence in the family. Security for the aged would eliminate an important source of tension. Since money symbolizes power and position, the meagre resources of the aged reflect their present impotence. We must find a way to right this inequality. How can this be done? By protecting what the older person now has, and by adding enough funds to meet his minimum physical and psychological needs. For many, the lameduck period at the end of life would not occur if we discouraged the practice of passing from father to son the control of property long before the father's death. Since this so often springs from a desire to avoid succession duties, wise legislation might help.

Where the aged need extra money to keep up a normal living standard, this should come from governments rather than from the offspring. More flexible and more generous pensions might accomplish this with a very helpful impact on the entire family —even if only an economist could predict the impact on the economy. For psychological reasons, pensions should, in part, come from the contributions made by the aged during their maximum earning period. Where funds must come from an offspring, this is best done unobtrusively—for instance by an automatic monthly bank transfer.

WHAT WORK SHOULD THE AGED DO?

Many problems which now plague the family would vanish if the aged could only find work. The shift from agriculture to industry decreased the potential earnings of most old people, and the mobility of today's wage earner handicaps him further. None would claim that we should, or could, roll back industrial advance, even to help the aged. However, I would strongly suggest that industry plan to include the elderly wage earner in its structure. Unfortunately, in some circles planning sounds like a bad word, especially when it states its goal as human betterment. Yet, we must plan, and wisely plan, to prevent the idle aged from disorganizing the family.

Employment for the elderly individual would help the family in many ways. Not only would it provide money for better incomes and better housing but also would increase family harmony by raising the morale of its elder members. Appropriate work is therapeutic; it relieves tensions and prevents illness. Those planning work for the aged should keep individual wants in mind. Some aged, especially old women with money, now have other activities which satisfy their needs.

How should we provide work for the elders of the family? Industries should protect those jobs of old people now threatened by automation or by industrial mobility. Flexible retirement plans should replace our present compulsory ones. Part-time jobs, with responsibilities diminishing as age increases, would help adjust to the inevitable slowing up from growing old. It also would delay the decline of the older person's powers. Sheltered workshops would help not only the sick aged but also those enfeebled by years. So an enlightened program of work for the aged would soften the impact of aging in the family.

SUMMARY

The family feels the impact of all the troubles of the aged. We help it best by doing something about the problems which now face old people.

All older persons in every culture suffer from being pushed into the background by advancing years. To lose power and prestige is never easy. Furthermore, illness always handicaps elderly people.

To these inevitable problems we've added the insults from recent social and economic change. Modern glamorizing of youth, today's speed and material progress have all made it tougher for the aged.

So now we must plan and act to deal with this crisis of age. We must meet health problems by setting up community-organized health services in a background of better hospitals—especially for the mentally sick. This also means improved ways of paying for health care.

We must help the aged regain the social prestige lost in the shuffle of the industrial revolution. We should meet the question of "where to live" by the general principle of, near but not with, married offspring. Economists must help families and governments solve the financial muddles of the older person. The aged need a fair share of the nation's earnings—and their new political power will help them get it.

Finally, we must fight the destructive idleness of the aged by jobs tailored to their needs either in industry or in sheltered workshops.

So the effect of aging in many families upsets all 3 generations. Fortunately we now know how to make things better. So now we must decide whether "Age is a dream that is dying or one that is coming to birth." It is up to us to help the aged help the family.

BIBLIOGRAPHY

1. Townsend, Peter: The Family Life of Old People, London, Routledge, 1957.
2. The Committee on The Family: Integration and conflict in family behaviour, Group Adv. Psychiat., Rep., Volume III report No. 27, 1954.
3. Burgess, E. W.: New Family Relationships, Fourth International Gerontological Conference.
4. Burgess, Locke: The Family, ed. 2, New York, Am. Bk. Co., 1956.
5. Home Care and Housing Needs of the Aged, New York State Division of Housing, 1956.
6. Committee on Labour and Public Welfare: Studies of the Aged and Aging, Selected Documents, Vol. II, Washington, D. C., U. S. Government Printing Office.
7. Ebaugh, F. G.: Age introduced stress into the family, Geriatrics 2:146, 1956.

7

The Individual, the Family
and the Community

Judd Marmor, M.D.

As a consequence of increasing interdisciplinary communication among psychiatry, psychology, sociology and anthropology, there has been a growing awareness in recent years of their complex inter-relatedness and interdependence. Social anthropologists who used to deal chiefly with descriptions of the external trappings of various cultures have become progressively concerned with understanding the origins and the meanings of diverse cultural institutions and their relationship to individual and family life within the culture. Conversely, psychiatrists and clinical psychologists have begun to realize that no awareness of individual psychodynamics truly can be complete without an understanding of the complex interpenetration of individual personality development with family structure, community organization and cultural mores. This is not to imply that these various disciplines are merging and losing their distinctive orientations, but, like specialists in the field of medicine, each is learning that its own complexities can be understood best only by a full awareness of all its interrelationships.

In the remarks which follow I shall endeavor to explore some present-day aspects of the intricate interdependency that exists among individual, family and community in America. We are living in an era of extraordinary change—change which is occuring at

103

a tempo that probably has no previous parallel in the entire history of man. The technologic innovations which scientists predict within the next 10 to 25 years in the uses of atomic energy, in automation, in technics of communication, in transportation over and under the surface of our planet and in travel into space, to mention but a few, stagger the imagination. What changes these ultimately will bring with them in the patterns of family life and in the personality of man can only be guessed at and perhaps, for the time being, are best left to the creative imaginations of the science fiction writers.

Indeed, if we try to survey the transformations that are going on even in contemporary community and family life the task is almost equally difficult. This is not merely because the rapidly changing scene is apt to make our conclusions obsolete almost before they are reached but also because ours is a particularly complex society. There is much less of the uniformity here that characterizes some of the societies of the Old World; and although the sharp cutting down of the stream of immigration in the past 4 decades has resulted in the melting pot of American life being stirred into greater homogeneity than ever before, the subcultural variations in mores are still quite considerable among different classes, races and ethnic groups within our country.

Therefore, generally, when social scientists talk about "the American family," they are discussing an abstraction—one based primarily on the native-born, white, urban, middle-class, which constitutes the largest single subcultural group within our country, and the one whose standards and aspirations form the prevailing ethos toward which most other groups tend to gravitate. With the understanding, then, that this is our focus also, let us turn to the consideration of the effects of the contemporary community on the prototypal American family and its individual members.

It is necessary at the outset that we clarify the specific sense in which we are using the term "community," since it is a vague concept which can refer to anything from a county to a continent. However, we shall use it in a much more limited sense to refer to *neighborhood*—the immediate physical and subcultural environment in which any particular family finds itself. In this sense the community traditionally always has been an important intermediary between the culture at large and the family, acting on both and, in turn, being acted upon by both.

The significant fact about the contemporary American community is that its specific influence on family life has been diminishing steadily over the past several decades. The chief reasons for this are to be found in the progressive changes that have taken place in communication and transportation. In earlier days people living together in a community were inevitably linked closely together by their geographic proximity. The barriers that stood in the way of communication with people in other areas, as well as the difficulties in transportation, held the members of the community together in bonds of mutual interest. They worked together, played together and prayed together. The community thus tended to have a high degree of stability and homogeneity which strengthened its ability to maintain social control. Each neighbor was known to every other one, and their mutual interdependency made it important to retain one another's good will, thus enhancing the pattern of mutual regulation. The social and the religious activities, the marriages, the births and the deaths were all shared experiences which led to common traditions and consistent community standards.

However, today, the prototypal community is no longer the integrated unit that it once was. The development of the telephone, the automobile, rapid transit and air transportation have dissolved almost completely the geographic barriers which once gave the community its stability. As often as not, especially in the large urban areas and upper middle-class suburban centers, people have no more than a nodding acquaintance with their neighbors. Their friends often live in widely separated areas, and the social, economic and recreational life of the family is no longer rooted in the geographic community.

The effect of this social fragmentation upon patterns of family life has been important and far-reaching. By making each family a relatively isolated unit in the community, it has contributed significantly to the deep inner feeling of isolation, the sense of loneliness within the crowd, that seems to characterize so many people in our time. This tendency towards isolation of the family, together with the over-all trend towards smaller family groups in recent decades, also has favored the development of intensified intrafamilial dependency patterns. "Family romances" and Oedipal ties, overprotectiveness, and "smother-love" all tend to flourish luxuriantly in the hothouse atmosphere of the isolated family. I believe that this has been a significant factor in the increased incidence of passive dependent immature personalities who are so often encountered in present-day psychiatric practice.

The dependency of today's child upon his parents is further accentuated by the actual physical aspects of modern community life. The street is no longer a safe place for play, and the growing scarcity of land in urban residential areas has led to the disappearance of the numerous neighborhood "vacant lots" which used to provide easily accessible and natural playgrounds for children. Playmates for the child often have to be specially arranged for by his parents and even imported from other neighborhoods. As a result, the child's life, outside of school, tends often to be a relatively isolated one. Thus the modern American child seems to be in danger of losing one of the important values that used to be derived from community living—the learning process that comes from belonging to a group of peers and acquiring a social identity in the course of growing up with them. The lessons of social competition, social adaptation, leadership and submission often were learned far more effectively in the peer society of the local community than anywhere else.

Still another effect of the isolation of the family within the fragmented community is the loss of the sense of trust toward others that an integrated community used to foster. A man's word is no longer considered his bond. Almost everyone outside of the family circle is regarded as a stranger, and the stranger is not to be trusted. A gesture of friendship, an act of kindness, is apt to be met with the suspicious inner feeling of "I wonder what he *wants* from me," or "What's in it for him?" The traditional American motto of "In God we trust," now tends to carry the cynical addendum "All others pay cash!" and the Golden Rule of "Do unto others as you would have them do unto you" is often replaced by the brassier one of "Do others before they do you!"

Despite the fact that the community was an important instrument of local social control, it also was a significant factor in individuation. Mark Twain's Hannibal, Missouri, Sherwood Anderson's Winesburg, Ohio, and Edgar Lee Masters' Spoon River all left their distinctive imprints upon their inhabitants. With the loss of cohesion of community life, people's personalities bear less and less of the stamp of the relatively individualized community subculture but instead become more and more responsive to the influences of the culture at large. One of the consequences of this has been the growing trend toward conformity, which numerous students of the current scene have noted with concern. Erich Fromm has described it in the context of "escape from freedom" and "alienation from the self," David Riesman has done so in terms of the tendency to "other-directedness," William Whyte, Jr., in terms of "the organization man," and Erik Erickson in terms of the "loss of personal identity," but they are all describing the same phenomenon. People all over the country read the same magazines and books, see the same movies and TV programs, hear the same broadcasts. They wear the same clothes, drive the same kinds of cars, strive to live in the same kinds of houses with the same kinds of appliances and furnishings. Indeed, even many of the differentiating aspects of urban and rural living which used to distinguish the "city slicker" from the "hayseed" are in the process of disappearing, as today's farmer is exposed to more and more of the same communication media as is the city dweller, and rapid transportation brings him nearer and nearer to the influences of urban life.

The fact that the pressures in our culture toward conformity grow steadily stronger is no accidental occurrence. They grow naturally out of the conditions of intensified industrialization and automation of our economy and represent the kind of "social character" which an increasingly complex technologic society requires. The artisan, the ruggedly independent individual who took pride in the product that he fashioned from beginning to end and was known to and patronized by his neighbors, is a rapidly disappearing phenomenon. In his place is the anonymous cog in the vast assembly line of production, where things are produced infinitely more efficiently and cheaply but necessarily alike. To fit into the assembly line, whether it be at the level of production or distribution, whether in the factory, in the clerical or the managerial ranks, or in the advertising apparatuses designed to create the needs that keep the assembly lines rolling, the human individual must submerge his individuality and his unique personal identity in terms of the group needs of the apparatus. The ideal of the inner-directed Renaissance Man has gradually given way to that of the other-directed Organization Man.

The remarkable rise in productive efficiency in contemporary American society has led to increasing emphasis on the need for increased domestic consumption to absorb the ever-increasing national product. To accomplish this, the technics of advertising and merchandising have reached levels of efficiency previously undreamed of. The manner in which they have employed all of the knowledge of human motivation provided by modern individual and group psychology has been ably documented by Vance Packard in his book "The Hidden Persuaders."[1] An important consequence of this societal need has been a significant change in the Protestant Ethic which had been the social character required by earlier industrial societies and dominated community life of the past. In a society of abundance, the social emphasis necessarily shifts from being thrifty to spending more and more. It is now more important to have credit at the bank than money in it. In our increasingly automated society, the emphasis is shifting from the sanctification of hard work to the glorification of leisure-time activities, in the support of which large industries have developed. As one of the high priests of the new marketing ethic, Dr. Ernest Dichter, has put it, "We are now confronted with the problem of permitting the average American to feel moral . . . even when he is spending, even when he is not saving, even when he is taking 2 vacations a year and buying a second and third car. One of the basic problems of this prosperity . . . is to give people the sanction and the justification to enjoy it and to demonstrate that the hedonistic approach to this life is a moral, not an immoral one."[2]

However, as one might gather from Dr. Dichter's remarks, this shift has not been achieved without inner cost to the individual. The residuals of religious teachings and old identifications create much guilt and conflict in many of those who try to conform to the new ethic of the Abundant Society. Although the conflict may not be conscious, their reactions at times seem to suggest, in the words of Elinor Wylie, that in "the Puritan marrow of (their) bones, there's something in this richness that (they) hate." The frenetic quality of the acquisitiveness and the pleasure-seeking that characterizes many modern Americans, like the forced gaiety of a New Year's Eve party, suggests a reaction-formation to deny the unconscious guilt, the emotional insecurity and the loss of personal identity that lie behind it.

The development of automation with its diminished need for huge labor forces, plus the increasing prosperity of the middle class, has led to another striking phenomenon which is having an important effect on family life—namely, the increasing move out of the large urban centers into the suburbs. Spectorsky[3] has described a syndrome which characterizes many families in the suburbs, to which he has given the name Destination Sickness. He describes it as "that constellation of physical and psychic disorders and discomforts which ensue on the artificial and premature attainment of deceptive and inadequate goals." The prototypal suburbanite whom he describes is surrounded by all the latest consumer gadgets and appliances—all bought on time. His house and car are mortgaged to the hilt but represent the best he can manage—on credit—in his and his wife's frantic effort to keep up with the Jones's. His work is in an organization in which he performs a specialized, fragmentary function which gives him little satisfaction and no real opportunity for self-expression. His inner emptiness plus the tensions from which he suffers in the process of trying to support the pyramid of possessions which he has been driven to purchase, leave him little capacity to enjoy the leisure which our society allows him. His weekends are apt to be a torment of boredom and emptiness, and he looks forward with relief to his assembly-line activities Monday morning. At least he knows what to do with himself then. His wife, too, suffers from a similar emptiness which she seeks to fill by acquiring new possessions. She makes a career of haunting the department stores, following the "sales," and daily adding to the conglomeration of possessions, many of which are never even used. Possession for possession's sake—the accumulation of things as symbols of security—is a passion which obsesses countless modern American families. The "conspicuous consumption" which Thorstein Veblen once attributed to the small, rich, "leisure class" has become the dominant goal of American society as a whole.

How do these facets of contemporary life become transmitted to the personality of the individual in our society? One of the most important ways is through the pattern of unconscious identification with the parents, by means of which much of the child's image of himself and his world is formed. In almost all societies parents are the most significant purveyors to their children of the values and mores of the culture. Parents who are anxious, confused, unsure of what they believe or do not believe, concerned only with social approval and material possessions, cannot help passing these attitudes on to their children. In an age of conformity, children learn very early from their parents that the most important thing in life is not to be one's self but rather "to be liked," to win the approval of others. As Riesman has put it:

> Approval itself, irrespective of content, becomes almost the only unequivocal good in this situation: one makes good when one is approved of. Thus all power, not merely some power, is in the hands of the actual or imaginary approving group, and the child learns from his parents' reactions to him that nothing in his character, no possession he owns, no inheritance of name or talent, no work he has done is valued for itself, but only for its effect on others.[4]

Success—and particularly material success—becomes the only criterion often of one's adequacy. Children grow up in an atmosphere of social competitiveness in which the home, the furniture and the family car become the measures by which they evaluate the status of their parents and themselves. Marital conflicts and tensions tend increasingly to revolve around these same issues. The "success" of a marriage, from the modern American woman's standpoint, often is gauged by the material status of her husband, rather than upon his worth as a human being. By the same token, the man is apt to appraise the worth of his own marital selection in terms of either the material status or the beauty of his wife, both external "marketing" values which bring him the approval of others, rather than upon her intrinsic worth as a human being, companion and mate. As motivation researcher Louis Cheskin has suggested, the contemporary American makes his selection of a mate on the basis of the size, shape and wrapping of the package instead of on its contents! Little wonder, when the factors that determine marital choice are so lacking in inner values and relatedness, that the process of communication breaks down or indeed never genuinely develops within the family.

Of couse, there are other factors involved. The working life of today's father is generally alien both to his wife and children in contrast with what it was in artisan days. In turn, the father usually has little real contact with the daily life of his wife or children. Thus, except for discussions of the trivia of everyday living, husband and wife, and parents and children, often have almost nothing to say to each other, and countless family evenings are spent in the semi-autism of television viewing. It would seem to be only a matter of time before each family member will have his own television set and be able to watch his own program in splendid isolation!

Admittedly, this is a rather extreme and bleak picture of the direction in which contemporary American family life seems to be moving. However, fortunately, this is not the entire picture. Just as in the individual neurosis regressive and reparative trends may exist side by side, the same seems to hold true at a sociologic level.

Social fragmentation creates inner feelings of loneliness and anxiety, and these in turn set into action strong motivational forces within the individual designed to counteract these feelings. Whether the social needs of human beings stem from biologically rooted "herd instincts," or whether they grow out of the early social conditioning dependent on the prolonged helplessness of the human infant, may be a debatable issue, but that these social needs exist and are powerful sources of human motivation is unquestionable. In the face of the decay of community influence and the trend towards family isolation that has ensued, there are indications that the group needs of people in our society are reasserting themselves in new ways.

Some of these indications in the postwar era have been the steady lowering of age at marriage, the decline in numbers of single adults, and the increase in birth rates. There are many others. For example, the widespread move to the suburbs that has been taking place in most large urban areas is not simply an automatic consequence of better transportation facilities. It also represents a genuine striving on the part of countless Americans to recapture the values of community living which they felt they were losing in the impersonal chaos of the big city. Indeed, to some extent, they do. If they fail, as often happens, it is not because their intentions were not good but because the community itself is no longer what it used to be. The competitive strivings and the pressures towards conformity pursue the suburban family even more fiercely than they did in the city, and the more leisurely pace of living which the family sought turns out to be an illusion. Between keeping the home and the garden up to the community's standards, and afternoons and evenings spent on boards, committees, PTA's, scout meetings, church work, entertaining the "right" people, etc., the suburban mother and father may get so submerged in their strivings for "belongingness" that they find themselves with even less leisure time than they had in the city, and the sense of inner emptiness may be even greater than ever.

Still another evidence of reparative strivings on the part of contemporary Americans has been the remarkable rise in the number and the influence of churches and synagogues in this country in the past 15 or 20 years. This is undoubtedly an overdetermined phenomenon, with a number of different factors involved. For some people, membership in a religious group seems to represent a search for emotional security in an age of uncertainties and anxieties. For others, it seems to reflect a rebellion against the overwhelming materialism and "marketing orientation" of our society and a search for spiritual, ethical, and moral values that would restore the inner dignity of man and give his life a meaningfulness that it seems to have lost. However, for many, the modern church or synagogue represents a center for group activities, group relationships and group values which can take the place once held by the community in this respect. The religious center of today is less and less a place where people merely gather together to worship their God. More and more it has become a social and educational center. The religious school has become an increasingly important instrument for teaching children not merely religious values and rituals but also the group identifications and group mores which the family and the community have become less able to transmit. Similarly, the social functions of the church—the dances, the picnics, the mens' clubs, the women's circles and sisterhoods, the teen-age groups and the little children's groups—embrace all members of the family and provide new centers for group-belongingness to combat the vacuum and loneliness that the disorganization of community influence has created.

There have been many other searchings for group membership in recent times—hobby clubs, fraternal and sororal lodges and organizations, country clubs and community centers, and while all have grown, none seems to have the vitality of the current religious revival.

Of course, in a broad sense, the entire trend toward what has been called "togetherness" represents not only an aspect of the "social character" which an increasingly complex technologic society requires of its membership but also reflects an inner striving for emotional security from *within* the individual in such a society; and the tendency to frown upon patterns of behavior associated with lower and upperclass mores—violence, eccentricity, irresponsibility—is part of this striving. Still another reflection of it is the postwar decline of all extremist political movements, both on the left and on the right. The values of the "beat generation" with its angry and despairing rejection of the material values of our time have found no significant answering echoes in the college youth of today.

The ego-reparative strivings of the individual in contemporary American society reflect themselves in many other ways also—some of them constructive, others not. The extraordinary popularity which self-help books in the area of mental health—from "dianetics" to the "power of positive thinking"—have enjoyed in the past decade is one of these. Other phenomena are indicative of efforts at the narcotization of loneliness and anxiety, notably the progressive rise in the consumption of alcohol. Less obvious but probably equally significant indications of this are the steady increase in cigarette smoking, cancer scares notwithstanding, and the enormous consumption of tranquillizing drugs.

Thus far we have directed our comments to the prototypal middle-class family primarily. However, no discussion of the relationship between community and family life in contemporary America can ignore the problem of the urban lower class and the impact of its community and family life upon some of our present social problems.

The natural history of the modern industrial city follows a fairly characteristic pattern. Leaving out that increase of population due to the surplus of births over deaths, its chief other source of growth is through the migration of unskilled persons who come from rural areas and smaller cities, and from other countries. Most of these come in at the bottom of the industrial totem pole and are apt to settle in the cheapest residential districts, the slum and semislum areas. *One of the most significant facts about these slum communities is that their imprint upon their inhabitants tends to be the same regardless of the nature of the groups which occupy them.* This is not to imply that different groups do not leave different imprints on such neighborhoods but merely to emphasize that the major effect seems to be in the reverse direction. Consider the example of juvenile delinquency. Its incidence varies markedly according to the community. Slum sections, inhabited by the lowest paid working people, have the highest rates. On the other hand, as one moves outward towards the middle and upper class residential communities, there is a progressive drop in delinquency rate. This drop cannot be explained in terms of different racial or ethnic origins of the groups involved. Studies have shown that delinquency and crime rates in large urban centers for such groups as German, Irish, Scandinavian, Italian, Slavic, Mexican, Negro and Puerto Rican groups have all been high when these groups occupied slum neighborhoods at various periods in their history, and all decreased when they moved to better residential districts. Although the delinquency rates for each national or racial group within the same areas were not identical, the *variation between these groups was much less than the variation between areas.* A study by Stofflet[5] revealed that the homicide rate of a particular group was low among first generation European immigrants, high in the second generation, low again in the third. This would seem to indicate that the generation whose personalities developed in the stable communities of the Old World had healthy super-egos; but that their children, raised in the slum communities to which they migrated, were much more prone to violent aggression. On the other hand, as the third generation began to improve its status and move out of the slums, the homicide rate again decreased. Therefore, it is clear that the behavioral patterns of these groups are influenced significantly by their community experiences and change when they move to other areas in which the aspects of community life and mores are different.

What is it about slum life that has this influence upon the individuals within it? The answer to this question is not a simple one and has many facets—social, economic and psychological. Sociologists used to attribute much of the area's problems to the assumption that the slums were essentially disorganized communities which lacked the inner cohesiveness and group sanctions which created social controls in other kinds of communities. More recent studies[6] have thrown doubt upon this assumption. The problem of the slum district often is not that it lacks organization but rather that its own organization and group standards fail to mesh with those of the dominant middle-class society around it. Since the slum's inhabitants are people who are regarded as inferiors and often exposed to discrimination by the rest of the community, it is not surprising that they develop group patterns of defensiveness and hostility. In addition, ignorance and prejudice often contribute to intense intracommunity hostilities between different racial, ethnic or religious groups within the slum area. The effect on family life of such a community is a disorganizing one. Parents, deprived of their traditional settings and values, find themselves confused in their relationship to their children in their new setting. In many underprivileged homes the influence of the father is seriously undermined by the circumstances of his life. If he is not working, his authority and status within the home suffer, and he is either bitter or defeated. If he is employed, his work is apt to be hard, long and unrewarding, and he returns home exhausted or incapable of evincing any interest in or control over his children—or else, bitter, tired and irritable, he may make them the scapegoats for the frustrations and indignities which he may have had to endure during the day. Often he stems from an authoritarian tradition which endeavors to impose harsh and strict disciplines upon the children. However, the absence of the group sanctions which existed in his original environment, plus the fact that in the new community he not only lacks social and economic status but may also be looked down on by his children as a foreigner, all serve to negate his efforts at such authoritarian control and to foster defiance and rebelliousness in his children. On the other hand, the mother often has to work also in order to supplement the family's meagre income; or else, as is frequently the case in lower income groups, she is harassed by a large brood of children and is weighed down by a never-ending mass of routine duties. In either event she often is unable to give any individual child the emotional guidance and warmth that he needs. Under these circumstances, it is not surprising that the "gang" on the corner becomes one of the centrally significant community institutions in the life of the child growing up in a slum district. The gang is the outlet through which his hurts and his hostilities, personal and social, can be discharged. It becomes his surrogate family, with its leader his surrogate father or big brother. It is to this group that he turns for acceptance, guidance and companionship. The goals and the sanctions of the gang become the most important guideposts in his development. It is little wonder, then, if the complex social and economic pressures within the community are such as to make racketeering, vandalism, narcotics or other forms of antisocial acting-out important patterns of gang behavior, that the young individual in the slum area will tend to become involved in these patterns, as a necessary means of achieving prestige and acceptance within his group. Here, then, we see another pattern of community life, one, unhappily, whose influence is not diminishing upon the families who reside within it, with profoundly serious consequences for society as a whole. Clearly, a recognition of the importance of community-family-individual interaction in the genesis of juvenile delinquency, narcotics and crime problems of our time is essential in endeavoring to find constructive solutions for them.

In conclusion, then, I have tried to indicate how in recent decades, as a result of vastly improved means of communication and transportation, the geographic barriers which once gave some degree of stability and homogeneity to neighborhood communities have broken down. As a consequence, neighborhood relationships have become more tenuous, and the individual family a relatively isolated unit, with resultant intensification of intrafamilial dependency patterns, and increased patterns of distrust, suspiciousness and insecurity extrafamilially. At the same time, through mass communication media and the needs of an increasingly complex technology, the social pressures towards conformity have accelerated tremendously, leading to loss of individuation and to inner feelings of emptiness and boredom. The frenetic drive towards the acquisition of material goods is seen as one consequence of this psychological void, with the old Protestant Ethic and its emphasis on work, thrift and reward in the hereafter giving way to a new Age-of-Anxiety Ethic, in which leisure, spending and pleasure in the here-and-now have become the goals of living.

However, that these new idols are failing to relieve the inner tensions of people is indicated by their almost frantic grasping at anything which promises relief from anxiety—from self-help books to tranquillizing drugs. It is speculated that the resurgence of religion in contemporary American life represents, in part at least, another aspect of this search for inner peace and meaningfulness.

What the society of the future will be like is difficult, if not impossible, to predict at this time. Indeed there is reason for some concern in this age of the nuclear bomb as to whether or not there will be *any* human society on this planet in decades to come; but one thing is certain. If man is to survive as a *human being,* he will have to devise social technics which will enable community, family and individual to interact in such a way as to preserve and enhance the uniqueness and dignity of the individual, instead of submerging or degrading it. If we fail to accomplish that, nothing else will matter.

REFERENCES

1. Packard, Vance: The Hidden Persuaders, New York, David McKay Co., Inc., 1957.
2. Quoted from Whyte, W. H., Jr.: The Organization Man, p. 19, New York, Simon & Schuster, Inc., 1956.
3. Spectorsky, A. C.: Destination Sickness *in* What's New, No. 204, 1958.
4. Riesman, David: The Lonely Crowd, Connecticut Yale, pp. 66, 1953.
5. Stofflet, E. H.: A study of national and cultural differences in criminal tendency, Arch. Psychol., N.Y., No. 185, pp. 1-60, 1935.
6. Whyte, W. F.: Street Corner Society, ed. 2, Chicago, Univ. Chicago Press, 1955.

8

The Individual, the Family
and the Boss

*Bertram Schaffner, M.D.**

INTRODUCTION

I WOULD LIKE TO BEGIN by congratulating Dr. Liebman on his choice of theme for this year's lecture series and to thank him for this opportunity to take part in it. The topics brought up for discussion in all the series have been most timely; this year's theme is no exception. The title, *Emotional Forces in the Family* brings together the original psychiatric concern with the well-being of the individual and the newer emphasis on the individual among all his interpersonal relationships, especially with the persons most significant to him—the members of his family. The speakers before me no doubt have pointed out the impossibility of comprehending the individual apart from his family. Nor can the family be comprehended apart from all the individuals in it. The family plays an essential part in the formation of the individual, and each individual plays an essential part in the formation of the family. The interpersonal transactions between them are not merely interesting facts, like plots in a novel or a play (although they may be as full of suspense); they are the vital

---

* With grateful acknowledgment for ideas and suggestions from Arthur Allen Schwartz, chief psychologist at the James Weldon Johnson Mental Health Clinic, New York City.

interplay of intellectual and emotional forces through which the personality and the character of a human being are formed, through means of which his mental health is either hindered or promoted. The implication is that in order to understand any individual person, we cannot do so in terms of himself alone; we must think in terms of the people with whom he has grown up, has played, has learned and later worked.

The subject assigned to me, "The Individual, the Family and the Boss," puzzled me at first. I wondered if I was to speak about the place of authority within the family, but it seemed to be likely that this would have been discussed some time earlier by Dr. Jessner or Dr. Ness. Later I thought it probable that Dr. Liebman, in preparing this series, had viewed "the boss" as an especially important personal and social force outside the family but one whose effect on the people within the family is potentially almost as great as that of a mother or a father. The significant influence of the boss on a family rarely has been singled out for psychiatric discussion and surely deserves attention today, when the destinies of many families and individuals depend largely on the destinies of huge organizations, such as universities, worldwide industries, national and international governmental agencies and far-flung military services. The boss sometimes may be a specific individual; sometimes the boss may be a corporation. In either case, he impinges forcefully on the breadwinner of the family and thereby impinges indirectly on every member of the family. In one case the effect may be personal, in another it may be called social. In years gone by, the boss's relationship to the breadwinner was an intimate one and fairly clearly defined; today it has undergone many changes and is generally less intimate. However, even a less personal boss can produce strong emotional effects, and for that reason alone he should not be ignored in a discussion of family life. In addition, the boss represents the employing organization; he enforces its values, expectations and standards and thereby exerts a social force on all the members of the employee's family. He sets the patterns of authority for those around him. Any of his words and behavior carry special weight because of his status and prestige and the confidence placed in him by authorities above him.

PERSONAL EFFECTS OF THE RELATIONSHIP WITH THE BOSS ON MEMBERS OF THE FAMILY

A boss necessarily plays a large role in the state of mind of the family's breadwinner, since the family's basic security depends on continuity of employment and income. In today's society, with most families living in cities, human beings are especially vulnerable to loss of employment. Consequently, there is a degree of apprehensiveness, even when not expressed, about the holding of a job. The job may depend not only on a man's competence but also specifically on the relationship between him and his employer. Even though labor unions and laws have reduced somewhat the danger of loss of employment due to personality conflicts, the resolutions of conflicts are still essentially up to the two people involved, who are not always equipped to resolve their quarrels.

The relationship with one's boss is bound to be tremendously charged emotionally. The boss is the appointed authority figure in the employee's environment, the person under whose supervision the employee spends at least 8 hours every day, the largest segment of his waking life. All of a man's attitudes toward his father in early life, all his early problems regarding father and other authorities, all his individual problems connected with competence, achievement, submission and the like, are prone to be rekindled and to manifest themselves again in relationship to the particular personality of the boss. Such reactions in the employee can be irrational and unrealistic, having nothing to do with the situation itself; or there may be troublesome reactions having little to do with a problem in the employee but arising from difficult behavior or real personality problems on the part of the boss.

For example, a man who has felt unable as a child to satisfy his father, or has felt that his father always favored an older brother, may come to experience the same feelings toward the boss of today, even though the boss does not behave like the man's father. The employee unconsciously may anticipate that today's authority figure will behave in the same way as yesterday's. In reverse, men who have had especially capable fathers may be genuinely surprised or even indignant to find that some men in positions of responsibility, are not as competent or mature as their own fathers (or themselves). Again, a man who as a child was particularly dependent upon his father for direction, guidance or approval, may find himself unprepared for a boss today who takes for granted that the men under him feel adequate and self-reliant.

Quite apart from an employee's unrealistic expectations, the real personality of the boss can play an important role. The boss may be a "good" authority figure, by which I mean a person experienced in dealing with people, one who is able to appreciate their needs and feelings as well as his own, able to assign duties, define situations and limits, without hurting the self-esteem of the other person. A "good" authority figure can recognize his own limitations without embarrassment, tolerate the inevitable frustrations of leadership, set sensible goals, and recognize the differences among his various employees. The boss sometimes turns out to be a poor authority figure, often described as the "authoritarian personality." The poor authority figure usually needs to present himself to others as omnipotent and omniscient, demanding complete faith in himself and unquestioning obedience. His underlying sense of weakness or inadequacy, the inner unsureness to which he is not reconciled, require him to behave in this authoritarian way. Unfortunately his behavior rarely reveals his true feelings of unsureness to his subordinates; it does not obtain for him the co-operation that he seeks, nor does it cause the employees to sympathize with or to assist him. On the contrary, especially in the United States, his subordinates are likely to become highly intolerant and rebellious. This in turns tends to increase his unfortunate authoritarian behavior, when he sees that he is failing with them. Such a poor authority figure cannot participate with his employees either in the solution of their practical work problems or in their personal and emotional problems at work. He is usually so preoccupied with his own uncomfortable feelings that he is not properly aware of what goes on inside other people.

The tension between an employee and his boss will have a powerful effect on both men's prevailing mood; it can play a major role in the way a man feels about his daily life as well as in the way he feels about himself as a human being.

A man's self-esteem must be good if he is to remain mentally healthy; his self-esteem is also crucial to his relationships with other persons. The psychological effects of the employee's relationship with his boss are inevitably reflected in his behavior toward his wife, his children—actually toward all of his associates but more forcefully among those who are emotionally closest to him. Therefore, the relationship with the boss deeply affects the mental health of the family as a whole.

The following description of the family of one of my patients illustrates the effects of the boss-employee relationship and its indirect effects upon other members of the employee's family:

The father in the family was a son of immigrant parents; he had a strong sense of duty toward his wife and children and a fierce urge to lift his family out of the poverty that he had known in his childhood. His own parents were hard-working but unhappy people, very stern and critical. Their goal had been to achieve some form of personal dignity through improving their economic status, but they never reached it. Their son felt quite inadequate at his work, partly due to lack of training and opportunity but chiefly because of a low estimate of his personal competence as a result of long-standing criticism from his parents. He married before he had any financial security and when he was emotionally too young, probably because of an enormous need to feel loved and appreciated.

Three children came in rapid succession; his salary was hardly adequate to take care of them. He sought to improve his work; he dreamed and schemed for promotions. Unfortunately his immediate superior, his boss, was very much like his own parents: hypercritical, easily dissatisfield, impatient with other people's discordant moods, and quite unaware of others' need for encouragement. The boss alternated between behaving irritably when with the man, and avoiding him altogether, and he never recommended a promotion. The employee never understood that the situation was due at least in part to their interaction upon each other. He felt that hard work alone entitled him to a salary increase and promotion and complained bitterly that he was being unfairly treated. Since he could do nothing about it, he felt weak and a failure. He became depressed, envied the other employees and increasingly resented the boss, the store in which he worked and the customers who patronized it. His sense of failure and inadequacy prevented him from seeking employment elsewhere.

Early in their marriage his wife sensed his unhappiness, and did her best to praise and to comfort him. She tried to compensate for his miserable days with a happy home life in the evenings, but as time went on, he continued to be most unhappy. Her efforts being of no avail, she began to feel that she was a failure as a wife, even though the unhappiness was not due primarily to shortcomings in herself. Meanwhile her own daily existence also had become more difficult; she was now taking care of 3 small children, while the family income remained the same. She began to take in sewing to have extra money but felt ashamed of this rather than proud; now she began to avoid her neighbors. Her husband, like his own parents, was so engrossed in his own problems that he could not be sympathetic with her. She began to nag him; instead of asking for the warmth and praise that she wanted, she reproached him for not bringing home more money. The anger against his boss now spilled over into the relationship with his wife, estranging them; his behavior had the undesired result of turning his wife into the very kind of person he had tried to escape from when he was growing up. Now he felt unhappy and unwanted both at work and at home. He resorted more and more to drinking and to staying by himself. When he would arrive home at night, there were frequent quarrels because now even less money was available for the family.

So far, this story illustrates the interaction among 3 people, 2 of whom never saw each other face-to-face, but whose behavior and emotions affected each other's lives. This is certainly not to say that the influence of a man's employer *alone* regularly determines the course of his marriage! Fortunately not. In this situation it was the man's particular background, his specific difficulty in relating to other people, his susceptibility to disapproval and lack of progress that permitted the personality of the employer to work such a hardship on him. In this situation neither husband, wife nor employer recognized that there was a psychological problem; consequently no help was sought, especially since both husband and wife were both strongly conditioned against asking for any kind of help whatsoever. Since they lacked ways of relieving the strain between them, the effect of the man's relationship with his boss was more significant in widening the gulf between man and wife. Had the boss been a different kind of person, open, encouraging and constructive in his dealings with the man, the employee might have been sufficiently sustained emotionally so as to improve in his work, enjoy his worklife, and function better in his human relations, especially with his family.

Actually the unhappiness of the boss, the employee and the wife affected their children too, in specific ways. The eldest child, a girl, was born while the mother and the father were still relatively hopeful and harmonious. To this day she retains a basically optimistic disposition; unfortunately superimposed upon it is an obsession for money, as she became mistakenly convinced from her mother's complaints that money alone was responsible for the unhappiness in their home. She is married to a man who is similarly oriented, and although they get along well, they lead lives that are emotionally impoverished.

The second child in the family, a girl, did not live to maturity. The third child, a boy, was about 4 when the unhappiness in the home became most acute. He seems to have suffered worse psychological consequences than his older sister. He witnessed the parents' vitriolic quarrels at a more impressionable age; to this day he has great fear of all anger and friction, and avoids fighting even when he ought to protect himself. He remembers his mother chiefly as demanding from his father things which could not be provided; today he is compulsively generous, especially to women. He has not married; however, his relations with women have not suffered quite as much as those with men. He remembers his mother as nagging and irritable but also as some one fighting in the interest of her children and trying to hold her home together.

His picture of men was derived from two sources, from his father as he saw him in the home and from the boss he heard described by his father. This picture was of particular significance with important bearing upon his conception of males and therefore his picture of himself.

His father appeared to him to be unable to cope with the ordinary problems in life; this alone started the youngster wondering whether or not he himself would be able to cope with them.

During the fights between his parents, he acquired a fear of his father. During the father's pervasive brooding, the boy got the impression that he meant nothing to his father as a person. His father's alcoholism set up in him the feeling that men were totally unpredictable, explosive people whom one should hold at a distance. As the father and mother eventually divorced when the boy was 10, this signified to him the hopelessness of marriage as such.

From his father's descriptions of his boss, he got distorted ideas about men in positions of power, ideas which for years impaired his capacity to get along with employers. He expected an employer to be indifferent, unkind, exploiting and unsympathetic, intractable and unapproachable. When he first went to work, he was ill at ease among his fellow workers and so hostile toward older men that he was often fired. He did not realize that his own attitude was provoking the unfavorable reaction in his employers and among the other workers.

This family history illustrates what might be called a chain reaction of vulnerability, beginning with the reciprocal effects that the boss and the breadwinner had upon each other, then the interaction between husband and wife, next the interaction between the parents and their elder daughter, and finally the specific effect upon their son. Thus we can consider the boss a dynamic force within the body of the family, even though he is technically outside its kinship structure.

ABDICATION OF AUTHORITY IN THE FAMILY AND AT WORK

Now let us turn our attention to another question involving the individual, the family and the boss: the matter of attitudes toward authority as such. There has been a vast change in the United States within the last 50 years which it is important to identify. It might be described as a complete swing from total reliance upon authority in problems of living, to total reliance upon the doctrine of "human relations" to solve the problems of living. As with any extreme swing or reversal, the result has been considerable confusion, with a loss of many original values. The defects of the new extremism in turn produced an extreme reaction in the opposite direction. At this very moment, articles are beginning to appear in professional journals on industrial management which take a strong stand against "human relations" policies in administration, and make a strong plea for a return to the "tough boss."

Let me make it clear that as a psychiatrist, I am not siding either with the protagonists of authority or with the defenders of the human relations school of thought. On the contrary, I believe that neither side sees the situation as a whole. I feel that the rebellion in our times against the doctrine of "spare the rod and spoil the child" had to come, since that point of view grossly ignored the natural developmental patterns of children, did not sufficiently take into account certain important needs of children and often covered up a profoundly self-centered point of view. The "spare the rod" school was deficient precisely because it did not know enough about human beings, human emotions and human relations. It also had some virtues.

On the other hand, I feel that the so-called "human relations" school of thought has gone too far in its direction. It did supply the long-needed emphasis on the healthy development and fulfillment of a child's potentials, on meeting the emotional needs of children. Unfortunately, it also carried with it an implied indictment of everything done by parents in the past, by all authority figures; this unhappily often came to mean that any parent who did not meet a child's need, or frustrated a child, or punished a child, was automatically a poor parent, guilty of a "crime" against human relations. Consequently, while formerly parents' needs often were met at the expense of children, today children's needs often are met at the expense of parents. With the unpopularity of the use of authority in child-rearing today, parents either tend to feel confused about it, to abdicate the use of authority altogether, or to use it most inconsistently and unpleasantly. As a result, it is often said about the prevailing United States family-pattern today that "the child is king" or that parents live "under the tyranny of their children."

What are the sources of this major change in the attitude toward authority? I do not know them all. It often has been claimed that our country originated in a rebellion against authority and that it has continued to have a distaste for authority in

government ever since. In all probability, the strongly democratic, egalitarian trend in our traditions militates against
authority. Certainly, in the area of government we have created
a most elaborate system of checks and balances to make sure that
authority cannot perpetuate itself or become too powerful. The
struggle against unlimited authority has not been purely a phenomenon in the United States. It has been part of a world-wide
rebellion against colonialism and against the possibility of the
domination of civilization by any one world sector. I believe that
the psychiatric profession has played a part in promoting the new
attitude toward authority, since it necessarily concerned itself
with the experiences of children and the relation of these experiences to mental illness. In the early years of psychiatry, there was
a tendency to place responsibility upon the parents, a tendency to
blame them as the culprits, while the "child could do no wrong."
In view of this, it behooves the psychiatric profession today to
review the subject of authority, from the point of view of the
psychological welfare of *both* children and parents in the home
and outside the home.

Much confusion apparently started with the instruction to the
parent that the young child needs the feeling of "security." It is
doubtless valid that the child is better off when it feels sure of a
person to depend upon, as long as it is still helpless and not yet
grown up. The fear of abandonment, and the anxiety which it
can arouse in the child about his own worth to others, can be
most destructive. Yet it need not mean that a child can *never* be
left alone, or *never* be allowed to feel helpless for a second, *never*
be refused something which it feels it needs. We know now that the
child's sense of security derives from many factors and not merely
the absense of frustration and fear, not merely the fulfilling of felt
needs. We now know that one of the major factors conducive to a
feeling of security in the young child is the sense of adequacy and
sureness within the authority figure on whom the child depends.
When the authority figure becomes unsure, vacillating, incon-
sistent, afraid to take a stand, then the child is sure to lose some of
the most important bases for its own sense of security.

To illustrate what I have in mind: It is the function of the
authority to define what is normally to be expected from each

person in a given situation. For example, in the home a father or mother defines what is expected from a young child, a school-child, an older child. Among my patients there have been individuals who never truly experienced the sense of being a child, defined as a child, respected for being still a child. They were treated at times like contemporaries of grown-ups, entitled to equal privileges, for which they actually were not ready; at other times they were treated like children, to which they then reacted with surprised indignation and a sense of humiliation. They had never known a period when it felt appropriate and honorable to be a child. On the same order, a child needs definition from the parent of what is to be regarded as adequate, in order that the child can be healthily satisfied with his own achievement, and the parent's definition needs to be established with due consideration of the child's individual capacity and stage of development. This may involve at times expecting more from the child than he is doing, while at others it may mean reducing the child's expectations of himself.

A further illustration of the function of authority is that of instructing the child what is expected of him or what is obligatory in our particular society, even when it is not necessarily pleasing to the child to carry it out. Too frequently it has been felt that the so-called "secure" child should naturally enjoy whatever he is doing or is asked to do. Realistically speaking, most of what the child tries to do has its difficulties; realistically, there are many duties that are truly chores, which must still be done for reasons of practicality, soundness or social custom. For the most part they are not enjoyable or interesting, although completing and getting them out of the way may have its satisfactions. It happens to be a source of considerable security to a person to know that he is able to tolerate the chores of life, the work-hours, for example, which are not always pleasant or always satisfying but cannot be avoided without cost.

Teaching a child what to expect and how to adjust to disappointment is another major task of the authority figure. In the recent confused picture of parent-child relations, some parents have taken the concept (to give a child security) to mean that the child should have every wish and need met, should not have the experience of being refused; this implies that the parent must supply whatever the child wishes, with the aim that the child, if so treated, will be frustration-free, although the poor parent (if he survives) may undergo either complete deprivation of all resources, frustration of all his own needs, or complete sense of parental inadequacy if the child should want more than the combined parents (and grandparents) can afford. The kind of child that results from years of such "security" is usually one so unaccustomed to disappointment and frustration that he is totally unprepared to endure it when it eventually comes, concluding that his parents do not love him, that he is unlovable or unworthy, that the world is unjust and unfair, since he "should" be receiving what he genuinely feels the need for. He may have learned through previous experience with his parents that he will have to force them to grant his wishes, as they have in the past succumbed to his pressure when he has kept it up long enough. However, in that case he will not enjoy what he receives, first because he has become accustomed to receiving and considers it only what is "coming to him," not a *gift*, received with pleasure; second, he will not enjoy it because he has finally achieved it after a struggle which he felt should never have had to take place; third, because the wish, finally granted by the parents with an air of exhaustion, anger or disapproval, removes any possible feeling of pleasure.

All this could be spared the child by a parental authority who can hear a child's request, understand the child's need, grant the wish if he deems it wise to do so, and is able and willing to grant it, or refuse the request for reasons of his own, kindly but firmly, with or without an explanation, and without the feeling that he is harming the child by his refusal. Much of the satisfaction a child feels is due not only to achieving or receiving what he wished for but also to his realization that the achievement was not something merely to be taken for granted. The enjoyment and the appreciation is in large part due to the sense that something has actually been added to one's life which would not have been part of it except for the efforts he himself made to obtain it or through the goodness of some other person.

It is time now to connect what I have just been saying about

parents, children and authority with "the boss." The principles
that the parental authority needs to be capable of teaching to a
child have a direct bearing on the child's future orientation to
employment, work, his relationship to his future employer and
his future job security. As I pointed out earlier, the boss rightly
can be considered a branch of the extended family group; hence
the attitudes learned at home in relation to authority, expectations, work and satisfaction will be carried over directly into the
future work situation with the future boss.

To illustrate, beginning with the same considerations that we
took up in regard to parents, first we have the sense of security.
We pointed out that the young child is most likely to feel secure
in proportion to the sense of adequacy and sureness of the parent;
the older child's security also will come with his own sense of
developing adequacy and competence. In the job situation,
much of the employee's sense of security must depend upon his
confidence in the intelligence, the judgment and the soundness
of his company or in the boss or the foreman who directs him in
his work; this is particularly true if his work is in any way potentially dangerous to life. We know that when management becomes frightened or unsure this is quickly reflected in the morale
of employees as manifested in the rate of illness, absenteeism
and decreased productivity.

Then, the second point mentioned, concerning the definition of
situations and making explicit what is expected from the person.
If a child has not been accustomed at home to knowing just what
is expected from him, he may not look for a defining of what is
expected from him on his job or what he is to expect from a job
and from an employer. Numbers of young people apply for work
today with definite ideas of what they hope to receive from the
job in the way of salary, title, prestige, increase in pay, promotion
but often with little or no picture of what is expected from them
in the way of contribution to the function of the organization,
with confused ideas as to rights and duties and confused ideas
about how proficient they need to be before asking for advancement. The boss of today is often at a loss to deal with the requests,
or sometimes demands, of the young employees. He may be confused, like the parents (the boss too is usually a parent in his own

home), and feel unsure about his own standards. He sometimes may find his personal standards at variance with those of the company; he may find himself ill-equipped to deal with the conflicting attitudes and emotions of his employees. If he is sure enough of himself and his authority, he may deal with situations by means of a simple statement of company policy; if that should fail, he may call on the services of the industrial psychologist. If no solution is possible, he may resort to discharging the employee, or the employee may quit.

If a child had never become accustomed at home to the so-called drudgery of chores as well as the activities that give pleasure, he may find it hard to work at those studies in high school or college which do not come easily or which require laborious plugging and repetition. He may find it equally hard to endure the routine of work, which by its very nature cannot be regularly easy or satisfying, and much of which is inevitably dull. He may feel that the dullness is due to the nature of the particular work that he has selected and yearn for any other type of employment available, not knowing that this too would turn out the same for him.

There is a corollary to be drawn from the parallel of family life and work life, which may be significant. The parent who is sure of his authority can refuse a child with kindness, set standards for the child's behavior, set limits for its actions and expect the child to be able to comply in time with these standards and within such limits. The child would be expected eventually to manage its own behavior without supervision, much as a child learns at first not to cross a street alone until it has learned to do so without being run over and later is able to protect himself automatically from traffic. That is, as the man matures, he will carry within himself the ego standards and ego controls which his parents had started in him.

In today's family situation, with parents eager to do what is best for their children but vacillating between what are still called "old-fashioned" ideas and what is "new and modern," many children receive no simple, clear concepts of behavior, values or standards; instead they receive many impressions, from differing social class contacts, from magazines, fiction, television. In this large country they also receive different values and concepts from the many cultural backgrounds which make up our heritage. There is hardly a person among us whose traditional cultural strains have not been well fused with some others.

It is not surprising that today's younger generation, with no single understandable code of expectable behavior, stirs about restlessly, from job to job and within the job. Like children for whom no limits or directions have been set, they push out toward this side or that, waiting till someone may have the wisdom or the courage to stop them or guide them. If we let ourselves recall the feelings we had as children, we may remember not only the joy we experienced when we could express ourselves freely but also those times when we became a bit too excited, too stimulated, when we sensed that we were going too far for our own benefit but were in no position to stop ourselves. What a relief to be brought to a definite halt by a strong but kind authority! Juvenile delinquency, I am sure, has numerous causes; it is too widespread in the world today for it to be a symptom of only one disturbing factor; but I feel strongly that one important cause of juvenile delinquency may be the confusions that follow when a family's life changes from rural to urban, from farming to industrial, from one city to another, from living in one kind of established cultural pattern and moving into a different culture, in short, from the enforced rapid adjustment to major changes, without either resolution of standards or the establishment of inner and outer controls for behavior.

We are living in a time of change and transition, when authorities—parents and bosses—are unclear as to their roles, values, rights or duties and when their children, large and small, are equally unclear about their roles and relationships. The result is much friction and confusion. Consequently, there is a new search for solutions, solutions which inevitably will restore some of the sense of authority from the past but without the excessive, authoritarian quality. There is a search going on for new forms of authority which will have far more sensitivity to the feelings and rights of others, with far greater tact and wisdom in resolving conflicts between those in authority and those for whom and to whom they are responsible.

9

The Disintegrating Impact of "Modern" Life on the Family in America and Its Explosive Repercussions

Lawrence S. Kubie, M.D.

USUALLY, IT IS TAKEN FOR GRANTED that the family is an essential and creative force in our cultural and social system. Indeed, many of our social institutions are based on this assumption. Unhappily, this overlooks the fact that forces at work in our culture are so deforming the family that it is in danger of becoming at least as destructive as creative. No thought is being given to the very drastic steps which must be taken if the creative potential of the family is to be preserved from these destructive influences. Our problem then is what are some of the forces which are deforming the family, what can be done about them and, in turn, what can be done about the destructive boomerang effects which emanate from the distorted family unit.

Everything which I have to say about the future of the family is predicated on the uncertain assumption that we will not blow ourselves to pieces before that future arrives. Yet, even apart from that, there is no minor or major prophet who has sufficient prescience to justify predictions about the future of the family. Of course, anyone can say with reasonable confidence that the family in some form is here to stay.[10] It is obvious that as long as the physical, intellectual and emotional maturation of the human being takes many years, infancy, childhood and even adolescence will require protective devices, built around either one parent or the other, or both. Although subject to extensive variations, this constitutes the essence of a family system. Furthermore, there is a constant lengthening of that phase of human life which is devoted to vocational and educational apprenticeship. Therefore, a family system or some derivative will not only continue but also will lengthen, even if it grows thinner in form and substance. Moreover, it will not be possible to shrink the duration of the long years of relative helplessness until methods are found to accelerate radically the rate of physical growth, of intellectual development and especially of the all-important process of emotional maturation. Any such acceleration is still in the realm of science fiction; but since the fiction of one generation becomes the reality of the next, it would not be safe to dismiss this possibility as being forever unattainable.

However, in the meantime, there is little about man's development in any known culture which gives occasion for complacency. The neurotic process is universal.[9] We have learned to use only a small fragment of our latent creative potential.[7,9] We do not know how to educate efficiently, nor how to make the scholar a wise man, nor even how to transmit the lessons learned painfully by one generation to its successors. Least of all do we know how to hand on the lessons of our own past errors and failures.[5,9] As parents, we do not know how to bring up psychologically healthy and happy young people. What is worse is that we are too timid to use even what is known because its use is opposed by deeply entrenched prejudices and vested interests.

Therefore, it would seem to be obvious that it is our responsibility to re-examine critically everything which used to be left to mother's or father's uninformed impulses, under such euphemistic clichés as "instinct" and "love," lest mother-love mask self-love and father-love mask unconscious impulses to destroy. All of this means that we still have much to learn about what family life could mean and even more to learn about the effective application of any knowledge that we gain. Such an unsparing re-examination of accepted premises is all the more vital because of the changes in family life which will be imposed by powerful and complex social, cultural, economic and population forces. Indeed, many such changes are already underway, and some of those which have crept up unnoted are frighteningly destructive. Only Pollyanna is pessimistic enough to deny this. Therefore, the question is not whether or not they are going to happen but whether we will allow them to happen to us passively or will have the courage and the imagination to guide these developments with some degree of intelligent planning.

Like the physician, we first must recognize the presence of illness, then attempt to diagnose its nature. In this discussion we shall not be able to go beyond a partial and approximate diagnosis of the nature of the ailment of the family as an institution. We may recognize several symptoms, but we shall not be certain whether the patient is suffering from one disease or from several, and we shall not make even this much progress if we continue to replace the eagle by the ostrich as the American emblem. For our country has become a land where no one allows himself to acknowledge or recall anything unpleasant, where no one is willing to turn humbly to the autopsy table of experience to learn from his mistakes. Nor do we seem to realize that Pollyanna is the ultimate pessimist: for in her heart-of-hearts Pollyanna believes that things are so desperate that she does not dare to acknowl-edge them, much less to do anything about them. Therefore, she pretends to herself that everything is rosy, with the childlike hope that what she does not put into words will not be true. In the world of reality, the optimist is the man who knows that things could hardly be worse, who is not even sure that there is a way out but is not frightened into silence by these hard facts and has the courage and the determination to seek a solution. This is the only kind of optimist who can save what is good in our culture. I am challenging you to be that kind of optimist, i.e., to have the courage, the humility, and the honesty that medicine showed when it first went to the autopsy table to study its failures.

It is in this spirit that we should consider the family, the nature and causes of its current failure and its possible future in our American culture.

I start with the assumption that family life as we know it today is a failure, that its failure is increasing, and that unless we do some hard thinking about it and some hard planning, its failure will destroy the culture and the civilization that we prize so highly.

Before undertaking to make a diagnostic survey of the family and more specifically of the American family, let us look briefly at a few facts about the world in which families exist. Surely the first important consideration is what has been so aptly called the Population Bomb.[12]

According to figures compiled by the United Nations, the world population in 1920 was *1,810,000,000*. In 1956 it was 2,734,000,000. The world population is growing now at the rate of 44 million a year, and the rate of increase is accelerating.[1,12]* Writing in 1959, Sir Charles Darwin, the grandson of the great biologist, gives evidence that there will be upwards of 5,000,000,000 in the world by the end of this century, i.e., in 40 years.[2]† This is not the rate of growth for our country but the anticipated population increase for the world as a whole, including those lands in which the economic level, the health level, the food level, the living level are far below ours.

To consider what this means merely in terms of food, water, space for living, fuel and raw materials gives us pause. I am reminded of the little boy of 5 who looked angrily at his new baby sister and asked his father, "And how do you *know* that there will be enough air to go around?" This hardly seems as funny as it did even a few years ago.

Against the general background of this world picture there also have been significant changes within our own country.

In the first place, in 1890 only 52 per cent of the men and 54

---

* Subsequent studies by the United Nations indicate more precisely that the world population in 1925 was 1,907,000,000. In 1957 it had become 2,790,000,000. The world population is now growing by more than 40,000,000 a year.

† In 1958 a report by the United Nations indicates that there may be 7 billion people in the world by the end of the century. (These figures are from Population Studies No. 28, entitled "1925 and 2000: The Future Growth of World Population," New York, 1958, pp. 23 and 70: and again from Population and Vital Statistics Report, New York, 1959, Series A, Volume II, No. 2, p. 3, 1957.)

per cent of the women were married, whereas in 1958 70 per cent of the men and 66 per cent of the women were married. Scarcely more than half the population were married in 1890, whereas nearly three quarters were married in 1958, an increase of 20 per cent. This should be kept in mind when anyone notes that the divorce rate of the population has risen by less than 2 per cent of the population in the same 60 years.[2,11] In this connection a further point has already been described in previous publications.[8] This is the fact that the family disruption rate in 1890 *due to early deaths* was higher than the family disruption rate in 1940 *through divorce*. In other words, the reshuffling of marriages primarily because of death occurred at a higher rate in 1890 than did the reshuffle rate through divorce in 1940. Thus the increase in the divorce rate is a direct consequence of two facts: (1) that a much larger percentage of the public are married[2,11]; and (2) that more married people remain alive longer.[8]

Furthermore, we are marrying at an earlier age.[2] Between 1890 and 1956 the average age of first marriages for men dropped from 27 to 22½, for women from 23 to 20. With this came the baby boom, earlier births, and a reduction of infant mortality by 50 per cent in the last 20 years. Recently there has been a slight increase in the number of children per family, which may be temporary only or may initiate a slight swing toward larger family groups.‡

---

‡ Since my manuscript was prepared, Paul H. Jacobson, Ph.D., Research Assistant in the Statistical Bureau of the Metropolitan Life Insurance Company, N. Y., has published a volume entitled *American Marriage and Divorce* (Rhinehart & Company, Inc., 1959). In this volume on page 90 and again on pages 138-144 the above statistics are given with greater precision and again are drawn from current Population Reports, Publication 20, No. 72, page 3, Table C, and No. 90, page 5, Table 7.

These indicate the following facts:

The contrast with the situation in 1890 becomes even clearer. At that time only 52 per cent of the men of 14 or over, and 55 per cent of the women were married; whereas in 1958 almost 70 per cent of the men and 66 per cent of the women were married. This means that scarcely more than half of the population were married in 1890, whereas about two thirds were married in 1958.

In this connection a further point should be noted, namely, that since 1890 divorce and annulment has increased from 3 to 9.3 per 1,000 existing marriages. At the same time, however, the rate of family dissolution through deaths has dropped by an even greater number. As a consequence, the annual rate of marital dissolutions from *all* causes is lower now than it was in 1890. (Further sources of these statistics are the 1950 census P-D1, p. 97, Table 46, and Current Population Reports T-20, No. 87,

Footnote continued on following page

After this hasty survey of the available statistics, let us consider some of the other changes that are happening to the family.

Although there are some recent indications that a trend toward slightly larger groups of progeny may be starting, the family unit still remains very small as compared with what it was in previous generations. Inevitably where few children are born and grow up in close succession, they experience a sharpened sibling rivalry, all of which is focused on the parental apex of a conical familial structure. In the larger family groups of brothers and sisters of previous generations the children also were born in close succession, but because these were interspersed with frequent deaths in infancy and early childhood, the survivors were spaced more widely.

Furthermore, there used to be several adults of assorted ages in each rambling home: an uncle and an aunt or so, an older cousin, 1 or more grandparents. There is no room for these any more. Few older siblings or parental surrogates live together under one gabled roof, to divide up the parental functions or to provide a special ally to buffer the situation for the child who is passing through an unhappy phase. The built-in baby sitter is gone. Moreover, nurses and servants, these other parental substitutes, are vanishing breed except for the numerically insignificant top economic layer. This means that there is no one other than the beleaguered parent to help a child. A remarkable intensification and sharpening of domestic rivalries results. One further paradoxical consequence is that within this conical microcosm of the family the more dedicated and devoted the parents are the more

---

Footnote continued from preceding page

p. 9.) The figures for males relates to the civilian population alone, plus those members of the armed forces who live off post, or with their families on post. Thus the percentage married among all males is somewhat lower than the selected sample.

Furthermore, as indicated above, we are marrying at an earlier age. Since 1890 the median age for first marriage for men has dropped from 26 to 23, for women from 22 to 20.

In a personal communication, Dr. Jacobson indicates that the estimates for the earlier years are gradually being revised as more complete coverage is being obtained from such areas as Africa and Asia.

Again, I am indebted to Dr. Jacobson, as I was in previous studies, for his invaluable and precise statistical data and for the valuable studies which are presented fully in his book which has just been published.

bitter becomes the rivalry among the children. This is a situation which by its very nature breeds hostility and illness.

The home is a stage-set on which the drama of human life evolves, and there are few things more destructive to this drama than overcrowding, no matter what the economic level or how pretentious the address. Yet families are living in ever smaller, more cramped and transitory quarters. Lack of space means lack of privacy, making it impossible for any member of the family to live with freedom or dignity, forcing an inescapable proximity on both child and adult, allowing no peace to eye or ear.

This is an assault on human dignity and human values. In our cities space has become a major health problem, as important as the control of poliomyelitis today or of typhoid fever and smallpox in previous generations. To allow real estate interests to pile more and more hapless human beings on a pinpoint of land for the sake of skyrocketting real estate values is literally criminal negligence. If this is creeping socialism, then in the name of common sense let all be creeping socialists.

Next, what about the economic changes as a result of which the family is no longer a source of economic security to its members but a threat? It has been a long time since the family was a producing unit. As producers, people work together, building allegiances and a sense of closeness and co-operation. However, today the family is only a consuming unit. This in itself has deleterious psychological consequences. As consumers, the same folks are pigs at the feeding trough. Among the different members the rivalry over consumer needs brings out and accentuates the worst in the natures of child and adult. Of course, this is exploited and played upon skillfully and destructively by the emphasis on possessing more things, more gadgets, more clothes, cosmetics and foods. The home is battered by high-pressure advertising by radio, TV, newspapers and periodicals. The family is the target of the country's advertising, turning it into a battleground of inflamed material greed. This too is happening to the family in America.

Furthermore, not only is the family no longer a center for creative and productive work, it is no longer a center even for the mutual interchange of ideas and entertainment. The family rarely entertains itself any more. Family games and sports are few. Instead, the home has become a place where one looks and listens while someone else performs. Of course, the quality and the content of what we look at or listen to makes a difference, but far greater are the consequences of being always and forever a spectator and never an actively creative participant. The entertainment industries with their sponsoring advertisers are making of the family a spectator culture, not a place for creative communication. Furthermore, the perfection of the facsimiles which can be piped into the home increases the confusion of the young, exploiting the limited capacity of the child to distinguish reality from fantasy. For many this makes it seem useless even to try to do anything. When one can "live" vicariously, why get out of bed at all!

Another distortion of values results from the family having become a motel for nomads on the go, with speed and distance as their only goal. The home has become the center of its own solar system, each member pursuing an irregular orbit in space on his own gasoline-propelled vehicle, with hot rods and radio but no purpose except to go as far as possible in as short a time as possible, whether speeding on the highways or into outer space.

It is useless to attempt to cope with forces of this kind by bland repetitions of romantic clichés, which never were true. The fantasy that the family was automatically a breeder of love was never real. If it bred love, it also bred hate; and unhappily, as is well known, hate can often be the stronger of the two. Love within the family is a goal given to few to achieve. Where it occurs, it is in spite of the forces which are operating in our present culture and not because of them.

At the same time, still other things are happening. There is a high survival rate in youth and again in old age; but there is no significant extension of the span of the average man's earning life, as the duration lengthens of the years of physical health. This increases still further the economic insecurity which the family has come to mean. Furthermore, because there are many early marriages before young folk are economically independent, because modern industrial technics and the world situation demand a higher degree of training for more people, and because industry cannot absorb so many healthy young and old people, the duration of the period in which the adolescent lives on a dole from the adult world becomes longer. Therefore, we mask unemployment by educating the young and fobbing off the old. . . .

In the midst of these forces, what is happening to the traditional picture of the family as an institution which functions under the powerful image of a working father and the constant influence of a home-dwelling, child-rearing, child-loving mother?

According to Ginzberg,[3] 28 million women were employed in the United States during 1955. This meant that there were in any one day 22½ million women in a total labor force of 66 million. This is 1 out of every 3 workers. Moreover, 3 out of every 5 working women were married; 40 per cent of all the mothers in the country with children below the age of 18 were in the labor force; 3 million mothers with children below the age of 6 are at work.

Add to this the fact that more girls graduate from high school than boys, but fewer graduate from college. They work instead. Finally, 93 of every 100 women get married and marry younger, have their children earlier, complete an average family of 3 by the time that they are 28, and in addition are at work by that time. To the child it does not matter whether the absent mother is scrubbing floors or playing bridge. An absent mother means a home without a woman; and restless mothers at work out of the home mean children and adolescents spending most of their waking hours without parents. It is not strange that in the absence of family ties the closeness that young people need but cannot find in the family is sought in gangs.

What are the effects of all of this on the psychology of women? Clearly we are making it harder to be a woman. There was a time when her activities within her home were to the woman a source of prideful distinction and of love. The woman who made the best apple pie was known for her pie, or she was known for the kind of home she kept, instead of her clothes. Today her self-esteem derives not from her home but from her activities outside of the home. When her heart and life were within the home she was compelled to realize that home was no place in which to ventilate anger, no matter how justified that anger or how skillfully disguised. This forced woman to attempt to find within herself the ability to live peacefully. She had been under pressure to resolve in her adult years the residues of nursery-bred anger, whereas man had been under pressure merely to harness that anger to some of the world's work. Whether in the process either became sick or remained well has been an issue to which until recently our culture has been strangely indifferent. Moreover, this confronts human education with its most urgent and difficult problem, one which we have quite failed to solve and to which we pay almost no attention. Instead, social and economic pressures are increasing on women to force them out into the unsatisfactory world of men, making them equally competitive in the "great" world. This makes it certain that woman will now displace onto her home and family the frustrations generated in her extra-domestic life, exactly as men have been doing since Cain and Abel. Indeed, it is strange to realize that people once were naive enough to think that women could remain uninfected by hate, in a culture in which their major goal is to prove that they can be women with one hand while outdistancing men (and also other women) with the other. We are developing a world of angry women to match a world of angry men; yet we wonder naively that this breeds angry children. The family has become a place in which competitiveness and possessiveness are bred into our culture, as destructive a development as anything could be.

Wars may have been won on the playing fields of Eton; but now more than ever before, wars and crimes and neuroses, man's universal pestilences, are bred in families. The family remains an autocratic system within a democratic culture; or as MacFie Campbell used to say, an autocracy ruled by its sickest member. Thus we face a situation in our American culture which is forcing the family to become almost wholly destructive.

Prolonged adolescence adds to this still another explosive force. Thus in every conceivable way it would seem that the family has become a source of individual and social instability and insecurity. Families are exploding in a world with an exploding population. Anyone who can sit and view this situation with equanimity is one of that strange species to which I referred above, the American Ostrich.

Let me summarize. Powerful influences arise out of our industrial organization to intensify the neurotic insatiability of the demand for things. Advertising media which taint the air we breathe deliberately inculcate neurotogenic envy, greed and discontent. A temporary abundance of leisure merges with the specter of unemployment. Unsolved psychological problems arise out of an unearned prosperity on the installment plan, turning the adult into a Get-Rich-Quick Wallingford: a child in long pants with "the Gimmies." Add to this the movement of women into industry, the population pressure, the concentration of more people into less space, the fantastic development of technics of transportation while transportation as a system breaks down, the almost totally destructive influence of the entertainment industries which turn our entire culture into a spectator culture—all of these and many more forces and vested interests are destroying the family as a creative way of life.

Yet we remain so blindly complacent that we will not make comparative studies of other patterns of homemaking and of child-rearing. We might at least have the humility to cease our fatuous glorification of the great American home in order to consider how we are destroying the souls of children in the debased family life to which we expose them. Perhaps with imagination and with courage we could then find some alternatives that might at least be worth a trial.

Let me cite one contrasting example[4]: "The Kibbutz is a kind of extended family—a community-parent-child relationship. The obvious affection of all the members for all the children came home to me that night when a young soldier in an ill-fitting uniform, a tall, awkward, unbelievably thin and gangling redheaded boy, with a childish, homely-humorous face, walked into the dining hall. Grinning with delight at being home for the weekend, he was hailed from every table, patted on the back, teased and embraced. Everyone looked on him with tender and amused pride. My neighbor at table said to me, 'Our first child to go into the army.' She did not say, 'The first child of the Kibbutz,' nor 'My first child,' but 'Our first child.' " Here is something on which to ponder, although it comes from a frontier culture where the externals of life are tough and the internals easy.

Actually I am not without hope that it lies in our power to make of family life within our own culture something healing and creative. Indeed we must, or the stirring experiments on the Palestinian frontier also will be destroyed as they attain a level of sophistication comparable to ours. However, such a spiritual rebirth will be possible only if we make use of everything that we have learned about the development of the infant and the child within the family structure. It will be a slow road and a hard one. Nor will we take that road unless there is a willingness to face old mistakes, to make new ones and to challenge all the interests which are vested in our economic structure, in education as it is, in the rigid structures of religious organization and in the romantic tradition itself. We will have to take up the challenge thrown at us years ago by G. Brock Chisholm: that although the most important thing any man does is to bring up his own children, he receives less training for this task than a farmer is given for raising cattle—and this in the proudest, most complacent and emptiest educational system in the world.

This does not mean that I have easy answers to offer. I know a thousand things which are wrong in our culture, but I cannot offer a single slick improvement. We can point to obvious and gross abuses, but the real problems are the subtle ones. However, the search for an answer will not begin until there is a frank initial acknowledgment that the problem exists.

Examples of a different kind of family can be found among a few sophisticated, cultured, sensitive, intelligent young folk, where a thoughtful and articulate interchange is fostered between the generations. In such rare families the child's fifth freedom is vigilantly guarded: his freedom to know all that he feels inside himself. Instead of surrounding the child with an elaborate conspiracy of silence, his right to self-knowledge and to be articulate about it is carefully nurtured.[6]

I think for instance of the family situation in which the parent never says Do or Don't to the child, yet never hesitates to say to a child, "We do it this way," or "We do not do it that way." A small yet vital difference. Such a child is guided firmly, yet never loses his close identification with the adult, never feels that he is pushed away by the command or that he is an inferior and second-class human who receives orders. He remains instead closely identified with the loved parental image.

Yet this can happen only in a family in which the parent is an image the child can love and with whom the child can identify. This depends not on what the parent says or does but on how he sounds and looks and moves, since how we live and how we sound exercise a deeper influence on the child than the content of what we say—and far more effect than any punishment or reward. This degree of self-direction and self-awareness can grow only out of self-study; and for this we need new ways of using tape recordings and film, to study how we actually project the images of ourselves to our children. To this end parent education must be cultivated continuously, a new depth and breadth of parent education to be developed by future Parent Study Associations, if we are to gain the humility to see ourselves as children see us. As a goal this is at least as important as nutritional health or campaigns against epidemics.

That this can be done, I would illustrate with one happy and moving example:

A little girl of 5 comes into her mother's bedroom when her father is away on a trip. The child looks thoughtfully at her mother's bed, as the mother is dressing for the evening, and finally says, "Mommy, are you going to be lonely tonight?" The mother puts an arm around her and says equally thoughtfully and equally seriously, "Yes dear, I guess I will be." The child answered, "Then, you know how I feel." Such a moment can determine the future shape of life for this child. In it she feels not only her mother's love for her but also her own compassionate love for her mother. The extension of her experience of loneliness to include her mother's breeds sympathy instead of rivalry, turning this tight little family unit into something spiritually creative. Such a moment of truth can determine a child's acceptance of her role as a woman and her future relationship to her own children, but this implies a family in which the minute but highly charged events of daily life are given thoughtful, penetrating and sensitive attention.

In that same family where love is carefully nurtured, when trouble stirs as it inevitably must, there will be no shamed hesitation to acknowledge that the child often needs an adult ally from outside the family circle to whom the child can talk without a sense of disloyalty to either parent or both, an adult ally who will be made available without any sense on the part of the parent that to call in such help is no more a confession of failure than to call in a pediatrician but rather an acknowledgment of the extraordinarily subtle and complex realities of human life.

I repeat that I am not here to offer any easy solutions. Medicine never comes up with quick cures. She does not even suggest any until she has faced the existence of a disease first and studied its dangers. If I have convinced you that the family in our culture is threatened by a deep and malignant illness, we will have taken one forward step.

REFERENCES

1. Darwin, Charles: The pressures of population *in* What's New, No. 210, pp. 2-5, North Chicago, Ill., Abbott Labs., 1959.
2. Davis, Kingsley: The early marriage trend pp. 2-6, *in* What's New, No. 207, North Chicago, Ill., Abbott Labs., 1958.
3. Ginzberg, Eli: The changing pattern of women's work: some psychological correlates, Am. J. Orthopsychiat. 28: (No. 2) 313-321, 1958.
4. Kubie, E. B.: Kibbutz Samir on the Syrian border, Israel Horizons, 12: (No. 1) 16-20 (from p. 19).
5. Kubie, L. S.: The concept of normality and neurosis *in* Heiman, M. (ed.): *Psychoanalysis and Social Work*, pp. 3-14, New York, Internat. Univ. Press., 1953.
6. ————: Is Preventive Psychiatry Possible?, Daedalus, Journal of the American Academy of Arts and Sciences, published jointly by the Academy and the Wesleyan University Press, Boston, 88: (No. 4) 646-669, 1959.
7. ————: Neurotic Distortion of the Creative Process *in* Porter Lectures, Series 22, Lawrence, Kan., University of Kansas Press, pp. 151, 1958.
8. ————: Psychoanalysis and marriage: practical and theoretical issues *in* Eisenstein, V. W. (ed.): Neurotic Interaction in Marriage, New York, Basic, pp. 10-43, 1956.
9. ————: Social forces and the neurotic process *in* Leighton, A. H. (ed): New York, Basic, 1957; J. Nerv. & Ment. Dis. 128: (No. 1) 65-80, 1959.

10. Opler, M. K.: History of the family as a social and cultural institution *in* Galdston, I. (ed) : The Family in Contemporary Society, pp. 23-38, New York, Internat. Univ. Press, 1958.

11. Parke, Davis and Company, New York: Marriage and the Family —Special Report in Patterns of Disease, Feb. 1959. (With best compilation of latest statistical data and sources.)

12. Population Council: Publications and Annual Reports. 1957, 3rd Printing by the Hugh Moore Fund, 51 E. 42nd St., New York 17, N. Y.

Index